AUDIO SIGNAL PROCESSING AND CODING

BICENTENNIAL 1807 WILEY 2007 BICENTENNIAL

THE WILEY BICENTENNIAL–KNOWLEDGE FOR GENERATIONS

Each generation has its unique needs and aspirations. When Charles Wiley first opened his small printing shop in lower Manhattan in 1807, it was a generation of boundless potential searching for an identity. And we were there, helping to define a new American literary tradition. Over half a century later, in the midst of the Second Industrial Revolution, it was a generation focused on building the future. Once again, we were there, supplying the critical scientific, technical, and engineering knowledge that helped frame the world. Throughout the 20th Century, and into the new millennium, nations began to reach out beyond their own borders and a new international community was born. Wiley was there, expanding its operations around the world to enable a global exchange of ideas, opinions, and know-how.

For 200 years, Wiley has been an integral part of each generation's journey, enabling the flow of information and understanding necessary to meet their needs and fulfill their aspirations. Today, bold new technologies are changing the way we live and learn. Wiley will be there, providing you the must-have knowledge you need to imagine new worlds, new possibilities, and new opportunities.

Generations come and go, but you can always count on Wiley to provide you the knowledge you need, when and where you need it!

William J. Pesce

WILLIAM J. PESCE
PRESIDENT AND CHIEF EXECUTIVE OFFICER

Peter Booth Wiley

PETER BOOTH WILEY
CHAIRMAN OF THE BOARD

AUDIO SIGNAL PROCESSING AND CODING

Andreas Spanias
Ted Painter
Venkatraman Atti

WILEY

Audio Signal Processing and Coding

First Indian Reprint: 2013

Authorized reprint by Wiley India Pvt. Ltd., 4435-36/7, Ansari Road, Daryaganj, New Delhi – 110002.

Published by John Wiley & Sons Inc.

Printed at: *Syndicate Print-n-Pack, NOIDA*

ISBN: 978-81-265-4102-7

To
Photini, John and Louis
Lizzy, Katie and Lee
Srinivasan, Sudha, Kavitha, Satish and Ammu

CONTENTS

PREFACE

Audio processing and recording has been part of telecommunication and entertainment systems for more than a century. Moreover bandwidth issues associated with audio recording, transmission, and storage occupied engineers from the very early stages in this field. A series of important technological developments paved the way from early phonographs to magnetic tape recording, and lately compact disk (CD), and super storage devices. In the following, we capture some of the main events and milestones that mark the history in audio recording and storage.[1]

Prototypes of phonographs appeared around 1877, and the first attempt to market cylinder-based gramophones was by the Columbia Phonograph Co. in 1889. Five years later, Marconi demonstrated the first radio transmission that marked the beginning of audio broadcasting. The Victor Talking Machine Company, with the little nipper dog as its trademark, was formed in 1901. The "telegraphone", a magnetic recorder for voice that used still wire, was patented in Denmark around the end of the nineteenth century. The Odeon and His Masters Voice (HMV) label produced and marketed music recordings in the early nineteen hundreds. The cabinet phonograph with a horn called "Victrola" appeared at about the same time. Diamond disk players were marketed in 1913 followed by efforts to produce sound-on-film for motion pictures. Other milestones include the first commercial transmission in Pittsburgh and the emergence of public address amplifiers. Electrically recorded material appeared in the 1920s and the first sound-on-film was demonstrated in the mid 1920s by Warner Brothers. Cinema applications in the 1930s promoted advances in loudspeaker technologies leading to the development of woofer, tweeter, and crossover network concepts. Juke boxes for music also appeared in the 1930s. Magnetic tape recording was demonstrated in Germany in the 1930s by BASF and AEG/Telefunken. The Ampex tape recorders appeared in the US in the late 1940s. The demonstration of stereo high-fidelity (Hi-Fi) sound in the late 1940s spurred the development of amplifiers, speakers, and reel-to-reel tape recorders for home use in the 1950s both in Europe and

Apple iPod®. (Courtesy of Apple Computer, Inc.) Apple iPod® is a registered trademark of Apple Computer, Inc.

the US. Meanwhile, Columbia produced the 33-rpm long play (LP) vinyl record, while its rival RCA Victor produced the compact 45-rpm format whose sales took off with the emergence of rock and roll music. Technological developments in the mid 1950s resulted in the emergence of compact transistor-based radios and soon after small tape players. In 1963, Philips introduced the compact cassette tape format with its EL3300 series portable players (marketed in the US as Norelco) which became an instant success with accessories for home, portable, and car use. Eight track cassettes became popular in the late 1960s mainly for car use. The Dolby system for compact cassette noise reduction was also a landmark in the audio signal processing field. Meanwhile, FM broadcasting, which had been invented earlier, took off in the 1960s and 1970s with stereo transmissions. Helical tape-head technologies invented in Japan in the 1960s provided high-bandwidth recording capabilities which enabled video tape recorders for home use in the 1970s (e.g., VHS and Beta formats). This technology was also used in the 1980s for audio PCM stereo recording. Laser compact disk technology was introduced in 1982 and by the late 1980s became the preferred format for Hi-Fi stereo recording. Analog compact cassette players, high-quality reel-to-reel recorders, expensive turntables, and virtually all analog recording devices started fading away by the late 1980s. The launch of the digital CD audio format in the 1980s coincided with the advent of personal computers, and took over in all aspects of music recording and distribution. CD playback soon dominated broadcasting, automobile, home stereo, and analog vinyl LP. The compact cassette formats became relics of an old era and eventually disappeared from music stores. Digital audio tape (DAT) systems enabled by helical tape head technology were also introduced in the 1980s but were commercially unsuccessful because of strict copyright laws and unusually large taxes.

Parallel developments in digital video formats for laser disk technologies included work in audio compression systems. Audio compression research papers started appearing mostly in the 1980s at IEEE ICASSP and Audio Engineering Society conferences by authors from several research and development labs including, Erlangen-Nuremburg University and Fraunhofer IIS, AT&T Bell Laboratories, and Dolby Laboratories. Audio compression or audio coding research, the art of representing an audio signal with the least number of information bits while maintaining its fidelity, went through quantum leaps in the late 1980s and 1990s. Although originally most audio compression algorithms were developed as part of the digital motion video compression standards, e.g., the MPEG series, these algorithms eventually became important as stand alone technologies for audio recording and playback. Progress in VLSI technologies, psychoacoustics and efficient time-frequency signal representations made possible a series of scalable real-time compression algorithms for use in audio and cinema applications. In the 1990s, we witnessed the emergence of the first products that used compressed audio formats such as the MiniDisc (MD) and the Digital Compact Cassette (DCC). The sound and video playing capabilities of the PC and the proliferation of multimedia content through the Internet had a profound impact on audio compression technologies. The MPEG-1/-2 layer III (MP3) algorithm became a defacto standard for Internet music downloads. Specialized web sites that feature music content changed the ways people buy and share music. Compact MP3 players appeared in the late 1990s. In the early 2000s, we had the emergence of the Apple iPod® player with a hard drive that supports MP3 and MPEG advanced audio coding (AAC) algorithms.

In order to enhance cinematic and home theater listening experiences and deliver greater realism than ever before, audio codec designers pursued sophisticated multichannel audio coding techniques. In the mid 1990s, techniques for encoding 5.1 separate channels of audio were standardized in MPEG-2 BC and later MPEG-2 AAC audio. Proprietary multichannel algorithms were also developed and commercialized by Dolby Laboratories (AC-3), Digital Theater System (DTS), Lucent (EPAC), Sony (SDDS), and Microsoft (WMA). Dolby Labs, DTS, Lexicon, and other companies also introduced 2:N channel upmix algorithms capable of synthesizing multichannel surround presentation from conventional stereo content (e.g., Dolby ProLogic II, DTS Neo6). The human auditory system is capable of localizing sound with greater spatial resolution than current multichannel audio systems offer, and as a result the quest continues to achieve the ultimate spatial fidelity in sound reproduction. Research involving spatial audio, real-time acoustic source localization, binaural cue coding, and application of head-related transfer functions (HRTF) towards rendering immersive audio has gained interest. Audiophiles appeared skeptical with the 44.1-kHz 16-bit CD stereo format and some were critical of the sound quality of compression formats. These ideas along with the need for copyright protection eventually gained momentum and new standards and formats appeared in the early 2000s. In particular, multichannel lossless coding such as the DVD-Audio (DVD-A) and the Super-Audio-CD (SACD) appeared. The standardization of these storage formats provided the audio codec designers with enormous storage capacity. This motivated *lossless* coding of digital audio.

The purpose of this book is to provide an in-depth treatment of audio compression algorithms and standards. The topic is currently occupying several communities in signal processing, multimedia, and audio engineering. The intended readership for this book includes at least three groups. At the highest level, any reader with a general scientific background will be able to gain an appreciation for the heuristics of perceptual coding. Secondly, readers with a general electrical and computer engineering background will become familiar with the essential signal processing techniques and perceptual models embedded in most audio coders. Finally, undergraduate and graduate students with focuses in multimedia, DSP, and computer music will gain important knowledge in signal analysis and audio coding algorithms. The vast body of literature provided and the tutorial aspects of the book make it an asset for audiophiles as well.

Organization

This book is in part the outcome of many years of research and teaching at Arizona State University. We opted to include exercises and computer problems and hence enable instructors to either use the content in existing DSP and multimedia courses, or to promote the creation of new courses with focus in audio and speech processing and coding. The book has twelve chapters and each chapter contains problems, proofs, and computer exercises. Chapter 1 introduces the readers to the field of audio signal processing and coding. In Chapter 2, we review the basic signal processing theory and emphasize concepts relevant to audio coding. Chapter 3 describes waveform quantization and entropy coding schemes. Chapter 4 covers linear predictive coding and its utility in speech and audio coding. Chapter 5 covers psychoacoustics and Chapter 6 explores filter bank design. Chapter 7 describes transform coding methodologies. Subband and sinusoidal coding algorithms are addressed in Chapters 8 and 9, respectively. Chapter 10 reviews several audio coding standards including the ISO/IEC MPEG family, the cinematic Sony SDDS, the Dolby AC-3, and the DTS-coherent acoustics (DTS-CA). Chapter 11 focuses on lossless audio coding and digital audio watermarking techniques. Chapter 12 provides information on subjective quality measures.

Use in Courses

For an undergraduate elective course with little or no background in DSP, the instructor can cover in detail Chapters 1, 2, 3, 4, and 5, then present select sections of Chapter 6, and describe in an expository and qualitative manner certain basic algorithms and standards from Chapters 7-11. A graduate class in audio coding with students that have background in DSP, can start from Chapter 5 and cover in detail Chapters 6 through Chapter 11. Audio coding practitioners and researchers that are interested mostly in qualitative descriptions of the standards and information on bibliography can start at Chapter 5 and proceed reading through Chapter 11.

Trademarks and Copyrights

Sony Dynamic Digital Sound, SDDS, ATRAC, and MiniDisc are trademarks of Sony Corporation. Dolby, Dolby Digital, AC-2, AC-3, DolbyFAX, Dolby Pro-Logic are trademarks of Dolby laboratories. The perceptual audio coder (PAC), EPAC, and MPAC are trademarks of AT&T and Lucent Technologies. The APT-x100 is trademark of Audio Processing Technology Inc. The DTS-CA is trademark of Digital Theater Systems Inc. Apple iPod® is a registered trademark of Apple Computer, Inc.

Acknowledgments

The authors have all spent time at Arizona State University (ASU) and Prof. Spanias is in fact still teaching and directing research in this area at ASU. The group of authors has worked on grants with Intel Corporation and would like to thank this organization for providing grants in scalable speech and audio coding that created opportunities for in-depth studies in these areas. Special thanks to our colleagues in Intel Corporation at that time including Brian Mears, Gopal Nair, Hedayat Daie, Mark Walker, Michael Deisher, and Tom Gardos. We also wish to acknowledge the support of current Intel colleagues Gang Liang, Mike Rosenzweig, and Jim Zhou, as well as Scott Peirce for proof reading some of the material. Thanks also to former doctoral students at ASU including Philip Loizou and Sassan Ahmadi for many useful discussions in speech and audio processing. We appreciate also discussions on narrowband vocoders with Bruce Fette in the late 1990s then with Motorola GEG and now with General Dynamics.

The authors also acknowledge the National Science Foundation (NSF) CCLI for grants in education that supported in part the preparation of several computer examples and paradigms in psychoacoustics and signal coding. Also some of the early work in coding of Dr. Spanias was supported by the Naval Research Laboratories (NRL) and we would like to thank that organization for providing ideas for projects that inspired future work in this area. We also wish to thank ASU and some of the faculty and administrators that provided moral and material support for work in this area. Thanks are extended to current ASU students Shibani Misra, Visar Berisha, and Mahesh Banavar for proofreading some of the material. We thank the Wiley Interscience production team George Telecki, Melissa Yanuzzi, and Rachel Witmer for their diligent efforts in copyediting, cover design, and typesetting. We also thank all the anonymous reviewers for their useful comments. Finally, we all wish to express our thanks to our families for their support.

The book content is used frequently in ASU online courses and industry short courses offered by Andreas Spanias. Contact Andreas Spanias (spanias@asu.edu / http://www.fulton.asu.edu/~spanias/) for details.

[1] Resources used for obtaining important dates in recording history include web sites at the University of San Diego, Arizona State University, and Wikipedia.

CHAPTER 1

INTRODUCTION

Audio coding or *audio compression* algorithms are used to obtain compact digital representations of high-fidelity (wideband) audio signals for the purpose of efficient transmission or storage. The central objective in audio coding is to represent the signal with a minimum number of bits while achieving transparent signal reproduction, i.e., generating output audio that cannot be distinguished from the original input, even by a sensitive listener ("golden ears"). This text gives an in-depth treatment of algorithms and standards for transparent coding of high-fidelity audio.

1.1 HISTORICAL PERSPECTIVE

The introduction of the compact disc (CD) in the early 1980s brought to the fore all of the advantages of digital audio representation, including true high-fidelity, dynamic range, and robustness. These advantages, however, came at the expense of high data rates. Conventional CD and digital audio tape (DAT) systems are typically sampled at either 44.1 or 48 kHz using pulse code modulation (PCM) with a 16-bit sample resolution. This results in uncompressed data rates of 705.6/768 kb/s for a monaural channel, or 1.41/1.54 Mb/s for a stereo-pair. Although these data rates were accommodated successfully in first-generation CD and DAT players, second-generation audio players and wirelessly connected systems are often subject to bandwidth constraints that are incompatible with high data rates. Because of the success enjoyed by the first-generation

Audio Signal Processing and Coding, by Andreas Spanias, Ted Painter, and Venkatraman Atti

systems, however, end users have come to expect "CD-quality" audio reproduction from any digital system. Therefore, new network and wireless multimedia digital audio systems must reduce data rates without compromising reproduction quality. Motivated by the need for compression algorithms that can satisfy simultaneously the conflicting demands of high compression ratios and transparent quality for high-fidelity audio signals, several coding methodologies have been established over the last two decades. Audio compression schemes, in general, employ design techniques that exploit both *perceptual irrelevancies* and *statistical redundancies*.

PCM was the primary audio encoding scheme employed until the early 1980s. PCM does not provide any mechanisms for redundancy removal. Quantization methods that exploit the signal correlation, such as differential PCM (DPCM), delta modulation [Jaya76] [Jaya84], and adaptive DPCM (ADPCM) were applied to audio compression later (e.g., PC audio cards). Owing to the need for drastic reduction in bit rates, researchers began to pursue new approaches for audio coding based on the *principles of psychoacoustics* [Zwic90] [Moor03]. Psychoacoustic notions in conjunction with the basic properties of signal quantization have led to the theory of *perceptual entropy* [John88a] [John88b]. Perceptual entropy is a quantitative estimate of the fundamental limit of transparent audio signal compression. Another key contribution to the field was the characterization of the auditory filter bank and particularly the time-frequency analysis capabilities of the inner ear [Moor83]. Over the years, several *filter-bank* structures that mimic the critical band structure of the auditory filter bank have been proposed. A filter bank is a parallel bank of bandpass filters covering the audio spectrum, which, when used in conjunction with a perceptual model, can play an important role in the identification of perceptual irrelevancies.

During the early 1990s, several workgroups and organizations such as the International Organization for Standardization/International Electro-technical Commission (ISO/IEC), the International Telecommunications Union (ITU), AT&T, Dolby Laboratories, Digital Theatre Systems (DTS), Lucent Technologies, Philips, and Sony were actively involved in developing perceptual audio coding algorithms and standards. Some of the popular commercial standards published in the early 1990s include Dolby's Audio Coder-3 (AC-3), the DTS Coherent Acoustics (DTS-CA), Lucent Technologies' Perceptual Audio Coder (PAC), Philips' Precision Adaptive Subband Coding (PASC), and Sony's Adaptive Transform Acoustic Coding (ATRAC). Table 1.1 lists chronologically some of the prominent audio coding standards. The commercial success enjoyed by these audio coding standards triggered the launch of several multimedia storage formats.

Table 1.2 lists some of the popular multimedia storage formats since the beginning of the CD era. High-performance stereo systems became quite common with the advent of CDs in the early 1980s. A compact-disc–read only memory (CD-ROM) can store data up to 700–800 MB in digital form as "microscopic-pits" that can be read by a laser beam off of a reflective surface or a medium. Three competing storage media – DAT, the digital compact cassette (DCC), and the

Table 1.1. List of perceptual and lossless audio coding standards/algorithms.

Standard/algorithm	Related references
1. ISO/IEC MPEG-1 audio	[ISOI92]
2. Philips' PASC (for DCC applications)	[Lokh92]
3. AT&T/Lucent PAC/EPAC	[John96c] [Sinh96]
4. Dolby AC-2	[Davi92] [Fiel91]
5. AC-3/Dolby Digital	[Davis93] [Fiel96]
6. ISO/IEC MPEG-2 (BC/LSF) audio	[ISOI94a]
7. Sony's ATRAC; (MiniDisc and SDDS)	[Yosh94] [Tsut96]
8. SHORTEN	[Robi94]
9. Audio processing technology – APT-x100	[Wyli96b]
10. ISO/IEC MPEG-2 AAC	[ISOI96]
11. DTS coherent acoustics	[Smyt96] [Smyt99]
12. The DVD Algorithm	[Crav96] [Crav97]
13. MUSICompress	[Wege97]
14. Lossless transform coding of audio (LTAC)	[Pura97]
15. AudioPaK	[Hans98b] [Hans01]
16. ISO/IEC MPEG-4 audio version 1	[ISOI99]
17. Meridian lossless packing (MLP)	[Gerz99]
18. ISO/IEC MPEG-4 audio version 2	[ISOI00]
19. Audio coding based on integer transforms	[Geig01] [Geig02]
20. Direct-stream digital (DSD) technology	[Reef01a] [Jans03]

Table 1.2. Some of the popular audio storage formats.

Audio storage format	Related references
1. Compact disc	[CD82] [IECA87]
2. Digital audio tape (DAT)	[Watk88] [Tan89]
3. Digital compact cassette (DCC)	[Lokh91] [Lokh92]
4. MiniDisc	[Yosh94] [Tsut96]
5. Digital versatile disc (DVD)	[DVD96]
6. DVD-audio (DVD-A)	[DVD01]
7. Super audio CD (SACD)	[SACD02]

MiniDisc (MD) – entered the commercial market during 1987–1992. Intended mainly for back-up high-density storage (~1.3 GB), the DAT became the primary source of mass data storage/transfer [Watk88] [Tan89]. In 1991–1992, Sony proposed a storage medium called the MiniDisc, primarily for audio storage. MD employs the ATRAC algorithm for compression. In 1991, Philips introduced the DCC, a successor of the analog compact cassette. Philips DCC employs a compression scheme called the PASC [Lokh91] [Lokh92] [Hoog94]. The DCC began as a potential competitor for DATs but was discontinued in 1996. The introduction of the digital versatile disc (DVD) in 1996 enabled both video and audio recording/storage as well as text-message programming. The DVD became one of the most successful storage media. With the improvements in the audio compression and DVD storage technologies, multichannel surround sound encoding formats gained interest [Bosi93] [Holm99] [Bosi00].

With the emergence of streaming audio applications, during the late 1990s, researchers pursued techniques such as combined speech and audio architectures, as well as joint source-channel coding algorithms that are optimized for the packet-switched Internet. The advent of ISO/IEC MPEG-4 standard (1996–2000) [ISOI99] [ISOI00] established new research goals for high-quality coding of audio at low bit rates. MPEG-4 audio encompasses more functionality than perceptual coding [Koen98] [Koen99]. It comprises an integrated family of algorithms with provisions for scalable, object-based speech and audio coding at bit rates from as low as 200 b/s up to 64 kb/s per channel.

The emergence of the DVD-audio and the super audio CD (SACD) provided designers with additional storage capacity, which motivated research in *lossless* audio coding [Crav96] [Gerz99] [Reef01a]. A lossless audio coding system is able to reconstruct perfectly a bit-for-bit representation of the original input audio. In contrast, a coding scheme incapable of perfect reconstruction is called *lossy*. For most audio program material, lossy schemes offer the advantage of lower bit rates (e.g., less than 1 bit per sample) relative to lossless schemes (e.g., 10 bits per sample). Delivering real-time lossless audio content to the network browser at low bit rates is the next grand challenge for codec designers.

1.2 A GENERAL PERCEPTUAL AUDIO CODING ARCHITECTURE

Over the last few years, researchers have proposed several efficient signal models (e.g., transform-based, subband-filter structures, wavelet-packet) and compression standards (Table 1.1) for high-quality digital audio reproduction. Most of these algorithms are based on the generic architecture shown in Figure 1.1.

The coders typically segment input signals into quasi-stationary frames ranging from 2 to 50 ms. Then, a time-frequency analysis section estimates the temporal and spectral components of each frame. The time-frequency mapping is usually matched to the analysis properties of the human auditory system. Either way, the ultimate objective is to extract from the input audio a set of time-frequency parameters that is amenable to quantization according to a *perceptual distortion metric*. Depending on the overall design objectives, the time-frequency analysis section usually contains one of the following:

- Unitary transform
- Time-invariant bank of critically sampled, uniform/nonuniform bandpass filters

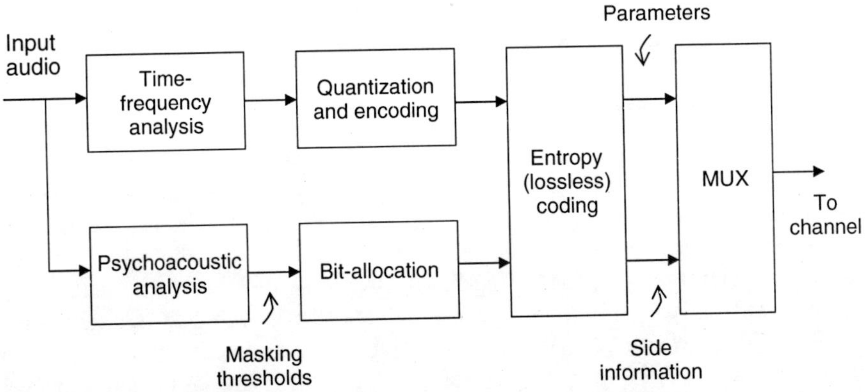

Figure 1.1. A generic perceptual audio encoder.

- Time-varying (signal-adaptive) bank of critically sampled, uniform/nonuniform bandpass filters
- Harmonic/sinusoidal analyzer
- Source-system analysis (LPC and multipulse excitation)
- Hybrid versions of the above.

The choice of time-frequency analysis methodology always involves a fundamental tradeoff between time and frequency resolution requirements. Perceptual distortion control is achieved by a psychoacoustic signal analysis section that estimates signal masking power based on psychoacoustic principles. The psychoacoustic model delivers masking thresholds that quantify the maximum amount of distortion at each point in the time-frequency plane such that quantization of the time-frequency parameters does not introduce audible artifacts. The psychoacoustic model therefore allows the quantization section to exploit perceptual irrelevancies. This section can also exploit statistical redundancies through classical techniques such as DPCM or ADPCM. Once a quantized compact parametric set has been formed, the remaining redundancies are typically removed through noiseless run-length (RL) and entropy coding techniques, e.g., Huffman [Cove91], arithmetic [Witt87], or Lempel-Ziv-Welch (LZW) [Ziv77] [Welc84]. Since the output of the psychoacoustic distortion control model is signal-dependent, most algorithms are inherently variable rate. Fixed channel rate requirements are usually satisfied through buffer feedback schemes, which often introduce encoding delays.

1.3 AUDIO CODER ATTRIBUTES

Perceptual audio coders are typically evaluated based on the following attributes: audio reproduction quality, operating bit rates, computational complexity, codec delay, and channel error robustness. The objective is to attain a high-quality (transparent) audio output at low bit rates (<32 kb/s), with an acceptable algorithmic delay (~5 to 20 ms), and with low computational complexity (~1 to 10 million instructions per second, or MIPS).

1.3.1 Audio Quality

Audio quality is of paramount importance when designing an audio coding algorithm. Successful strides have been made since the development of simple near-transparent perceptual coders. Typically, classical objective measures of signal fidelity such as the signal to noise ratio (SNR) and the total harmonic distortion (THD) are inadequate [Ryde96]. As the field of perceptual audio coding matured rapidly and created greater demand for listening tests, there was a corresponding growth of interest in perceptual measurement schemes. Several subjective and objective quality measures have been proposed and standardized during the last decade. Some of these schemes include the noise-to-mask ratio (NMR, 1987) [Bran87a] the perceptual audio quality measure (PAQM, 1991) [Beer91], the perceptual evaluation (PERCEVAL, 1992) [Pail92], the perceptual objective measure (POM, 1995) [Colo95], and the objective audio signal evaluation (OASE, 1997) [Spor97]. We will address these and several other quality assessment schemes in detail in Chapter 12.

1.3.2 Bit Rates

From a codec designer's point of view, one of the key challenges is to represent high-fidelity audio with a minimum number of bits. For instance, if a 5-ms audio frame sampled at 48 kHz (240 samples per frame) is represented using 80 bits, then the encoding bit rate would be 80 bits/5 ms = 16 kb/s. Low bit rates imply high compression ratios and generally low reproduction quality. Early coders such as the ISO/IEC MPEG-1 (32–448 kb/s), the Dolby AC-3 (32–384 kb/s), the Sony ATRAC (256 kb/s), and the Philips PASC (192 kb/s) employ high bit rates for obtaining transparent audio reproduction. However, the development of several sophisticated audio coding tools (e.g., MPEG-4 audio tools) created ways for efficient transmission or storage of audio at rates between 8 and 32 kb/s. Future audio coding algorithms promise to offer reasonable quality at low rates along with the ability to scale both *rate* and *quality* to match different requirements such as time-varying channel capacity.

1.3.3 Complexity

Reduced computational complexity not only enables real-time implementation but may also decrease the power consumption and extend battery life. Computational complexity is usually measured in terms of millions of instructions per second (MIPS). Complexity estimates are processor-dependent. For example, the complexity associated with Dolby's AC-3 decoder was estimated at approximately 27 MIPS using the Zoran ZR38001 general-purpose DSP core [Vern95]; for the Motorola DSP56002 processor, the complexity was estimated at 45 MIPS [Vern95]. Usually, most of the audio codecs rely on the so-called asymmetric encoding principle. This means that the codec complexity is not evenly shared between the encoder and the decoder (typically, encoder 80% and decoder 20% complexity), with more emphasis on reducing the decoder complexity.

1.3.4 Codec Delay

Many of the network applications for high-fidelity audio (streaming audio, audio-on-demand) are delay tolerant (up to 100–200 ms), providing the opportunity to exploit long-term signal properties in order to achieve high coding gain. However, in two-way real-time communication and voice-over Internet protocol (VoIP) applications, low-delay encoding (10–20 ms) is important. Consider the example described before, i.e., an audio coder operating on frames of 5 ms at a 48 kHz sampling frequency. In an ideal encoding scenario, the minimum amount of delay should be 5 ms at the encoder and 5 ms at the decoder (same as the frame length). However, other factors such as analysis-synthesis filter bank window, the look-ahead, the bit-reservoir, and the channel delay contribute to additional delays. Employing shorter analysis-synthesis windows, avoiding look-ahead, and re-structuring the bit-reservoir functions could result in low-delay encoding, nonetheless, with reduced coding efficiencies.

1.3.5 Error Robustness

The increasing popularity of streaming audio over packet-switched and wireless networks such as the Internet implies that any algorithm intended for such applications must be able to deal with a noisy time-varying channel. In particular, provisions for error robustness and error protection must be incorporated at the encoder in order to achieve reliable transmission of digital audio over error-prone channels. One simple idea could be to provide better protection to the error-sensitive and priority (important) bits. For instance, the audio frame header requires the maximum error robustness; otherwise, transmission errors in the header will seriously impair the entire audio frame. Several error detecting/correcting codes [Lin82] [Wick95] [Bayl97] [Swee02] [Zara02] can also be employed. Inclusion of error correcting codes in the bitstream might help to obtain error-free reproduction of the input audio, however, with increased complexity and bit rates.

From the discussion in the previous sections, it is evident that several tradeoffs must be considered in designing an algorithm for a particular application. For this reason, audio coding standards consist of several tools that enable the design of scalable algorithms. For example, MPEG-4 provides tools to design algorithms that satisfy a variety of bit rate, delay, complexity, and robustness requirements.

1.4 TYPES OF AUDIO CODERS – AN OVERVIEW

Based on the signal model or the analysis-synthesis technique employed to encode audio signals, audio coders can be broadly classified as follows:

- Linear predictive
- Transform

- Subband
- Sinusoidal.

Algorithms are also classified based on the lossy or the lossless nature of audio coding. Lossy audio coding schemes achieve compression by exploiting perceptually irrelevant information. Some examples of lossy audio coding schemes include the ISO/IEC MPEG codec series, the Dolby AC-3, and the DTS CA. In lossless audio coding, the audio data is merely "packed" to obtain a bit-for-bit representation of the original. The meridian lossless packing (MLP) [Gerz99] and the direct stream digital (DSD) techniques [Brue97] [Reef01a] form a class of high-end lossless compression algorithms that are embedded in the DVD-audio [DVD01] and the SACD [SACD02] storage formats, respectively. Lossless audio coding techniques, in general yield high-quality digital audio without any artifacts at high rates. For instance, perceptual audio coding yields compression ratios from 10:1 to 25:1, while lossless audio coding can achieve compression ratios from 2:1 to 4:1.

1.5 ORGANIZATION OF THE BOOK

This book is organized as follows. In Chapter 2, we review basic signal processing concepts associated with audio coding. Chapter 3 provides introductory material to waveform quantization and entropy coding schemes. Some of the key topics covered in this chapter include scalar quantization, uniform/nonuniform quantization, pulse code modulation (PCM), differential PCM (DPCM), adaptive DPCM (ADPCM), vector quantization (VQ), bit-allocation techniques, and entropy coding schemes (Huffman, Rice, and arithmetic).

Chapter 4 provides information on linear prediction and its application in narrow and wideband coding. First, we address the utility of LP analysis/synthesis approach in speech applications. Next, we describe the open-loop analysis-synthesis LP and closed-loop analysis-by-synthesis LP techniques.

In Chapter 5, psychoacoustic principles are described. Johnston's notion of perceptual entropy is presented as a measure of the fundamental limit of transparent compression for audio. The ISO/IEC 11172-3 MPEG-1 psychoacoustic analysis model 1 is used to describe the five important steps associated with the global masking threshold computation. Chapter 6 explores filter bank design issues and algorithms, with a particular emphasis placed on the modified discrete cosine transform (MDCT) that is widely used in several perceptual audio coding algorithms. Chapter 6 also addresses pre-echo artifacts and control strategies.

Chapters 7, 8, and 9 review established and emerging techniques for transparent coding of FM and CD-quality audio signals, including several algorithms that have become international standards. Transform coding methodologies are described in Chapter 7, subband coding algorithms are addressed in Chapter 8, and sinusoidal algorithms are presented in Chapter 9. In addition to methods based on uniform bandwidth filter banks, Chapter 8 covers coding methods that utilize discrete wavelet transforms (DWT), discrete wavelet packet transforms (DWPT), and other nonuniform filter banks. Examples of hybrid algorithms that make use of more than one signal model appear throughout Chapters 7, 8, and 9.

Chapter 10 is concerned with standardization activities in audio coding. It describes coding standards and products such as the ISO/IEC MPEG family (−1 ".MP1/2/3", −2, −4, −7, and −21), the Sony Minidisc (ATRAC), the cinematic Sony SDDS, the Lucent Technologies PAC/EPAC/MPAC, the Dolby AC-2/AC-3, the Audio Processing Technology APT-x100, and the DTS-coherent acoustics (DTS-CA). Details on the MP3 and MPEG-4 AAC algorithms that are popular in Web and in handheld media applications, e.g., Apple iPod, are provided.

Chapter 11 focuses on lossless audio coding and digital audio watermarking techniques. In particular, the SHORTEN, the DVD algorithm, the MUSICompress, the AudioPaK, the C-LPAC, the LTAC, and the IntMDCT lossless coding schemes are described in detail. Chapter 11 also addresses the two popular high-end storage formats, i.e., the SACD and the DVD-Audio. The MLP and the DSD techniques for lossless audio coding are also presented.

Chapter 12 provides information on subjective quality measures for perceptual codecs. The five-point absolute and differential subjective quality scales are addressed, as well as the subjective test methodologies specified in the ITU-R Recommendation BS.1116. A set of subjective benchmarks is provided for the various standards in both stereophonic and multichannel modes to facilitate algorithm comparisons.

1.6 NOTATIONAL CONVENTIONS

Unless otherwise specified, bit rates correspond to single-channel or monaural coding throughout this text. Subjective quality measurements are specified in terms of either the five-point mean opinion score (MOS, Table 1.3) or the 41-point subjective difference grade (SDG, Chapter 12, Table 12.1). Table 1.4 lists some of the symbols and notation used in the book.

Table 1.3. Mean opinion score (MOS) scale.

MOS	Perceptual quality
1	Bad
2	Poor
3	Average
4	Good
5	Excellent

Table 1.4. Symbols and notation used in the book.

Symbol/notation	Description
t, n	Time index/sample index
ω, Ω	Frequency index (analog domain, discrete domain)
$f(=\omega/2\pi)$	Frequency (Hz)
F_s, T_s	Sampling frequency, sampling period
$x(t) \leftrightarrow X(\omega)$	Continuous-time Fourier transform (FT) pair
$x(n) \leftrightarrow X(\Omega)$	Discrete-time Fourier transform (DTFT) pair
$x(n) \leftrightarrow X(z)$	z transform pair
$s[.]$	Indicates a particular element in a coefficient array
$h(n) \leftrightarrow H(\Omega)$	Impulse-frequency response pair of a discrete time system
$e(n)$	Error/prediction residual
$H(z) = \dfrac{B(z)}{A(z)} = \dfrac{1 + b_1 z^{-1} + \ldots + b_L z^{-L}}{1 + a_1 z^{-1} + \ldots + a_M z^{-M}}$	Transfer function consisting of numerator-polynomial and denominator-polynomial (corresponding to b-coefficients and a-coefficients)
$\tilde{s}(n) = \sum_{i=0}^{M} a_i s(n-i)$	Predicted signal
$Q\langle s(n)\rangle = \hat{s}(n)$	Quantization/approximation operator or estimated/encoded value
$s^{[.]}$	Square brackets in the superscript denote recursion
$s^{(.)}$	Parenthesis superscript; time dependency
N, N_f, N_{sf}	Total number of samples, samples per frame, samples per subframe
log(.), ln(.), $\log_p$(.)	Log to the base-10, log to the base-e, log to the base-p
E[.]	Expectation operator
ε	Mean squared error (MSE)
μ_x, σ_x^2	Mean and the variance of the signal, $x(n)$
$r_{xx}(m)$	Autocorrelation of the signal, $x(n)$
$r_{xy}(m)$	Cross-correlation of $x(n)$ and $y(n)$
$R_{xx}(e^{j\Omega})$	Power spectral density of the signal, $x(n)$
Bit rate	Number of bits per second (b/s, kb/s, or Mb/s)
dB, SPL	Decibels, sound pressure level

PROBLEMS

The objective of these introductory problems are to introduce the novice to simple relations between the sampling rate and the bit rate in PCM coded sequences.

1.1. Consider an audio signal, $s(n)$, sampled at 44.1 kHz and digitized using a) 8-bit, b) 24-bit, and c) 32-bit resolution. Compute the data rates for the cases (a)–(c). Give the number of samples within a 16-ms frame and compute the number of bits per frame.

1.2. List some of the typical data rates (in kb/s) and sampling rates (in kHz) employed in applications such as a) video streaming, b) audio streaming, c) digital audio broadcasting, d) digital compact cassette, e) MiniDisc, f) DVD, g) DVD-audio, h) SACD, i) MP3, j) MP4, k) video conferencing, and l) cellular telephony.

COMPUTER EXERCISES

The objective of this exercise is to familiarize the reader with the handling of sound files using MATLAB and to expose the novice to perceptual attributes of sampling rate and bit resolution.

1.3. For this computer exercise, use MATLAB workspace ***ch1pb1.mat*** from the website.

Load the workspace ***ch1pb1.mat*** using,

```
>> load('ch1pb1.mat');
```

Use `whos` command to view the variables in the workspace. The data-vector `'audio_in'` contains 44,100 samples of audio data. Perform the following in MATLAB:

```
>> wavwrite(audio_in,44100,16,'pb1_aud44_16.wav');
>> wavwrite(audio_in,10000,16,'pb1_aud10_16.wav');
>> wavwrite(audio_in,44100,8,'pb1_aud44_08.wav');
```

Listen to the wave files ***pb1_aud44_16.wav, pb1_aud10_16.wav***, and ***pb1_aud44_08.wav*** using a media player. Comment on the perceptual quality of the three wave files.

1.4. Down-sample the data-vector `'audio_in'` in problem 1.3 using

```
>> aud_down_4 = downsample(audio_in, 4);
```

Use the following commands to listen to `audio_in` and `aud_down_4`. Comment on the perceptual quality of the data vectors in each of the cases below:

```
>> sound(audio_in, fs);
>> sound(aud_down_4, fs);
>> sound(aud_down_4, fs/4);
```

CHAPTER 2

SIGNAL PROCESSING ESSENTIALS

2.1 INTRODUCTION

The signal processing theory described here will be restricted only to the concepts that are relevant to audio coding. Because of the limited scope of this chapter, we provide mostly qualitative descriptions and establish only the essential mathematical formulas. First, we briefly review the basics of continuous-time (analog) signals and systems and the methods used to characterize the frequency spectrum of analog signals. We then present the basics of analog filters and subsequently describe discrete-time signals. Coverage of the basics of discrete-time signals includes: the fundamentals of transforms that represent the spectra of digital sequences and the theory of digital filters. The essentials of random and multirate signal processing are also reviewed in this chapter.

2.2 SPECTRA OF ANALOG SIGNALS

The frequency spectrum of an analog signal is described in terms of the continuous Fourier transform (CFT). The CFT of a continuous-time signal, $x(t)$, is given by

$$X(\omega) = \int_{-\infty}^{\infty} x(t)e^{-j\omega t}\,dt, \qquad (2.1)$$

where ω is the frequency in radians per second (rad/s). Note that $\omega = 2\pi f$, where f is the frequency in Hz. The complex-valued function, $X(\omega)$, describes the CFT

Audio Signal Processing and Coding, by Andreas Spanias, Ted Painter, and Venkatraman Atti

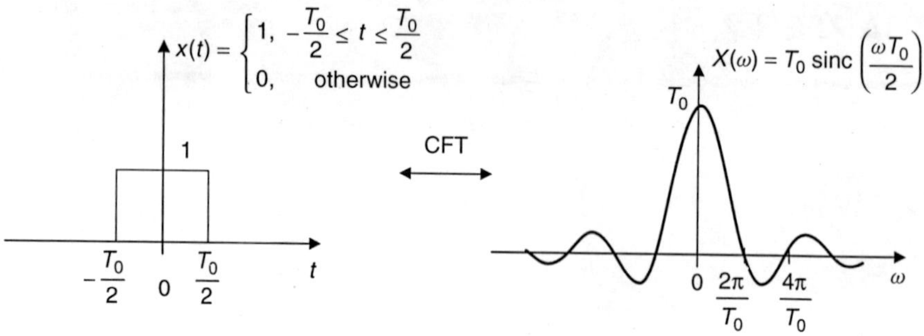

Figure 2.1. The pulse-sinc CFT pair.

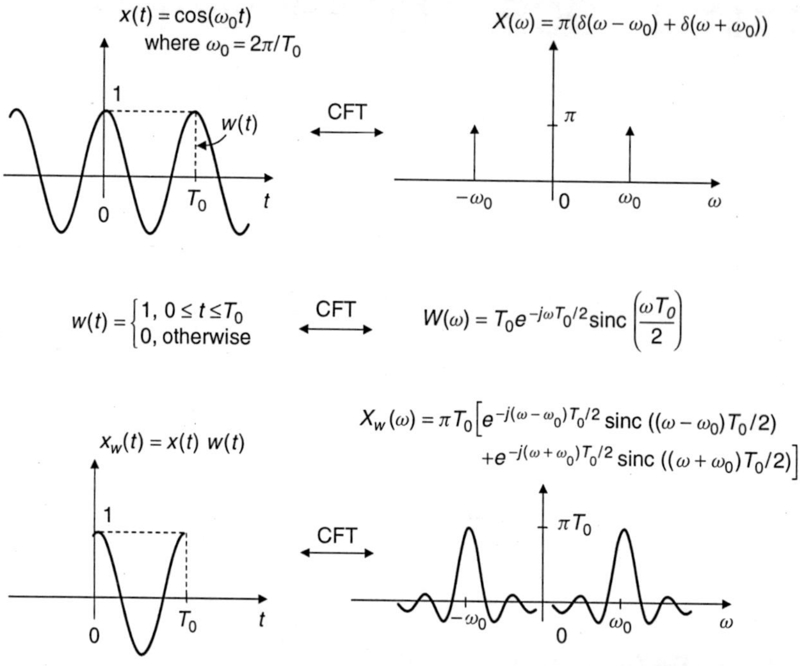

Figure 2.2. CFT of a sinusoid and a truncated sinusoid.

magnitude and phase spectrum of the signal. The inverse CFT is given by

$$x(t) = \frac{1}{2\pi}\int_{-\infty}^{\infty} X(\omega)e^{j\omega t}\,d\omega. \tag{2.2}$$

The inverse CFT is also known as the synthesis formula because it describes the time-domain signal, $x(t)$, in terms of complex sinusoids. In CFT theory, $x(t)$ and $X(\omega)$ are called a transform pair, i.e.,

$$x(t) \leftrightarrow X(\omega). \tag{2.3}$$

The pulse-sinc pair shown in Figure 2.1 is useful in explaining the effects of time-domain truncation on the spectra. For example, when a sinusoid is truncated then there is loss of resolution and spectral leakage as shown in Figure 2.2.

In real-life signal processing, all signals have finite length, and hence time-domain truncation always occurs. The truncation of an audio segment by a rectangular window is shown in Figure 2.3. To smooth out frame transitions and control spectral leakage effects, the signal is often tapered prior to truncation using window functions such as the Hamming, the Bartlett, and the trapezoidal windows. A tapered window avoids the sharp discontinuities at the edges of the truncated time-domain frame. This in turn reduces the spectral leakage in the frequency spectrum of the truncated signal. This reduction of spectral leakage is attributed to the reduced level of the sidelobes associated with tapered windows. The reduced sidelobe effects come at the expense of a modest loss of spectral

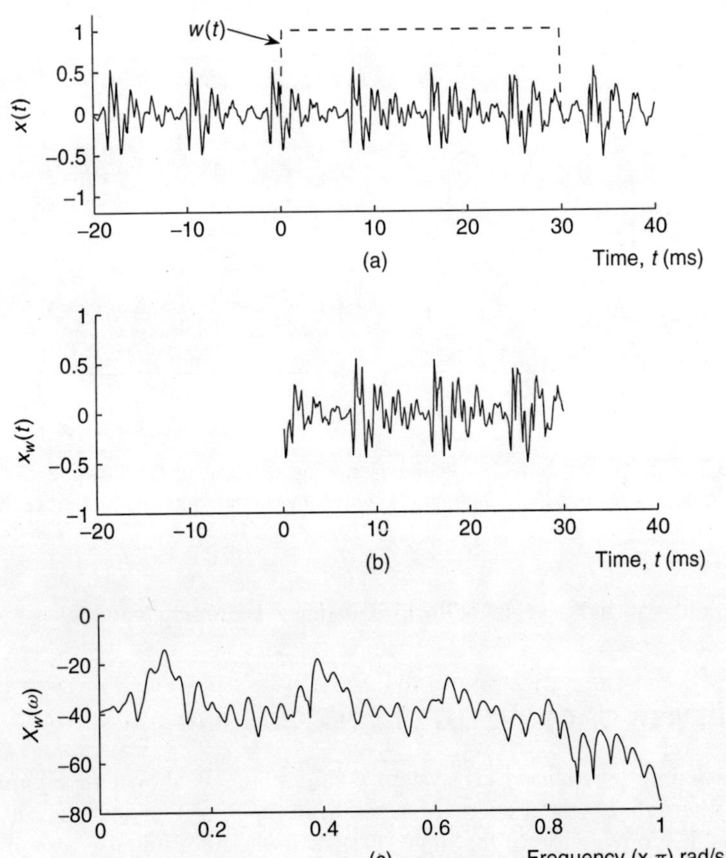

Figure 2.3. (a) Audio signal, $x(t)$ and a rectangular window, $w(t)$ (shown in dashed line); (b) truncated audio signal, $x_w(t)$; and (c) frequency-domain representation, $X_w(\omega)$, of the truncated audio.

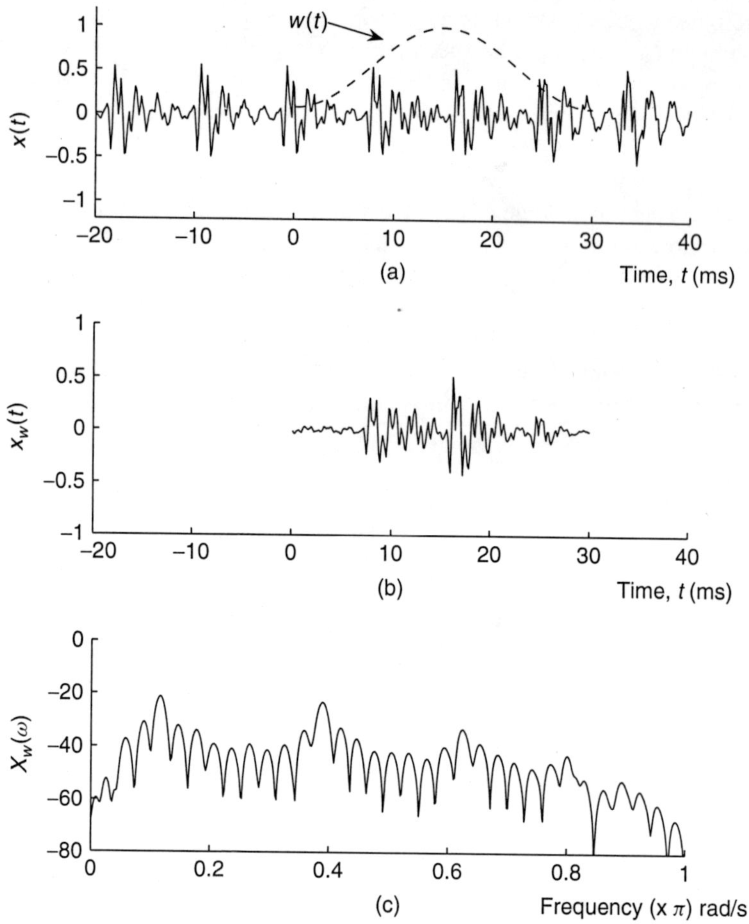

Figure 2.4. (a) Audio signal, $x(t)$ and a Hamming window, $w(t)$ (shown in dashed line); (b) truncated audio signal, $x_w(t)$; and (c) frequency-domain representation, $X_w(\omega)$, of the truncated audio.

resolution. An audio segment formed using a Hamming window is shown in Figure 2.4.

2.3 REVIEW OF CONVOLUTION AND FILTERING

A linear time-invariant (LTI) system configuration is shown in Figure 2.5. A linear filter satisfies the property of generalized superposition and hence its output, $y(t)$, is the convolution of the input, $x(t)$, with the filter impulse response, $h(t)$. Mathematically, convolution is represented by the integral in Eq. (2.4):

$$y(t) = \int_{-\infty}^{\infty} h(\tau)x(t-\tau)d\tau = h(t) * x(t). \tag{2.4}$$

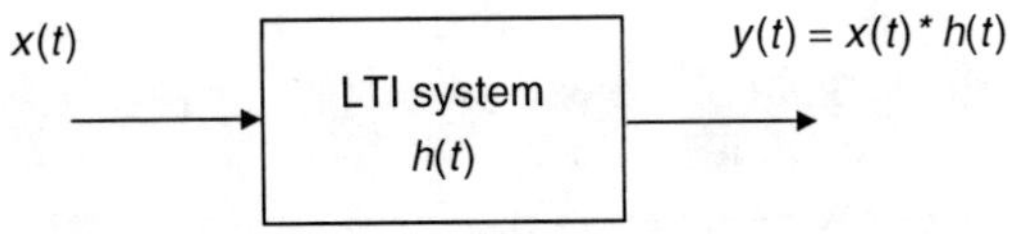

Figure 2.5. A linear time-invariant (LTI) system and convolution operation.

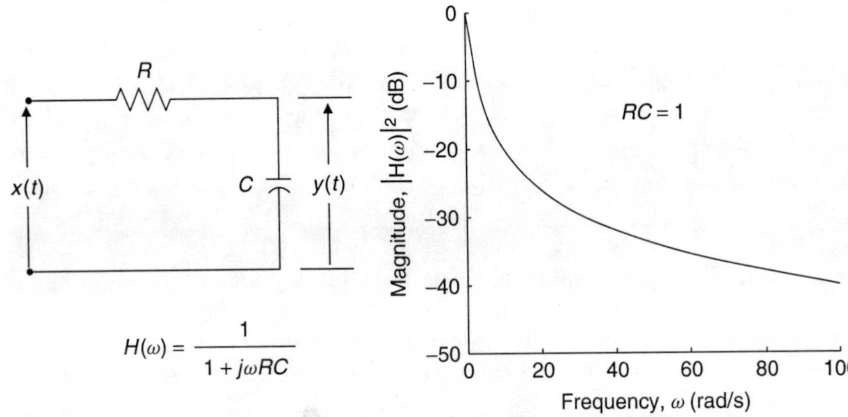

Figure 2.6. A simple RC low-pass filter.

The symbol * between the impulse response, $h(t)$, and the input, $x(t)$, is often used to denote the convolution operation.

The CFT of the impulse response, $h(t)$, is the frequency response of the filter, i.e.,

$$h(t) \leftrightarrow H(\omega). \tag{2.5}$$

As an example for reviewing these fundamental concepts in linear systems, we present in Figure 2.6 a simple first-order RC circuit that corresponds to a low-pass filter. The impulse response for this RC filter is a decaying exponential, and its frequency response is given by a simple first-order rational function, $H(\omega)$. This function is complex-valued and its magnitude represents the gain of the filter with respect to frequency at steady state. If a sinusoidal signal drives the linear filter, the steady-state output is also a sinusoid with the same frequency. However, its amplitude is scaled and phase is shifted in a manner consistent with the magnitude and phase of the frequency response function, respectively.

2.4 UNIFORM SAMPLING

In all of our subsequent discussions, we will be treating audio signals and associated systems in discrete time. The rules for uniform sampling of analog speech/audio are provided by the sampling theorem [Shan48]. This theorem states that a signal that is strictly bandlimited to a bandwidth of B rad/s can be uniquely represented by its sampled values spaced at uniform intervals that are not more than π/B seconds apart. In other words, if we denote the sampling period as T_s, then the sampling theorem states that $T_s \leqslant \pi/B$. In the frequency domain, and with the sampling frequency defined as $\omega_s = 2\pi f_s = 2\pi/T_s$, this condition can be stated as,

$$\omega_s \geqslant 2B(\text{rad/s}) \quad \text{or} \quad f_s \geqslant \frac{B}{\pi}. \tag{2.6}$$

Mathematically, the sampling process is represented by time-domain multiplication of the analog signal, $x(t)$, with an impulse train, $p(t)$, as shown in Figure 2.7.

Since multiplication in time is convolution in frequency, the CFT of the sampled signal, $x_s(t)$, corresponds to the CFT of the original analog signal, $x(t)$, convolved with the CFT of the impulse train, $p(t)$. The CFT of the impulses is also a train of uniformly spaced impulses in frequency that are spaced $1/T_s$ Hz apart. The CFT of the sampled signal is therefore a periodic extension of the CFT of the analog signal as shown in Figure 2.8. In Figure 2.8, the analog signal was considered to be ideally bandlimited and the sampling frequency, ω_s, was chosen to be more than $2B$ to avoid aliasing. The CFT of the sampled signal is

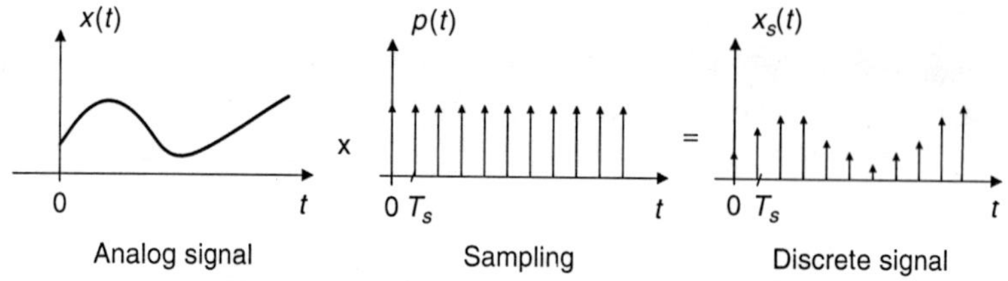

Figure 2.7. Uniform sampling of analog signals.

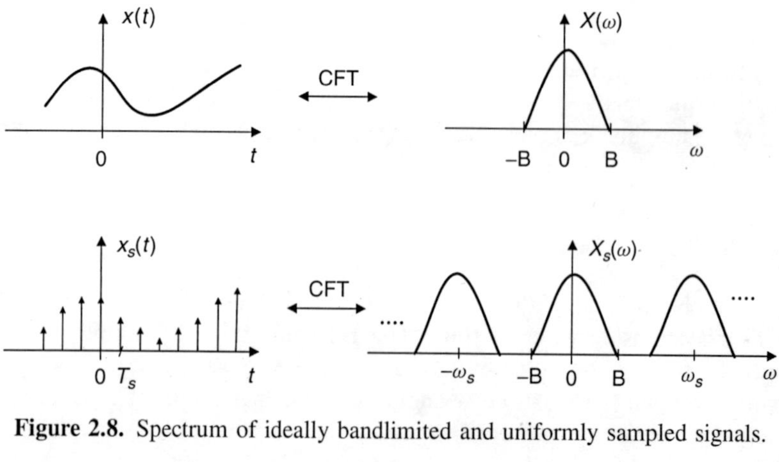

Figure 2.8. Spectrum of ideally bandlimited and uniformly sampled signals.

given by,

$$X_s(\omega) = \frac{1}{T_s} \sum_{k=-\infty}^{\infty} X(\omega - k\omega_s). \tag{2.7}$$

Note that the spectrum of the sampled signal in Figure 2.8 is such that an ideal low-pass filter (LPF) can recover the baseband of the signal and hence perfectly reconstruct the analog signal from the digital signal. The reconstruction process is shown in Figure 2.9. This reconstruction LPF essentially interpolates between the samples and reproduces the analog signal from the digital signal. The interpolation process becomes evident once the filtering operation is interpreted in the time domain as convolution. Reconstruction occurs by interpolating with the sinc function, which is the impulse response of the ideal low-pass filter. The reconstruction process for $\omega_s = 2B$ is given by,

$$x(t) = \sum_{n=-\infty}^{\infty} x(nT_s)\text{sinc}(B(t - nT_s)). \tag{2.8}$$

Note that if the sampling frequency is less than $2B$, then aliasing will occur, and therefore the signal can no longer be reconstructed perfectly. Figure 2.10 illustrates aliasing.

In real-life applications, the analog signal is not ideally bandlimited and the sampling process is not perfect, i.e., sampling pulses have finite amplitude and finite duration. Therefore, some level of aliasing is always present. To reduce

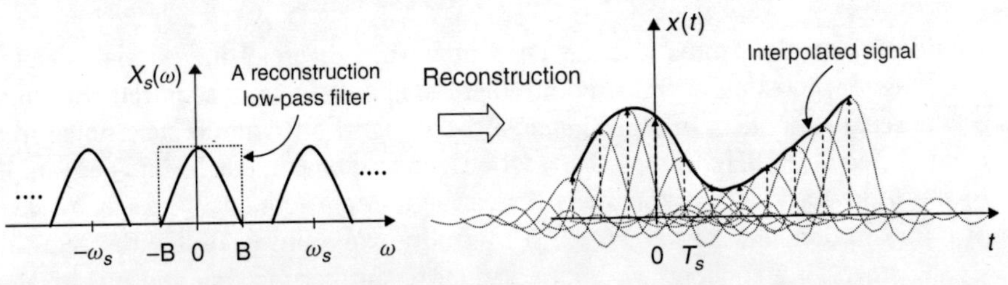

Figure 2.9. Reconstruction (interpolation) using a low-pass filter.

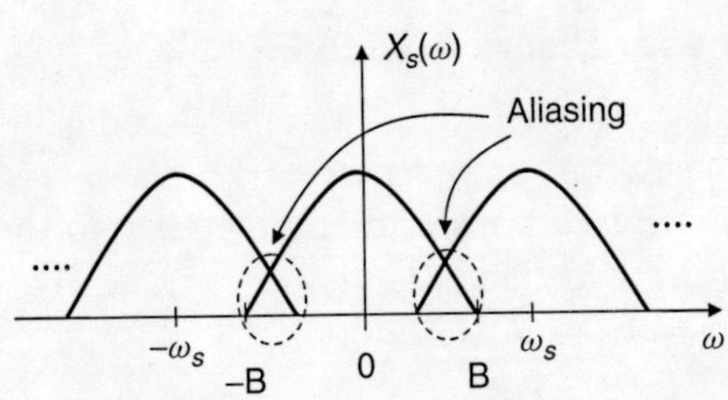

Figure 2.10. Aliasing when $\omega_s < 2B$.

Table 2.1. Sampling rates and bandwidth specifications.

Format	Bandwidth	Sampling frequency
Telephony	3.2 kHz	8 kHz
Wideband audio	7 kHz	16 kHz
High-fidelity, CD	20 kHz	44.1 kHz
Digital audio tape (DAT)	20 kHz	48 kHz
Super audio CD (SACD)	100 kHz	2.8224 MHz
DVD audio (DVD-A)	96 kHz	44.1, 48, 88.2, 96, 176.4, or 192 kHz

aliasing, the signal is prefiltered by an anti-aliasing low-pass filter and usually over-sampled ($\omega_s > 2B$). The degree of over-sampling depends also on the choice of the analog anti-aliasing filter. For high-quality reconstruction and modest over-sampling, the anti-aliasing filter must have good rejection characteristics. On the other hand, over-sampling by a large factor relaxes the requirements on the analog anti-aliasing filter and hence simplifies analog hardware at the expense of a higher data rate. Nowadays, over-sampling is practiced often even in high-fidelity systems. In fact, the use of inexpensive Sigma-Delta ($\Sigma\Delta$) analog-to-digital (A/D) converters, in conjunction with down-sampling in the digital domain, is a common practice. Details on $\Sigma\Delta$ A/D conversion and some over-sampling schemes tailored for high-fidelity audio will be presented in Chapter 11. Standard sampling rates for the different grades of speech and audio are given in Table 2.1.

2.5 DISCRETE-TIME SIGNAL PROCESSING

Audio coding algorithms operate on a quantized discrete-time signal. Prior to compression, most algorithms require that the audio signal is acquired with high-fidelity characteristics. In typical standardized algorithms, audio is assumed to be bandlimited at 20 kHz, sampled at 44.1 kHz, and quantized at 16 bits per sample. In the following discussion, we will treat audio as a sequence, i.e., as a stream of numbers denoted $x(n) = x(t)|_{t=nT_s}$. Initially, we will review the discrete-time signal processing concepts without considering further aliasing and quantization effects. Quantization effects will be discussed later during the description of specific coding algorithms.

2.5.1 Transforms for Discrete-Time Signals

Discrete-time signals are described in the transform domain using the z-transform and the discrete-time Fourier transform (DTFT). These two transformations have similar roles as the Laplace transform and the CFT for analog signals, respectively. The z-transform is defined as

$$X(z) = \sum_{n=-\infty}^{\infty} x(n)z^{-n}, \tag{2.9}$$

where z is a complex variable. Note that if the z-transform is evaluated on the unit circle, i.e., for

$$z = e^{j\Omega}, \Omega = 2\pi f T_s \tag{2.10}$$

then the z-transform becomes the discrete-time Fourier transform (DTFT). The DTFT is given by,

$$X(e^{j\Omega}) = \sum_{n=-\infty}^{\infty} x(n)e^{-jn\Omega}. \tag{2.11}$$

The DTFT is discrete in time and continuous in frequency. As expected, the frequency spectrum associated with the DTFT is periodic with period 2π rads.

Example 2.1

Consider the DTFT of a finite-length pulse:

$$\begin{aligned} x(n) &= 1, \quad \text{for } 0 \leqslant n \leqslant N-1 \\ &= 0, \quad \text{else.} \end{aligned}$$

Using geometric series results and trigonometric identities on the DTFT sum,

$$X(e^{j\Omega}) = \sum_{n=0}^{N-1} e^{-jn\Omega}$$

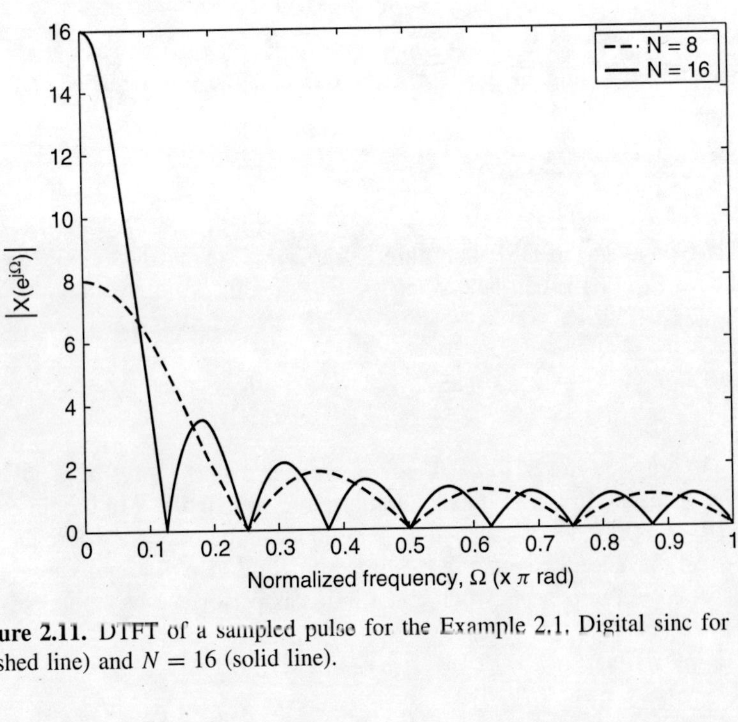

Figure 2.11. DTFT of a sampled pulse for the Example 2.1. Digital sinc for $N = 8$ (dashed line) and $N = 16$ (solid line).

$$= \frac{1 - e^{-jN\Omega}}{1 - e^{-j\Omega}}$$

$$= e^{-j(N-1)\Omega/2} \frac{\sin(N\Omega/2)}{\sin(\Omega/2)}. \tag{2.12}$$

The ratio of sinusoidal functions in Eq. (2.12) is known as the *Dirichlet* function or as a *digital sinc* function. Figure 2.11 shows the DTFT of a finite-length pulse. The digital sinc is quite similar to the continuous-time sinc function except that it is periodic with period 2π and has a finite number of sidelobes within a period.

2.5.2 The Discrete and the Fast Fourier Transform

A computational tool for Fourier transforms is developed by starting from the DTFT analysis expression (2.11), and considering a finite length signal consisting of N points, i.e.,

$$X(e^{j\Omega}) = \sum_{n=0}^{N-1} x(n)e^{-jn\Omega}. \tag{2.13}$$

Furthermore, the frequency-domain signal is sampled uniformly at N points within one period, $\Omega = 0$ to 2π, i.e.,

$$\Omega \Rightarrow \Omega_k = \frac{2\pi}{N}k, k = 0, 1, \ldots, N-1. \tag{2.14}$$

With the sampling in the frequency domain, the Fourier sum of Eq. (2.13) becomes

$$X(e^{j\Omega_k}) = \sum_{n=0}^{N-1} x(n)e^{-jn\Omega_k}. \tag{2.15}$$

It is typical in the DSP literature to replace Ω_k with the frequency index k and hence Eq. (2.15) can be written as,

$$X(k) = \sum_{n=0}^{N-1} x(n)e^{-j2\pi kn/N}, \quad k = 0, 1, 2, \ldots, N-1. \tag{2.16}$$

The expression in (2.16) is called the *discrete Fourier transform* (DFT). Note that the sampling in the frequency domain forces periodicity in the time domain, i.e., $x(n) = x(n+N)$. We also have periodicity in the frequency domain, $X(k) = X(k+N)$, because the signal in the time domain is also discrete. These periodicities create circular effects when convolution is performed by frequency-domain multiplication, i.e.,

$$x(n) \otimes h(n) \leftrightarrow X(k)H(k), \tag{2.17}$$

where

$$x(n) \otimes h(n) = \sum_{m=0}^{N-1} h(m)\, x((n-m)_{\mathrm{mod}N}). \tag{2.18}$$

The symbol $\otimes$ stands for circular or periodic convolution; and mod N implies modulo N subtraction of indices. The DFT is a one-to-one transformation whose basis functions are orthogonal. With the proper normalization, the DFT matrix can be written as a unitary matrix. The N-point inverse DFT (IDFT) is written as

$$x(n) = \frac{1}{N}\sum_{k=0}^{N-1} X(k)e^{j2\pi kn/N}, n = 0, 1, 2, \ldots, N-1. \tag{2.19}$$

The DFT transform pair is represented by the following notation:

$$x(n) \leftrightarrow X(k). \tag{2.20}$$

The DFT can be computed efficiently using the fast Fourier transform (FFT). The FFT takes advantage of redundancies in the DFT sum by decimating the sequence into subsequences with even and odd indices. It can be shown that if N is a radix-2 integer, the N-point DFT can be computed using a series of *butterfly* stages. The complexity associated with the DFT algorithm is of the order of N^2 computations. In contrast, the number of computations associated with the FFT algorithm is roughly of the order of $N \log_2 N$. This is a significant reduction in computational complexity and FFTs are almost always used in lieu of a DFT.

2.5.3 The Discrete Cosine Transform

The discrete cosine transform (DCT) of $x(n)$ can be defined as

$$X(k) = c(k)\sqrt{\frac{2}{N}}\sum_{n=0}^{N-1} x(n)\cos\left[\frac{\pi}{N}\left(n+\frac{1}{2}\right)k\right], 0 \leqslant k \leqslant N-1, \tag{2.21}$$

where $c(0) = 1/\sqrt{2}$, and $c(k) = 1$ for $1 \leqslant k \leqslant N-1$. Depending on the periodicity and the symmetry of the input signal, $x(n)$, the DCT can be computed using different orthonormal transforms (usually DCT-1, DCT-2, DCT-3, and DCT-4). More details on the DCT and the modified DCT (MDCT) [Malv91] are given in Chapter 6.

2.5.4 The Short-Time Fourier Transform

Spectral analysis of nonstationary signals cannot be accommodated by the classical Fourier transform since the signal has time-varying characteristics. Instead, a time-frequency transformation is required. Time-varying spectral

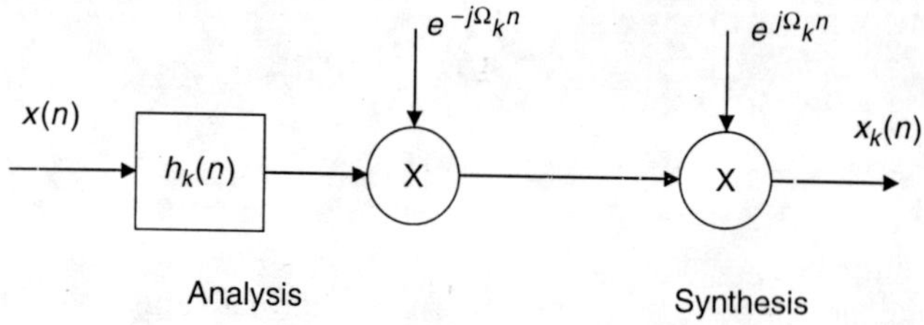

Figure 2.12. The k-th channel of the analysis-synthesis filterbank (after [Rabi78]).

analysis [Silv74] [Alle77] [Port81a] can be performed using the short-time Fourier transform (STFT). The analysis expression for the STFT is given by

$$X(n, \Omega) = \sum_{m=-\infty}^{\infty} x(m)h(n-m)e^{-j\Omega m} = h(n) * x(n)e^{-j\Omega n}, \tag{2.22}$$

where $\Omega = \omega T = 2\pi f T$ is the normalized frequency in radians, and $h(n)$ is the sliding analysis window. The synthesis expression (inverse transform) is given by

$$h(n-m)x(m) = \frac{1}{2\pi}\int_{-\pi}^{\pi} X(n, \Omega)e^{j\Omega m}\, d\Omega. \tag{2.23}$$

Note that if $n = m$ and $h(0) = 1$ [Rabi78] [Port80], then $x(n)$ can be obtained from Eq. (2.23). The basic assumption in this type of analysis-synthesis is that the signal is slowly time-varying and can be modeled by its short-time spectrum. The temporal and spectral resolution of the STFT are controlled by the length and shape of the analysis window. For speech and audio signals, the length of the window is often constrained to be about 5–20 ms and hence spectral resolution is sacrificed. The sequence, $h(n)$, can also be viewed as the impulse response of a LTI filter, which is excited by a frequency-shifted signal (see Eq. (2.22)). The latter leads to the filter-bank interpretation of the STFT, i.e., for a discrete frequency variable $\Omega_k = k(\Delta\Omega)$, $k = 0, 1, \ldots N-1$ and $\Delta\Omega$ and N chosen such that the speech band is covered. Then the analysis expression is written as

$$X(n, \Omega_k) = \sum_{m=-\infty}^{\infty} x(m)h(n-m)e^{-j\Omega_k m} = h(n) * x(n)e^{-j\Omega_k n} \tag{2.24}$$

and the synthesis expression is

$$\tilde{x}_{STFT}(n) = \sum_{k=0}^{N-1} X(n, \Omega_k)e^{j\Omega_k n}, \tag{2.25}$$

where $\tilde{x}_{STFT}(n)$ is the signal reconstructed within the band of interest. If $h(n)$, $\Delta\Omega$, and N are chosen carefully [Scha75], the reconstruction by Eq. (2.25) can be exact. The k-th channel analysis-synthesis scheme is depicted in Figure 2.12, where $h_k(n) = h(n)e^{j\Omega_k n}$.

2.6 DIFFERENCE EQUATIONS AND DIGITAL FILTERS

Digital filters are characterized by difference equations of the form

$$y(n) = \sum_{i=0}^{L} b_i x(n-i) - \sum_{i=1}^{M} a_i y(n-i). \quad (2.26)$$

In the input-output difference equation above, the output $y(n)$ is given as the linear combination of present and past inputs minus a linear combination of past outputs (feedback term). The parameters a_i and b_i are the filter coefficients or filter taps and they control the frequency response characteristics of the digital filter. Filter coefficients are programmable and can be made adaptive (time-varying). A direct-form realization of the digital filter is shown in Figure 2.13.

The filter in the Eq. (2.26) is referred to as an infinite-length impulse response (IIR) filter. The impulse response, $h(n)$, of the filter shown in Figure 2.13 is given by

$$h(n) = \sum_{i=0}^{L} b_i \delta(n-i) - \sum_{i=0}^{M} a_i h(n-i). \quad (2.27)$$

The IIR classification stems from the fact that, when the feedback coefficients are non-zero, the impulse response is infinitely long. In a statistical signal representation, Eq. (2.26) is referred to as a time-series model. That is, if the input of this filter is white noise then $y(n)$ is called an autoregressive moving average (ARMA) process. The feedback coefficients, a_i, are chosen such that the filter is stable, i.e., a bounded input gives a bounded output (BIBO). An input-output equation of a causal filter can also be written in terms of the impulse response of the filter, i.e.,

$$y(n) = \sum_{m=0}^{\infty} h(m)x(n-m). \quad (2.28)$$

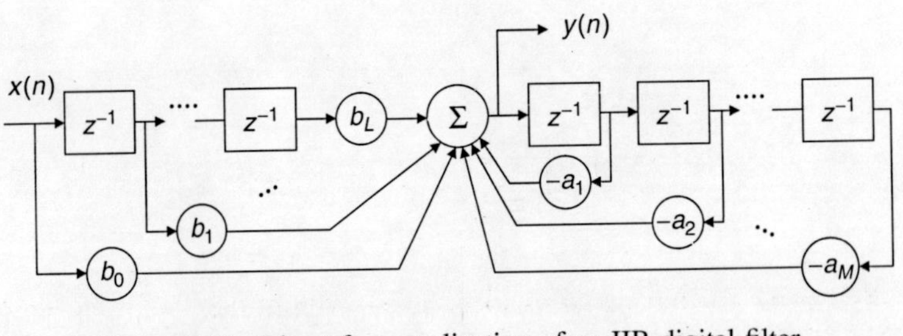

Figure 2.13. Direct-form realization of an IIR digital filter.

The impulse response of the filter is associated with its coefficients and can be computed explicitly by programming the difference equation. It can also be obtained in closed form by solving the difference equation.

Example 2.2

Consider the first-order IIR filter shown in Figure 2.14. The difference equation of this digital filter is given by

$$y(n) = 0.2x(n) + 0.8y(n-1).$$

The coefficient $b_0 = 0.2$ and $a_1 = -0.8$. The impulse response of this filter is given by

$$h(n) = 0.2\delta(n) + 0.8h(n-1).$$

The impulse response can be determined in closed-form by solving the above difference equation. Note that $h(0) = 0.2$ and $h(1) = 0.16$. Therefore, the closed-form expression for the impulse response is $h(n) = 0.2(0.8)^n u(n)$. Note also that this first-order IIR filter is BIBO stable because

$$\sum_{n=-\infty}^{\infty} |h(n)| < \infty. \quad (2.29)$$

Digital filters with finite-length impulse response (FIR) are realized by setting the feedback coefficients, $a_i = 0$, for $i = 1, 2, \ldots, M$. FIR filters, Figure 2.15, are inherently BIBO stable because their impulse response is always absolutely summable. The output of an FIR filter is a weighted moving average of the input. The simplest FIR filter is the so-called *averaging filter* that is used in some simple estimation applications. The input-output equation of the averaging filter is given by

$$y(n) = \frac{1}{L+1}\sum_{i=0}^{L} x(n-i). \quad (2.30)$$

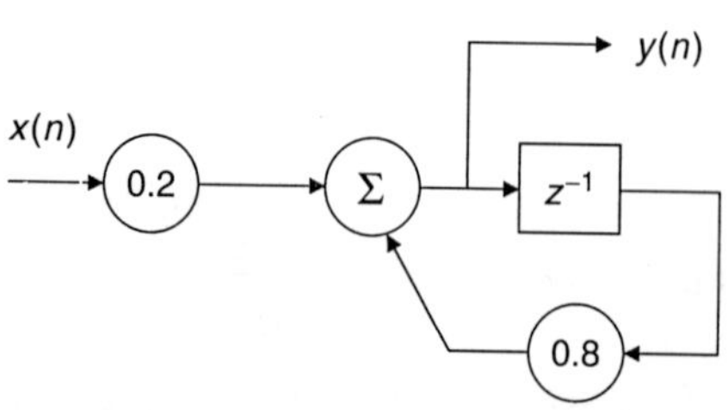

Figure 2.14. A first-order IIR digital filter.

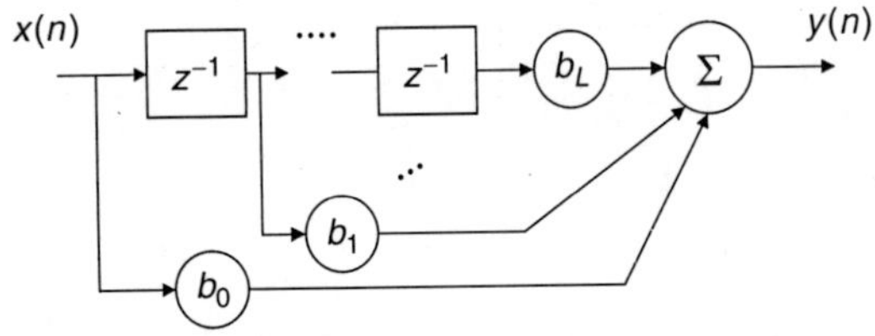

Figure 2.15. An FIR digital filter.

The impulse response of this filter is equal to $h(n) = 1/(L+1)$ for $n = 0, 1, \ldots, L$. The frequency response of the averaging filter is the DTFT of its impulse response, $h(n)$. Therefore, frequency responses of averaging filters for $L = 7$ and 15 are normalized versions of the DTFT spectra shown in Figure 2.11(a) and Figure 2.11(b), respectively.

2.7 THE TRANSFER AND THE FREQUENCY RESPONSE FUNCTIONS

The z-transform of the impulse response of a filter is called the transfer function and is given by

$$H(z) = \sum_{n=-\infty}^{\infty} h(n) z^{-n}. \tag{2.31}$$

Considering the difference equation, we can also obtain the transfer function in terms of filter parameters, i.e.,

$$X(z)\left(\sum_{i=0}^{L} b_i z^{-i}\right) = Y(z)\left(1 + \sum_{i=1}^{M} a_i z^{-i}\right). \tag{2.32}$$

The ratio of output over input in the z domain gives the transfer function in terms of the filter coefficients

$$H(z) = \frac{Y(z)}{X(z)} = \frac{b_0 + b_1 z^{-1} + \ldots + b_L z^{-L}}{1 + a_1 z^{-1} + \ldots + a_M z^{-M}}. \tag{2.33}$$

For an FIR filter, the transfer function is given by

$$H(z) = \sum_{i=0}^{L} b_i z^{-i}. \tag{2.34}$$

The frequency response function is a special case of the transfer function of the filter. That is for $z = e^{j\Omega}$, then

$$H(e^{j\Omega}) = \sum_{n=-\infty}^{\infty} h(n) e^{-jn\Omega}.$$

By considering the difference equation associated with the LTI digital filter, the frequency response can be written as the ratio of two polynomials, i.e.,

$$H(e^{j\Omega}) = \frac{b_0 + b_1 e^{-j\Omega} + b_2 e^{-j2\Omega} + \ldots + b_L e^{-jL\Omega}}{1 + a_1 e^{-j\Omega} + a_2 e^{-j2\Omega} + \ldots + a_M e^{-jM\Omega}}.$$

Note that for an FIR filter the frequency response becomes

$$H(e^{j\Omega}) = b_0 + b_1 e^{-j\Omega} + b_2 e^{-j2\Omega} + \ldots + b_L e^{-jL\Omega}.$$

Example 2.3

Frequency responses of four different first-order filters are shown in Figure 2.16. The frequency responses in Figure 2.16 are plotted up to the fold-over frequency, which is half the sampling frequency. Note from Figure 2.16 that low-pass and high-pass filters can be realized as either FIR or as IIR filters. The location of the root of the polynomial of the FIR filter determines where the notch in the frequency response occurs. Therefore, in the top two figures that correspond to the FIR filters, the low-pass filter (top left) has a notch at π rads (zero at $z = -1$), while the high-pass filter has a notch at

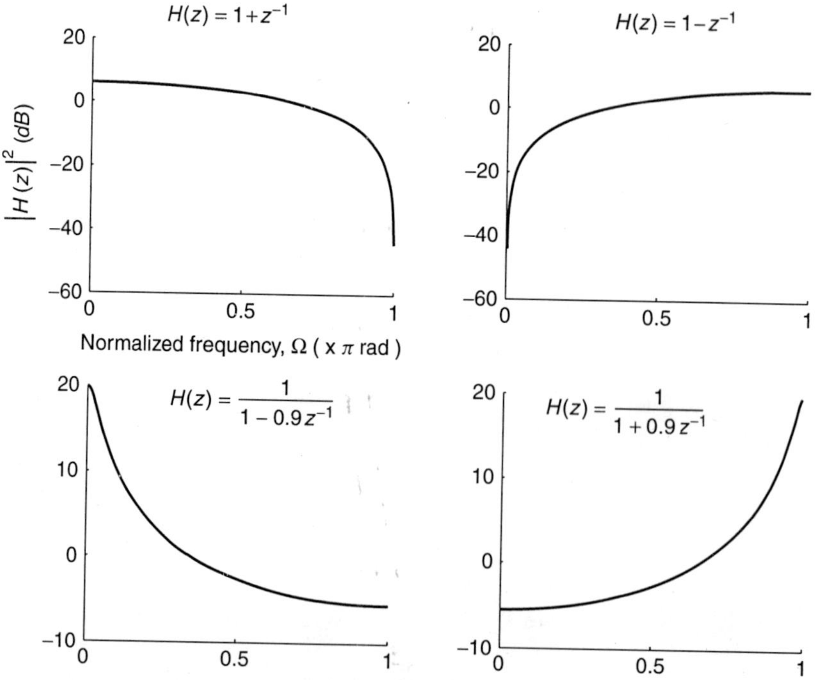

Figure 2.16. Frequency responses of first-order FIR and IIR digital filters.

0 rads (zero at $z = 1$). The bottom two IIR filters have a pole at $z = 0.9$ (peak at 0 rads) and $z = -0.9$ (peak at π rads) for the low-pass and high-pass filters, respectively.

2.7.1 Poles, Zeros, and Frequency Response

A z domain function, $H(z)$, can be written in terms of its poles and zeros as follows:

$$H(z) = G\frac{(z-\zeta_1)(z-\zeta_2)\ldots(z-\zeta_L)}{(z-p_1)(z-p_2)\ldots(z-p_M)} = G\frac{\prod_{i=1}^{L}(z-\zeta_i)}{\prod_{i=1}^{M}(z-p_i)}, \qquad (2.35)$$

where ζ_i and p_i are the zeros and poles of $H(z)$, respectively, and G is a constant. The locations of the poles and zeros affect the shape of the frequency response. The magnitude of the frequency response can be written as

$$|H(e^{j\Omega})| = G\frac{\prod_{i=1}^{L}|e^{j\Omega}-\zeta_i|}{\prod_{i=1}^{M}|e^{j\Omega}-p_i|}. \qquad (2.36)$$

It is therefore evident that when an isolated zero is close to the unit circle, then the magnitude frequency response will assume a small value at that frequency. When an isolated pole is close to unit circle it will give rise to a peak in the magnitude frequency response at that frequency. In speech processing, the presence of poles in z domain representations of the vocal tract, has been associated with the speech *formants* [Rabi78]. In fact, formant synthesizers use the pole locations to form synthesis filters for certain phonemes. On the other hand, the presence of zeros has been associated with the coupling of the nasal tract. For example, zeros associate with nasal sounds such as m and n [Span94].

Example 2.4

For the second-order system below, find the poles and zeros, give a z-domain diagram with the pole and zeros, and sketch the frequency response:

$$H(z) = \frac{1 - 1.3435z^{-1} + 0.9025z^{-2}}{1 - 0.45z^{-1} + 0.55z^{-2}}.$$

The poles and zeros appear in conjugate pairs because the coefficients of $H(z)$ are real-valued:

$$H(z) = \frac{(z - .95e^{j45^o})(z - .95e^{-j45^o})}{(z - .7416e^{j72.34^o})(z - .7416e^{-j72.34^o})}.$$

The pole zero diagram and the frequency response are shown in Figure 2.17. Poles give rise to spectral peaks and zeros create spectral valleys in the magnitude of the frequency response. The symmetry around π is due to the fact that roots appear in complex conjugate pairs.

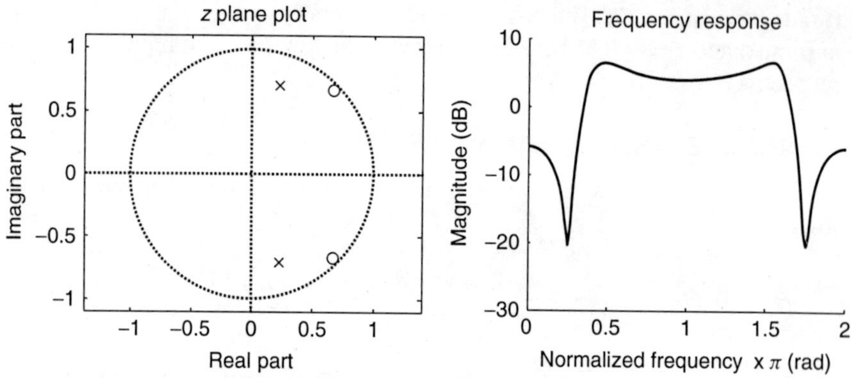

Figure 2.17. z domain and frequency response plots of the second-order system given in Example 2.4.

2.7.2 Examples of Digital Filters for Audio Applications

There are several standard designs for digital filters that are targeted specifically for audio-type applications. These designs include the so-called *shelving* filters, *peaking* filters, *cross-over* filters, and quadrature mirror filter (QMF) bank filters. Low-pass and high-pass shelving filters are used for bass and treble tone controls, respectively, in stereo systems.

Example 2.5

The transfer function of a low-pass shelving filter can be expressed as

$$H_{lp}(z) = C_{lp}\left(\frac{1 - b_1 z^{-1}}{1 - a_1 z^{-1}}\right), \tag{2.37}$$

where

$$C_{lp} = \frac{1 + k\mu}{1 + k},\ b_1 = \left(\frac{1 - k\mu}{1 + k\mu}\right),\ a_1 = \left(\frac{1 - k}{1 + k}\right)$$

$$k = \left(\frac{4}{1 + \mu}\right)\tan\left(\frac{\Omega_c}{2}\right) \text{ and } \mu = 10^{g/20}.$$

Note also that $\Omega_c = 2\pi f_c/f_s$ is the normalized cutoff frequency and g is the gain in decibels (dB).

Example 2.6

The transfer function of a high-pass shelving filter is given by

$$H_{hp}(z) = C_{hp}\left(\frac{1 - b_1 z^{-1}}{1 - a_1 z^{-1}}\right), \tag{2.38}$$

where

$$C_{hp} = \frac{\mu + p}{1 + p}, b_1 = \left(\frac{\mu - p}{\mu + p}\right), a_1 = \left(\frac{1 - p}{1 + p}\right)$$

$$p = \left(\frac{1 + \mu}{4}\right) \tan\left(\frac{\Omega_c}{2}\right) \text{ and } \mu = 10^{g/20}.$$

Again $\Omega_c = 2\pi f_c/f_s$ is the normalized cutoff frequency and g is the gain in dB. More complex tone controls that operate as graphic equalizers are accomplished using bandpass *peaking* filters.

Example 2.7

The transfer function of a peaking filter is given by

$$H_{pk}(z) = C_{pk}\left(\frac{1 + b_1 z^{-1} + b_2 z^{-2}}{1 + a_1 z^{-1} + a_2 z^{-2}}\right), \tag{2.39}$$

where

$$C_{pk} = \left(\frac{1 + k_q \mu}{1 + k_q}\right)$$

$$b_1 = \frac{-2\cos(\Omega_c)}{1 + k_q \mu}, \quad b_2 = \frac{1 - k_q \mu}{1 + k_q \mu},$$

$$a_1 = \frac{-2\cos(\Omega_c)}{1 + k_q}, \quad a_2 = \frac{1 - k_q}{1 + k_q},$$

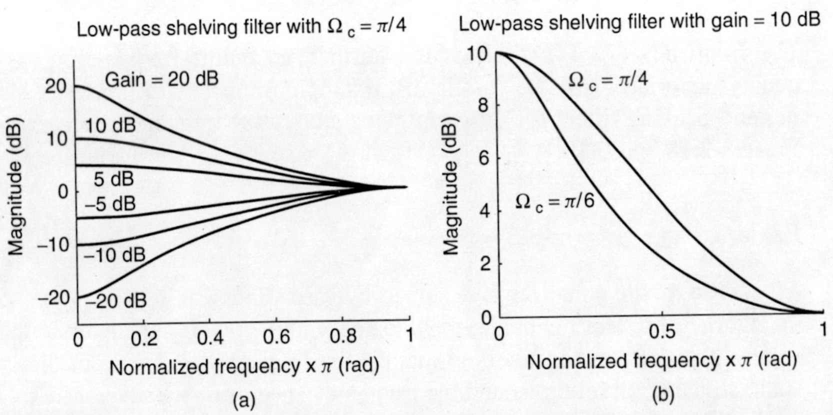

Figure 2.18. Frequency responses of a low-pass shelving filter: (a) for different gains and (b) for different cutoff frequencies, $\Omega_c = \pi/6$ and $\pi/4$.

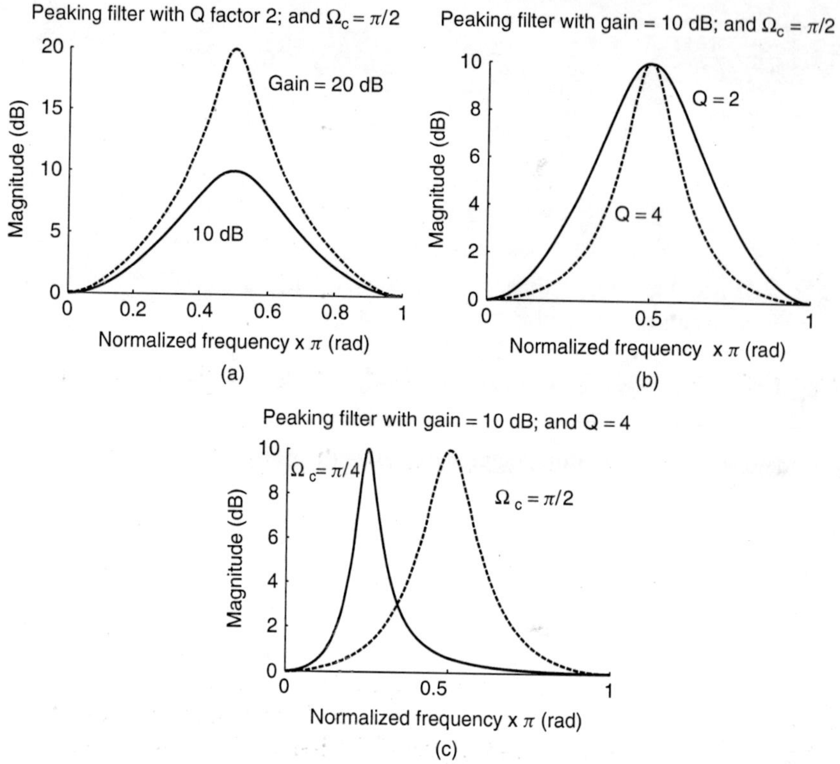

Figure 2.19. Frequency responses of a peaking filter: (a) for different gains, g = 10 dB and 20 dB; (b) for different quality factors, Q = 2 and 4; and (c) for different cutoff frequencies, $\Omega_c = \pi/4$ and $\pi/2$.

$$k_q = \left(\frac{4}{1+\mu}\right)\tan\left(\frac{\Omega_c}{2Q}\right), \quad \text{and } \mu = 10^{g/20}.$$

The frequency $\Omega_c = 2\pi f_c/f_s$ is the normalized cutoff frequency, Q is the quality factor, and g is the gain in dB. Example frequency responses of shelving and peaking filters for different gains and cutoff frequencies are given in Figures 2.18 and 2.19.

Example 2.8

An audio graphic equalizer is designed by cascading peaking filters as shown in Figure 2.20. The main idea behind the audio graphic equalizer is that it applies a set of peaking filters to modify the frequency spectrum of the input audio signal by dividing its audible frequency spectrum into several frequency bands. Then the frequency response of each band can be controlled by varying the corresponding peaking filter's gain.

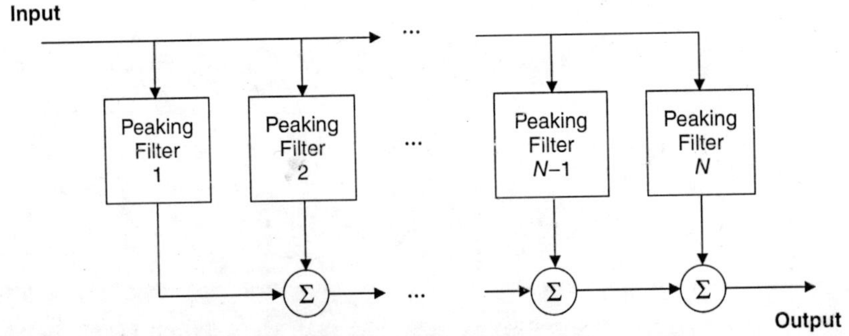

Figure 2.20. A cascaded setup of peaking filters to design an audio graphic equalizer.

2.8 REVIEW OF MULTIRATE SIGNAL PROCESSING

Multirate signal processing (MSP) involves the change of the sampling rate while the signal is in the digital domain. Sampling rate changes have been popular in DSP and audio applications. Depending on the application, changes in the sampling rate may reduce algorithmic and hardware complexity or increase resolution in certain signal processing operations by introducing additional signal samples. Perhaps the most popular application of MSP is over-sampling analog-to-digital (A/D) and digital-to-analog (D/A) conversions. In over-sampling A/D, the signal is over-sampled thereby relaxing the anti-aliasing filter design requirements, and, hence, the hardware complexity. The additional time-resolution in the over-sampled signal allows a simple 1-bit delta modulation (DM) quantizer to deliver a digital signal with sufficient resolution even for high-fidelity audio applications. This reduction of analog hardware complexity comes at the expense of a data rate increase. Therefore, a down-sampling operation is subsequently performed using a DSP chip to reduce the data rate. This reduction in the sampling rate requires a high precision anti-aliasing digital low-pass filter along with some other correcting DSP algorithmic steps that are of appreciable complexity. Therefore, the over-sampling analog-to-digital (A/D) conversion, or as otherwise called Delta-Sigma A/D conversion, involves a process where complexity is transferred from the analog hardware domain to the digital software domain. The reduction of analog hardware complexity is also important in D/A conversion. In that case, the signal is up-sampled and interpolated in the digital domain, thereby, reducing the requirements on the analog reconstruction (interpolation) filter.

2.8.1 Down-sampling by an Integer

Multirate signal processing is characterized by two basic operations, namely, up-sampling and down-sampling. Down-sampling involves increasing the sampling period and hence decreasing the sampling frequency and data rate of the digital signal. A sampling rate reduction by integer L is represented by

$$x_d(n) = x(nL). \tag{2.40}$$

Given the DTFT transform pairs

$$x(n) \overset{DTFT}{\longleftrightarrow} X(e^{j\Omega}) \quad \text{and} \quad x_d(n) \overset{DTFT}{\longleftrightarrow} X_d(e^{j\Omega}), \tag{2.41}$$

it can be shown [Oppe99] that the DTFT of the original and decimated signal are related by

$$X_d(e^{j\Omega}) = \frac{1}{L}\sum_{l=0}^{L-1} X(e^{j(\Omega-2\pi l)/L}). \tag{2.42}$$

Therefore, down-sampling introduces L copies of the original DTFT that are both amplitude and frequency scaled by L. It is clear that the additional copies may introduce aliasing. Aliasing can be eliminated if the DTFT of the original signal is bandlimited to a frequency π/L, i.e.,

$$X(e^{j\Omega}) = 0, \quad \frac{\pi}{L} \leqslant |\Omega| \leqslant \pi. \tag{2.43}$$

An example of the DTFTs of the signal during the down-sampling process is shown Figure 2.21. To approximate the condition in Eq. (2.43), a digital anti-aliasing filter is used. The down-sampling process is illustrated in Figure 2.22.

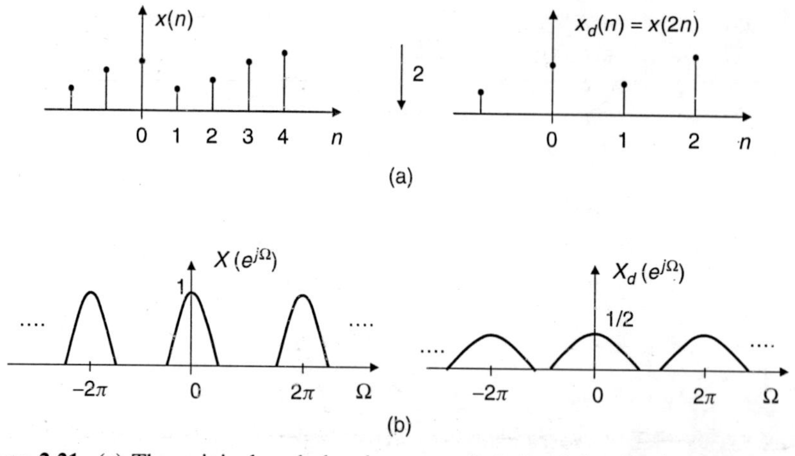

Figure 2.21. (a) The original and the down-sampled signal in the time-domain; and (b) the corresponding DTFTs.

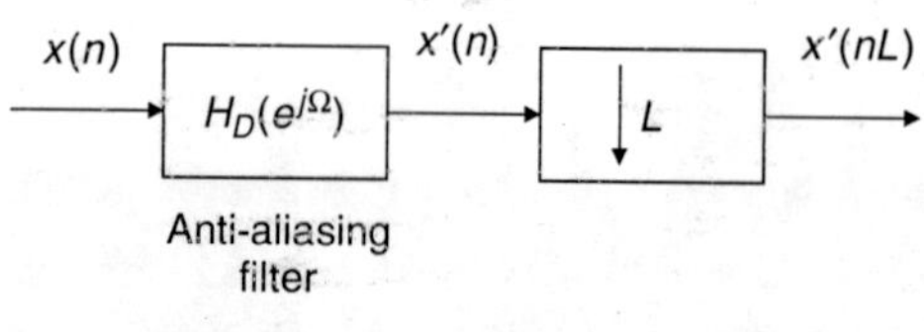

Figure 2.22. Down-sampling by an integer L.

In Figure 2.22, $H_D(e^{j\Omega})$ is given by

$$H_D(e^{j\Omega}) = \begin{cases} 1, & 0 \leqslant |\Omega| \leqslant \pi/L \\ 0, & \pi/L \leqslant |\Omega| \leqslant \pi \end{cases}. \tag{2.44}$$

2.8.2 Up-sampling by an Integer

Up-sampling involves reducing the sampling period by introducing additional regularly spaced samples in the signal sequence

$$x_u(n) = \sum_{m=-\infty}^{\infty} x(m)\delta(n - mM) = \begin{cases} x(n/M), & n = 0, \pm M, \pm 2M \ldots \\ 0, & else \end{cases}. \tag{2.45}$$

The introduction of zero-valued samples in the up-sampled signal, $x_u(n)$, increases the sampling rate of the signal. The DTFT of the up-sampled signal relates to the DTFT of the original signal as follows:

$$X_u(e^{j\Omega}) = X(e^{j\Omega M}). \tag{2.46}$$

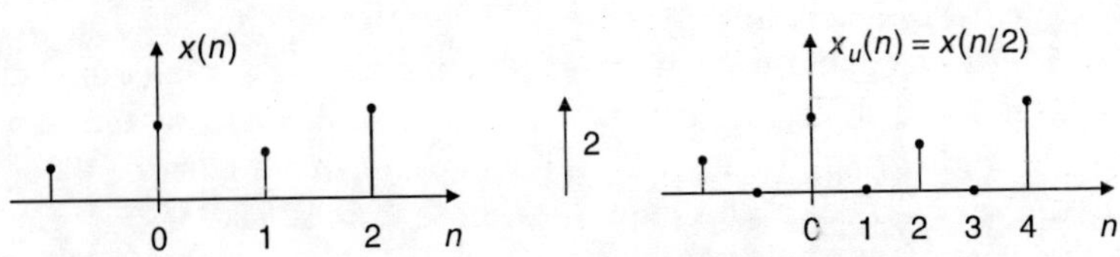

(a)

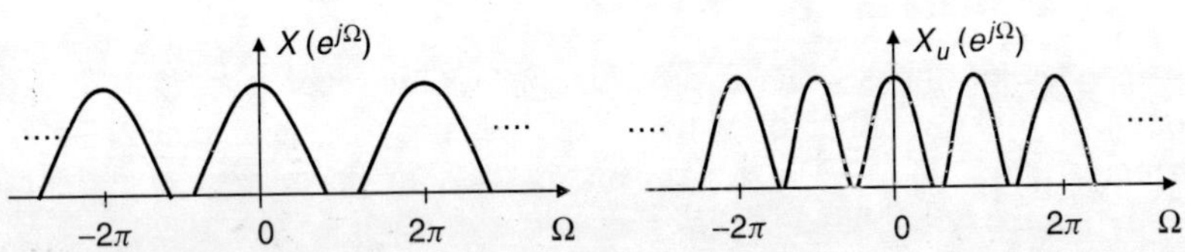

(b)

Figure 2.23. (a) The original and the up-sampled signal in the time-domain; and (b) the corresponding DTFTs.

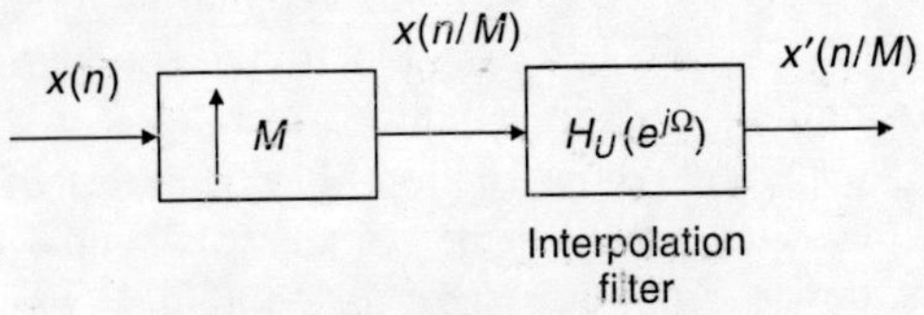

Figure 2.24. Up-sampling by an integer M.

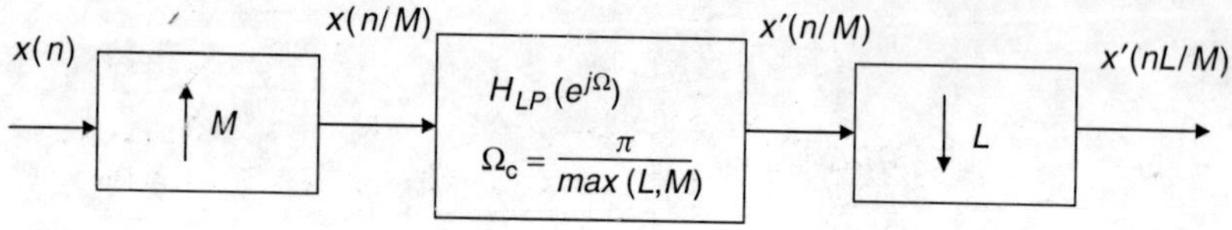

Figure 2.25. Sampling rate changes by a noninteger factor.

Therefore, the DTFT of the up-sampled signal, $X_u(e^{j\Omega})$, is described by a series of frequency compressed images of the DTFT of the original signal located at integer multiples of $2\pi/M$ rads (see Figure 2.23). To complete the up-sampling process, an interpolation stage is required that fills appropriate values in the time-domain to replace the artificial zero-valued samples introduced by the sampling. In Figure 2.24, $H_U(e^{j\Omega})$ is given by

$$H_U(e^{j\Omega}) = \begin{cases} M, & 0 \leqslant |\Omega| \leqslant \pi/M \\ 0, & \pi/M \leqslant |\Omega| \leqslant \pi \end{cases}. \tag{2.47}$$

2.8.3 Sampling Rate Changes by Noninteger Factors

Sampling rate by noninteger factors can be accomplished by cascading up-sampling and down-sampling operations. The up-sampling stage precedes the down-sampling stage and the low-pass interpolation and anti-aliasing filters are combined into one filter whose bandwidth is the minimum of the two filters, Figure 2.25. For example, if we want a noninteger sampling period modification such that $T_{new} = 12T/5$. In this case, we choose $L = 12$ and $M = 5$. Hence, the bandwidth of the low-pass filter is the minimum of $\pi/12$ and $\pi/5$.

2.8.4 Quadrature Mirror Filter Banks

The analysis of the signal in a perceptual audio coding system is usually accomplished using either filter banks or frequency-domain transformations or a combination of both. The filter bank is used to decompose the signal into several frequency subbands. Different coding strategies are then derived and implemented in each subband. The technique is known as *subband coding* in the coding literature Figure 2.26.

One of the important aspects of subband decomposition is the aliasing between the different subbands because of the imperfect frequency responses of the digital filters, Figure 2.27. These aliasing problems prevented early analysis-synthesis filter banks from perfectly reconstructing the original input signal in the absence of quantization effects. In 1977, a solution to this problem was provided by combining down-sampling and up-sampling operations with appropriate filter designs [Este77]. The perfect reconstruction filter bank design came to be known as a quadrature mirror filter (QMF) bank. An analysis-synthesis QMF consists of anti-aliasing filters, down-sampling stages, up-sampling stages, and interpolation filters. A two-band QMF structure is shown in Figure 2.28

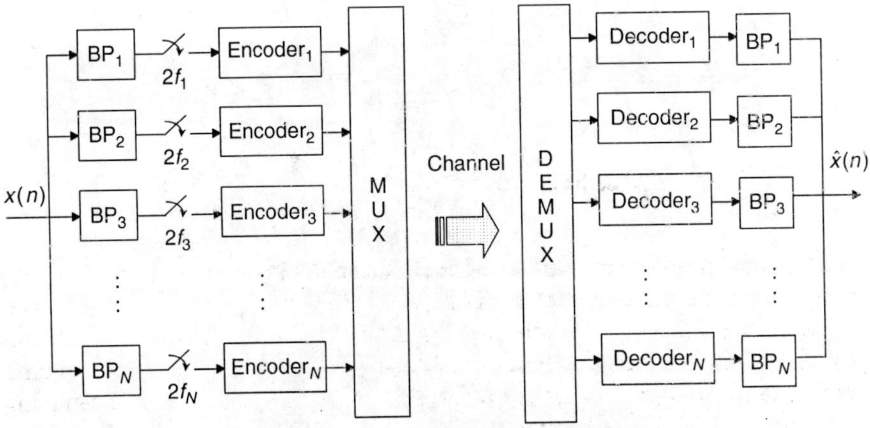

Figure 2.26. Signal coding in subbands.

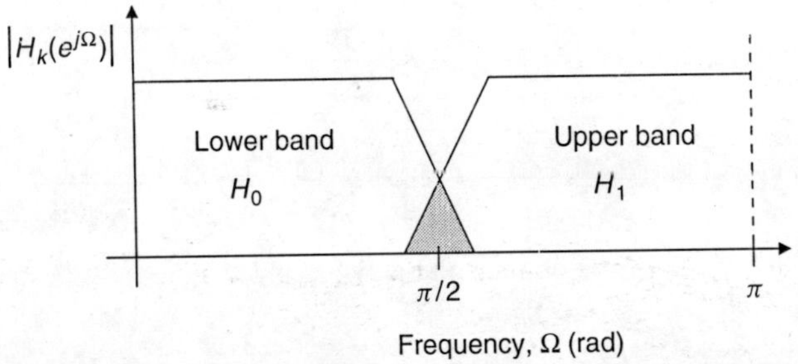

Figure 2.27. Aliasing effects in a two-band filter bank.

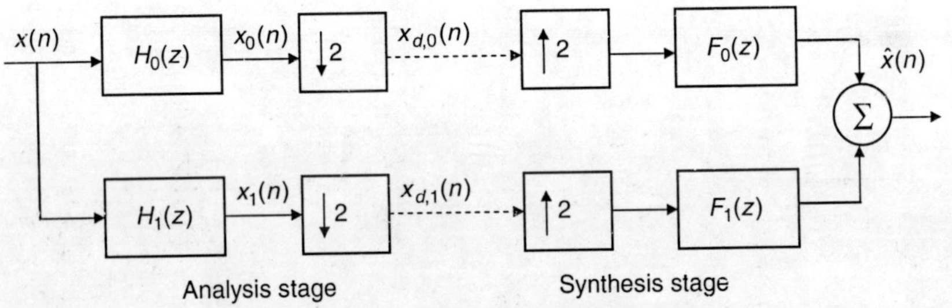

Figure 2.28. A two-band QMF structure.

The analysis stage consists of the filters $H_0(z)$ and $H_1(z)$ and down-sampling operations. The synthesis stage includes up-sampling stages and the filters $F_0(z)$ and $F_1(z)$. If the process includes quantizers, those will be placed after the down-sampling stages. We first examine the filter bank without the quantization stage. The input signal, $x(n)$, is first filtered and then down-sampled. The DTFT of the down-sampled signal can be shown to be

$$X_{d,k}(e^{j\Omega}) = \frac{1}{2}(X_k(e^{j\Omega/2}) + X_k(e^{j(\Omega-2\pi)/2})), k = 0, 1. \quad (2.48)$$

Figure 2.29 presents plots of the DTFTs of the original and down-sampled signals. It can be seen that an aliasing term is present.

The reconstructed signal, $\hat{x}(n)$, is derived by adding the contributions from the up-sampling and interpolations of the low and the high band. It can be shown

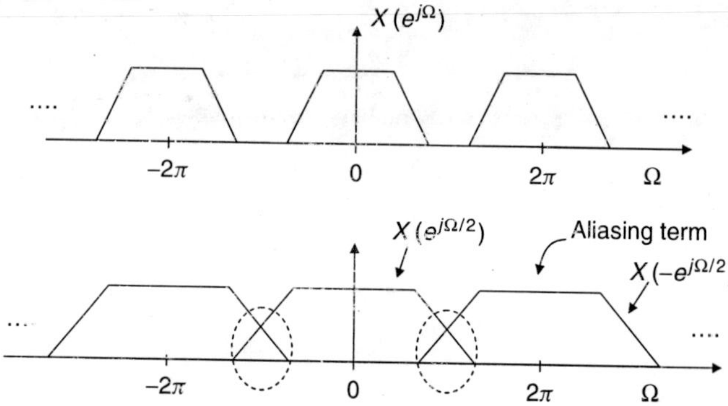

Figure 2.29. DTFTs of the original and down-sampled signals to illustrate aliasing.

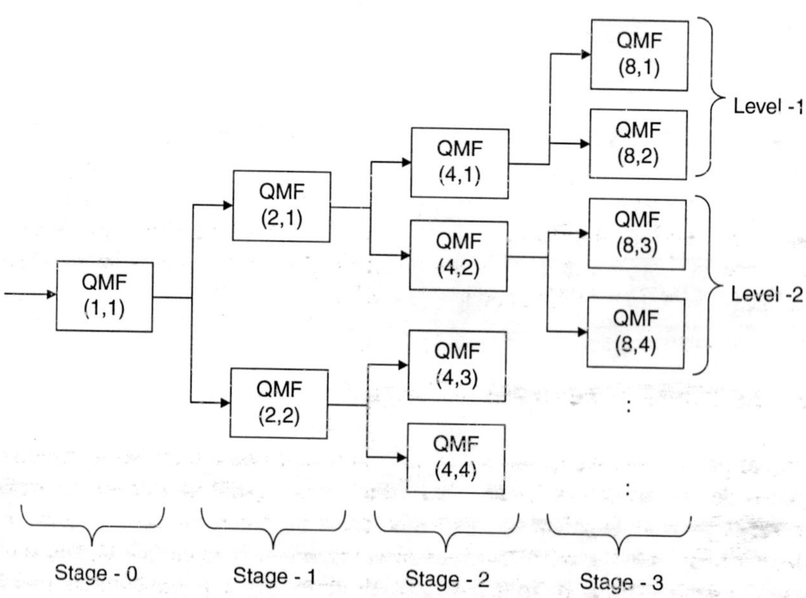

Figure 2.30. Tree-structured QMF bank.

that the reconstructed signal in the z-domain has the form

$$\hat{X}(z) = \frac{1}{2}(H_0(z)F_0(z) + H_1(z)F_1(z))X(z) + \frac{1}{2}(H_0(-z)F_0(z) + H_1(-z)F_1(z))X(-z). \quad (2.49)$$

The signal $X(-z)$ in Eq. (2.49) is associated with the aliasing term. The aliasing term can be cancelled by designing filters to have the following *mirror* symmetries:

$$F_0(z) = H_1(-z) \qquad F_1(z) = -H_0(-z). \quad (2.50)$$

Under these conditions, the overall transfer function of the filter bank can then be written as

$$T(z) = \frac{1}{2}(H_0(z)F_0(z) + H_1(z)F_1(z)). \quad (2.51)$$

If $T(z) = 1$, then the filter bank allows perfect reconstruction. Perfect delay-less reconstruction is not realizable, but an all-pass filter bank with linear phase characteristics can be designed easily. For example, the choice of first-order FIR filters

$$H_0(z) = 1 + z^{-1} \qquad H_1(z) = 1 - z^{-1} \quad (2.52)$$

results in alias-free reconstruction. The overall transfer function of the QMF in this case is

$$T(z) = \frac{1}{2}((1 + z^{-1})^2 - (1 - z^{-1})^2) = 2z^{-1}. \quad (2.53)$$

Therefore, the signal is reconstructed within a delay of one sample and with an overall gain of 2. QMF filter banks can be cascaded to form tree structures. If we represent the analysis stage of a filter bank as a block that divides the signal in low and high frequency subbands, then by cascading several such blocks, we can divide the signal into smaller subbands. This is shown in Figure 2.30. QMF banks are part of many subband and hybrid subband/transform audio and image/video coding standards [Thei87] [Stoll88] [Veld89] [John96]. Note that the theory of quadrature mirror filterbanks has been associated with wavelet transform theory [Wick94] [Akan96] [Stra96].

2.9 DISCRETE-TIME RANDOM SIGNALS

In signal processing, we generally classify signals as deterministic or random. A signal is defined as *deterministic* if its values at any point in time can be defined precisely by a mathematical equation. For example, the signal $x(n) = \sin(\pi n/4)$ is deterministic. On the other hand, *random* signals have uncertain values and are usually described using their statistics. A discrete-time random process involves an ensemble of sequences $x(n,m)$ where m is the index of the m-th sequence in the ensemble and n is the time index. In practice, one does not have access to all possible sample signals of a random process. Therefore, the determination of the statistical structure of a random process is often done from the observed waveform. This approach becomes valid and simplifies random signal analysis if the random process at hand is *ergodic*. *Ergodicity* implies that the statistics of a random process can be determined using time-averaging operations on a single observed signal. Ergodicity requires that the statistics of the signal are independent of the time of observation. Random signals whose statistical structure is independent of time of origin are generally called *stationary*. More specifically, a random process is said to be *widesense stationary* if its statistics, upto the second order, are independent of time. Although it is difficult to show analytically that signals with various statistical distributions are ergodic, it can be shown that a stationary zero-mean Gaussian process is ergodic up to second order. In many practical applications involving a stationary or quasi-stationary process, it is assumed that the process is also ergodic. The definitions of signal statistics presented henceforth will focus on real-valued, stationary processes that are ergodic.

The mean value, μ_x, of the discrete-time, wide sense stationary signal, $x(n)$, is a first-order statistic that is defined as the expected value of $x(n)$, i.e.,

$$\mu_x = E[x(n)] = \lim_{N\to\infty} \frac{1}{2N+1} \sum_{n=-N}^{N} x(n), \tag{2.54}$$

where $E[]$ denotes the statistical expectation. The assumption of ergodicity allows us to determine the mean value with a time-averaging process shown on the right-hand side of Eq. (2.54). The mean value can be viewed as the D.C. component in the signal. In many applications involving speech and audio signals, the D.C. component does not carry any useful information and is either ignored or filtered out.

The variance, σ_x^2, is a second-order signal statistic and is a measure of signal dispersion from its mean value. The variance is defined as

$$\sigma_x^2 = E[(x(n) - \mu_x)(x(n) - \mu_x)] = E[x^2(n)] - \mu_x^2. \tag{2.55}$$

The square root of the variance is the *standard deviation* of the signal. For a zero-mean signal, the variance is simply $E[x^2(n)]$. The autocorrelation of a signal is a second-order statistic defined by

$$r_{xx}(m) = E[x(n+m)x(n)] = \lim_{N\to\infty} \frac{1}{2N+1} \sum_{n=-N}^{N} x(n+m)x(n), \tag{2.56}$$

where m is called the autocorrelation lag index. The autocorrelation can be viewed as a measure of predictability of the signal in the sense that a future value of a correlated signal can be predicted by processing information associated with its past values. For example, speech is a correlated waveform, and, hence, it can be modeled by linear prediction mechanisms that predict its current value from a linear combination of past values. Correlation can also be viewed as a measure of redundancy in the signal, in that correlated waveforms can be parameterized in terms of statistical time-series models; and, hence, represented by a reduced number of information bits.

The autocorrelation sequence, $r_{xx}(m)$, is symmetric and positive definite, i.e.,

$$r_{xx}(-m) = r_{xx}(m) \quad r_{xx}(0) \geqslant |r_{xx}(m)|. \tag{2.57}$$

Example 2.9

The autocorrelation of a white noise signal is

$$r_{ww}(m) = \sigma_w^2 \delta(m),$$

where σ_w^2 is the variance of the noise. The fact that the autocorrelation of white noise is the unit impulse implies that white noise is an uncorrelated signal.

Example 2.10

The autocorrelation of the output of a second-order FIR digital filter, $H(z)$, (in Figure 2.31) to a white noise input of zero mean and unit variance is

$$\begin{aligned} r_{yy}(m) &= E[y(n+m)y(n)] \\ &= \delta(m+2) + 2\delta(m+1) + 3\delta(m) + 2\delta(m-1) + \delta(m-2). \end{aligned}$$

Cross-correlation is a measure of similarity between two signals. The cross-correlation of a signal, $x(n)$, relative to a signal, $y(n)$, is given by

$$r_{xy}(m) = E[x(n+m)y(n)]. \tag{2.58}$$

Similarly, cross-correlation of a signal, $y(n)$, relative to a signal, $x(n)$, is given by

$$r_{yx}(m) = E[y(n+m)x(n)]. \tag{2.59}$$

Note that the symmetry property of the cross-correlation is

$$r_{yx}(m) = r_{xy}(-m). \tag{2.60}$$

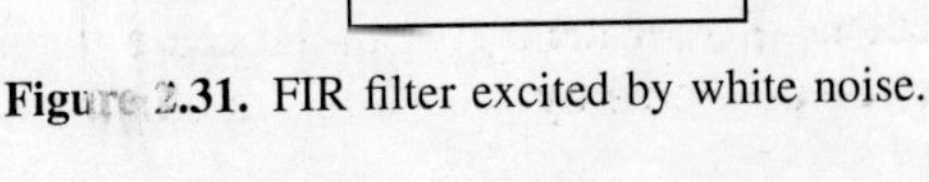

Figure 2.31. FIR filter excited by white noise.

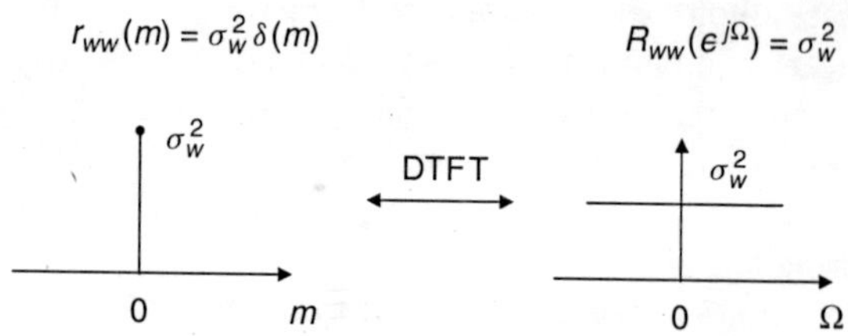

Figure 2.32. The PSD of white noise.

The power spectral density (PSD) of a random signal is defined as the DTFT of the autocorrelation sequence,

$$R_{xx}(e^{j\Omega}) = \sum_{m=-\infty}^{\infty} r_{xx}(m) e^{-j\Omega m}. \tag{2.61}$$

The PSD is real-valued and positive and describes how the power of the random process is distributed across frequency.

Example 2.11

The PSD of a white noise signal (see Figure 2.32) is

$$R_{ww}(e^{j\Omega}) = \sigma_w^2 \sum_{m=-\infty}^{\infty} \delta(m) e^{-j\Omega m} = \sigma_w^2.$$

2.9.1 Random Signals Processed by LTI Digital Filters

In Example 2.10, we determined the autocorrelation of the output of a second-order FIR digital filter when the excitation is white noise. In this section, we review briefly the characterization of the statistics of the output of a causal LTI digital filter that is excited by a random signal. The output of a causal digital filter can be computed by convolving the input with its impulse response, i.e.,

$$y(n) = \sum_{i=0}^{\infty} h(i) x(n-i). \tag{2.62}$$

Based on the convolution sum we can derive the following expressions for the mean, the autocorrelation, the cross-correlation, and the power spectral density of the steady-state output of an LTI digital filter:

$$\mu_y = \sum_{k=0}^{\infty} h(k) \mu_x = H(e^{j\Omega})|_{\Omega=0} \mu_x \tag{2.63}$$

$$r_{yy}(m) = \sum_{k=0}^{\infty} \sum_{i=0}^{\infty} h(k) h(i) r_{xx}(m-k+i) \tag{2.64}$$

$$r_{yx}(m) = \sum_{i=0}^{\infty} h(i) r_{xx}(m-i) \tag{2.65}$$

$$R_{yy}(e^{j\Omega}) = |H(e^{j\Omega})|^2 R_{xx}(e^{j\Omega}). \tag{2.66}$$

These equations describe the statistical behavior of the output at steady state. During the transient state of the filter, the output is essentially nonstationary, i.e.,

$$\mu_y(n) = \sum_{k=0}^{n} h(k) \mu_x. \tag{2.67}$$

Example 2.12

Determine the output variance of an LTI digital filter excited by white noise of zero mean and unit variance:

$$\sigma_y^2 = r_{yy}(0) = \sum_{k=0}^{\infty} \sum_{i=0}^{\infty} h(k) h(i) \delta(i-k) = \sum_{k=0}^{\infty} h^2(k).$$

Example 2.13

Determine the variance, the autocorrelation, and the PSD of the output of the digital filter in Figure 2.33 when its input is white noise of zero mean and unit variance.

The impulse response and transfer function of this first-order IIR filter is

$$h(n) = 0.8^n u(n) \qquad H(z) = \frac{1}{1 - 0.8z^{-1}}.$$

The variance of the output at steady state is

$$\sigma_y^2 = \sum_{k=0}^{\infty} h^2(k) \sigma_x^2 = \sum_{k=0}^{\infty} 0.64^k = \frac{1}{1 - 0.64} = 2.78.$$

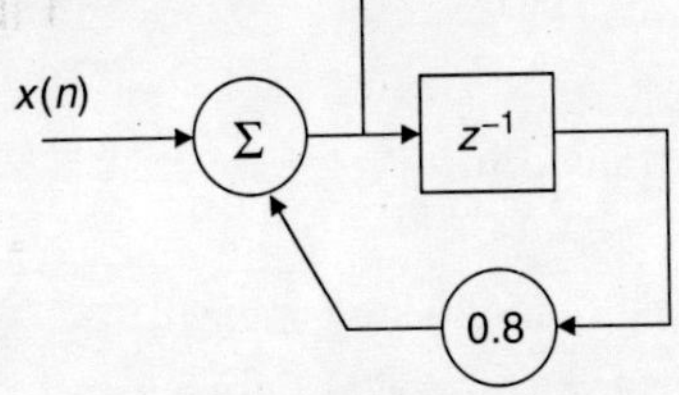

Figure 2.33. An example of an IIR filter.

The autocorrelation of the output is given by

$$r_{yy}(m) = \sum_{k=0}^{\infty}\sum_{i=0}^{\infty} 0.8^{k+i}\delta_{xx}(m-k+i).$$

It is easy to see that the unit impulse will be non-zero only for $k = m + i$ and hence

$$r_{yy}(m) = \sum_{i=0}^{\infty} 0.8^{m+2i} \quad m \geqslant 0.$$

And, taking into account the autocorrelation symmetry,

$$r_{yy}(m) = r_{yy}(-m) = 2.77(0.8)^{|m|}\forall m.$$

Finally, the PSD is given by

$$R_{yy}(e^{j\Omega}) = |H(e^{j\Omega})|^2 R_{xx}(e^{j\Omega}) = \frac{1}{|1-0.8e^{-j\Omega}|^2}.$$

2.9.2 Autocorrelation Estimation from Finite-Length Data

Estimators of signal statistics given N observations are based on the assumption of stationarity and ergodicity. The following is an estimator of the autocorrelation (based on sample averaging) of a signal, $x(n)$,

$$\hat{r}_{xx}(m) = \frac{1}{N}\sum_{n=0}^{N-m-1} x(n+m)x(n),\, m = 0, 1, 2, \ldots, N-1. \quad (2.68)$$

Correlations for negative lags can be taken using the symmetry property in Eq. (2.57). The estimator above is asymptotically unbiased (fixed m and $N >> m$) but for small N it is biased.

2.10 SUMMARY

A brief review of some of the essentials of signal processing techniques were described in this chapter. In particular, some of the important concepts covered in this chapter include:

- Continuous Fourier transform
- Spectral leakage effects
- Convolution, sampling, and aliasing issues
- Discrete-time Fourier transform and z-transform
- The DFT, FFT, DCT, and STFT basics

- IIR and FIR filter representations
- Pole/zero and frequency response interpretations
- Shelving and peaking filters, audio graphic equalizers
- Down-sampling and up-sampling
- QMF banks and alias-free reconstruction
- Discrete-time random signal processing review.

PROBLEMS

2.1. Determine the continuous Fourier transform (CFT) of a pulse described by

$$x(t) = u(t+1) - u(t-1),$$

where $u(t)$ is the unit step function.

2.2. State and derive the CFT properties of duality, time shift, modulation, and convolution.

2.3. For the circuit shown in Figure 2.34(a) and for RC = 1,

a. Write the input output differential equation.

b. Determine the impulse response in closed-form by solving the differential equation.

c. Write the frequency response function.

d. Determine the steady state response, $y(t)$, for $x(t) = \sin(10t)$.

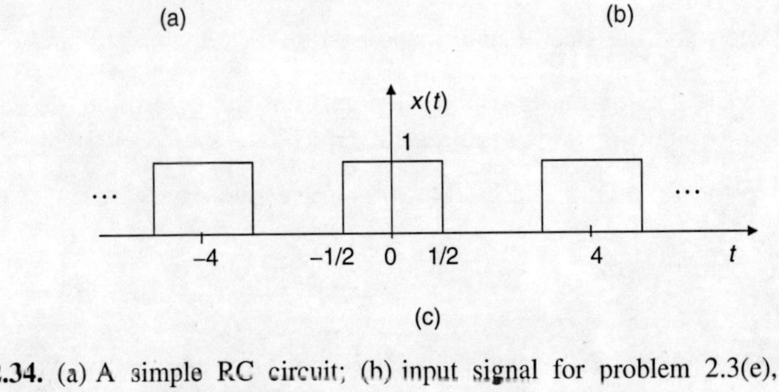

Figure 2.34. (a) A simple RC circuit; (b) input signal for problem 2.3(e), $x(t)$; and (c) input signal for problem 2.3(f).

e. Given $x(t)$ as shown in Figure 2.34(b), find the circuit output, $y(t)$, using convolution.

f. Determine the Fourier series of the output, $y(t)$, of the RC circuit for the input shown in Figure 2.34(c).

2.4. Determine the CFT of $p(t) = \sum_{n=-\infty}^{\infty} \delta(t - nT_s)$. Given, $x_s(t) = x(t)p(t)$, derive the following,

$$X_s(\omega) = \frac{1}{T_s} \sum_{k=-\infty}^{\infty} X(\omega - k\omega_s)$$

$$x(t) = \sum_{n=-\infty}^{\infty} x(nT_s)\,\mathrm{sinc}(B(t - nT_s)),$$

where $X(\omega)$ and $X_s(\omega)$ are the spectra of ideally bandlimited and uniformly sampled signals, respectively, and $\omega_s = 2\pi/T_s$. (Refer to Figure 2.8 for variable definitions.)

2.5. Determine the z-transforms of the following causal signals:

a. $\sin(\Omega n)$

b. $\delta(n) + \delta(n-1)$

c. $p^n \sin(\Omega n)$

d. $u(n) - u(n-9)$

2.6. Determine the impulse and frequency responses of the averaging filter

$$h(n) = \frac{1}{L+1}, n = 0, 1, \ldots, L, \text{ for } L = 9.$$

2.7. Show that the IDFT can be derived as a least squares signal matching problem, where N points in the time domain are matched by a linear combination of N sampled complex sinusoids.

2.8. Given the transfer function $H(z) = (z-1)^2/(z^2 + 0.81)$,

a. Determine the impulse response, $h(n)$.

b. Determine the steady state response due to the sinusoid $\sin\left(\frac{\pi n}{4}\right)$.

2.9. Derive the decimation-in-time FFT algorithm and determine the number of complex multiplications required for an FFT size of $N = 1024$.

2.10. Derive the following expression in a simple two-band QMF

$$X_{d,k}(e^{j\Omega}) = \frac{1}{2}(X_k(e^{j\Omega/2}) + X_k(e^{j(\Omega-2\pi)/2})), \quad k = 0, 1.$$

Refer to Figure 2.28 for variable definitions. Give and justify the conditions for alias-free reconstruction in a simple QMF bank.

2.11. Design a tree-structured uniform QMF bank that will divide the spectrum of 0-20 kHz into eight uniform subbands. Give appropriate figures and denote on the branches the range of frequencies.

2.12. Modify your design in problem 2.11 and give one possible realization of a simple nonuniform tree structured QMF bank that will divide the of 0-20 kHz spectrum into eight subbands whose bandwidth increases with the center frequency.

2.13. Design a low-pass shelving filter for the following specifications: $f_s =$ 16 kHz, $f_c = 4$ kHz, and $g = 10$ dB.

2.14. Design peaking filters for the following cases: a) $\Omega_c = \pi/4$, $Q = 2$, $g =$ 10 dB and b) $\Omega_c = \pi/2$, $Q = 3$, $g = 5$ dB. Give frequency responses of the designed peaking filters.

2.15. Design a five-band digital audio equalizer using the concept of peaking digital filters. Select center frequencies, f_c, as follows: 500 Hz, 1500 Hz, 4000 Hz, 10 kHz, and 16 kHz; sampling frequency, $f_s = 44.1$ kHz and the corresponding peaking filter gains as 10 dB. Choose a constant Q for all the peaking filters.

2.16. Derive equations (2.63) and (2.64).

2.17. Derive equation (2.66).

2.18. Show that the PSD is real-valued and positive.

2.19. Show that the estimator (2.68) of the autocorrelation is biased.

2.20. Show that the estimator (2.68) provides autocorrelations such that $r_{xx}(0) \geqslant |r_{xx}(m)|$.

2.21. Provide an unbiased autocorrelation estimator by modifying the estimator in (2.68).

2.22. A digital filter with impulse response $h(n) = 0.7^n u(n)$ is excited by white Gaussian noise of zero mean and unit variance. Determine the mean and variance of the output of the digital filter in closed-form during the transient and steady state.

2.23. The filter $H(z) = z/(z - 0.8)$ is excited by white noise of zero mean and unit variance.

a. Determine all the autocorrelation values at the output of the filter at steady state.

b. Determine the PSD at the output.

COMPUTER EXERCISES

Use the speech file '***Ch2speech.wav***' from the Book Website for all the computer exercises in this chapter.

2.24. Consider the 2-band QMF bank shown in Figure 2.35. In this figure, $x(n)$ denotes speech frames of 256 samples and $\hat{x}(n)$ denotes the synthesized speech frames.

a. Design the transfer functions, $F_0(z)$ and $F_1(z)$, such that aliasing is cancelled. Also calculate the overall delay of the QMF bank.

b. Select an arbitrary voiced speech frame from Ch2speech.wav. Give time-domain and frequency-domain plots of $x_{d0}(n)$ and $x_{d1}(n)$ for that particular frame. Comment on the frequency-domain plots with regard to low-pass/high-pass band-splitting.

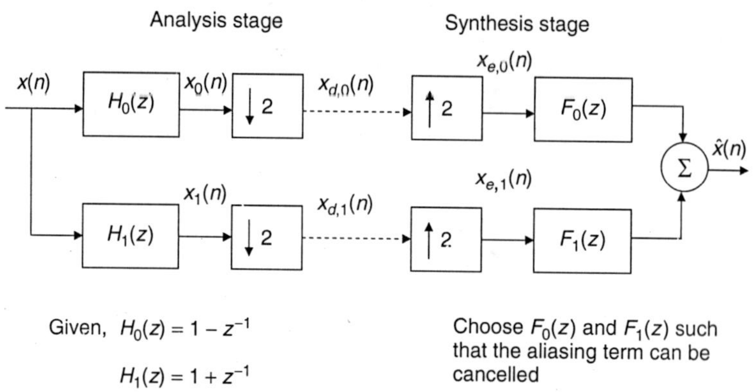

Figure 2.35. A two-band QMF bank.

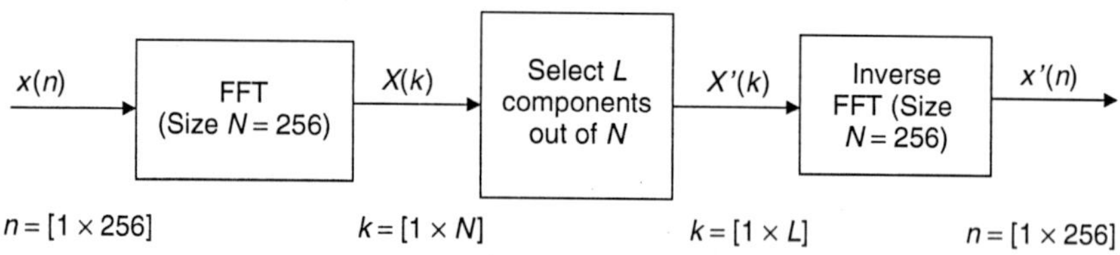

Figure 2.36. Speech synthesis from a select number (subset) of FFT components.

Table 2.2. Signal-to-noise ratio (SNR) and MOS values.

Number of FFT components, L	$SNR_{overall}$	Subjective evaluation MOS (mean opinion score) in a scale of 1–5 for the entire speech record
16		
128		

c. Repeat step (b) for $x_1(n)$ and $x_{d1}(n)$ in order to compare the signals before and after the downsampling stage.

d. Calculate the SNR between the input speech record, $x(n)$, and the synthesized speech record, $\hat{x}(n)$. Use the following equation to compute the SNR,

$$\mathrm{SNR} = 10\log_{10}\left(\frac{\sum_n x^2(n)}{\sum_n (x(n)-\hat{x}(n))^2}\right)\ \text{(dB)}$$

Listen to the synthesized speech record and comment on its quality.

e. Choose a low-pass $F_0(z)$ and a high-pass $F_1(z)$, such that aliasing occurs. Compute the SNR. Listen to the synthesized speech and describe its perceptual quality relative to the output speech in step(d). (Hint: Use first-order IIR filters.)

2.25. In Figure 2.36, $x(n)$ denotes speech frames of 256 samples and $x'(n)$ denotes the synthesized speech frames. For $L = N(= 256)$, the synthesized speech will be identical to the input speech. In this computer exercise, you need to perform speech synthesis on a frame-by-frame basis from a select number (subset) of FFT components, i.e., $L < N$. We will use two methods for the FFT component selection, (*i*) Method 1: selecting the first L components including their conjugate-symmetric ones out of a total of N components; and (*ii*) Method 2: the least-squares method (peak-picking method that selects the L components that minimize the sum of squares error.)

a. Use $L = 64$ and Method 1 for component selection. Perform speech synthesis and give time-domain plots of both input and output speech records.

b. Repeat the above step using the peak-picking Method 2 (choose L peaks including symmetric components in the FFT magnitude spectrum). List the SNR values in both the cases. Listen to the output files corresponding to (a) and (b) and provide a subjective evaluation (on a MOS scale 1–5). To calibrate the process think of a wireline telephone quality (toll) as 4, cellphone quality around 3.7.

c. Perform speech synthesis for (i) $L = 16$ and (ii) $L = 128$. Use the peak-picking Method 2 for the FFT component selection. Compute the overall SNR values and provide a subjective evaluation of the output speech for both the cases. Tabulate your results in Table 2.2.

CHAPTER 3

QUANTIZATION AND ENTROPY CODING

3.1 INTRODUCTION

This chapter provides an introduction to waveform quantization, and entropy coding algorithms. Waveform quantization deals with the digital or, more specifically, binary representation of signals. All the audio encoding algorithms typically include a quantization module. Theoretical aspects of waveform quantization methods were established about fifty years ago [Shan48]. Waveform quantization can be: *i*) *memoryless* or *with memory*, depending upon whether the encoding rules rely on past samples; and *ii*) *uniform* or *nonuniform* based on the step-size or the quantization (discretization) levels employed. Pulse code modulation (PCM) [Oliv48] [Jaya76] [Jaya84] [Span94] is a memoryless method for discrete-time, discrete-amplitude quantization of analog waveforms. On the other hand, Differential PCM (DPCM), delta modulation (DM), and adaptive DPCM (ADPCM) have memory.

Waveform quantization can also be classified as *scalar* or *vector*. In scalar quantization, each sample is quantized individually, as opposed to vector quantization, where a block of samples is quantized jointly. Scalar quantization [Jaya84] methods include PCM, DPCM, DM, and their adaptive versions. Several vector quantization (VQ) schemes have been proposed, including the VQ [Lind80], the split-VQ [Pali91] [Pali93], and the conjugate structure-VQ [Kata93] [Kata96]. Quantization can be *parametric* or *nonparametric*. In nonparametric quantization, the actual signal is quantized. Parametric representations are generally based on signal transformations (often unitary) or on source-system signal models.

A bit allocation algorithm is typically employed to compute the number of quantization bits required to encode an audio segment. Several bit allocation schemes have been proposed over the years; these include bit allocation based on the noise-to-mask-ratio (NMR) and masking thresholds [Bran87a] [John88a] [ISOI92], perceptually motivated bit allocation [Vora97] [Naja00], and dynamic bit allocation based on signal statistics [Jaya84] [Rams86] [Shoh88] [West88] [Beat89] [Madi97]. Note that the NMR-based perceptual bit allocation scheme [Bran87a] is one of the most popular techniques and is embedded in several audio coding standards (e.g., ISO/IEC MPEG codec series, etc.).

In audio compression, entropy coding techniques are employed in conjunction with the quantization and bit-allocation modules in order to obtain improved coding efficiencies. Unlike the DPCM and the ADPCM techniques that remove the redundancy by exploiting the correlation of the signal, while entropy coding schemes exploit the likelihood of the symbol occurrence [Cove91]. Entropy is a measure of uncertainty of a random variable. For example, consider two random variables, x and y; and two random events, A and B. For the random variable x, let the probability of occurrence of the event A be $p_x(A) = 0.5$ and the event B be $p_x(B) = 0.5$. Similarly, define $p_y(A) = 0.99999$ and $p_y(B) = 0.00001$. The random variable x has a high uncertainty measure, i.e., it is very hard to predict whether event A or B is likely to occur. On the contrary, in the case of the random variable y, the event A is more likely to occur, and, therefore, we have less uncertainty relative to the random variable x. In entropy coding, the information symbols are mapped into codes based on the frequency of each symbol. Several entropy-coding schemes have been proposed including Huffman coding [Huff52], Rice coding [Rice79], Golomb coding [Golo66], arithmetic coding [Riss79] [Howa94], and Lempel-Ziv coding [Ziv77]. These entropy coding schemes are typically called *noiseless*. A noiseless coding system is able to reconstruct the signal perfectly from its coded representation. In contrast, a coding scheme incapable of perfect reconstruction is called *lossy*.

In the rest of the chapter we provide an overview of the quantization-bit allocation-entropy coding (QBE) framework. We also provide background on the probabilistic signal structures and we show how they are exploited in quantization algorithms. Finally, we introduce vector quantization basics.

3.1.1 The Quantization–Bit Allocation–Entropy Coding Module

After perceptual irrelevancies in an audio frame are exploited, a quantization–bit allocation–entropy coding (QBE) module is employed to exploit statistical correlation. In Figure 3.1, typical output parameters from stage I include the transform coefficients, the scale factors, and the residual error. These parameters are first quantized using one of the aforementioned PCM, DPCM, or VQ schemes. The number of bits allocated per frame is typically specified by a bit-allocation module that uses perceptual masking thresholds.

The quantized parameters are entropy coded using an explicit noiseless coding stage for final redundancy reduction. Huffman or Rice codes are available in the form of look-up tables at the entropy coding stage. Entropy coders are employed

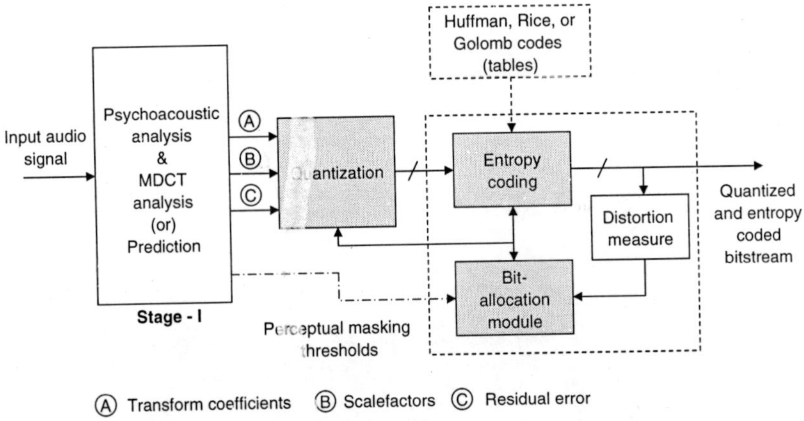

Figure 3.1. A typical QBE module employed in audio coding.

in scenarios where the objective is to achieve maximum coding efficiency. In entropy coders, more probable symbols (i.e., frequently occurring amplitudes) are encoded with shorter codewords, and vice versa. This will essentially reduce the average data rate. Next, a distortion measure between the input and the encoded parameters is computed and compared against an established threshold. If the distortion metric is greater than the specified threshold, the bit-allocation module supplies additional bits in order to reduce the quantization error. The above procedure is repeated until the distortion falls below the threshold.

3.2 DENSITY FUNCTIONS AND QUANTIZATION

In this section, we discuss the characterization of a random process in terms of its probability density function (PDF). This approach will help us derive the quantization noise equations for different quantization schemes. A random process is characterized by its PDF, which is a non-negative function, $p(x)$, whose properties are

$$\int_{-\infty}^{\infty} p(x)dx = 1 \tag{3.1}$$

and

$$\int_{x_1}^{x_2} p(x)dx = \Pr(x_1 < X \leqslant x_2). \tag{3.2}$$

From the above equations, it is evident that the PDF area from x_1 to x_2 is the probability that the random variable X is observed in this range. Since X lies somewhere in $[-\infty, \infty]$, the total area under $p(x)$ is one. The mean and the variance of the random variable X are defined as

$$\mu_x = E[X] = \int_{-\infty}^{\infty} xp(x)dx \tag{3.3}$$

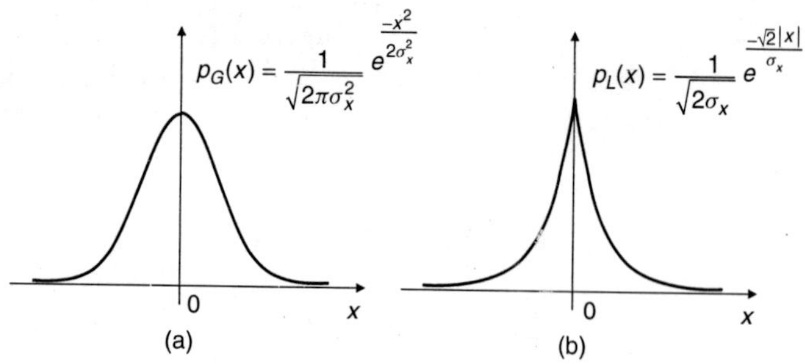

Figure 3.2. (a) The Gaussian PDF and (b) The Laplacian PDF.

$$\sigma_x^2 = \int_{-\infty}^{\infty} (x - \mu_x)^2 p(x)dx = E[(X - \mu_x)^2] \tag{3.4}$$

Note that the expectation is computed either as a weighted average (3.3) or under ergodicity assumptions as a time average (Chapter 2, Eq. 2.54). PDFs are useful in the design of optimal signal quantizers as they can be used to determine the assignment of optimal quantization levels. PDFs often used to design or analyze quantizers include the zero-mean uniform ($p_U(x)$), the Gaussian ($p_G(x)$) (Figure 3.2a), and the Laplacian ($p_L(x)$) These are given in that order below:

$$p_U(x) = \frac{1}{2S}, -S \leqslant x \leqslant S \tag{3.5}$$

$$p_G(x) = \frac{1}{\sqrt{2\pi\sigma_x^2}} e^{-\frac{x^2}{2\sigma_x^2}} \tag{3.6}$$

$$p_L(x) = \frac{1}{\sqrt{2\sigma_x}} e^{-\frac{\sqrt{2}|x|}{\sigma_x}}, \tag{3.7}$$

where S is some arbitrary non-zero real number and σ_x^2 is the variance of the random variable X. Readers are referred to Papoulis' classical book on probability and random variables [Papo91] for an in-depth treatment of random processes.

3.3 SCALAR QUANTIZATION

In this section, we describe the various scalar quantization schemes. In particular, we review uniform and nonuniform quantization, and then we present differential PCM coding methods and their adaptive versions.

3.3.1 Uniform Quantization

Uniform PCM is a memoryless process that quantizes amplitudes by rounding off each sample to one of a set of discrete values (Figure 3.3). The difference between adjacent quantization levels, i.e., the step size, Δ, is constant in non-adaptive uniform PCM. The number of quantization levels, Q, in uniform PCM

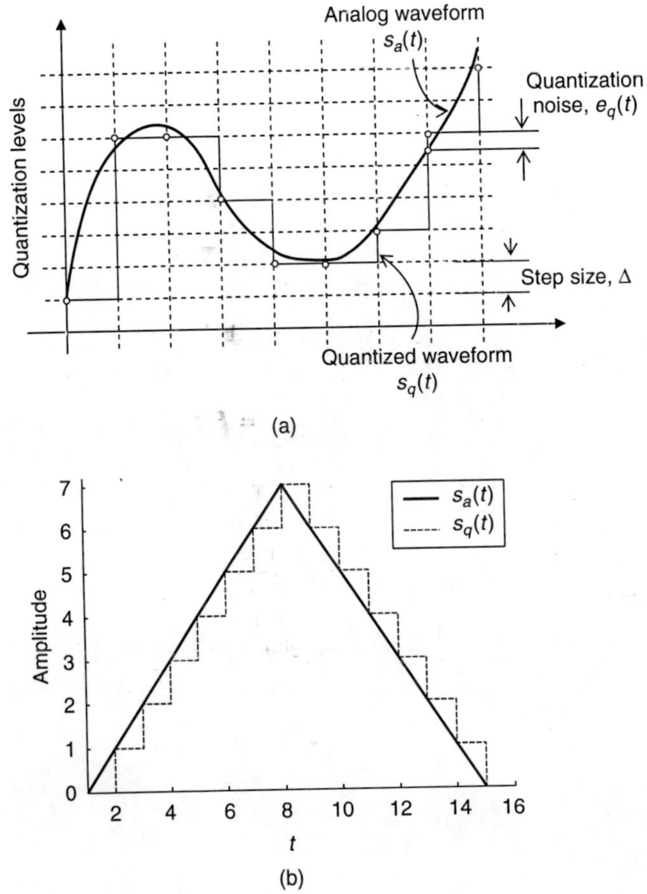

Figure 3.3. (a) Uniform PCM and (b) uniform quantization of a triangular waveform. From the figure, $R_b = 3$ bits; $Q = 8$ uniform quantizer levels.

binary representations is $Q = 2^{R_b}$, where R_b denotes the number of bits. The performance of uniform PCM can be described in terms of the signal-to-noise ratio (SNR). Consider that the signal, s, is to be quantized and its values lie in the interval

$$s \in (-s_{\max}, s_{\max}). \tag{3.8}$$

A uniform step size can then be determined by

$$\Delta = \frac{2s_{\max}}{2^{R_b}}. \tag{3.9}$$

Let us assume that the quantization noise, e_q, has a uniform PDF, i.e.,

$$-\frac{\Delta}{2} \leqslant e_q \leqslant \frac{\Delta}{2} \tag{3.10}$$

$$p_{e_q}(e_q) = \frac{1}{\Delta}, \text{ for } |e_q| \leqslant \frac{\Delta}{2}. \tag{3.11}$$

From (3.9), (3.10), and (3.11), the variance of the quantization noise can be shown [Jaya84] to be

$$\sigma_{e_q}^2 = \frac{\Delta^2}{12} = \frac{s_{\max}^2 2^{-2R_b}}{3}. \tag{3.12}$$

Therefore, if the input signal is bounded, an increase by 1 bit reduces the noise variance by a factor of four. In other words, the SNR for uniform PCM will

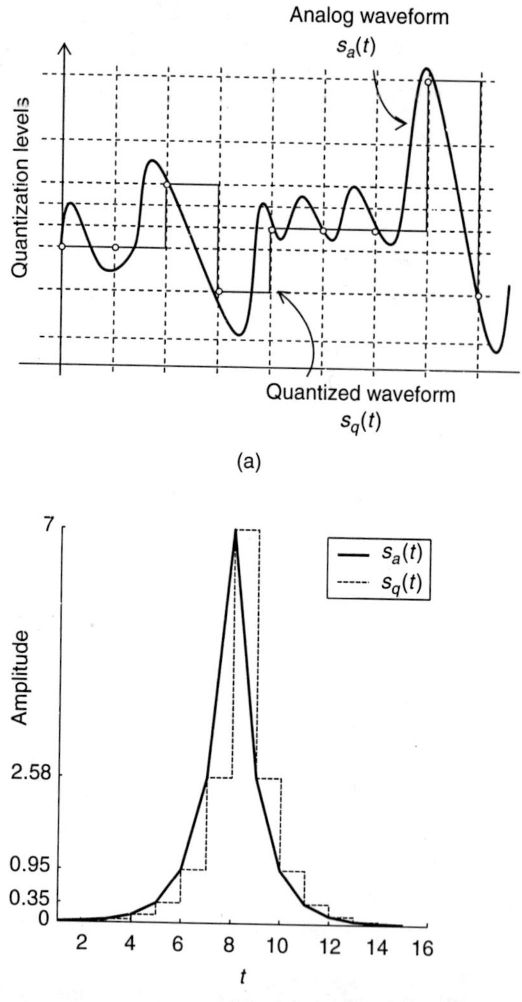

Figure 3.4. (a) Nonuniform PCM and (b) nonuniform quantization of a decaying-exponential waveform. From the figure, $R_b = 3$ bits; $Q = 8$ nonuniform quantizer levels.

improve approximately by 6 dB per bit, i.e.,

$$SNR_{PCM} = 6.02R_b + K_1(\text{dB}). \tag{3.13}$$

The factor K_1 is a constant that accounts for the step size and loading factors. For telephone speech, $K_1 = -10$ [Jaya84].

3.3.2 Nonuniform Quantization

Uniform nonadaptive PCM has no mechanism for exploiting signal redundancy. Moreover, uniform quantizers are optimal in the mean square error (MSE) sense for signals with uniform PDF. Nonuniform PCM quantizers use a nonuniform step size (Figure 3.4) that can be determined from the statistical structure of the signal.

PDF-optimized PCM uses fine step sizes for frequently occurring amplitudes and coarse step sizes for less frequently occurring amplitudes. The step sizes can be optimally designed by exploiting the shape of the signal's PDF. A signal with a Gaussian PDF (Figure 3.5), for instance, can be quantized more efficiently in terms of the overall MSE by computing the quantization step sizes and the corresponding centroids such that the mean square quantization noise is minimized [Scha79].

Another class of nonuniform PCM relies on log-quantizers that are quite common in telephony applications [Scha79] [Jaya84]. In Figure 3.6, a nonuniform quantizer is realized by using a nonlinear mapping function, $g(.)$, that maps nonuniform step sizes to uniform step sizes such that a simple linear quantizer is used. An example of the mapping function is given in Figure 3.7. The decoder uses an expansion function, $g^{-1}(.)$, to recover the signal.

Two telephony standards have been developed based on logarithmic companding, i.e., the μ-law and the A-law. The μ-law companding function is used in

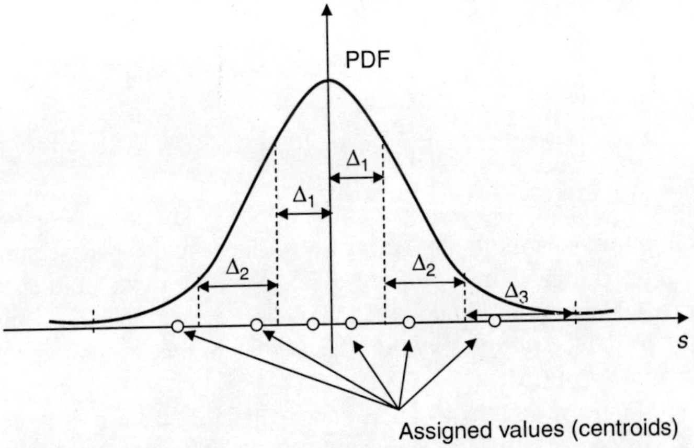

Figure 3.5. PDF-optimized PCM for signals with Gaussian distribution. Quantization levels are on the horizontal axis.

Figure 3.6. Nonuniform PCM via compressor and expansion functions.

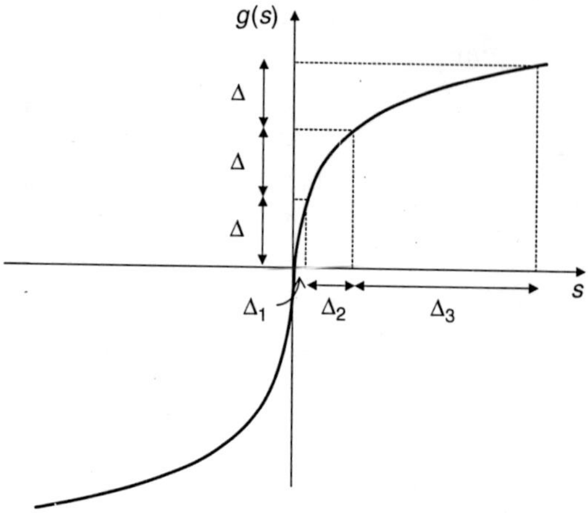

Figure 3.7. Companding function for nonuniform PCM.

the North American PCM standard ($\mu = 255$). The μ-law is given by

$$|g(s)| = \frac{\log(1 + \mu|s/s_{\max}|)}{\log(1 + \mu)}. \tag{3.14}$$

For $\mu = 255$, (3.14) gives approximately linear mapping for small amplitudes and logarithmic mapping for larger amplitudes. The European A-law companding standard is slightly different and is based on the mapping

$$|g(s)| = \begin{cases} \dfrac{A|s/s_{\max}|}{1 + \log(A)}, & \text{for } 0 < |s/s_{\max}| < 1/A \\ \dfrac{1 + \log(A|s/s_{\max}|)}{1 + \log(A)}, & \text{for } 1/A < |s/s_{\max}| < 1. \end{cases} \tag{3.15}$$

The idea with A-law companding is similar with μ-law in that again for signals with small amplitudes the mapping is almost linear and for large amplitudes the transformation is logarithmic. Both of these techniques can yield superior SNRs particularly for small amplitudes. In telephony, the companding schemes have been found to reduce bit rates, without degradation, by as much as 4 bits/sample relative to uniform PCM.

Dynamic range variations in PCM can be handled by using an adaptive step size. A PCM system with an adaptive step-size is called adaptive PCM (APCM). The step size in a feed forward system is transmitted as side information while in a feedback system the step size is estimated from past coded samples, Figure 3.8. In

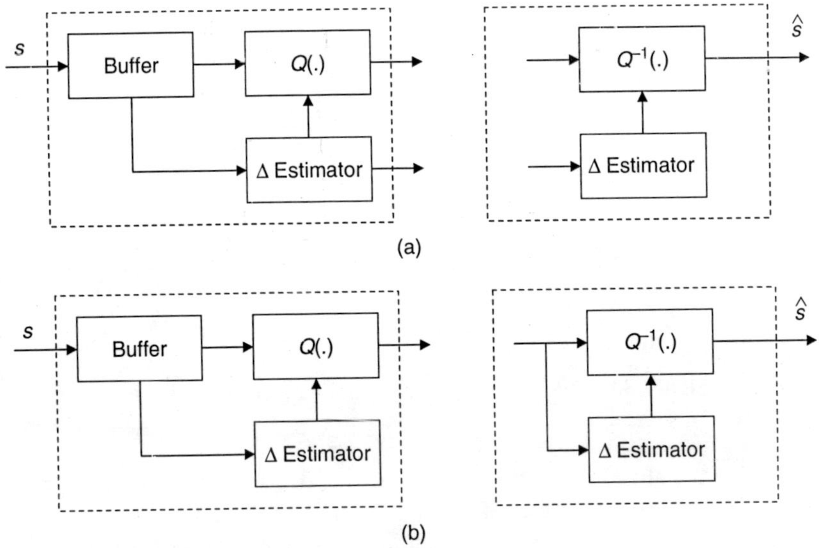

Figure 3.8. Adaptive PCM with (a) forward estimation of step size and (b) backward estimation of step size.

this figure, Q represents either uniform or nonuniform quantization (compression) scheme, and Δ corresponds to the stepsize.

3.3.3 Differential PCM

A more efficient scalar quantizer is the differential PCM (DPCM) that removes the redundancy in the audio waveform by exploiting the correlation between adjacent samples. In its simplest form, a DPCM transmitter encodes only the difference between successive samples and the receiver recovers the signal by integration. Practical DPCM schemes incorporate a time-invariant short-term prediction process, $A(z)$. This is given by

$$A(z) = \sum_{i=1}^{p} a_i z^{-i}, \tag{3.16}$$

where a_i are the prediction coefficients and z is the complex variable of the z-transform. This DPCM scheme is also called *predictive differential coding* (Figure 3.9) and reduces the quantization error variance by reducing the variance of the quantizer input. An example of a representative DPCM waveform, $e_q(n)$, along with the associated analog and PCM quantized waveforms, $s(t)$ and $s(n)$, respectively, is given in Figure 3.10.

The DPCM system (Figure 3.9) works as follows. The sample $\tilde{s}'(n)$ is the estimate of the current sample, $s(n)$, and is obtained from past sample values. The prediction error, $e(n)$, is then quantized (i.e., $e_q(n)$) and transmitted to the

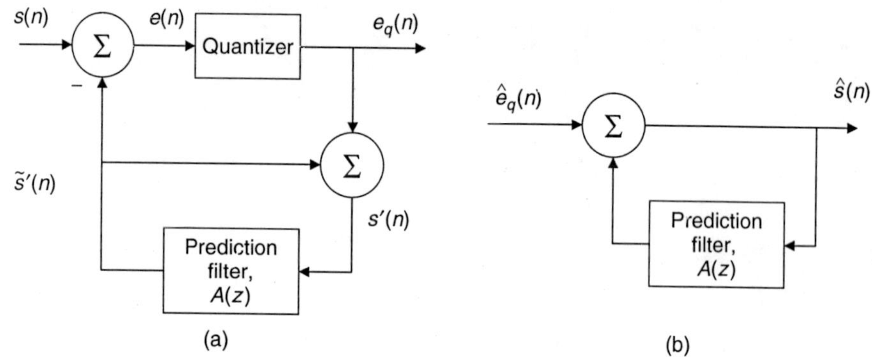

Figure 3.9. DPCM system (a) transmitter and (b) receiver.

receiver. The quantized prediction error is also added to $\tilde{s}'(n)$ in order to reconstruct the sample $s'(n)$. In the absence of channel errors, $s'(n) = \hat{s}(n)$. In the simplest case, $A(z)$ is a first-order polynomial. In Figure 3.9, $A(z)$ is given by

$$A(z) = \sum_{i=1}^{p} a_i z^{-i} \tag{3.17}$$

and the predicted signal is given by

$$\tilde{s}'(n) = \sum_{i=1}^{p} a_i s'(n-i). \tag{3.18}$$

The prediction coefficients are usually determined by solving the autocorrelation equations,

$$r_{ss}(m) - \sum_{i=1}^{p} a_i r_{ss}(m-i) = 0 \quad \text{for} \quad m = 1, 2, .., p, \tag{3.19}$$

where $r_{ss}(m)$ are the autocorrelation samples of $s(n)$. The details of the equation above will be discussed in Chapter 4. Two other types of scalar coders are the delta modulation (DM) and the adaptive DPCM (ADPCM) coders [Cumm73] [Gibs74] [Gibs78] [Yatr88]. DM can be viewed as a special case of DPCM where the difference (prediction error) is encoded with one bit. DM typically operates at sampling rates much higher than the rates commonly used with DPCM. The step size in DM may also be adaptive.

In an adaptive differential PCM (ADPCM) system, both the step size and the predictor are allowed to adapt and track the time-varying statistics of the input signal [Span94]. The predictor can be either *forward adaptive* or *backward adaptive*. In forward adaptation, the prediction parameters are estimated from the current data, which are not available at the receiver. Therefore, the

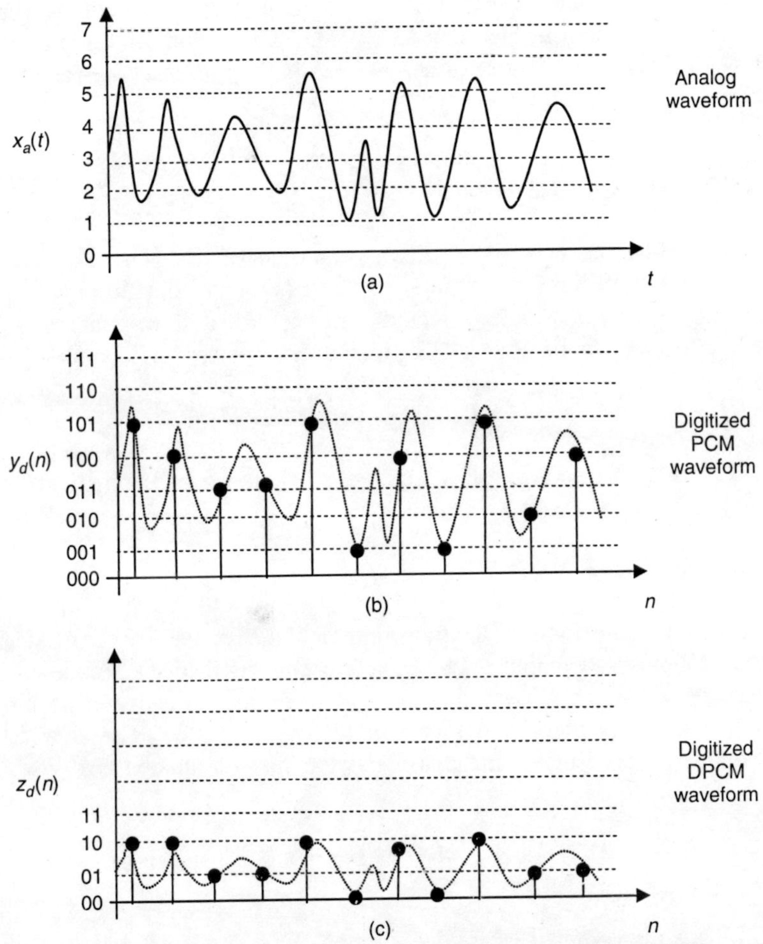

Figure 3.10. Uniform quantization: (a) Analog input signal; (b) PCM waveform (3-bit digitization of the analog signal). Output after quantization: [101, 100, 011, 011, 101, 001, 100, 001, 101, 010, 100]. Total number of bits in PCM digitization = 33. (c) Differential PCM (2-bit digitization of the analog signal). Output after quantization: [10, 10, 01, 01, 10, 00, 10, 00, 10, 01, 01]. Total number of bits in DPCM digitization = 22. As an aside, it can be noted that, relative to the PCM, the DPCM reduces the number of bits for encoding by reducing the variance of the input signal. The dynamic range of the input signal can be reduced by exploiting the redundancy present within the adjacent samples of the signal.

prediction parameters must be encoded and transmitted separately in order to reconstruct the signal at the receiver. In backward adaptation, the parameters are estimated from past data, which are already available at the receiver. Therefore, the prediction parameters can be estimated locally at the receiver. Backward predictor adaptation is amenable to low-delay coding [Gibs90] [Chen92]. ADPCM encoders with pole-zero decoder filters have proved to be particularly versatile in speech applications. In fact, the ADPCM 32 kb/s algorithm adopted for the ITU-T G.726 [G726] standard (formerly G.721 [G721]) uses a pole-zero adaptive predictor.

3.4 VECTOR QUANTIZATION

Data compression via vector quantization (VQ) is achieved by encoding a data-set jointly in block or vector form. Figure 3.11(a) shows an N-dimensional quantizer and a codebook. The incoming vectors can be formed from consecutive data samples or from model parameters. The quantizer maps the i-th incoming $[N \times 1]$ vector given by

$$\mathbf{s}_i = [s_i(0), s_i(1), \ldots, s_i(N-1)]^T \tag{3.20}$$

to a n-th channel symbol u_n, $n = 1, 2, \ldots, L$ as shown in Figure 3.11(a). The codebook consists of L code vectors,

$$\hat{\mathbf{s}}_n = [\hat{s}_n(0), \hat{s}_n(1), \ldots, \hat{s}_n(N-1)]^T, n = 1, 2, .., L, \tag{3.21}$$

which reside in the memory of the transmitter and the receiver. A vector quantizer works as follows. The input vectors, $\mathbf{s}_i$, are compared to each codeword, $\hat{\mathbf{s}}_n$, and the address of the closest codeword, with respect to a distortion measure $\varepsilon(\mathbf{s}_i, \hat{\mathbf{s}}_n)$, determines the channel symbol to be transmitted. The simplest and most commonly used distortion measure is the sum of squared errors which is given by

$$\varepsilon(\mathbf{s}_i, \hat{\mathbf{s}}_n) = \sum_{k=0}^{N-1} (s_i(k) - \hat{s}_n(k))^2. \tag{3.22}$$

The L $[N \times 1]$ real-valued vectors are entries of the codebook and are designed by dividing the vector space into L nonoverlapping cells, c_n, as shown in Figure 3.11(b). Each cell, c_n, is associated with a template vector $\hat{\mathbf{s}}_n$. The quantizer assigns the channel symbol, u_n, to the vector $\mathbf{s}_i$, if $\mathbf{s}_i$ belongs to c_n. The channel symbol u_n is usually a binary representation of the codebook index of $\hat{\mathbf{s}}_n$.

A vector quantizer can be considered as a generalization of the scalar PCM and, in fact, Gersho [Gers83] calls it vector PCM (VPCM). In VPCM, the codebook is fully searched and the number of bits per sample is given by

$$B = \frac{1}{N} \log_2 L. \tag{3.23}$$

The signal-to-noise ratio for VPCM is given by

$$SNR_N = 6B + K_N (\text{dB}). \tag{3.24}$$

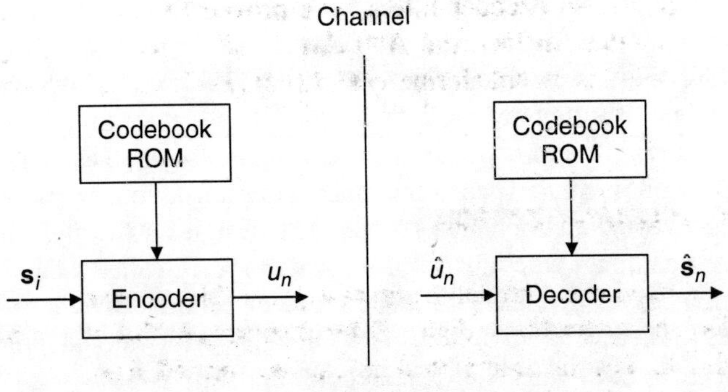

(a)

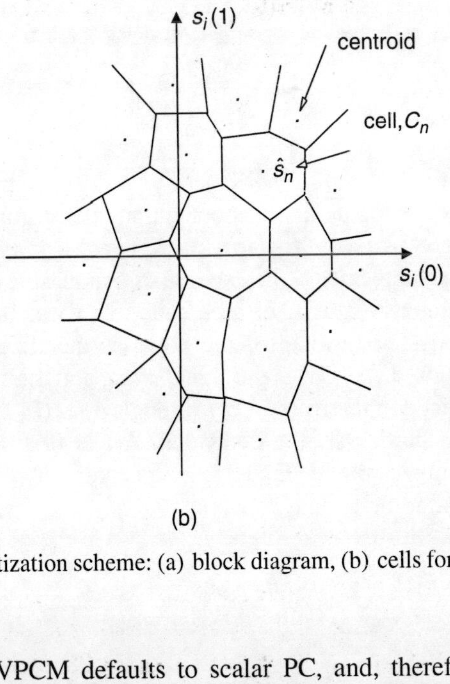

(b)

Figure 3.11. Vector quantization scheme: (a) block diagram, (b) cells for two-dimensional VQ, i.e., $N = 2$.

Note that for $N = 1$, VPCM defaults to scalar PC, and, therefore, (3.13) is a special case of (3.24). Although the two equations are quite similar, VPCM yields improved SNR (reflected in K_N), since it exploits the correlation within the vectors. VQ offers significant coding gain by increasing N and L. However, the memory and the computational complexity required grows exponentially with N for a given rate. In general, the benefits of VQ are realized at rates of 1 bit per sample or less. The codebook design process, also known as the training or populating process, can be fixed or adaptive.

Fixed codebooks are designed *a priori* and the basic design procedure involves an initial guess for the codebook and then iterative improvement by using a large number of training vectors. An iterative codebook design algorithm that works for a large class of distortion measures was given by Linde, Buzo, and Gray [Lind80]. This is essentially an extension of Lloyd's [Lloy82] scalar quantizer design and is often referred to as the "LBG algorithm." Typically, the number of training vectors per code vector must be at least ten and preferably fifty [Makh85]. Since speech and audio are nonstationary signals, one may also wish to adapt the codebooks ("codebook design on the fly") to the signal statistics. A quantizer with an adaptive codebook is called adaptive VQ (A-VQ) and applications to speech coding have been reported in [Paul82] [Cupe85] and [Cupe89]. There are two types of A-VQ, namely, forward adaptive and backward adaptive. In backward A-VQ, codebook updating is based on past data that is also available at the decoder. Forward A-VQ updates the codebooks based on current (or sometimes future) data and as such additional information must be encoded.

3.4.1 Structured VQ

The complexity in high-dimensionality VQ can be reduced significantly with the use of structured codebooks that allow for efficient search. Tree-structured [Buzo80] and multi-step [Juan82] vector quantizers are associated with lower encoding complexity at the expense of a modest loss of performance. Multi-step vector quantizers consist of a cascade of two or more quantizers each one encoding the error or residual of the previous quantizer. In Figure 3.12(a), the first VQ codebook, $\boldsymbol{L}_1$, encodes the signal, $s(k)$, and the subsequent VQ stages, $\boldsymbol{L}_2$ through $\boldsymbol{L}_M$, encode the errors, $e_1(k)$ through $e_{M-1}(k)$ from the previous stages, respectively. In particular, the codebook, $\boldsymbol{L}_1$, is first searched and the vector, $\hat{s}_{l_1}(k)$, that minimizes the MSE (3.25) is selected; where l_1 is the codebook index associated with the first-stage codebook:

$$\varepsilon_l = \sum_{k=0}^{N-1}(s(k) - \hat{s}_l(k))^2, \text{ for } l = 1, 2, 3, \ldots, L_1. \tag{3.25}$$

Next, the difference between the original input, $s(k)$, and the first-stage codeword, $\hat{s}_{l_1}(k)$ is computed as shown in Figure 3.12 (a). This is given by

$$e_1(k) = s(k) - \hat{s}_{l_1}(k), \text{ for } k = 0, 1, 2, 3, \ldots, N-1. \tag{3.26}$$

The residual, $e_1(k)$, is used in the second stage as the reference signal to be approximated. Codebooks $\boldsymbol{L}_2, \boldsymbol{L}_3, \ldots \boldsymbol{L}_M$ are searched sequentially, and the code vectors $\hat{e}_{l_2,1}(k), \hat{e}_{l_3,2}(k), \ldots, \hat{e}_{l_M,M-1}(k)$ that result in the minimum MSE (3.27) are chosen as the codewords. Note that $l_2, l_3, \ldots, l_M$ are the codebook indices

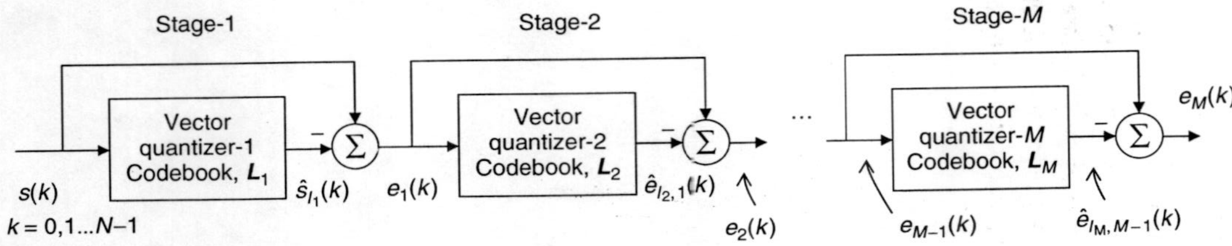

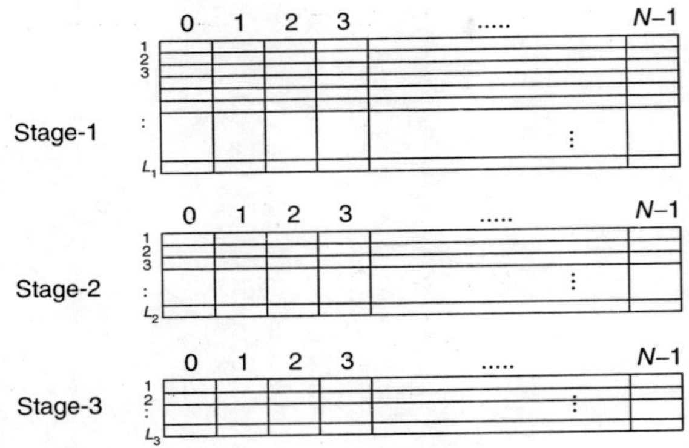

Figure 3.12. (a) Multi-step M-stage VQ block diagram. At the decoder, the signal, $s(k)$, can be reconstructed as $\tilde{s}(k) = \hat{s}_{l_1}(k) + \hat{e}_{l_2,1}(k) + \ldots + \hat{e}_{l_M,M-1}(k)$, for $k = 0, 1, 2, \ldots N - 1$. (b) An example multi-step codebook structure ($M = 3$). The first stage N-dimensional codebook consists of L_1 code vectors. Similarly, the number of entries in the second- and third-stage codebooks include L_2 and L_3, respectively. Usually, in order to reduce the computational complexity, the number of codevectors is as follows: $L_1 > L_2 > L_3$.

associated with the 2nd, 3rd, ..., M-th-stage codebooks, respectively.

$$\begin{aligned} &\textit{For codebook } \mathbf{L_2} \\ &\varepsilon_{l,2} = \sum_{k=0}^{N-1} (e_1(k) - \hat{e}_{l,1}(k))^2, \text{ for } l = 1, 2, 3, \ldots, L_2 \\ &\quad :: \\ &\quad :: \\ &\textit{For codebook } \mathbf{L_M} \\ &\varepsilon_{l,M} = \sum_{k=0}^{N-1} (e_{M-1}(k) - \hat{e}_{l,M-1}(k))^2, \text{ for } l = 1, 2, 3, \ldots, L_M \end{aligned} \tag{3.27}$$

At the decoder, the transmitted codeword, $\tilde{s}(k)$, can be reconstructed as follows:

$$\tilde{s}(k) = \hat{s}_{l_1}(k) + \hat{e}_{l_2,1}(k) + \ldots + \hat{e}_{l_M,M-1}(k), \quad \text{for } k = 0, 1, 2, \ldots N-1. \tag{3.28}$$

The complexity of VQ can also be reduced by normalizing the vectors of the codebook and encoding the gain separately. The technique is called gain/shape VQ (GS-VQ) and has been introduced by Buzo *et al.* [Buzo80] and later studied by Sabin and Gray [Sabi82]. The waveform shape is represented by a code vector from the shape codebook while the gain can be encoded from the gain codebook, Figure 3.13. The idea of encoding the gain separately allows for the encoding of

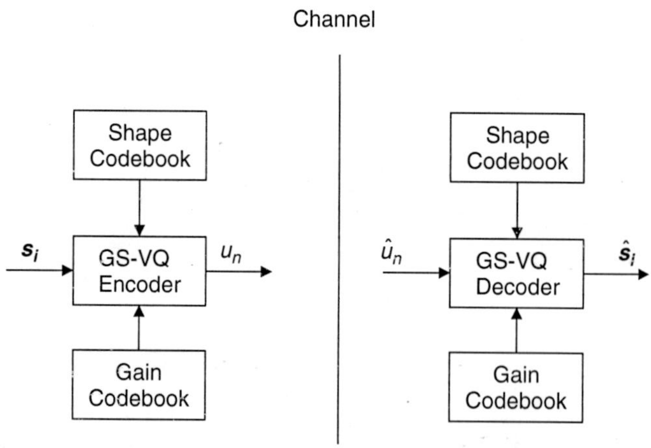

Figure 3.13. Gain/shape (GS)-VQ encoder and decoder. In the GS-VQ, the idea is to encode the waveform shape and the gain separately using the shape and gain codebooks, respectively.

vectors of high dimensionality with manageable complexity and is being widely used in encoding the excitation signal in code-excited linear predictive (CELP) coders [Atal90]. An alternative method for building highly structured codebooks consists of forming the code vectors by linearly combining a small set of basis vectors [Gers90b].

3.4.2 Split-VQ

In split-VQ, it is typical to employ two stages of VQs; multiple codebooks of smaller dimensions are used in the second stage. A two-stage split-VQ block diagram is given in Figure 3.14(a) and the corresponding split-VQ codebook structure is shown in Figure 3.14(b). From Figure 3.14(b), note that the first-stage codebook, $\mathbf{L}_1$, employs an N-dimensional VQ and consists of L_1 code vectors. The second-stage codebook is implemented as a split-VQ and consists of a combination of two $N/2$-dimensional codebooks, $\mathbf{L}_2$ and $\mathbf{L}_3$. The number of entries in these two codebooks are L_2 and L_3, respectively. First, the codebook $\mathbf{L}_1$ is searched and the vector, $\hat{s}_{l_1}(k)$, that minimizes the MSE (3.29) is selected; where l_1 is the index of the first-stage codebook:

$$\varepsilon_l = \sum_{k=0}^{N-1}(s(k) - \hat{s}_l(k))^2, \text{ for } l = 1, 2, 3, \ldots, L_1. \tag{3.29}$$

Next, the difference between the original input, $s(k)$, and the first-stage codeword, $\hat{s}_{l_1}(k)$, is computed as shown in Figure 3.14(a):

$$e(k) = s(k) - \hat{s}_{l_1}(k), \text{ for } k = 0, 1, 2, 3, \ldots, N-1. \tag{3.30}$$

The residual, $e(k)$, is used in the second stage as the reference signal to be approximated. Codebooks $\mathbf{L}_2$ and $\mathbf{L}_3$ are searched separately and the code vectors $\hat{\mathbf{e}}_{l_2,low}$ and $\hat{\mathbf{e}}_{l_3,upp}$ that result in the minimum MSE (3.31) and (3.32), respectively, are chosen as the codewords. Note that l_2 and l_3 are the codebook indices associated with the second- and third-stage codebooks:

For codebook $\mathbf{L}_2$

$$\varepsilon_{l,low} = \sum_{k=0}^{N/2-1}(e(k) - \hat{e}_{l,low}(k))^2, \text{ for } l = 1, 2, 3, \ldots, L_2 \tag{3.31}$$

For codebook $\mathbf{L}_3$

$$\varepsilon_{l,upp} = \sum_{k=N/2}^{N-1}(e(k) - \hat{e}_{l,upp}(k - N/2))^2, \text{ for } l = 1, 2, 3, \ldots, L_3. \tag{3.32}$$

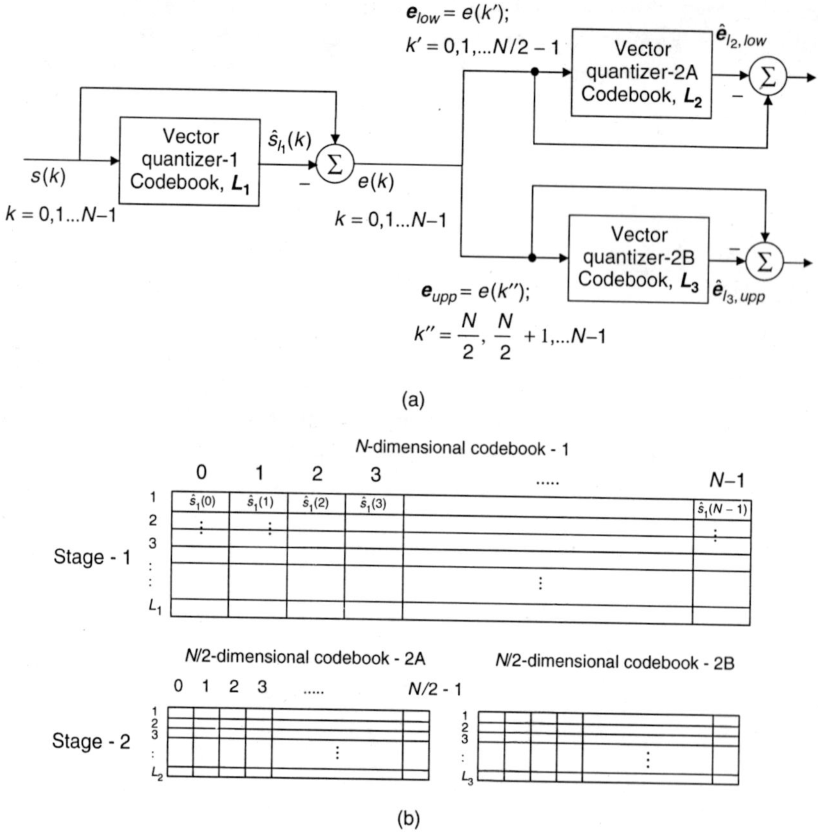

Figure 3.14. Split VQ: (a) A two-stage split-VQ block diagram; (b) the split-VQ codebook structure. In the split-VQ, the codevector search is performed by "dividing" the codebook into smaller dimension codebooks. In Figure 3.14(b), the second stage N-dimensional VQ has been divided into two $N/2$-dimensional split-VQ. Note that the first-stage codebook, $\boldsymbol{L}_1$, employs an N-dimensional VQ and consists of L_1 entries. The second-stage codebook is implemented as a split-VQ and consists of a combination of two $N/2$-dimensional codebooks, $\boldsymbol{L}_2$ and $\boldsymbol{L}_3$. The number of entries in these two codebooks are L_2 and L_3, respectively. The codebook indices, i.e., l_1, l_2, and l_3 will be encoded and transmitted. At the decoder, these codebook indices are used to reconstruct the transmitted codeword, $\tilde{s}(k)$, Eq. (3.33).

At the decoder, the transmitted codeword, $\tilde{s}(k)$, can be reconstructed as follows:

$$\tilde{s}(k) = \begin{cases} \hat{s}_{l_1}(k) + \hat{e}_{l_2,low}(k), & \text{for } k = 0, 1, 2, \ldots, N/2 - 1 \\ \hat{s}_{l_1}(k) + \hat{e}_{l_3,upp}(k - N/2), & \text{for } k = N/2, N/2 + 1, \ldots, N - 1. \end{cases} \tag{3.33}$$

Split-VQ offers high coding accuracy, however, with increased computational complexity and with a slight drop in the coding gain. Paliwal and Atal discuss these issues in [Pali91] [Pali93] while presenting an algorithm for vector quantization of speech LPC parameters at 24 bits/frame. Despite the aforementioned shortcomings, split-VQ techniques are very efficient when it comes to encoding line spectrum prediction parameters in several speech and audio coding standards, such as the ITU-T G.729 CS-ACELP standard [G729], the IS-893 Selectable Mode Vocoder (SMV) [IS-893], and the MPEG-4 General Audio (GA) Coding-Twin VQ tool [ISOI99].

3.4.3 Conjugate-Structure VQ

Conjugate structure-VQ (CS-VQ) [Kata93] [Kata96] enables joint quantization of two or more parameters. The CS-VQ works as follows. Let $s(n)$ be the target vector that has to be approximated, and the MSE ε to be minimized is given by

$$\varepsilon = \frac{1}{N}\sum_{n=0}^{N-1}|e(n)|^2 = \frac{1}{N}\sum_{n=0}^{N-1}|(s(n) - g_1u(n) - g_2v(n))|^2, \tag{3.34}$$

where $u(n)$ and $v(n)$ are some arbitrary vectors, and g_1 and g_2 are the gains to be vector quantized using the CS codebook given in Figure 3.15. From this figure, codebooks A and B contain P and Q entries, respectively. In both codebooks, the first-column element corresponds to parameter 1, i.e., g_1 and the second-column element represents parameter 2, i.e., g_2. The optimum combination of g_1 and g_2 that results in the minimum MSE (3.34) is computed from PQ permutations as follows:

$$g_1(i, j) = g_{A1,i} + g_{B1,j} i \in [1, 2, \ldots, P], j \in [1, 2, \ldots, Q] \tag{3.35}$$

$$g_2(i, j) = g_{A2,i} + g_{B2,j} i \in [1, 2, \ldots, P], j \in [1, 2, \ldots, Q]. \tag{3.36}$$

Index	Codebook – A	
1	$g_{A1,1}$	$g_{A2,1}$
2	$g_{A1,2}$	$g_{A2,2}$
3	$g_{A1,3}$	$g_{A2,3}$
⋮	⋮	⋮
P	$g_{A1,P}$	$g_{A2,P}$

Index	Codebook – B	
1	$g_{B1,1}$	$g_{B2,1}$
2	$g_{B1,2}$	$g_{B2,2}$
3	$g_{B1,3}$	$g_{B2,3}$
⋮	⋮	⋮
Q	$g_{B1,Q}$	$g_{B2,Q}$

Figure 3.15. An example CS-VQ. In this figure, the codebooks 'A' and 'B' are conjugate.

CS-VQ codebooks are particularly handy in scenarios that involve joint quantization of excitation gains. Second-generation near-toll-quality CELP codecs (e.g., ITU-T G.729) and third-generation (3G) CELP standards for cellular applications (e.g., TIA/IS-893 Selectable Mode Vocoder) employ the CS-VQ codebooks to encode the adaptive and stochastic excitation gains [Atal90] [Sala98] [G729] [IS-893]. A CS-VQ is used to vector quantize the transformed spectral coefficients in the MPEG-4 Twin-VQ encoder.

3.5 BIT-ALLOCATION ALGORITHMS

Until now, we discussed various scalar and vector quantization algorithms without emphasizing how the number of quantization levels are determined. In this section, we review some of the fundamental bit allocation techniques. A *bit-allocation algorithm* determines the number of bits required to quantize an audio frame with reduced audible distortions. Bit-allocation can be based on certain perceptual rules or spectral characteristics. From Figure 3.1, parameters typically quantized include the transform coefficients, **x**, scale factors, **S**, and the residual error, **e**. For now, let us consider that the transform coefficients, **x**, are to be quantized, i.e.,

$$\mathbf{x} = [x_1, x_2, x_3, \ldots, x_{N_f}]^{\mathrm{T}}, \tag{3.37}$$

where N_f represents the total number of transform coefficients. Let the total number of bits available to quantize the transform coefficients be N bits. Our objective is to find an optimum way of distributing the available N bits across the individual transform coefficients, such that a distortion measure, D, is minimized. The distortion, D, is given by

$$D = \frac{1}{N_f}\sum_{i=1}^{N_f} \mathrm{E}[(x_i - \hat{x}_i)^2] = \frac{1}{N_f}\sum_{i=1}^{N_f} d_i, \tag{3.38}$$

where x_i and $\hat{x}_i$ denote the i-th unquantized and quantized transform coefficients, respectively; and E[.] is the expectation operator. Let n_i be the number of bits assigned to the coefficient x_i for quantization, such that,

$$\sum_{i=1}^{N_f} n_i \leqslant N. \tag{3.39}$$

Note that if x_i are uniformly distributed $\forall i$, then we can employ a simple *uniform* bit-allocation across all the transform coefficients, i.e.,

$$n_i = \left[\frac{N}{N_f}\right], \forall i \in [1, N_f]. \tag{3.40}$$

However, in practice, the transform coefficients, **x**, may not have uniform probability distribution. Therefore, employing an equal number of bits for both large and small amplitudes may result in spending extra bits for smaller amplitudes. Moreover, in such scenarios, for a given N, the distortion, D, can be very high.

Example 3.1

An example of the aforementioned discussion is presented in Table 3.1. Uniform bit-allocation is employed to quantize both the uniformly distributed and Gaussian-distributed transform coefficients. In this example, we assume that a total number of $N = 64$ bits are available for quantization; and $N_f = 16$ samples. Therefore, $n_i = 4, \forall i \in [1, 16]$. Note that the input vectors, $\mathbf{x}_u$ and $\mathbf{x}_g$, have been randomly generated in MATLAB™ using *rand(1, N_f)* and *randn(1, N_f)* functions, respectively. The distortions, D_u and D_g, are computed using (3.38) and are given by 0.00023927 and 0.00042573, respectively. From Example 3.1, we note that the uniform bit-allocation is not optimal for all the cases, especially when the distribution of the unquantized vector, $\mathbf{x}$, is not uniform. Therefore, we must have some cost function available that minimizes the distortion, d_i, subject to the constraint given in (3.39) is met. This is given by

$$\min_{n_i}\{D\} = \min_{n_i}\left\{\frac{1}{N_f}\sum_{i=1}^{N_f} \mathrm{E}[(x_i - \hat{x}_i)^2]\right\} = \min_{n_i}\left\{\frac{1}{N_f}\sum_{i=1}^{N_f}\sigma_i^2\right\}, \tag{3.41}$$

where σ_i^2 is the variance. Note that the above minimization problem can be simplified if the quantization noise has a uniform PDF [Jaya84]. From (3.12),

$$\sigma_i^2 = \frac{x_i^2}{3(2^{2n_i})}. \tag{3.42}$$

Table 3.1. Uniform bit-allocation scheme, where $n_i = [N/N_f]$, $\forall i \in [1, N_f]$.

Uniformly distributed coefficients		Gaussian-distributed coefficients	
Input vector, $\mathbf{x}_u$	Quantized vector, $\hat{\mathbf{x}}_u$	Input vector, $\mathbf{x}_g$	Quantized vector, $\hat{\mathbf{x}}_g$
[0.6029, 0.3806, 0.56222, 0.12649, 0.26904, 0.47535, 0.4553, 0.38398, 0.41811, 0.35213, 0.23434, 0.32256, 0.31352, 0.3026, 0.32179, 0.16496]	[0.625, 0.375, 0.5625, 0.125, 0.25, 0.5, 0.4375, 0.375, 0.4375, 0.375, 0.25, 0.3125, 0.3125, 0.3125, 0.3125, 0.1875]	[0.5199, 2.4205, −0.94578, −0.0081113, −0.42986, −0.87688, 1.1553, −0.82724, −1.345, −0.15859, −0.23544, 0.85353, 0.016574, −2.0292, 1.2702, 0.28333]	[0.5, 2.4375, −0.9375, 0, −0.4375, −0.875, 1.125, −0.8125, −1.375, −0.1875, −0.25, 0.875, 0, −2, 1.25, 0.3125]

$D = \frac{1}{N_f}\sum_{i=1}^{N_f} \mathrm{E}[(x_i - \hat{x}_i)^2]$, $n_i = 4 \forall i \in [1, 16]$, $D_u = 0.00023927$ and $D_g = 0.00042573$

Substituting (3.42) in (3.41) and minimizing w.r.t. n_i,

$$\frac{\partial D}{\partial n_i} = \frac{x_i^2}{3}(-2)2^{(-2n_i)} \ln 2 + K_1 = 0 \tag{3.43}$$

$$n_i = \frac{1}{2} \log_2 x_i^2 + K. \tag{3.44}$$

From (3.44) and (3.39),

$$\sum_{i=1}^{N_f} n_i = \sum_{i=1}^{N_f} \left(\frac{1}{2} \log_2 x_i^2 + K \right) = N \tag{3.45}$$

$$K = \frac{N}{N_f} - \frac{1}{2N_f} \sum_{i=1}^{N_f} \log_2 x_i^2 = \frac{N}{N_f} - \frac{1}{2N_f} \log_2 (\prod_{i=1}^{N_f} x_i^2). \tag{3.46}$$

Substituting (3.46) in (3.44), we can obtain the optimum bit-allocation, $n_i^{optimum}$, as

$$n_i^{optimum} = \frac{N}{N_f} + \frac{1}{2} \log_2 \left(\frac{x_i^2}{(\prod_{i=1}^{N_f} x_i^2)^{\frac{1}{N_f}}} \right). \tag{3.47}$$

Table 3.2 presents the optimum bit assignment for both uniformly distributed and Gaussian-distributed transform coefficients considered in the previous example (Table 3.1). From Table 3.2, note that the resulting optimal bit-allocation for Gaussian-distributed transform coefficients resulted in two negative integers. Several techniques have been proposed to avoid this scenario, namely, the sequential bit-allocation method [Rams82] and the Segall's method [Sega76]. For more detailed descriptions, readers are referred to [Jaya84] [Madi97]. Note that the bit-allocation scheme given by (3.47) may not be optimal either in the perceptual sense or in the SNR sense. This is because the minimization of (3.41) is performed without considering either the perceptual noise masking thresholds or the dependence of the signal-to-noise power on the optimal number of bits, $n_i^{optimum}$. Also, note that the distortion, D_u, in the case of optimal bit-assignment (Table 3.2) is slightly greater than in the case of uniform bit-allocation (Table 3.1). This can be attributed to the fact that fewer quantization levels must have been assigned to the low-powered transform coefficients relative to the number of levels implied by (3.47). Moreover, a maximum coding gain can be achieved when the audio signal spectrum is non-flat. One of the important remarks presented in [Jaya84] relevant to the on-going discussion is that when the geometric mean of x_i^2 is less than the arithmetic mean of x_i^2, then the optimal bit-allocation scheme performs better than the uniform bit-allocation. The ratio

Table 3.2. Optimal bit-allocation scheme, where $n_i^{optimum} = \frac{N}{N_f} + \frac{1}{2}\log_2 x_i^2 - \frac{1}{2}\log_2\left(\prod_{i=1}^{N_f} x_i^2\right)^{\frac{1}{N_f}}$.

Uniformly distributed coefficients		Gaussian-distributed coefficients	
Input vector, $\mathbf{x}_u$	Quantized vector, $\hat{\mathbf{x}}_u$	Input vector, $\mathbf{x}_g$	Quantized vector, $\hat{\mathbf{x}}_g$
[0.6029, 0.3806, 0.56222, 0.12649, 0.26904, 0.47535, 0.4553, 0.38398, 0.41811, 0.35213, 0.23434, 0.32256, 0.31352, 0.3026, 0.32179, 0.16496]	[0.59375, 0.375, 0.5625, 0.125, 0.25, 0.46875, 0.4375, 0.375, 0.4375, 0.375, 0.25, 0.3125, 0.3125, 0.3125, 0.3125, 0.125]	[0.5199, 2.4205, −0.94578, −0.0081113, −0.42986, −0.87688, 1.1553, −0.82724, −1.345, −0.15859, −0.23544, 0.85353, 0.016574, −2.0292, 1.2702, 0.28333]	[0.5, 2.4219, −0.9375, 0, −0.4375, −0.875, 1.1563, −0.8125, −1.3438, −0.125, −0.25, 0.84375, 0, −2.0313, 1.2656, 0.25]

Bits allocated, $n =$ [5 4 5 3 4 5 4 4 4 4 4 4 4 4 4 3] Distortion, $D_u = 0.0002468$

Bits allocated, $n =$ [4 6 5 −2 4 5 5 5 6 3 3 5 −1 6 6 3] Distortion, $D_g = 0.00022878$

of the two means is captured in the *spectral flatness measure (SFM)*, i.e.,

$$\text{Geometric Mean, } G_M = \left(\prod_{i=1}^{N_f} x_i^2\right)^{\frac{1}{N_f}}; \text{Arithemetic Mean, } A_M = \frac{1}{N_f}\sum_{i=1}^{N_f} x_i^2$$

$$SFM = \frac{G_M}{A_M}; \text{ and } SFM \in [0\ 1]. \tag{3.48}$$

Other important considerations, in addition to SNR and spectral flatness measures, are the perceptual noise masking, the noise-to-mask ratio (NMR), and the signal-to-mask ratio (SMR). All these measures are used in *perceptual bit allocation* methods. Since, at this point, readers are not introduced to the *principles of psychoacoustics* and the concepts of SMR and NMR; we defer a discussion of perceptually based bit allocation to Chapter 5, Section 5.8.

3.6 ENTROPY CODING

It is worthwhile to consider the theoretical limits for the minimum number of bits required to represent an audio sample. Shannon, in his mathematical theory of communication [Shan48], proved that the minimum number of bits required to encode a message, X, is given by the entropy, $H_e(X)$. The entropy of an input signal can be defined as follows. Let $X = [x_1, x_2, \ldots, x_N]$ be the input data vector of length N and p_i be the probability that i-th symbol (over the symbol set, $V = [v_1, v_2, \ldots, v_K]$) is transmitted. The entropy, $H_e(X)$, is given by

$$H_e(X) = -\sum_{i=1}^{K} p_i \log_2(p_i). \tag{3.49}$$

In simple terms, entropy is a *measure of uncertainty* of a random variable. For example, let the input bitstream to be encoded be $X = [4\ 5\ 6\ 6\ 2\ 5\ 4\ 4\ 5\ 4\ 4]$, i.e., $N = 11$; symbol set, $V = [2\ 4\ 5\ 6]$ and the corresponding probabilities are $\left[\frac{1}{11}, \frac{5}{11}, \frac{3}{11}, \frac{2}{11}\right]$, respectively, with $K = 4$. The entropy, $H_e(X)$, can be computed as follows:

$$\begin{aligned} H_e(X) &= -\sum_{i=1}^{K} p_i \log_2(p_i) \\ &= -\left\{\frac{1}{11}\log_2\left(\frac{1}{11}\right) + \frac{5}{11}\log_2\left(\frac{5}{11}\right) + \frac{3}{11}\log_2\left(\frac{3}{11}\right) + \frac{2}{11}\log_2\left(\frac{2}{11}\right)\right\} \\ &= 1.7899. \end{aligned} \tag{3.50}$$

Table 3.3. An example entropy code for Example 3.2.

Swimmer	Probability of winning	Binary string or the identifier
S_1	1/2	0
S_2	1/4	10
S_3	1/8	110
S_4	1/16	1110
S_5	1/64	111100
S_6	1/64	111101
S_7	1/64	111110
S_8	1/64	111111

Example 3.2

Consider eight swimmers {S_1, S_2, S_3, S_4, S_5, S_6, S_7, and S_8} in a race with win probabilities {1/2, 1/4, 1/8, 1/16, 1/64, 1/64, 1/64, and 1/64}, respectively. The entropy of the message announcing the winner can be computed as

$$
\begin{aligned}
H_e(X) &= -\sum_{i=1}^{8} p_i \log_2(p_i) \\
&= -\left\{\frac{1}{2}\log_2\left(\frac{1}{2}\right) + \frac{1}{4}\log_2\left(\frac{1}{4}\right) + \frac{1}{8}\log_2\left(\frac{1}{8}\right) + \frac{1}{16}\log_2\left(\frac{1}{16}\right)\right. \\
&\quad \left. + \frac{4}{64}\log_2\left(\frac{1}{64}\right)\right\} \\
&= 2.
\end{aligned}
$$

An example of the entropy code for the above message can be obtained by associating binary strings w.r.t. the swimmers' probability of winning as shown in Table 3.3. The average length of the example entropy code given in Table 3.3 is 2 bits, in contrast with 3 bits for a uniform code.

The statistical entropy alone does not provide a good measure of compressibility in the case of audio coding, since several other factors, i.e., quantization noise, masking thresholds, and tone- and noise-masking effects, must be accounted for in order to achieve efficiency. Johnston, in 1988, proposed a theoretical limit on compressibility for audio signals ($\sim$ 2.1 bits/sample) based on the measure of perceptually relevant information content. The limit is obtained based on both the psychoacoustic signal analysis and the statistical entropy and is called the *perceptual entropy* [John88a] [John88b]. The various steps involved in the perceptual entropy estimation are described later in Chapter 5.

In all the entropy coding schemes, the objective is to construct an ensemble code for each message, such that the code is uniquely decodable, prefix-free, and optimum in the sense that it provides minimum-redundancy encoding. In particular, some basic restrictions and design considerations imposed on a source-coding process include:

- *Condition 1*: Each message should be assigned a unique code (see Example 3.3).
- *Condition 2*: The codes must be prefix-free. For example, consider the following two code sequences (CS): CS-1 = {00, 11, 10, 011, 001} and CS-2 = {00, 11, 10, 011, 010}. Consider the output sequence to be decoded to be {001100011...}. At the decoder, if CS-1 is employed, the decoded sequence can be either {00, 11, 00, 011, ...} or {001, 10, 001, ...} or {001, 10, 00, 11, ...}, etc. This is due to the confusion at the decoder whether to select '00' or '001' from the output sequence. This confusion is avoided using CS-2, where the decoded sequence is unique and is given by {00, 11, 00, 011...}. Therefore, in a valid, no code in its entirety can be found as a prefix of another code.
- *Condition 3*: Additional information regarding the beginning- and the end-point of a message source will not usually be available at the decoder (once synchronization occurs).
- *Condition 4*: A necessary condition for a code to be prefix-free is given by the Kraft inequality [Cove91]:

$$K_I = \sum_{i=1}^{N} 2^{-L_i} \leqslant 1, \tag{3.51}$$

where L_i is the codeword length of the i-th symbol. In the example discussed in condition 2, the Kraft inequality K_I for both CS-1 and CS-2 is '1'. Although the Kraft inequality for CS-1 is satisfied, the encoding sequence is not prefix-free. Note that (3.51) is not a sufficient condition for a code to be prefix-free.

- *Condition 5*: To obtain a minimum-redundancy code, the compression rate, R, must be minimized and (3.51) must be satisfied:

$$R = \sum_{i=1}^{N} p_i L_i. \tag{3.52}$$

Example 3.3

Let the input bitstream $X = [4\ 5\ 6\ 6\ 2\ 5\ 4\ 4\ 1\ 4\ 4]$ be chosen over a data set $V = [0\ 12\ 3\ 4\ 5\ 6\ 7]$. Here, $N = 11$, $K = 8$, and the probabilities $p_i = \left[0, \frac{1}{11}, \frac{1}{11}, 0, \frac{5}{11}, \frac{2}{11}, \frac{2}{11}, 0\right] i \in V$.

In Figure 3.16(a), a simple binary representation with equal-length code is used. The code length of each symbol is given by $l = int(\log_2 K)$, i.e., $l = int(\log_2 8) = 3$. Therefore, a possible binary mapping would be, $1 \rightarrow 001, 2 \rightarrow 010, 4 \rightarrow 100, 5 \rightarrow 101, 6 \rightarrow 110$ and the total length, $L_b = 33$bits. Figure 3.16(b) depicts the Shannon-Fano coding procedure. Each symbol is encoded using a unary representation based on the symbol probability.

Input bitstream, [4 5 6 6 2 5 4 4 1 4 4]

(*a*) *Encoding based on the binary equal - length code* :
[100 101 110 110 010 101 100 100 001 100 100]
Total length, $L_b = 33$ *bits*

(*b*) *Encoding based on the Shannon - Fano Coding* :
4's-5 *times*, 5's-2 *times*, 6's-2 *twice*, 2's-*once*, *and* 1's-*once*
Probabilities, $p_4 = 5/11$, $p_5 = 2/11$, $p_6 = 2/11$, $p_2 = 1/11$, *and* $p_1 = 1/11$
Representations : 4 → 0, 5 → 10, 6 → 110, 1 → 1110, 2 → 11110
[0 10 110 110 11110 10 0 0 1110 0 0]
Total length, $L_{SF} = 24$ *bits*

Figure 3.16. Entropy coding schemes: (a) binary equal-length code, (b) Shannon-Fano coding.

3.6.1 Huffman Coding

Huffman proposed a technique to construct minimum redundancy codes [Huff52]. Huffman codes found applications in audio and video encoding due to their simplicity and efficiency. Moreover, Huffman coding is regarded as the most effective compression method, provided that the codes designed using a specific set of symbol frequencies match the input symbol frequencies. PDFs of audio signals of shorter frame-lengths are better described by the Gaussian distribution, while the long-time PDFs of audio can be characterized by the Laplacian or gamma densities [Rabi78] [Jaya84]. Hence, for example, Huffman codes designed based on the Gaussian or Laplacian PDFs can provide minimum redundancy entropy codes for audio encoding. Moreover, depending upon the symbol frequencies, a series of Huffman codetables can also be employed for entropy coding, e.g., the MPEG-1 Layer-III employs 32 Huffman codetables [ISOI94].

Example 3.4

Figure 3.17 depicts the Huffman coding procedure for the numerical Example 3.3. The input symbols, i.e., 1, 2, 4, 5, and 6, are first arranged in ascending order w.r.t their probabilities. Next, the two symbols with the smallest probabilities are combined to form a binary tree. The left tree is assigned a "0", and the right tree is represented by a "1." The probability of the resulting node is obtained by adding the two probabilities of the previous nodes as shown in Figure 3.17. The above procedure is continued until all the input symbol nodes are used. Finally, Huffman codes for each input symbol is formed by reading the bits along the tree. For example, the Huffman codeword for the input symbol "1" is given by "0000." The resulting Huffman bit-mapping is given in Table 3.4, and the total length of the encoded bitstream is $L_{HF} = 23$ bits. Note that, depending on the node selection for the code tree formation,

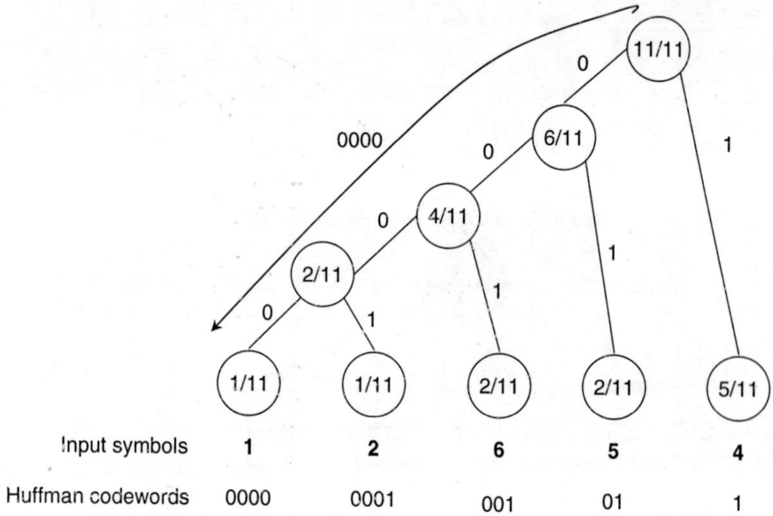

Figure 3.17. A possible Huffman coding tree for Example 3.4.

Table 3.4. Huffman codetable for the input bitstream, X = [4 5 6 6 2 5 4 4 1 4 4].

Input symbol	Probability	Huffman codeword
4	5/11	1
5	2/11	01
6	2/11	001
2	1/11	0001
1	1/11	0000

several Huffman bit-mappings can be possible, for example, $4 \rightarrow 1$, $5 \rightarrow 011$, $6 \rightarrow 010$, $2 \rightarrow 001$, and $1 \rightarrow 000$, as shown in Figure 3.18. However, the total number of bits remain the same, i.e., $L_{HF} = 23$ bits.

Example 3.5

The entropy of the input bitstream, $X = [4\ 5\ 6\ 6\ 2\ 5\ 4\ 4\ 1\ 4\ 4]$, is given by

$$\begin{aligned} H_e(X) &= -\sum_{i=1}^{K} p_i \log_2(p_i) \\ &= -\left\{\frac{2}{11}\log_2\left(\frac{1}{11}\right) + \frac{4}{11}\log_2\left(\frac{2}{11}\right) + \frac{5}{11}\log_2\left(\frac{5}{11}\right)\right\} \qquad (3.53) \\ &= 2.04. \end{aligned}$$

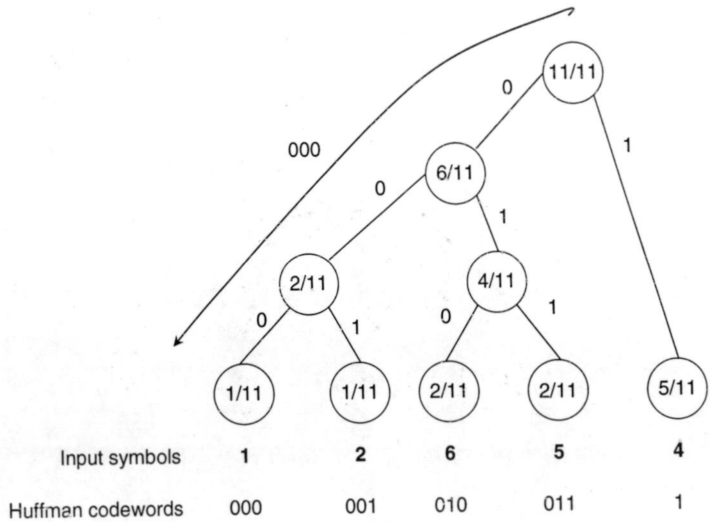

Figure 3.18. Huffman coding tree for Example 3.4.

From Figures 3.16 and 3.17, the compression rate, R, is obtained using

(a) the uniform binary representation $= 33/11 = 3$ bits/symbol,
(b) Shannon-Fano coding $= 24/11 = 2.18$ bits/symbol, and
(c) Huffman coding $= 23/11 = 2.09$ bits/symbol.

In the case of Huffman coding, entropy, $H_e(X)$, and the compression rate, R, can be related using the entropy bounds [Cove91]. This is given by

$$H_e(X) \leqslant R \leqslant H_e(X) + 1. \tag{3.54}$$

It is interesting to note that the compression rate for the Huffman code will be equal to the lower entropy bound, i.e., $R = H_e(X)$, if the input symbol frequencies are radix 2 (see Example 3.2).

Example 3.6

The Huffman code table for a different input symbol frequency than the one given in Example 3.4. Consider the input bitstream $Y = [2\ 5\ 6\ 6\ 2\ 5\ 5\ 4\ 1\ 4\ 4]$, chosen over a data set $V = [0\ 1\ 2\ 3\ 4\ 5\ 6\ 7]$; and the probabilities $p_i = \left[0, \frac{1}{11}, \frac{2}{11}, 0, \frac{3}{11}, \frac{3}{11}, \frac{2}{11}, 0\right]$ $i \in V$. Using the design procedure described above, a Huffman code tree can be formed as shown in Figure 3.19. Table 3.5 presents the resulting Huffman code table. Total length of the Huffman encoded bitstream is given by $L_{HF} =$ 25 bits.

Depending on the Huffman code table design procedure employed, three different encoding approaches can be possible. First, entropy coding based on

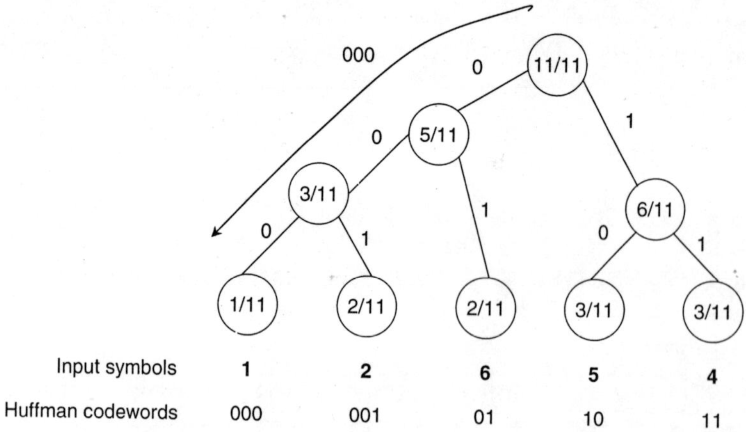

Figure 3.19. Huffman coding tree for Example 3.6.

Table 3.5. Huffman codetable for the input bitstream Y = [2 5 6 6 2 5 5 4 1 4 4].

Input symbol	Probability	Huffman codeword
4	3/11	11
5	3/11	10
6	2/11	01
2	2/11	001
1	1/11	000

the Huffman codes designed beforehand, i.e., *nonadaptive Huffman coding*. In particular, a training process involving a large database of input symbols is employed to design Huffman codes. These Huffman code tables will be available both at the encoder and at the decoder. It is important to note that this approach may not (always) result in minimum redundancy encoding. For example, if the Huffman bitmapping given in Table 3.5 is used to encode the input bitstream $X = [4\ 5\ 6\ 6\ 2\ 5\ 4\ 4\ 1\ 4\ 4]$ given in Example 3.3, the resulting total number of bits is $L_{HF} = 24$ bits, i.e., one bit more compared to Example 3.4. Therefore, in order to obtain better compression, a reliable symbol-frequency model is necessary. A series of Huffman code tables (in the range of 10–32) based on the symbol probabilities is usually employed in order to overcome the aforementioned shortcomings. The non-adaptive Huffman coding method is typically employed in a variety of audio coding standards [ISOI92] [ISOI94] [ISOI96] [John96]. Second, Huffman coding based on an iterative design/encode procedure, i.e., *semi-adaptive Huffman coding*. In the entropy coding literature, this approach is typically called the "two-pass" encoding scheme. In the first pass, a Huffman codetable is designed based on the input symbol statistics. In the second pass, entropy coding is performed using the designed Huffman codetable (similar to Examples 3.4 and 3.6). In this approach, note that the designed Huffman codetables must also be transmitted along with the entropy coded bitstream. This results in reduced coding efficiency, however, with an improved symbol-frequency modeling. Third, *adaptive Huffman coding* based on the symbol frequencies computed dynamically from the previous samples. Adaptive Huffman coding schemes based on the input quantization step size have also been proposed in order to accommodate for wide range of input word lengths [Crav96] [Crav97].

3.6.2 Rice Coding

Rice, in 1979, proposed a method for constructing practical noiseless codes [Rice79]. Rice codes are usually employed when the input signal, x, exhibits the Laplacian distribution, i.e.,

$$p_L(x) = \frac{1}{\sqrt{2}\sigma_x} e^{-\frac{\sqrt{2}|x|}{\sigma_x}}. \tag{3.55}$$

A Rice code can be considered as a Huffman code for the Laplacian PDF. Several efficient algorithms are available to form Rice codes [Rice79] [Cove91]. A simple method to represent the integer, I, as a Rice code is to divide the integer into four parts, i.e., a sign bit, m low-order bits (LSBs), and the number corresponding to the remaining MSBs of I as zeros, followed by a stop bit '1.' The parameter 'm' characterizes the Rice code, and is given by [Robi94]

$$m = \log_2(\log_e(2)E(|x|)). \tag{3.56}$$

For example, the Rice code for $I = 69$ and $m = 4$ is given by [0 0101 0000 1].

Example 3.7

Rice coding for Example 3.3. Input bitstream = [4 5 6 6 2 5 4 4 1 4 4]

Input symbol	Binary representation	Rice code ($m = 2$)
4	100	0 00 0 1
5	101	0 01 0 1
6	110	0 10 0 1
2	010	0 10 1
1	001	0 011

3.6.3 Golomb Coding

Golomb codes [Golo66] are optimal for exponentially decaying probability distributions of positive integers. Golomb codes are prefix codes that can be characterized by a unique parameter "m." An integer "I" can be encoded using a Golomb code as follows. The code consists of two parts: a binary representation of (m mod I) and a unary representation of $\left[\frac{I}{m}\right]$. For example, consider $I = 69$ and $m = 16$. The Golomb code will be [010111100] as explained in Figure 3.20. In Method 1, the positive integer "I" is divided in two parts, i.e., binary and unary bits along with a stop bit. On the other hand, in Method 2, if $m = 2^k$, the codeword for "I" consists of "k" LSBs of "I," followed by the number formed by the remaining MSBs of "I" in unary representation and with a stop bit. Therefore, the length of the code is $k + \left[\frac{I}{2^k}\right] + 1$.

Example 3.8

Consider the input bitstream, $X = [4\ 4\ 4\ 2\ 2\ 4\ 4\ 4\ 4\ 4\ 4\ 2\ 4\ 4\ 4\ 4]$, chosen over the data set $V = [2\ 4]$. The run-length encoding scheme [Golo66] can be employed to efficiently encode X. Note that "4" is the most frequently occurring symbol in X. The number of consecutive occurrences of "4" is called the run length, n. The run lengths are monitored and encoded, i.e., [3, 0, 6, 4]. Here "0" represents the consecutive occurrence of "2". The probability of occurrence of "4" and "2" are given by $p(4) = p = 13/16$ and $p(2) = (1 - p) = q = 3/16$, respectively. Note that $p >> q$. For this case,

Method – 1:

$n = 69,\ m = 16$

First part: $Binary(m \bmod n) = Binary(16 \bmod 69) = Binary(5) = 0101$

Second part: $Unary\left(\left[\frac{n}{m}\right]\right) = Unary(4) = 1110$

Stop bit $= 0$

Golomb Code : *First part + Second part + Stop bit*
[0101 1110 0]

Method – 2:

$n = 69,\ m = 16$, i.e., $k = 4$ (where, $m = 2^k$)

First part: *k LSBs of n* = 4 *LSBs of* [1000101] = 0101

Second part: *Unary (rest of MSBs)* = *unary*(4) = 1110

Stop bit $= 0$

Golomb Code : *First part + Second part + Stop bit*
[0101 1110 0]

Figure 3.20. Golomb coding.

Huffman coding of X results in 16 bits. Moreover, the PDFs of the run lengths are better described using an exponential distribution, i.e., the probability of a run length of n is given by, qp^n, which is an exponential distribution. Rice coding [Rice79] or the Golomb coding [Golo66] can be employed to efficiently encode the exponentially distributed run lengths. Furthermore, both Golomb and Rice codes are fast prefix codes that allow for practical implementation.

3.6.4 Arithmetic Coding

Arithmetic coding [Riss79] [Witt87] [Howa94] deals with encoding a sequence of input symbols as a *large binary fraction* known as a "codeword." For example, let $V = [v_1, v_2, \ldots, v_K]$ be the data set; let p_i be the probability that the i-th symbol is transmitted; and let $X = [x_1, x_2, \ldots, x_N]$ be the input data vector of length N. The main idea behind an arithmetic coder is to encode the input data stream, X, as one codeword that corresponds to a rational number in the half-open unit interval [0 1). Arithmetic coding is particularly useful when dealing with adaptive encoding and with highly skewed symbols [Witt87].

Example 3.9

Arithmetic coding of the input stream $X = 1\ 0 - 1\ 0\ 1 \ldots.$ chosen over a data set $V = [-1\ 0\ 1]$. Here, $N = 5$, $K = 3$. We will use the following symbol probabilities $p_i = \left[\frac{1}{3}, \frac{1}{2}, \frac{1}{6}\right] i \in V$.

Step 1

The probabilities associated with the data set $V = [-1\ 0\ 1]$ are arranged as intervals on a scale of [0, 1) as shown in Figure 3.21.

Step 2

The first input symbol in the data stream, X, is '1.' Therefore, the interval $\left[\frac{5}{6}, 1\right)$ is chosen as the target range.

Step 3

The second input symbol in the data stream, X, is '0.' Now, the interval $\left[\frac{5}{6}, 1\right)$ is partitioned according to the symbol probabilities, 1/3, 1/2, and 1/6. The resulting interval ranges are given in Figure 3.21. For example, the interval range for symbol '-1' can be computed as $\left[\frac{5}{6}, \frac{5}{6} + \frac{1}{6}\frac{1}{3}\right) = \left[\frac{5}{6}, \frac{16}{18}\right)$, and for symbol '0' the interval ranges is given by, $\left[\frac{16}{18}, \frac{5}{6} + \frac{1}{6}\frac{1}{3} + \frac{1}{6}\frac{1}{2}\right) = \left[\frac{16}{18}, \frac{35}{36}\right)$.

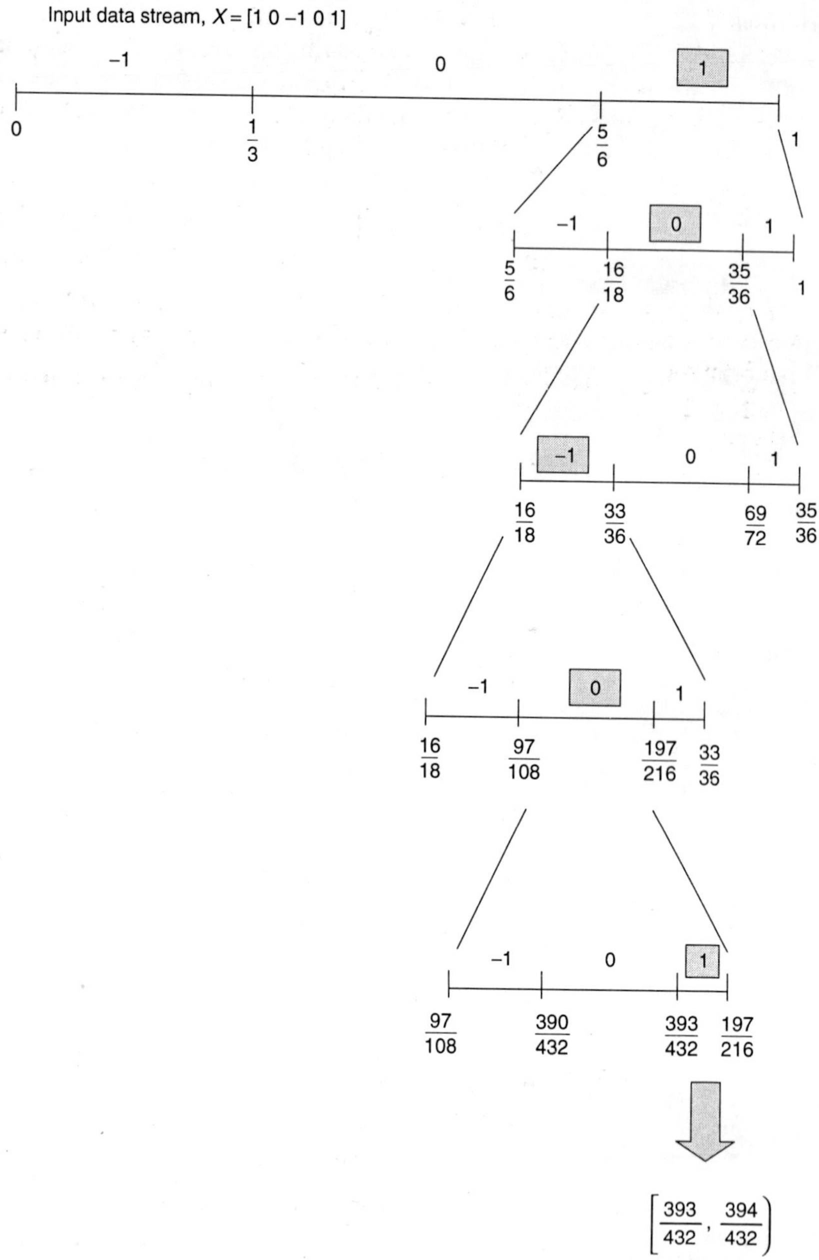

Figure 3.21. Arithmetic coding. First, the probabilities associated with the data set $V = [-1\ 0\ 1]$ are arranged as intervals on a scale of [0, 1). Next, in step 2, an interval is chosen that corresponds to the probability of the input symbol, '1', in the data sequence, X, i.e., [5/6, 1). In step 3, the interval [5/6, 1) is partitioned according to the probabilities, 1/3, 1/2, and 1/6; and the range corresponding to the input symbol, '0' is chosen, i.e., [16/18,35/36). This procedure is repeated for the rest of the input symbols, and the final interval range (typically a rational number) is encoded in the binary form.

Step 4

The third input symbol in the data stream, X, is '−1.' The interval $\left[\frac{16}{18}, \frac{35}{36}\right)$ is partitioned according to the symbol probabilities, 1/3, 1/2, and 1/6. The resulting interval ranges are given in Figure 3.21.

The above procedure is repeated for the rest of the input symbols, and an interval range is obtained, i.e., $\left[\frac{393}{432}, \frac{394}{432}\right) \simeq [0.9097222, 0.912037)$. In the binary form, the interval is given by $[0.1110100011\ldots, 0.1110100101\ldots)$. Since all binary numbers that begin with 0.1110100100 are within the interval $\left[\frac{393}{432}, \frac{394}{432}\right)$, the binary codeword 1110100100 uniquely represents the input data stream $X = [1\ 0\ \ -1\ 0\ 1]$.

3.7 SUMMARY

This chapter covered quantization essentials and provided background on PCM, DPCM, vector quantization, bit allocation, and entropy coding algorithms. A quantization–bit allocation–entropy coding (QBE) framework that is part of most of the audio coding standards was described. Some of the important concepts addressed in this chapter include:

- Uniform and nonuniform quantization
- PCM, DPCM, and ADPCM techniques
- Vector quantization, structured VQ, split VQ, and conjugate structure VQ
- Bit-allocation strategies
- Source coding principles
- Lossless (entropy) coders – Huffman coding, Rice coding, Golomb coding, and arithmetic coding.

PROBLEMS

3.1. Derive the PCM 6 dB per bit rule when the quantization error has a uniform probability density function.

3.2. For a signal with Gaussian distribution (zero mean and unit variance)

   a. Design a uniform PCM quantizer with four levels.

   b. Design a nonuniform four-level quantizer that is optimized for the signal PDF. Compare with the uniform PCM in terms of SNR.

3.3. For the PDF $p(x) = \frac{1}{2}e^{-|x|}$, determine the mean, the variance, and the probability that a random variable will fall within $\pm\sigma_x$ of the mean value.

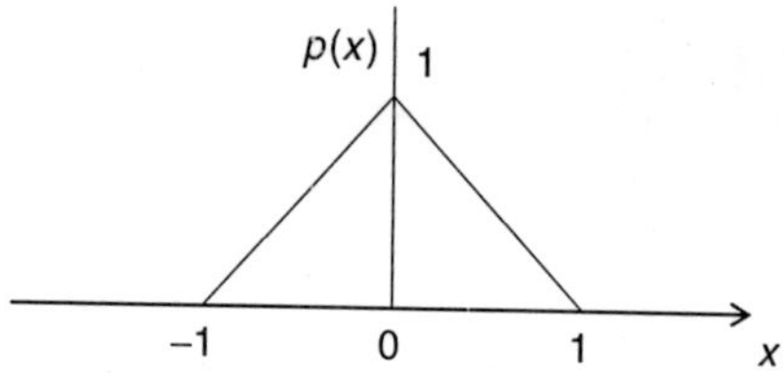

Figure 3.22. An example PDF.

3.4. For the PDF, $p(x)$, given in Figure 3.22, design a four-level PDF-optimized PCM and compare to uniform PCM in terms of SNR.

3.5. Give and justify a formula for the number of bits in simple vector quantization with $N \times 1$ vectors and L template vectors.

3.6. Give in terms of L and N the order of complexity in a VQ codebook search, where L is the number of codebook entries and N is the codebook dimension. Consider the following cases: (i) a simple VQ, (ii) a multi-step VQ, and (iii) a split VQ. For (ii) and (iii), use configurations given in Figure 3.12 and Figure 3.14, respectively.

COMPUTER EXERCISES

3.7. Design a DPCM coder for a stationary random signal with power spectral density $S(e^{j\Omega}) = \dfrac{1}{|1 + 0.8e^{j\Omega}|^2}$. Use a first-order predictor. Give a block diagram and all pertinent equations. Write a program that implements the DPCM coder and evaluate the MSE at the receiver. Compare the SNR (for the same data) for a PCM system operating at the same bit rate.

3.8. In this problem, you will write a computer program to design a vector quantizer and generate a codebook of size $[L \times N]$. Here, L is the number of codebook entries and N is the codebook dimension.

Step 1

Generate a training set, T_{in}, of size $[Ln \times N]$, where n is the number of training vectors per codevector. Assume $L = 16$, $N = 4$, and $n = 10$. Denote the training set elements as $t_{in}(i, j)$, for $i = 0, 1, \ldots, 159$ and $j = 0, 1, 2, 3$. Use Gaussian vectors of zero mean and unit variance for training.

Step 2

Using the LBG algorithm [Lind80], design a vector quantizer and generate a codebook, $\mathbf{C}$, of size $[L \times N]$, i.e., $[16 \times 4]$. In the LBG VQ design, choose the distortion threshold as 0.001. Label the codevectors as $c(i, j)$, for $i = 0, 1, \ldots, 15$ and $j = 0, 1, 2, 3$.

Table 3.6. Segmental and overall SNR values for a test signal within and outside the training sequence for different number of codebook entries, L, different codebook dimensions, N, and $\varepsilon = 0.001$.

Training vectors per entry, n	Codebook dimension, N	SNR for a test signal **within** the training sequence (dB)				SNR for a test signal **outside** the training sequence (dB)			
		Segmental		Overall		Segmental		Overall	
		$L = 16$	$L = 64$	$L = 16$	$L = 64$	$L = 16$	$L = 64$	$L = 16$	$L = 64$
10	2								
	4								
100	2								
	4								

Figure 3.23. Four-bit VQ design specifications for Problem 3.9.

Step 3

Similar to Step 1 generate another training set, T_{out} of size $[160 \times 4]$ that we will use for testing the VQ performance. Label these training set values as $t_{out}(i, j)$, for $i = 0, 1, \ldots, 159$ and $j = 0, 1, 2, 3$.

Step 4

Using the codebook, $\boldsymbol{C}$, designed in Step 2, perform vector quantization of $t_{in}(i, j)$ and $t_{out}(i, j)$. Let us denote the VQ results as $\hat{t}_{in}(i, j)$ and $\hat{t}_{out}(i, j)$, respectively.

a. When the test vectors are *within* the training sequence, compute the over-all SNR and segmental SNR values as follows,

$$SNR_{overall} = \frac{\sum_{i=0}^{Ln-1} \sum_{j=0}^{N-1} t_{in}^2(i, j)}{\sum_{i=0}^{Ln-1} \sum_{j=0}^{N-1} (t_{in}(i, j) - \hat{t}_{in}(i, j))^2} \quad (3.57)$$

$$SNR_{segmental} = \frac{1}{Ln} \sum_{i=0}^{Ln-1} \frac{\sum_{j=0}^{N-1} t_{in}^2(i, j)}{\sum_{j=0}^{N-1} (t_{in}(i, j) - \hat{t}_{in}(i, j))^2} \quad (3.58)$$

b. Compute the over-all SNR and segmental SNR values when the test vectors are different from the training ones, i.e., replace $t_{in}(i, j)$ with $t_{out}(i, j)$ and $\hat{t}_{in}(i, j)$ with $\hat{t}_{out}(i, j)$ in (3.57) and (3.58).

c. List in Table 3.6 the overall and segmental SNR values for different number of codebook entries and different codebook dimensions. Explain the effects of choosing different values of L, n, and N on the SNR values.

d. Compute the MSE, $\varepsilon(t_{in}, \hat{t}_{in}) = \frac{1}{Ln}\frac{1}{N} \sum_{i=0}^{Ln-1} \sum_{j=0}^{N-1} (t_{in}(i, j) - \hat{t}_{in}(i, j))^2$ for different cases, e.g., $L = 16, 64$, $n = 10, 100, 1000$, $N = 2, 8$. Explain how the MSE varies for different values of L, n, and N.

3.9. Write a program to design a 4-bit VQ codebook $\boldsymbol{L}_1$ (i.e., use $L = 16$ codebook entries) with codebook dimension, $N = 4$ (Figure 3.23). Use $n = 100$ training vectors per codebook entry and a distortion threshold, $\varepsilon = 0.00001$. For VQ-training choose zero mean and unit variance Gaussian vectors.

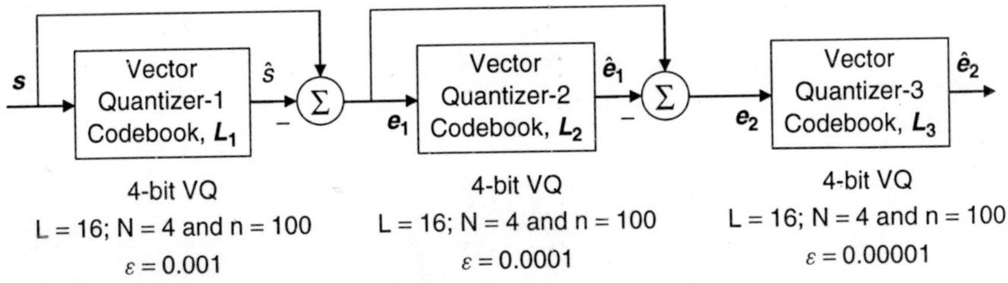

Figure 3.24. A three-stage vector quantizer.

3.10. Extend the VQ design in problem 3.9 to a multi-step VQ (see Figure 3.24 for an example multi-step VQ configuration). Use a total of three stages in your VQ design. Choose the MSE distortion thresholds in each of the stages as $\varepsilon_1 = 0.001$, $\varepsilon_2 = 0.0001$, and $\varepsilon_3 = 0.00001$. Comment on the MSE convergence in each of the stages. How would you compare the multi-step VQ with a simple VQ in terms of the segmental and overall SNR values. (Note: In Figure 3.24, the first VQ codebook ($\boldsymbol{L}_1$) encodes the signal $\boldsymbol{s}$ and the subsequent VQ stages $\boldsymbol{L}_2$ and $\boldsymbol{L}_3$ encode the error from the previous stage. At the decoder, the signal, s', can be reconstructed as, $s' = \hat{s} + \hat{e}_1 + \hat{e}_2$).

3.11. Design a two-stage split-VQ. Choose $L = 16$ and $n = 100$. Implement the first-stage as a 4-dimensional VQ and the second-stage as two 2-dimensional VQs. Select the distortion thresholds as follows: for the first stage, $\varepsilon_1 = 0.001$, and for the second stage, $\varepsilon_2 = 0.00001$. Compare the coding accuracy in terms of a distance measure and the coding gain in terms of the number of bits/sample of the split-VQ with respect to the simple VQ in problem 3.9 and the multi-step VQ in problem 3.10. (See Figure 3.14).

3.12. Given the input data stream $X = [1\ 0\ 2\ 1\ 0\ 1\ 2\ 1\ 0\ 2\ 0\ 1\ 1]$ chosen over a data set $V = [0\ 1\ 2]$:

a. Write a program to compute the entropy, $H_e(X)$.

b. Compute the symbol probabilities $p_i, i \in V$ for the input data stream, X.

c. Write a program to encode X using Huffman codes. Employ an appropriate Huffman bit-mapping. Give the length of the output bitstream. (Hint: See Example 3.4 and Example 3.6).

d. Use arithmetic coding to encode the input data stream, X. Give the final codeword interval range in the binary form. Give also the length of the output bitstream. (See Example 3.9.)

CHAPTER 4

LINEAR PREDICTION IN NARROWBAND AND WIDEBAND CODING

4.1 INTRODUCTION

Linear predictive coders are embedded in several telephony and multimedia standards [G.729] [G.723.1] [IS-893] [ISOI99]. Linear predictive coding (LPC) [Kroo95] is mostly used for source coding of speech signals and the dominant application of LPC is cellular telephony. Recently linear prediction (LP) analysis/synthesis has also been integrated in some of the wideband speech coding standards [G.722] [G.722.2] [Bess02] and in audio modeling [Iwak96] [Mori96] [Harm97a] [Harm97b] [Bola98] [ISOI00].

LP analysis/synthesis exploits the short- and long-term correlation to parameterize the signal in terms of a source-system representation. LP analysis can be open loop or closed loop. In closed-loop analysis, also called analysis-by-synthesis, the LP parameters are estimated by minimizing the "perceptually weighted" difference between the original and reconstructed signal. Speech coding standards use a perceptual weighting filter (PWF) to shape the quantization noise according to the masking properties of the human ear [Schr79] [Kroo95] [Sala98]. Although the PWF has been successful in speech coding, audio coding requires a more sophisticated strategy to exploit perceptual redundancies. To this end, several extensions [Bess02] [G.722.2] to the conventional LPC have been proposed. Hybrid transform/predictive coding techniques have also been employed for high-quality, low-bit-rate coding [Ramp98] [Ramp99] [Rong99] [ISOI99] [ISOI00]. Other LP methods that make use of perceptual constrains and

---

auditory psychophysics include the perceptual LP (PLP) [Herm90], the warped LP (WLP) [Stru80] [Harm96] [Harm01], and the perceptually-motivated all-pole (PMAP) modeling [Atti05]. In the PLP, a perceptually based auditory spectrum is obtained by filtering the signal using a filter bank that mimics the auditory filter bank. An all-pole filter that approximates the auditory spectrum is then computed using the autocorrelation method [Makh75]. On the other hand, in the WLP, the main idea is to warp the frequency axis, according to a Bark scale prior to performing LP analysis. The PMAP modeling employs an auditory excitation pattern matching-method to directly estimate the perceptually-relevant pole locations. The estimated "perceptual poles" are then used to construct an all-pole filter for speech analysis/synthesis.

Whether or not LPC is amenable for audio modeling depends on the signal properties. For example, a code-excited linear predictive (CELP) coder seems to be more adequate than a sinusoidal coder for telephone speech, while the sinusoidal coder seems to be more promising for music.

4.2 LP-BASED SOURCE-SYSTEM MODELING FOR SPEECH

Speech is produced by the interaction of the vocal tract with the vocal chords. The LP analysis/synthesis framework (Figures 4.1 and 4.2) has been successful for speech coding because it fits well the source-system paradigm for speech [Makh75] [Mark76]. In particular, the slowly time-varying spectral characteristics of the upper vocal tract (system) are modeled by an all-pole filter, while the prediction residual captures the voiced, unvoiced, or mixed excitation signal. The LP analysis filter, $A(z)$, in Figure 4.1 is given by

$$A(z) = 1 - \sum_{i=1}^{L} a_i z^{-i}, \tag{4.1}$$

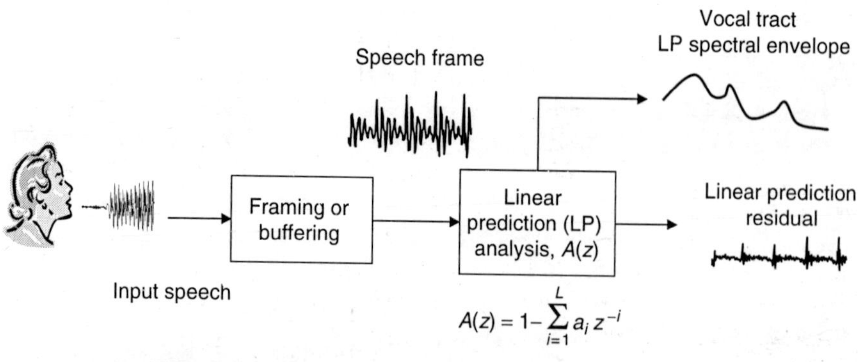

Figure 4.1. Parameter estimation using linear prediction.

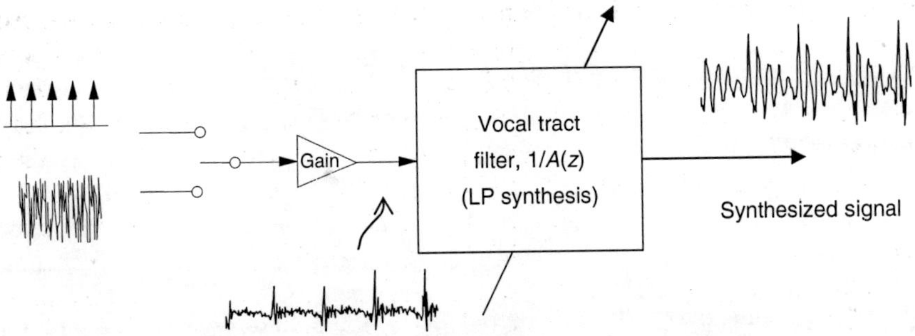

Figure 4.2. Engineering model for speech synthesis.

where L is the order of the linear predictor. Figure 4.2 depicts a simple speech synthesis model where a time-varying digital filter is excited by quasi-periodic waveforms when speech is voiced (e.g., as in steady vowels) and random waveforms for unvoiced speech (e.g., as in consonants). The inverse filter, $1/A(z)$, shown in Figure 4.2, is an all-pole LP synthesis filter

$$H(z) = \frac{G}{A(z)} = \frac{G}{1 - \sum_{i=1}^{L} a_i z^{-i}}, \tag{4.2}$$

where G represents the gain. Note that the term all-pole is used loosely since (4.2) has zeros at $z = 0$. The frequency response associated with the LP synthesis filter, i.e., the LPC spectrum, represents the formant structure of the speech signal (Figure 4.3). In this figure, F_1, F_2, F_3, and F_4 represent the four formants.

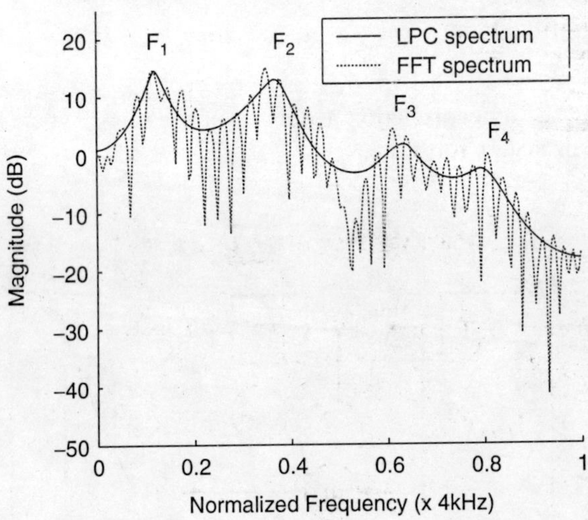

Figure 4.3. The LPC and FFT spectra (dotted line). The formants represent the resonant modes of the vocal tract.

4.3 SHORT-TERM LINEAR PREDICTION

Figure 4.4 presents a typical L-th order FIR linear predictor. During forward linear prediction of, $s(n)$, an estimated value, $\hat{s}(n)$, is computed as a linear combination of the previous samples, i.e.,

$$\hat{s}(n) = \sum_{i=1}^{L} a_i s(n-i), \tag{4.3}$$

where the weights, a_i, are the LP coefficients. The output of the LP analysis filter, $A(z)$, is called the *prediction residual*, $e(n) = s(n) - \hat{s}(n)$. This is given by

$$e(n) = s(n) - \sum_{i=1}^{L} a_i s(n-i). \tag{4.4}$$

Because only short-term delays are considered in (4.4), the linear predictor in Figure 4.4 is also referred to as the *short-term linear predictor*. The linear predictor coefficients, a_i, are estimated using least-square minimization of the prediction error, i.e.,

$$\varepsilon = E[e^2(n)] = E\left[\left(s(n) - \sum_{i=1}^{L} a_i s(n-i)\right)^2\right]. \tag{4.5}$$

The minimization of ε in (4.5) with respect to a_i, i.e., $\partial\varepsilon/\partial a_i = 0$, for $i = 1, 2, \ldots, L$, yields a set of equations involving autocorrelations

$$r_{ss}(m) - \sum_{i=1}^{L} a_i r_{ss}(m-i) = 0, \quad \text{for} \quad m = 1, 2, \ldots, L, \tag{4.6}$$

where $r_{ss}(m)$ is the autocorrelation sequence of the signal $s(n)$. Equation (4.6) can be written in matrix form, i.e.,

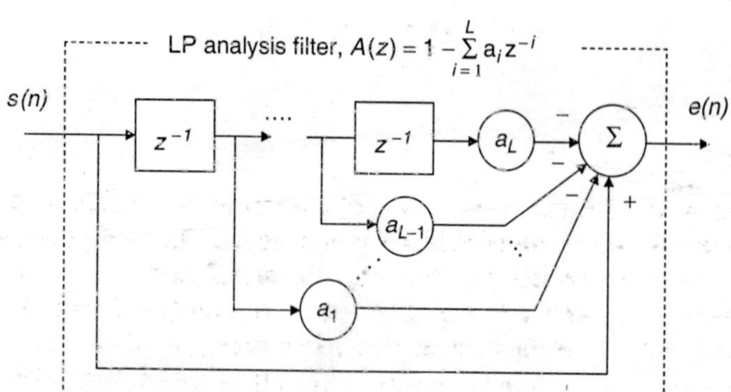

Figure 4.4. Linear prediction (LP) analysis.

$$\begin{bmatrix} a_1 \\ a_2 \\ a_3 \\ \cdot \\ \cdot \\ a_L \end{bmatrix} = \begin{bmatrix} r_{ss}(0) & r_{ss}(-1) & r_{ss}(-2) & \dots & r_{ss}(1-L) \\ r_{ss}(1) & r_{ss}(0) & r_{ss}(-1) & \dots & r_{ss}(2-L) \\ r_{ss}(2) & r_{ss}(1) & r_{ss}(0) & \dots & r_{ss}(3-L) \\ \cdot & \cdot & \cdot & \dots & \cdot \\ \cdot & \cdot & \cdot & \dots & \cdot \\ r_{ss}(L-1) & r_{ss}(L-2) & r_{ss}(L-3) & \dots & r_{ss}(0) \end{bmatrix}^{-1} \begin{bmatrix} r_{ss}(1) \\ r_{ss}(2) \\ r_{ss}(3) \\ \cdot \\ \cdot \\ r_{ss}(L) \end{bmatrix} \tag{4.7}$$

or more compactly,

$$\mathbf{a} = \mathbf{R}_{ss}^{-1}\mathbf{r}_{ss}, \tag{4.8}$$

where $\mathbf{a}$ is the LP coefficient vector, $\mathbf{r}_{ss}$ is the autocorrelation vector, and $\mathbf{R}_{ss}$ is the autocorrelation matrix. Note that $\mathbf{R}_{ss}$ has a Toeplitz and symmetric structure. Efficient algorithms [Makh75] [Mark76] [Marp87] are available for inverting the autocorrelation matrix, $\mathbf{R}_{ss}$, including algorithms tailored to work well with finite precision arithmetic [Gers90]. Typically, the Levinson-Durbin recursive algorithm [Makh75] is used to compute the LP coefficients. Preconditioning of the input sequence, $s(n)$, and autocorrelation data, $r_{ss}(m)$, using tapered windows improves the numerical behavior of these algorithms [Klei95] [Kroo95]. In addition, bandwidth expansion or scaling of the LP coefficients is typical in LPC as it reduces distortion during synthesis.

In low-bit-rate coding, the prediction coefficients and the residual must be efficiently quantized. Because the direct-form LP coefficients, a_i, do not have adequate quantization properties, transformed coefficients are typically quantized. First-generation voice coders (vocoders) such as the LPC10e [FS1015] and the IS-54 VSELP [IS-54] quantize *reflection coefficients* that are a by-product of the Levinson-Durbin recursion. Transformation of the reflection coefficients can lead to a set of parameters that are also less sensitive to quantization. In particular, the *log area ratios* and the *inverse sine* transformation have been used in the early GSM 6.10 algorithm [GSM89] and in the skyphone standard [Boyd88]. Recent LP-based cellular standards quantize *line spectrum pairs* (LSPs). The main advantage of the LSPs is that they relate directly to frequency-domain, and, hence, they can be encoded using perceptual criteria.

4.3.1 Long-Term Prediction

Long-term prediction (LTP), as opposed to short-term prediction, is a process that captures the long-term correlation in the signal. The LTP provides a mechanism for representing the periodicity in the signal and as such it represents the *fine* harmonic structure in the short-term spectrum (see Eq. (4.1)). LTP synthesis, (4.9), requires estimation of two parameters, i.e., the delay, τ, and the gain parameter, a_τ. For strongly voiced segments of speech, the delay is usually an integer that approximates the pitch period. A transfer function of a simple LTP synthesis filter, $H_\tau(z)$, is given in (4.9). More complex LTP filters involve multiple parameters and noninteger delays [Kroo90]:

$$H_\tau(z) = \frac{1}{A_L(z)} = \frac{1}{1 - a_\tau z^{-\tau}}. \tag{4.9}$$

The LTP can be implemented by open loop or closed loop analysis. The open-loop LTP parameters are typically obtained by searching the autocorrelation sequence. The gain is simply obtained by $a_\tau = r_{ss}(\tau)/r_{ss}(0)$. In closed-loop LTP search, the signal is synthesized for a range of candidate LTP lags and the lag that produces the best waveform matching is chosen. Because of the intensive computations in full-search, closed-loop LTP, recent algorithms use open-loop LTP to establish an initial LTP lag that is then refined using closed-loop search around the neighborhood of the initial estimate. In order to further reduce the complexity, LTP searches are often carried in every other subframe.

4.3.2 ADPCM Using Linear Prediction

One of the simplest compression schemes that uses the short-term LP analysis-synthesis is the adaptive differential pulse code modulation (ADPCM) coder [Bene86] [G.726]. ADPCM algorithms encode the difference between the current and the predicted speech samples. The block diagram of the ITU-T G.726 32 kb/s ADPCM encoder [Bene86] is shown in Figure 4.5. The algorithm consists of an adaptive quantizer and an adaptive pole-zero predictor. The prediction parameters are obtained by backward estimation, i.e., from quantized data using a gradient algorithm at the decoder. From Figure 4.5, it can be noted that the decoder is embedded in the encoder. The pole-zero predictor (2 poles and 6 zeros) estimates the input signal and hence it reduces the variance of $e(n)$. The quantizer encodes the error, $e(n)$, into a sequence of 4-bit words. The ITU-T G.726 also accommodates 16, 24, and 40 kb/s with individually optimized quantizers.

4.4 OPEN-LOOP ANALYSIS-SYNTHESIS LINEAR PREDICTION

In almost all LP-based speech codecs, speech is approximated on short analysis intervals, typically in the neighborhood of 20 ms. As shown in Figure 4.6, a set of LP synthesis parameters is estimated on each analysis frame to capture the shape of the vocal tract envelope and to model the excitation.

Some of the typical synthesis parameters encoded and transmitted in the open-loop LP include the prediction coefficients, the pitch information, the frame energy, and the voicing. At the receiver, the transmitted "source" parameters are used to form the excitation. The excitation, $e(n)$, is then used to excite the LP synthesis filter, $1/A(z)$, to reconstruct the speech signal. Some of the standardized open-loop analysis-synthesis LP algorithms include the LPC10e Federal Standard FS-1015 [FS1015] [Trem82] [Camp86] and the Mixed Excitation LP (MELP) [McCr91]. The LPC10e FS-1015 uses a tenth-order predictor to estimate the vocal tract parameters and a two-state voiced or unvoiced excitation model for residual modeling. Mixed excitation schemes in conjunction with LPC were

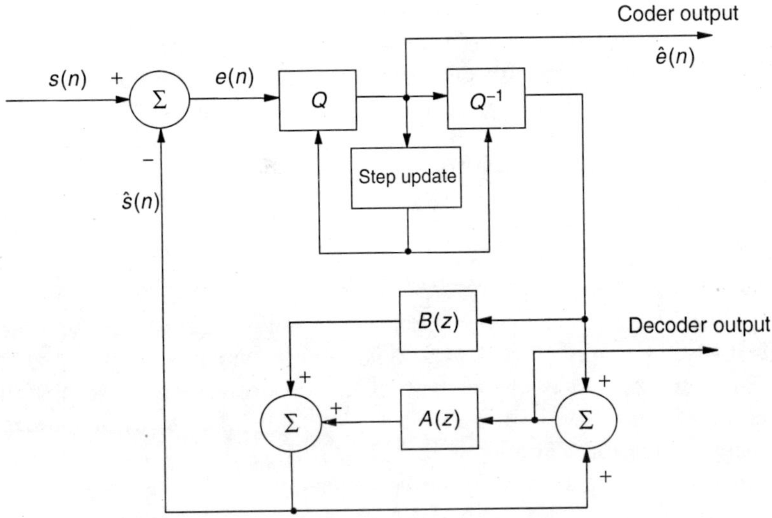

Figure 4.5. The ADPCM ITU-T G.726 encoder.

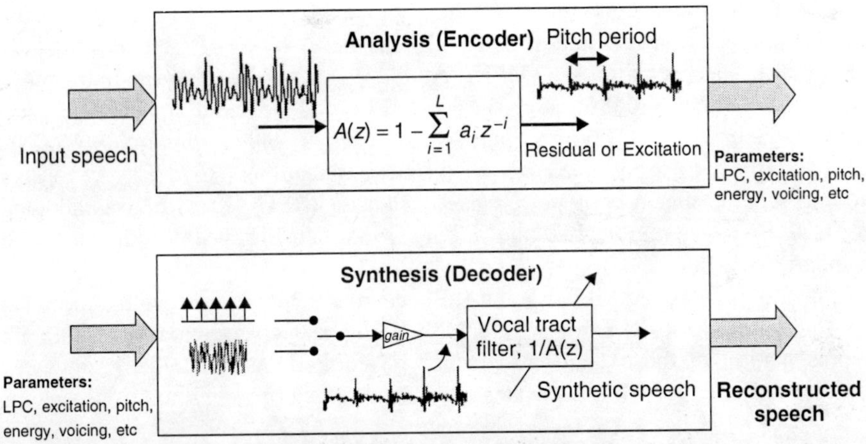

Figure 4.6. Open-loop analysis-synthesis LP.

proposed by Makhoul *et al.* [Makh78] and were later revisited by McCree and Barnwell [McCr91] [McCr93].

4.5 ANALYSIS-BY-SYNTHESIS LINEAR PREDICTION

In closed-loop source-system coders (Figure 4.7), the excitation source is determined by closed-loop or analysis-by-synthesis (A-by-S) optimization. The optimization process determines an excitation sequence that minimizes the perceptually weighted mean-square-error (MSE) between the input speech and reconstructed speech [Atal82b] [Sing84] [Schr85]. The closed-loop LP combines the spectral modeling properties of vocoders with the waveform matching attributes of waveform coders; and, hence, the A-by-S LP coders are also called *hybrid* LP coders. The system consists of a short-term LP synthesis filter, $1/A(z)$, and a LTP synthesis filter, $1/A_L(z)$, shown in Figure 4.7. The perceptual weighting filter (PWF), $W(z)$, shapes the error such that quantization noise is masked by the high-energy formants. The PWF is given by

$$W(z) = \frac{A(z/\gamma_1)}{A(z/\gamma_2)} = \frac{1 - \sum_{i=1}^{L} \gamma_1^i a_i z^{-i}}{1 - \sum_{i=1}^{L} \gamma_2^i a_i z^{-i}}, \quad 0 < \gamma_2 < \gamma_1 < 1, \tag{4.10}$$

where γ_1 and γ_2 are the adaptive weights and L is the order of the linear predictor. Typically, γ_1 ranges from 0.94 to 0.98, and γ_2 varies between 0.4 and 0.7, depending upon the tilt or the flatness characteristics associated with the LPC spectral envelope [Sala98] [Bess02]. The role of $W(z)$ is to de-emphasize the error energy in the formant regions [Schr79]. This de-emphasis strategy is based on the fact that quantization noise in the formant regions is partially masked by speech. From Figure 4.7, note that a gain factor, g, scales the excitation vector, $\mathbf{x}$, and the excitation samples are filtered by the long-term and short-term synthesis filters.

The three most common excitation models typically embedded in the excitation generator module (Figure 4.7) in the A-by-S LP schemes include the multi-pulse excitation (MPE) [Atal82b] [Sing84], the regular pulse excitation (RPE) [Kroo86], and the vector or code excited linear prediction (CELP) [Schr85]. A 9.6 kb/s *multi-pulse excited linear prediction* (MPE-LP) algorithm is used in *Skyphone* airline applications [Boyd88]. A 13 kb/s coding scheme that uses *regular pulse excitation* (RPE) [Kroo86] was adopted for the first generation full-rate ETSI GSM Pan-European digital cellular standard [GSM89].

The aforementioned MPE-LP and RPE schemes achieve high-quality speech at medium rates (13 kb/s). For low-rate, high-quality speech coding, a more efficient representation of the excitation sequence is required. Atal [Atal82a] suggested that high-quality speech at low rates may be produced by using noninstantaneous

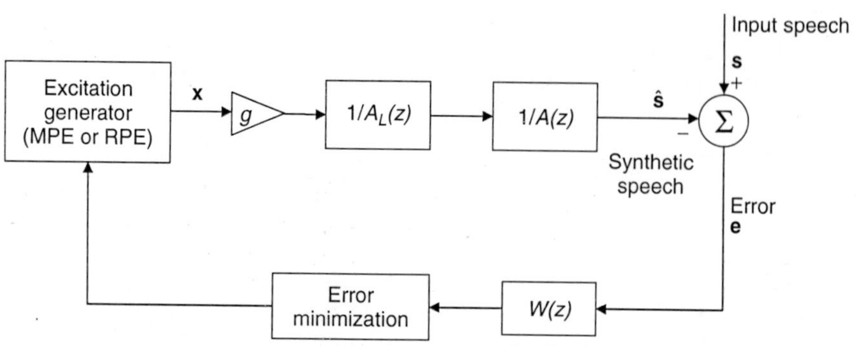

Figure 4.7. A typical source-system model employed in the analysis-by-synthesis LP.

(delayed decision) coding of Gaussian excitation sequences in conjunction with A-by-S linear prediction and perceptual weighting. In the mid-1980s, Atal and Schroeder [Atal84] [Schr85] proposed a vector or *code excited linear prediction* (CELP) algorithm for A-by-S linear predictive coding.

We provide further details on CELP in this section because of its recent use in wideband coding standards. The excitation codebook search process in CELP can be explained by considering the A-by-S scheme shown in Figure 4.8. The $N \times 1$ error vector, **e**, associated with the k-th excitation vector, can be written as

$$\mathbf{e}[k] = \mathbf{s}_w - \hat{\mathbf{s}}_w^0 - g_k \hat{\mathbf{s}}_w[k] \tag{4.11}$$

where $\mathbf{s}_w$ is the $N \times 1$ vector that contains the perceptually-weighted speech samples, $\hat{\mathbf{s}}_w^0$ is the vector that contains the output due to the initial filter state, $\hat{\mathbf{s}}_w[k]$ is the filtered synthetic speech vector associated with the k-th excitation vector, and g_k is the gain factor. Minimizing $\varepsilon_k = \mathbf{e}^T[k]\mathbf{e}[k]$ w.r.t. g_k, we obtain

$$g_k = \frac{\overline{\mathbf{s}}_w^T \hat{\mathbf{s}}_w[k]}{\hat{\mathbf{s}}_w^T[k] \hat{\mathbf{s}}_w[k]}, \tag{4.12}$$

where $\overline{\mathbf{s}}_w = \mathbf{s}_w - \mathbf{s}_w^0$, and T represents the transpose operator. From (4.12), ε_k can be written as

$$\varepsilon_k = \overline{\mathbf{s}}_w^T \overline{\mathbf{s}}_w - \frac{(\overline{\mathbf{s}}_w^T \hat{\mathbf{s}}_w[k])^2}{\hat{\mathbf{s}}_w^T[k] \hat{\mathbf{s}}_w[k]}. \tag{4.13}$$

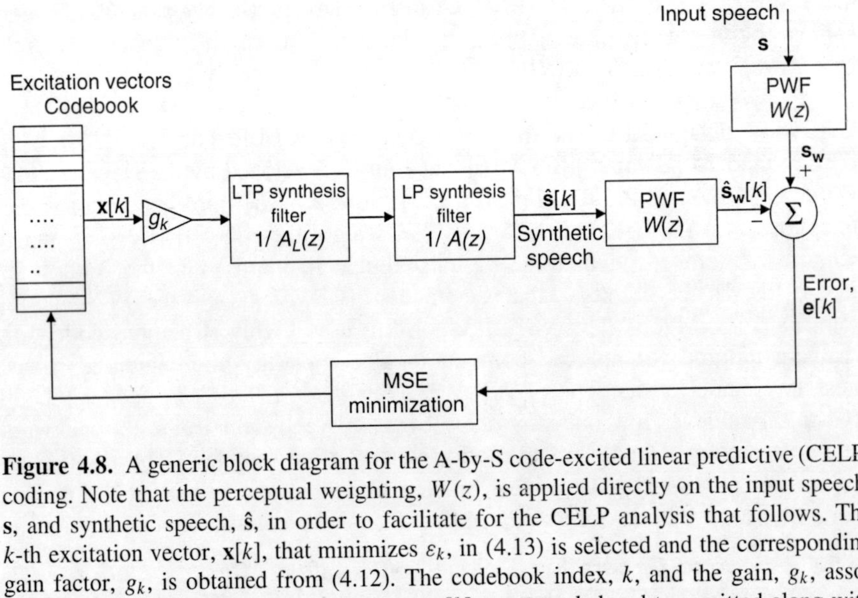

Figure 4.8. A generic block diagram for the A-by-S code-excited linear predictive (CELP) coding. Note that the perceptual weighting, $W(z)$, is applied directly on the input speech, **s**, and synthetic speech, $\hat{\mathbf{s}}$, in order to facilitate for the CELP analysis that follows. The k-th excitation vector, $\mathbf{x}[k]$, that minimizes ε_k, in (4.13) is selected and the corresponding gain factor, g_k, is obtained from (4.12). The codebook index, k, and the gain, g_k, associated with the candidate excitation vector, $\mathbf{x}[k]$, are encoded and transmitted along with the short-term and long-term prediction filter parameters.

The k-th excitation vector, $\mathbf{x}[k]$, that minimizes (4.13) is selected and the corresponding gain factor, g_k, is obtained from (4.12).

One of the disadvantages of the original CELP algorithm is the large computational complexity required for the codebook search [Schr85]. This problem motivated a great deal of work focused upon developing structured codebooks [Davi86] [Klei90a] and fast search procedures [Tran90]. In particular, Davidson and Gersho [Davi86] proposed sparse codebooks and Kleijn *et al.* [Klei90a] proposed a fast algorithm for searching stochastic codebooks with overlapping vectors. In addition, Gerson and Jasiuk [Gers90] [Gers91] proposed a vector sum excited linear predictive (VSELP) coder, which is associated with fast codebook search and robustness to channel errors. Other implementation issues associated with CELP include the quantization of the CELP parameters, the effects of channel errors on CELP coders, and the operation of the algorithm on finite-precision and fixed-point machines. A study on the effects of parameter quantization on the performance of CELP was presented in [Kroo90], and the issues associated with the channel coding of the CELP parameters were discussed by Kleijn [Klei90b]. Some of the problems associated with the fixed-point implementation of CELP algorithms were presented in [Span92].

4.5.1 Code-Excited Linear Prediction Algorithms

In this section, we taxonomize CELP algorithms into three categories that are consistent with the chronology of their development, i.e., first-generation CELP (1986–1992), second-generation CELP (1993–1998), and third-generation CELP (1999–present).

4.5.1.1 First-Generation CELP Coders The first-generation CELP algorithms operate at bit rates between 5.8 kb/s and 16 kb/s. These are generally high complexity and non-toll-quality algorithms. Some of the first-generation CELP algorithms include the FS-1016 CELP, the IS-54 VSELP, the ITU-T G.728 low delay-CELP, and the IS-96 Qualcomm CELP. The *FS-1016 4.8 kb/s CELP* standard [Camp90] [FS1016] was jointly developed by the Department of Defense (DoD) and the Bell Labs for possible use in the third-generation secure telephone unit (STU-III). The *IS-54 VSELP* algorithm [IS-54] [Gers90] and its variants are embedded in three digital cellular standards, i.e., the 8 kb/s TIA IS-54 [IS-54], the 6.3 kb/s Japanese standard [GSM96a], and the 5.6 kb/s half-rate GSM [GSM96b]. The VSELP algorithm uses highly structured codebooks that are tailored for reduced computational complexity and increased robustness to channel errors. The *ITU-T G.728 low-delay (LD) CELP* coder [G.728] [Chen92] achieves low one-way delay by using very short frames, a backward-adaptive predictor, and short excitation vectors (five samples). The *IS-96 Qualcomm CELP* [IS-96] is a variable bit rate algorithm and is part of the original code division multiple access (CDMA) standard for cellular communications.

4.5.1.2 Second-Generation Near-Toll-Quality CELP Coders The second-generation CELP algorithms are targeted for TDMA and CDMA cellphones, Internet audio streaming, voice-over-Internet-protocol (VoIP), and secure communications. Second-generation CELP algorithms include the ITU-T G.723.1 dual-rate speech codec [G.723.1], the GSM enhanced full rate (EFR) [GSM96a] [IS-641], the IS-127 Relaxed CELP (RCELP) [IS-127] [Klei92], and the ITU-T G.729 CS-ACELP [G.729] [Sala98].

The coding gain improvements in second-generation CELP coders can be attributed, partly, to the use of algebraic codebooks in excitation coding [Adou87] [Lee90] [Sala98] [G.729]. The term *algebraic CELP* refers to the structure of the excitation codebooks. Various algebraic codebook structures have been proposed [Adou87] [Lafl90], but the most popular is the *interleaved pulse permutation* code. In this codebook, the code vector consists of a set of interleaved permutation codes containing only few non-zero elements. This is given by

$$p_i = i + jd,\ j = 0, 1, \ldots, 2^M - 1, \tag{4.14}$$

where p_i is the pulse position, i is the pulse number, and d is the interleaving depth. The integer M represents the number of bits describing the pulse positions. Table 4.1 shows an example ACELP codebook structure, where the interleaving depth, $d = 5$, the number of pulses or tracks equal to 5, and the number of bits to represent the pulse positions, $M = 3$. From (4.14), $p_i = i + j5$, where $i = 0, 1, 2, 3, 4$, $j = 0, 1, 2, \ldots, 7$.

For a given value of i, the set defined by (4.14) is known as 'track,' and the value of j defines the pulse position. From the codebook structure shown in Table 4.1, the codevector, $x(n)$, is given by

$$x(n) = \sum_{i=0}^{4} \alpha_i \delta(n - p_i), \quad n = 0, 1, \ldots, 39, \tag{4.15}$$

where $\delta(n)$ is the unit impulse, α_i are the pulse amplitudes (± 1), and p_i are the pulse positions. In particular, the codebook vector, $x(n)$, is computed by placing the 5-unit pulses at the determined locations, p_i, multiplied with their signs (± 1). The pulse position indices and the signs are coded and transmitted. Note that the algebraic codebooks do not require any storage.

4.5.1.3 Third-Generation CELP for 3G Cellular Standards

The third-generation (3G) CELP algorithms are multimodal and accommodate several

Table 4.1. An example algebraic codebook structure: tracks and pulse positions.

Track (i)	Pulse positions (p_i)
0	P_0: 0, 5, 10, 15, 20, 25, 30, 35
1	P_1: 1, 6, 11, 16, 21, 26, 31, 36
2	P_2: 2, 7, 12, 17, 22, 27, 32, 37
3	P_3: 3, 8, 13,18, 23, 28, 33, 38
4	P_4: 4, 9, 14, 19, 24, 29, 34, 39

different bit rates. This is consistent with the vision on wideband wireless standards [Knis98] that operate in different modes including low-mobility, high-mobility, indoor, etc. There are at least two algorithms that have been developed and standardized for these applications. In Europe, GSM standardized the adaptive multi-rate coder [ETSI98] [Ekud99] and, in the United States, the TIA has tested the selectable mode vocoder (SMV) [Gao01a] [Gao01b] [IS-893]. In particular, the adaptive multirate coder [ETSI98] [Ekud99] has been adopted by ETSI for use in the GSM network. This is an algebraic CELP algorithm that operates at multiple rates: 12.2, 10.2, 7.95, 6.7, 5.9, 5.15, and 5.75 kb/s. The bit rate is adjusted according to the traffic conditions. The SMV algorithm (IS-893) was developed to provide higher quality, flexibility, and capacity over the existing IS-96 QCELP and IS-127 enhanced variable rate coding (EVRC) CDMA algorithms. The SMV is based on 4 codecs: full-rate at 8.5 kb/s, half-rate at 4 kb/s, quarter-rate at 2 kb/s, and eighth-rate at 0.8 kb/s. The rate and mode selections in SMV are based on the frame voicing characteristics and the network conditions. Efforts to establish wideband cellular standards continue to drive further the research and development towards algorithms that work at multiple rates and deliver enhanced speech quality.

4.6 LINEAR PREDICTION IN WIDEBAND CODING

Until now, we discussed the use of LP in narrowband coding with signal bandwidth limited to 150–3400 Hz. Signal bandwidth in wideband speech coding spans 50 Hz to 7 kHz; which substantially improves the quality of signal reconstruction, intelligibility, and naturalness. In particular, the introduction of the low-frequency components improves the naturalness, while the higher frequency extension provides more adequate speech intelligibility. In case of high-fidelity audio, it is typical to consider sampling rates of 44.1 kHz and signal bandwidth can range from 20 Hz to 20 kHz. Some of the recent super high-fidelity audio storage formats (Chapter 11) such as the DVD-audio and the super audio CD (SACD) consider signal bandwidths up to 100 kHz.

4.6.1 Wideband Speech Coding

Over the last few years, several wideband speech coding algorithms have been proposed [Orde91] [Jaya92] [Lafl93] [Adou95]. Some of the coding principles associated with these algorithms have been successfully integrated into several speech coding standards, for example, the ITU-T G.722 subband ADPCM standard and the ITU-T G.722.2 AMR-WB codec.

4.6.1.1 The ITU-T G.722 Codec The ITU-T G.722 standard (Figure 4.9) uses a combination of both subband and ADPCM (SB-ADPCM) techniques [G.722] [Merm88] [Span94] [Pain00]. The input signal is sampled at 16 kHz and decomposed into two subbands of equal bandwidth using quadrature mirror filter (QMF) banks. The subband filters $h_{low}(n)$ and $h_{high}(n)$ should satisfy,

$$h_{high}(n) = (-1)^n h_{low}(n) \text{and} |H_{low}(z)|^2 + |H_{high}(z)|^2 = 1. \tag{4.16}$$

The low-frequency subband is typically quantized at 48 kb/s while the high-frequency subband is coded at 16 kb/s. The G.722 coder includes an adaptive bit allocation scheme and an auxiliary data channel. Moreover, provisions for quantizing the low-frequency subband at 40 or at 32 kb/s are available. In particular, the G.722 algorithm is multimodal and can operate in three different modes, i.e., 48, 56, and 64 kb/s by varying the bits used to represent the lower band signal. The MOS at 64 kb/s is greater than four for speech and slightly less than four for music signals [Jaya90], and the analysis-synthesis QMF banks introduce a delay of less than 3 ms. Details on the real-time implementation of this coder are given in [Taka88].

4.6.1.2 The ITU-T G.722.2 AMR-WB Codec The ITU-T G.722.2 [G.772.2] [Bess02] is an adaptive multi-rate wideband (AMR-WB) codec that operates at bit rates ranging from 6.6 to 23.85 kb/s. The G.722 AMR-WB standard is primarily targeted for the voice-over IP (VoIP), 3G wireless communications, ISDN wideband telephony, and audio/video teleconferencing. It is important to note that the AMR-WB codec has also been adopted by the third-generation partnership project (3GPP) for GSM and the 3G WCDMA systems for wideband mobile communications [Bess02]. This, in fact, brought to the fore all the interoperability-related advantages for wideband voice applications across wireline and wireless communications. The ITU-T G.722.2 AMR-WB codec is based on the ACELP coder and operates on audio frames of 20 ms sampled at 16 kHz. The codec supports the following nine bit rates: 23.85, 23.05, 19.85, 18.25, 15.85, 14.25, 12.65, 8.85, and 6.6 kb/s. Excepting the two lowest modes, i.e., the 8.85 kb/s and the

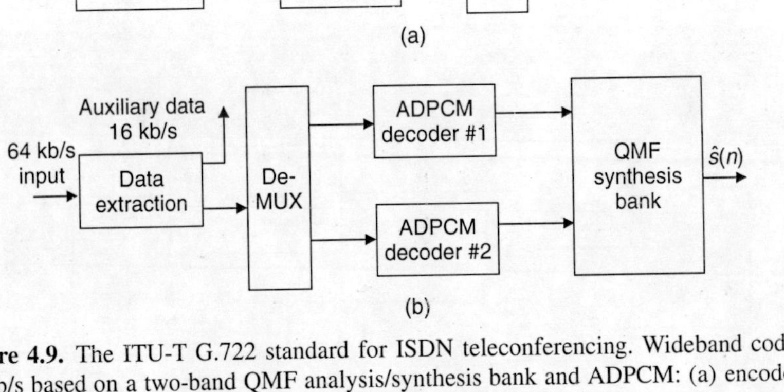

Figure 4.9. The ITU-T G.722 standard for ISDN teleconferencing. Wideband coding at 64 kb/s based on a two-band QMF analysis/synthesis bank and ADPCM: (a) encoder and (b) decoder. Note that the low-frequency band is encoded at 32 kb/s in order to allow for an auxiliary data channel at 16 kb/s.

6.6 kb/s that are intended for transmission over noisy time-varying channels, other encoding modes, i.e., 23.85 through 12.65 kb/s, offer high-quality signal reconstruction. The G.722 AMR-WB embeds several innovative techniques [Bess02] such as *i*) a modified perceptual weighting filter that decouples the formant weighting from the spectrum tilt, *ii*) an enhanced closed-loop pitch search to better accommodate the variations in the voicing level, and *iii*) efficient codebook structures for fast searches. The codec also includes a voice activity detection (VAD) scheme that activates a comfort noise generator module (1–2 kb/s) in case of discontinuous transmission.

4.6.2 Wideband Audio Coding

Motivated by the need to reduce the computational complexity associated with the CELP-based excitation source coding, researchers have proposed several hybrid (LP + subband/transform) coders [Lefe94] [Ramp98] [Rong99]. In this section, we consider LP-based wideband coding methods that encode the prediction residual based upon the transform, or subband, or sinusoidal coding techniques.

4.6.2.1 Multipulse Excitation Model Singhal at Bell Labs [Sing90] reported that analysis-by-synthesis multipulse excitation, with sufficient pulse density, can be applied to correct for LP envelope errors introduced by bandwidth expansion and quantization (Figure 4.10). This algorithm uses a 24th-order LPC synthesis filter, while optimizing pulse positions and amplitudes to minimize perceptually weighted reconstruction errors. Singhal determined that densities of approximately 1 pulse per 4 output samples of each excitation subframe are required to achieve near transparent quality.

Spectral coefficients are transformed to inverse sine reflection coefficients, then differentially encoded and quantized using PDF-optimized Max quantizers. Entropy (Huffman) codes are also used. Pulse locations are differentially encoded relative to the location of the first pulse. Pulse amplitudes are fractionally encoded relative to the largest pulse and then quantized using a Max quantizer. The proposed MPLPC audio coder achieved output SNRs of 35–40 dB at a bit rate of 128 kb/s. Other MPLPC audio coders have also been proposed [Lin91],

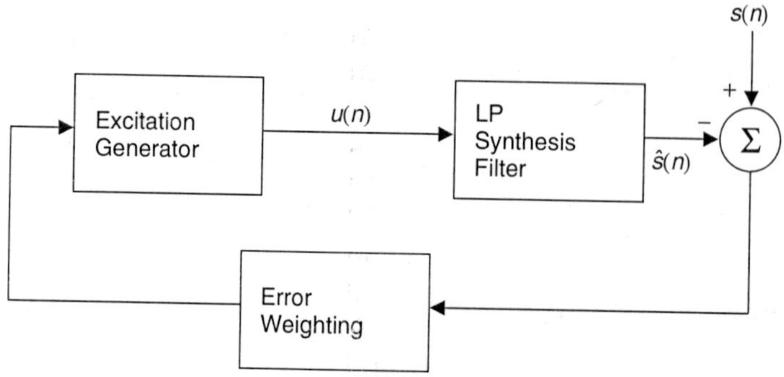

Figure 4.10. Multipulse excitation model used in [Sing90].

including a scheme based on MPLPC in conjunction with the discrete wavelet transform [Bola95].

4.6.2.2 *Discrete Wavelet Excitation Coding* While most of the successful speech codecs nowadays use some form of closed-loop time-domain analysis-by-synthesis such as MPLPC, high-performance LP-based perceptual audio coding has been realized with alternative frequency-domain excitation models. For instance, Boland and Deriche reported output quality comparable to MPEG-1, Layer II at 128 kb/s for an LPC audio coder operating at 96 kb/s [Bola98] in which the prediction residual was transform coded using a three-level discrete-wavelet-transform (DWT) (see also Section 8.2) based on a four-band uniform filter bank. At each level of the DWT, the lowest subband of the previous level was decomposed into four uniform bands. This 10-band nonuniform structure was intended to mimic critical bandwidths to a certain extent. A perceptual bit allocation according to MPEG-1, psychoacoustic model-2 was applied to the transform coefficients.

4.6.2.3 *Sinusoidal Excitation Coding* Excitation sequences modeled as a sum of sinusoids were investigated in [Chan96]. This form of excitation is based on the tendency of the prediction residuals to be spectrally impulsive rather than flat for high-fidelity audio. In coding experiments using 32-kHz-sampled input audio, subjective and objective quality improvements relative to the MPLPC coders were reported for the sinusoidal excitation schemes, with high-quality output audio reported at 72 kb/s. In the experiments reported in [Chan97], a set of ten LP coefficients is estimated on 9.4 ms analysis frames and split-vector quantized using 24 bits. Then, the prediction residual is analyzed and sinusoidal parameters are estimated for the seven best out of a candidate set of thirteen sinusoids for each of six subframes. The masked threshold is estimated and used to form a time-varying bit allocation for the amplitudes, frequencies, and phases on each subframe. Given a frame allocation of 675, a total of 573, 78, and 24 bits, respectively, are allocated to the sinusoidal, bit allocation side information, and LP coefficients. Sinusoidal excitation coding when used in conjunction with a masking-threshold adapted weighting filter, resulted in improved quality relative to MPEG-1 layer I at a bit rate of 96 kb/s [Chan96] for selected test material.

4.6.2.4 *Frequency-Warped LP* Beyond the performance improvements realized through the use of different excitation models, there has been interest in warping the frequency axis before LP analysis to effectively provide better resolution at certain frequencies. In the context of perceptual coding, it is naturally of interest to achieve a Bark-scale warping. Frequency axis warping to achieve nonuniform FFT resolution was first introduced by Oppenheim, Johnson, and Steiglitz [Oppe71] [Oppe72] using a network of cascaded first-order all-pass sections for frequency warping of the signal, followed by a standard FFT. The idea was later extended to warped linear prediction (WLP) by Strube [Stru80], and was ultimately applied to an ADPCM codec [Krug88]. Cascaded first-order all-pass sections were used to warp the signal, and then the LP autocorrelation analysis was performed on the warped autocorrelation sequence. In this scenario, a single-parameter warping of the frequency axis can be introduced into the LP analysis by replacing the delay elements in the FIR analysis filter (Figure 4.4) with all-pass sections. This is done by replacing the complex variable, z^{-1}, of the FIR system function with another filter, $H(z)$, of the form

$$H(z) = \frac{z^{-1} - \lambda}{1 - \lambda z^{-1}}. \tag{4.17}$$

Thus, the predicted sample value is not produced from a combination of past samples as in Eq. (4.4), but rather from the samples of a warped signal. In fact, it has been shown [Smit95] [Smit99] that selecting the value of 0.723 for the parameter λ leads to a frequency warp that approximates well the Bark frequency scale. A WLP-based audio codec [Harm96] was recently proposed. The inherent Bark frequency resolution of the WLP residual yields a perceptually shaped quantization noise without the use of an explicit perceptual model or time-varying bit allocation. In this system, a 40-th order WLP synthesis filter is combined with differential encoding of the prediction residual. A fixed rate of 2 bits per sample (88.2 kb/s) is allocated to the residual sequence, and 5 bits per coefficient are allocated to the prediction coefficients on an analysis frame of 800 samples, or 18 ms. This translates to a bit rate of 99.2 kb/s per channel. In objective terms, an auditory error measure showed considerable improvement for the WLP coding error in comparison to a conventional LP coding error when the same number of bits was allocated to the prediction residuals. Subjectively, the algorithm was reported to achieve transparent quality for some material but it also had difficulty with transients at the frame boundaries.

The algorithm was later extended to handle stereophonic signals [Harm97a] by forming a complex-valued representation of the two channels and then using a version of WLP modified for complex signals (CWLP). It was suggested that significant quality improvement could be realized for the WLPC audio coder by using a closed-loop analysis-by-synthesis procedure [Harm97b]. One of the shortcomings of the original WLP coder was inadequate attention to temporal effects. As a result, further experiments were reported [Harm98] in which WLP was combined with temporal noise shaping (TNS) to realize additional quality improvement.

4.7 SUMMARY

In this Chapter, we presented the LP-based source-system model and described its applications in narrowband and wideband coding. Some of the topics presented in this chapter include:

- Short-term linear prediction
- Conventional LP analysis-synthesis

- Closed-loop analysis-by-synthesis hybrid coders
- Code-excited linear prediction (CELP) speech standards
- Linear prediction in wideband coding
- Frequency-warped LP.

PROBLEMS

4.1. The following autocorrelation sequence is given, $r_{ss}(m) = \dfrac{0.8^{|m|}}{0.36}$. Describe a source-system mechanism that will generate a signal with this autocorrelation.

4.2. Sketch the magnitude frequency response of the following filter function

$$H(z) = \frac{1}{1 - 0.9z^{-10}}$$

4.3. The autocorrelation sequence in Figure 4.11 corresponds to a strongly voiced speech. Show with an arrow which autocorrelation sample relates to the pitch period of the voiced signal. Estimate the pitch period from the graph.

4.4. Consider Figure 4.12 with a white Gaussian input signal.

a. Determine analytically the LP coefficients for a first-order predictor and for $H(z) = 1/(1 - 0.8z^{-1})$.

b. Determine analytically the LP coefficients for a second-order predictor and for $H(z) = 1 + z^{-1} + z^{-2}$.

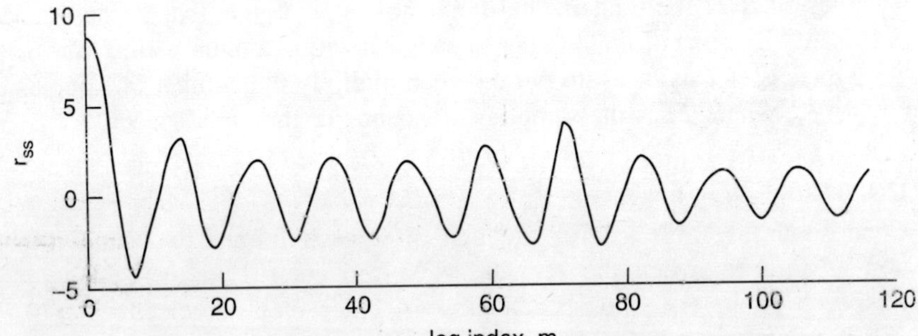

Figure 4.11. Autocorrelation of a voiced speech segment.

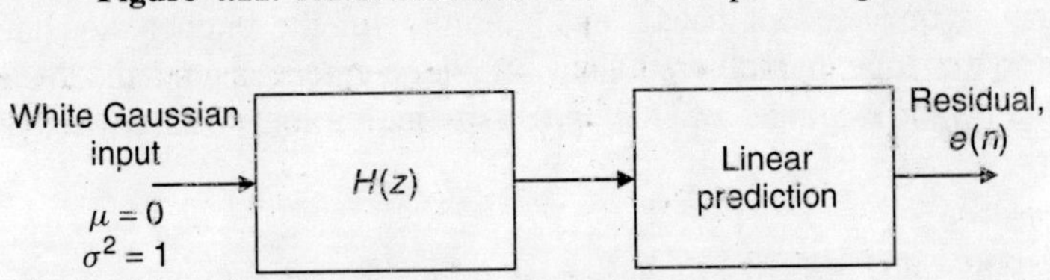

Figure 4.12. LP coefficients estimation.

COMPUTER EXERCISES

The necessary MATLAB software for computer simulations and the speech/audio files (*Ch4Sp8.wav, Ch4Au8.wav*, and *Ch4Au16.wav*) can be obtained from the Book website.

4.5. Linear predictive coding (LPC)

a. Write a MATLAB program to load, display, and play back speech files. Use *Ch4Sp8.wav* for this computer exercise.

b. Include a framing module in your program and set the frame size to 256 samples. Every frame should be read in a 256×1 real vector called *Stime*. Compute the fast Fourier transform (FFT) of this vector, i.e., $Sfreq = fft(Stime)$. Next, compute the magnitude of the complex vector *Sfreq* and plot its magnitude in dB up to the fold-over frequency. This computation should be part of your frame-by-frame speech processing program.

Deliverable 1:

Present at least one plot of time and one corresponding plot of frequency-domain data for a voiced, unvoiced, and a mixed speech segment. (A total of six plots – use the subplot command.)

c. *Pitch period and voicing estimation*: The period of a strongly voiced speech signal is associated in a reciprocal manner to the fundamental frequency of the corresponding harmonic spectrum. That is, if the pitch period is T, the fundamental frequency is $1/T$. Note that T can be measured in terms of the number of samples within a pitch period for voiced speech. If T is measured in ms, then multiply the number of samples by $1/F_s$, where F_s is the sampling frequency of the input speech.

Deliverable 2:

Create and fill Table 4.2 for the first 30 speech frames by visual inspection as follows: when the segment is voiced enter 1 in the 2nd column. If speech pause (i.e., no speech present) enter 0, if unvoiced enter 0.25, and if mixed enter 0.5. Measure the pitch period visually from the time-domain plot in terms of the number of samples in a pitch period. If the segment is unvoiced or pause, enter infinity for the pitch period and hence zero for the fundamental frequency. If the segment is mixed, do your best to obtain an estimate of the pitch if it is not possible set pitch to infinity.

Deliverable 3:

From Table 4.2, plot the fundamental frequency as a function of the frame number for all thirty frames. This is called the pitch frequency contour.

Table 4.2. Pitch period, voicing, and frame energy measurements.

Speech frame number	Voiced/unvoiced/ mixed/pause	Pitch (number of samples)	Frame energy	Fundamental frequency (Hz)
1				
2				
:				
30				

Deliverable 4:

From Table 4.2, plot also *i*) the voicing, and *ii*) frame energy (in dB) as a function of the frame number for all thirty frames.

d. *The FFT and LP spectra*: Write a MATLAB program to implement the Levinson-Durbin recursion. Assume a tenth-order LP analysis and estimate the LP coefficients (*lp_coeff*) for each speech frame.

Deliverable 5:

Compute the LP spectra as follows: *H_allpole* = *freqz(1, lp_coeff)*. Superimpose the LP spectra (*H_allpole*) with the FFT speech spectra (*Sfreq*) for a voiced segment and an unvoiced segment. Plot the spectral magnitudes in dB up to the foldover frequency. Note that the LPC spectra look like a smoothed version of the FFT spectra. Divide *Sfreq* by *H_allpole*. Plot the magnitude of the result in dB up to the fold-over frequency. What does the resulting spectrum represent?

Deliverable 6:

From the LP spectra, measure (visually) the frequencies of the first three formants, F_1, F_2, and F_3. Give these frequencies in Hz (Table 4.3). Plot the three formants across the frame number. These will be the formant contours. Use different line types or colors to discriminate the three contours.

e. *LP analysis-synthesis*: Using the prediction coefficients (*lp_coeff*) from part **(d)**, perform LP analysis. Use the mathematical formulation given in Sections 4.2 and 4.3. Quantize both the LP coefficients and the prediction

Table 4.3. Formants F_1, F_2, and F_3.

Speech frame number	F_1 (Hz)	F_2 (Hz)	F_3 (Hz)
1			
2			
:			
30			

residual using a 3-bit (i.e., 8 levels) scalar quantizer. Next, perform LP synthesis and reconstruct the speech signal.

Deliverable 7:

Plot the quantized residual and its corresponding dB spectrum for a voiced and an unvoiced frame. Provide plots of the original and reconstructed speech. Compute the SNR in dB for the entire reconstructed signal relative to the original record. Listen to the reconstructed signal and provide a subjective score on a scale of 1 to 5. Repeat this step when a 8-bit scalar quantizer is employed. In your simulation, when the LP coefficients were quantized using a 3-bit scalar quantizer, the LP synthesis filter will become unstable for certain frames. What are the consequences of this?

4.6. The FS-1016 CELP standard.

a. Obtain the MATLAB software for the FS-1016 CELP from the Book website. Use the following wave files: *Ch4Sp8.wav* and *Ch4Au8.wav*.

Deliverable 1:

Give the plots of the entire input and the FS-1016 synthesized output for the two wave files. Comment on the quality of the two synthesized wave files. In particular, give more emphasis on the *Ch4Au8.wav* and give specific reasons why the FS-1016 does not synthesize *Ch4Au8.wav* with high quality. Listen to the output files and provide a subjective evaluation. The CELP FS1016 coder scored 3.2 out of 5 in government tests on a MOS scale. How would you rate its performance in terms of MOS for *Ch4Sp8.wav* and *Ch4Au8.wav*? Also give segmental SNR values for a voiced/unvoiced/mixed frame and overall SNR for the entire record. Present your results as Table 4.4 with appropriate caption.

b. Spectrum analysis: File *CELPANAL.M*, from lines 134 to 159. Valuable comments describing the variable names, globals, and inputs/outputs are provided at the beginning of the MATLAB file CELPANAL.M to further assist you with understanding the MATLAB

Table 4.4. FS-1016 CELP subjective and objective evaluation.

	Speech frame number	Segmental SNR for the chosen frame (dB)
	Voiced #_____	
Ch4Sp8.wav	Unvoiced #_____	
	Mixed #_____	
	Over-all SNR for the entire speech record =_____ (dB)	
	MOS in a scale of 1–5 =_____	
Ch4Au8.wav	Over-all SNR for the entire music record =_____ (dB)	
	MOS in a scale of 1–5 =_____	

program. Specifically, some of the useful variables in the MATLAB code include snew-input speech buffer, fcn-LP filter coefficients of $1/A(z)$, rcn-reflection coefficients, newfreq-LSP frequencies, unqfreq-unquantized LSPs, newfreq-quantized LSPs, and lsp-interpolated LSPs for each subframe.

Deliverable 2:

Choose *Ch4Sp8.wav* and use the voiced/unvoiced/mixed frames selected in the previous step. Indicate the frame numbers. FS-1016 employs 30-ms speech frames, so the frame size is fixed = 240 samples. Give time-domain (variable in the code '*snew*') and frequency-domain plots (use FFT size 512; include commands as necessary in the program to obtain the FFT) of the selected voiced/unvoiced/mixed frame; Also plot the LPC spectrum using

figure, freqz(1, fcn)

Study the interlacing property of the LSPs on the unit circle. Note that in the FS-1016 standard, the LSPs are encoded using scalar quantization. Plot the LPC spectra obtained from the unquantized LSPs and quantized LSPs. You have to convert LSPs to LPCs in both the cases (unquantized and quantized) and use the freqz command to plot the LPC spectra. Give a $z =$ domain plot (of a voiced and an unvoiced frame) containing the pole locations (show as crosses 'x') of the LPC spectra, and the roots of the symmetric (show as black circles 'o') and asymmetric (show as red circles 'o') LSP polynomials. Note the interlacing nature of black and red circles, they always lie on the unit circle. Also note that if a pole is close to the unit circle, the corresponding LSPs will be close to each other. In the file *CELPANAL.M; line 124,* high-pass filtering is performed to eliminate the undesired low frequencies. Experiment with and without a high-pass filter to note the presence of humming and low-frequency noise in the synthesized speech.

c. Pitch analysis: File: *CELPANAL.M;* from lines *162 to 191.*

Deliverable 3:

What are the key advantages of employing subframes in speech coding (e.g., interpolation, pitch prediction?). Explain, in general, the differences between long-term prediction and short-term prediction. Give the necessary transfer functions. In particular, describe what aspects of speech each of the two predictors captures.

Deliverable 4:

Insert in file *CELPANAL.M*; after line 182

tauptr = 75;

Perform an evaluation of the perceptual quality of synthesis speech and give your remarks. How does the speech quality change by forcing a pitch to a predetermined value? (Choose different tauptr values, 40, 75, and 110.)

4.7. The warped LP for audio analysis-synthesis.

a. Write a MATLAB program to perform analysis-synthesis using *i*) the conventional LP, and *ii*) the warped LP. Use *Ch4Au8.wav* as the input wave file.

b. Perform a tenth-order LP analysis, and use a 5-bit scalar quantizer to quantize the LP and the WLP coefficients and a 3-bit scalar quantizer for the excitation vector. Perform audio synthesis using the quantized LP and WLP analysis parameters. Compute the warping coefficient [Smit99] [Harm01] using

$$\lambda = 1.0674\left(\frac{2}{\pi}\arctan\left(\frac{0.06583F_s}{1000}\right)\right)^{1/2} - 0.1916,$$

where F_s is the sampling frequency of the input audio. Comment on the quality of the synthesized audio from the LP and WLP analysis-synthesis. Repeat this step for *Ch4Au16.wav*. Refer to [Harm00] for implementation of WLP synthesis filters.

CHAPTER 5

PSYCHOACOUSTIC PRINCIPLES

5.1 INTRODUCTION

The field of psychoacoustics [Flet40] [Gree61] [Zwis65] [Scha70] [Hell72] [Zwic90] [Zwic91] has made significant progress toward characterizing human auditory perception and particularly the time-frequency analysis capabilities of the inner ear. Although applying perceptual rules to signal coding is not a new idea [Schr79], most current audio coders achieve compression by exploiting the fact that "irrelevant" signal information is not detectable by even a well-trained or sensitive listener. Irrelevant information is identified during signal analysis by incorporating into the coder several psychoacoustic principles, including absolute hearing thresholds, critical band frequency analysis, simultaneous masking, the spread of masking along the basilar membrane, and temporal masking. Combining these psychoacoustic notions with basic properties of signal quantization has also led to the theory of *perceptual entropy* [John88b], a quantitative estimate of the fundamental limit of transparent audio signal compression.

This chapter reviews psychoacoustic fundamentals and perceptual entropy and then gives as an application example some details of the ISO/MPEG psychoacoustic model 1. Before proceeding, however, it is necessary to define the *sound pressure level* (SPL), a standard metric that quantifies the intensity of an acoustical stimulus [Zwic90]. Nearly all of the auditory psychophysical phenomena addressed in this book are treated in terms of SPL. The SPL gives the level (intensity) of sound pressure in decibels (dB) relative to an internationally defined reference level, i.e., $L_{SPL} = 20\log_{10}(p/p_0)$ dB, where L_{SPL} is the SPL of a stimulus, p is the sound pressure of the stimulus in Pascals (Pa, equivalent to Newtons

Audio Signal Processing and Coding, by Andreas Spanias, Ted Painter, and Venkatraman Atti

per square meter (N/m^2)), and p_0 is the standard reference level of 20 μPa, or 2×10^{-5} N/m^2 [Moor77]. About 150 dB SPL spans the dynamic range of intensity for the human auditory system, from the limits of detection for low-intensity (quiet) stimuli up to the threshold of pain for high-intensity (loud) stimuli. The SPL reference level is calibrated such that the frequency-dependent absolute threshold of hearing in quiet (Section 5.2) tends to measure in the vicinity of 0 dB SPL. On the other hand, a stimulus level of 140 dB SPL is typically at or above the threshold of pain.

5.2 ABSOLUTE THRESHOLD OF HEARING

The absolute threshold of hearing characterizes the amount of energy needed in a pure tone such that it can be detected by a listener in a noiseless environment. The absolute threshold is typically expressed in terms of dB SPL. The frequency dependence of this threshold was quantified as early as 1940, when Fletcher [Flet40] reported test results for a range of listeners that were generated in a National Institutes of Health (NIH) study of typical American hearing acuity. The quiet threshold is well approximated [Terh79] by the non linear function

$$T_q(f) = 3.64(f/1000)^{-0.8} - 6.5e^{-0.6(f/1000-3.3)^2} + 10^{-3}(f/1000)^4 (\text{dB SPL}), \quad (5.1)$$

which is representative of a young listener with acute hearing. When applied to signal compression, $T_q(f)$ could be interpreted naively as a maximum allowable energy level for coding distortions introduced in the frequency domain (Figure 5.1).

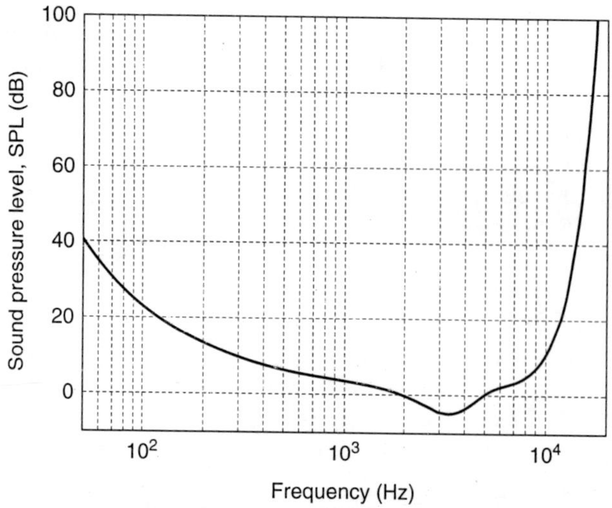

Figure 5.1. The absolute threshold of hearing in quiet.

At least two caveats must govern this practice, however. First, whereas the thresholds captured in Figure 5.1 are associated with pure tone stimuli, the quantization noise in perceptual coders tends to be spectrally complex rather than tonal. Secondly, it is important to realize that algorithm designers have no *a priori* knowledge regarding actual playback levels (SPL), and therefore the curve is often referenced to the coding system by equating the lowest point (i.e., near 4 kHz) to the energy in +/− 1 bit of signal amplitude. In other words, it is assumed that the playback level (volume control) on a typical decoder will be set such that the smallest possible output signal will be presented close to 0 dB SPL. This assumption is conservative for quiet to moderate listening levels in uncontrolled open-air listening environments, and therefore this referencing practice is commonly found in algorithms that utilize the absolute threshold of hearing. We note that the absolute hearing threshold is related to a commonly encountered acoustical metric other than SPL, namely, dB sensation level (dB SL). *Sensation level (SL)* denotes the intensity level difference for a stimulus relative to a listener's individual unmasked detection threshold for the stimulus [Moor77]. Hence, "equal SL" signal components may have markedly different absolute SPLs, but all equal SL components will have equal supra-threshold margins. The motivation for the use of SL measurements is that SL quantifies listener-specific audibility rather than an absolute level. Whether the target metric is SPL or SL, perceptual coders must eventually reference the internal PCM data to a physical scale. A detailed example of this referencing for SPL is given in Section 5.7 of this chapter.

5.3 CRITICAL BANDS

Using the absolute threshold of hearing to shape the coding distortion spectrum represents the first step towards perceptual coding. Considered on its own, however, the absolute threshold is of limited value in coding. The detection threshold for spectrally complex quantization noise is a modified version of the absolute threshold, with its shape determined by the stimuli present at any given time. Since stimuli are in general time-varying, the detection threshold is also a time-varying function of the input signal. In order to estimate this threshold, we must first understand how the ear performs spectral analysis. A frequency-to-place transformation takes place in the cochlea (inner ear), along the basilar membrane [Zwic90].

The transformation works as follows. A sound wave generated by an acoustic stimulus moves the eardrum and the attached ossicular bones, which in turn transfer the mechanical vibrations to the cochlea, a spiral-shaped, fluid-filled structure that contains the coiled basilar membrane. Once excited by mechanical vibrations at its oval window (the input), the cochlear structure induces traveling waves along the length of the basilar membrane. Neural receptors are connected along the length of the basilar membrane. The traveling waves generate peak responses at frequency-specific membrane positions, and therefore different neural receptors are effectively "tuned" to different frequency bands according to their locations. For sinusoidal stimuli, the traveling wave on the basilar membrane propagates from the oval window until it nears the region with a resonant frequency near that of the stimulus frequency. The wave then slows and the magnitude increases to a peak. The wave decays rapidly beyond the peak. The location of the peak is referred to as the "best place" or "characteristic place" for the stimulus frequency, and the frequency that best excites a particular place [Beke60] [Gree90] is called the "best frequency" or "characteristic frequency." Thus, a frequency-to-place transformation occurs. An example is given in Figure 5.2 for a three-tone stimulus. The interested reader can also find online a number of high-quality animations demonstrating this aspect of cochlear mechanics [Twve99].

As a result of the frequency-to-place transformation, the cochlea can be viewed from a signal processing perspective as a bank of highly overlapping bandpass filters. The magnitude responses are asymmetric and nonlinear (level-dependent). Moreover, the cochlear filter passbands are of nonuniform bandwidth, and the bandwidths increase with increasing frequency. The "critical bandwidth" is a function of frequency that quantifies the cochlear filter passbands. Empirical work by several observers led to the modern notion of critical bands [Flet40] [Gree61] [Zwis65] [Scha70]. We will consider two typical examples.

In one scenario, the loudness (perceived intensity) remains constant for a narrowband noise source presented at a constant SPL even as the noise bandwidth is increased up to the critical bandwidth. For any increase beyond the critical bandwidth, the loudness then begins to increase. In this case, one can imagine that loudness remains constant as long as the noise energy is present within only one cochlear "channel" (critical bandwidth), and then that the loudness increases as soon as the noise energy is forced into adjacent cochlear "channels." Critical bandwidth can also be viewed as the result of auditory detection efficacy in terms of a signal-to-noise ratio (SNR) criterion. The power spectrum model

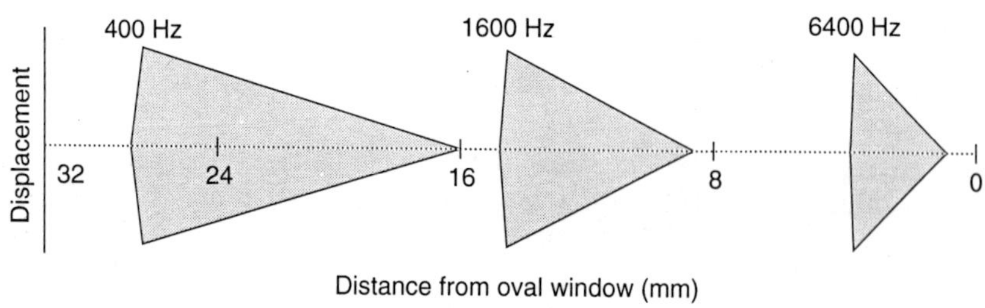

Figure 5.2. The frequency-to-place transformation along the basilar membrane. The picture gives a schematic representation of the traveling wave envelopes (measured in terms of vertical membrane displacement) that occur in response to an acoustic tone complex containing sinusoids of 400, 1600, and 6400 Hz. Peak responses for each sinusoid are localized along the membrane surface, with each peak occurring at a particular distance from the oval window (cochlear "input"). Thus, each component of the complex stimulus evokes strong responses only from the neural receptors associated with frequency-specific loci (after [Zwic90]).

of hearing assumes that masked threshold for a given listener will occur at a constant, listener-specific SNR [Moor96]. In the critical bandwidth measurement experiments, the detection threshold for a narrowband noise source presented between two masking tones remains constant as long as the frequency separation between the tones remains within a critical bandwidth (Figure 5.3a). Beyond this bandwidth, the threshold rapidly decreases (Figure 5.3c). From the SNR viewpoint, one can imagine that as long as the masking tones are presented within the passband of the auditory filter (critical bandwidth) that is tuned to the probe noise, the SNR presented to the auditory system remains constant, and hence the detection threshold does not change. As the tones spread further apart and transition into the filter stopband, however, the SNR presented to the auditory system improves, and hence the detection task becomes easier. In order to maintain a constant SNR at threshold for a particular listener, the power spectrum model calls for a reduction in the probe noise commensurate with the reduction in the energy of the masking tones as they transition out of the auditory filter passband. Thus, beyond critical bandwidth, the detection threshold for the probe tones decreases, and the threshold SNR remains constant. A notched-noise experiment with a similar interpretation can be constructed with masker and maskee roles reversed (Figure 5.3, b and d). Critical bandwidth tends to remain constant (about 100 Hz) up to 500 Hz, and increases to approximately 20% of the center frequency above 500 Hz. For an average listener, critical bandwidth (Figure 5.3b) is conveniently approximated [Zwic90] by

$$BW_c(f) = 25 + 75[1 + 1.4(f/1000)^2]^{0.69}(\text{Hz}). \tag{5.2}$$

Although the function BW_c is continuous, it is useful when building practical systems to treat the ear as a discrete set of bandpass filters that conforms to (5.2). The function [Zwic90]

$$Z_b(f) = 13 \arctan(0.00076 f) + 3.5 \arctan\left[\left(\frac{f}{7500}\right)^2\right] (\text{Bark}) \tag{5.3}$$

is often used to convert from frequency in Hertz to the Bark scale, Figure 5.4 (a). Corresponding to the center frequencies of the Table 5.1 filter bank, the numbered points in Figure 5.4 (a) illustrate that the nonuniform Hertz spacing of the filter bank (Figure 5.5) is actually uniform on a Bark scale. Thus, one critical bandwidth (CB) comprises one Bark. Table 5.1 gives an idealized filter bank that corresponds to the discrete points labeled on the curves in Figure 5.4(a, b). A distance of 1 critical band is commonly referred to as *one Bark* in the literature.

Although the critical bandwidth captured in Eq. (5.2) is widely used in perceptual models for audio coding, we note that there are alternative expressions. In particular, the equivalent rectangular bandwidth (ERB) scale emerged from research directed towards measurement of auditory filter shapes. Experimental data is obtained typically from notched noise masking procedures. Then, the masking data is fitted with parametric weighting functions that represent the spectral shaping properties of the

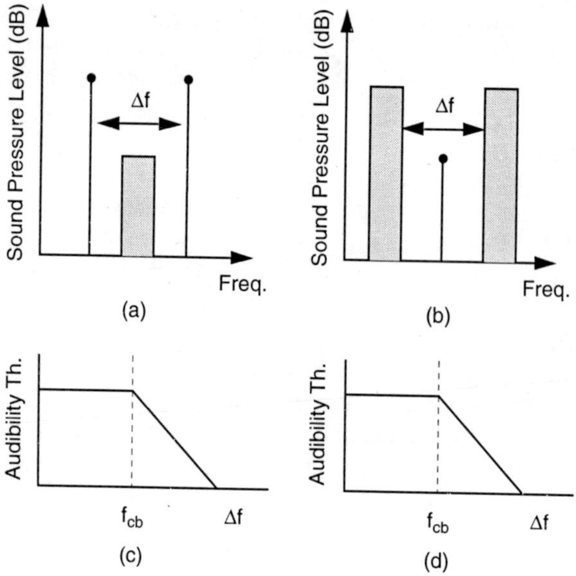

Figure 5.3. Critical band measurement methods. (a,c) Detection threshold decreases as masking tones transition from auditory filter passband into stopband, thus improving detection SNR. (b,d) Same interpretation with roles reversed (after [Zwic90]).

auditory filters [Moor96]. Rounded exponential models with one or two free parameters are popular. For example, the single-parameter *roex(p)* model is given by

$$W(g) = (1 + pg)e^{-pg}, \tag{5.4}$$

where $g = |f - f_0|/f_0$ is the normalized frequency, f_0 is the center frequency of the filter, and f represents frequency, in Hz. Although the *roex(p)* model does not capture filter asymmetry, asymmetric filter shapes are possible if two *roex(p)* models are used independently for the high- and low-frequency filter skirts. Two parameter models such as the *roex(p, r)* are also used to gain additional degrees of freedom [Moor96] in order to improve the accuracy of the filter shape estimates. After curve-fitting, an ERB estimate is obtained directly from the parametric filter shape. For the *roex(p)* model, it can be shown easily that the equivalent rectangular bandwidth is given by

$$ERB_{roex(p)} = \frac{4f_0}{p} \tag{5.5}$$

We note that some texts denote ERB by *equivalent noise bandwidth*. An example is given in Figure 5.6. The solid line in Figure 5.6 (a) shows an example *roex(p)* filter estimated for a center frequency of 1 kHz, while the dashed line shows the ERB associated with the given *roex(p)* filter shape.

In [Moor83] and [Glas90], Moore and Glasberg summarized experimental ERB measurements for *roex(p,r)* models obtained over a period of several years

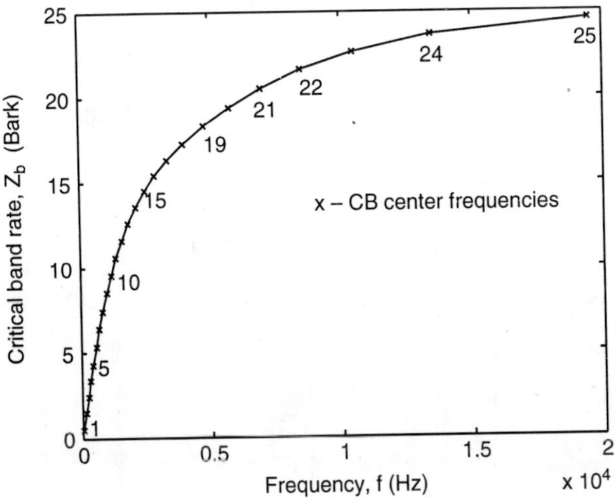

(a)

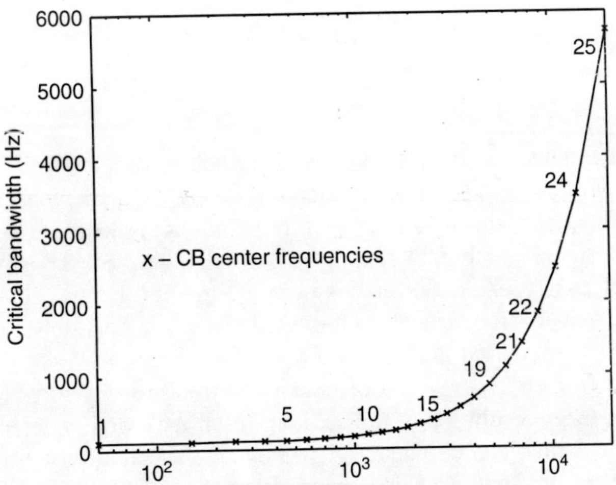

(b)

Figure 5.4. Two views of critical bandwidth. (a) Critical band rate, $Z_b(f)$, maps from Hertz to Barks, and (b) critical bandwidth, $BW_c(f)$ expresses critical bandwidth as a function of center frequency, in Hertz. The "Xs" denote the center frequencies of the idealized critical band filter bank given in Table 5.1.

by a number of different investigators. Given a collection of ERB measurements on center frequencies across the audio spectrum, a curve fitting on the data set yielded the following expression for ERB as a function of center frequency

$$ERB(f) = 24.7(4.37(f/1000) + 1). \tag{5.6}$$

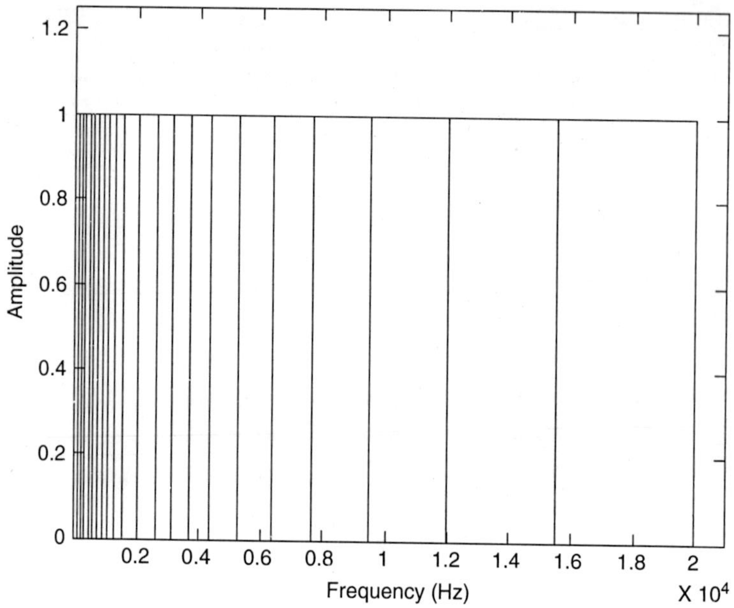

Figure 5.5. Idealized critical band filter bank.

As shown in Figure 5.6 (b), the function specified by Eq. (5.6) differs from the critical bandwidth of Eq. (5.2). Of particular interest for perceptual codec designers, the ERB scale implies that auditory filter bandwidths decrease below 500 Hz, whereas the critical bandwidth remains essentially flat. The apparent increased frequency selectivity of the auditory system below 500 Hz has implications for optimal filter-bank design, as well as for perceptual bit allocation strategies. These implications are addressed later in the book.

Regardless of whether it is best characterized in terms of critical bandwidth or ERB, the frequency resolution of the auditory filter bank largely determines which portions of a signal are perceptually irrelevant. The auditory time-frequency analysis that occurs in the critical band filter bank induces simultaneous and nonsimultaneous masking phenomena that are routinely used by modern audio coders to shape the coding distortion spectrum. In particular, the perceptual models allocate bits for signal components such that the quantization noise is shaped to exploit the detection thresholds for a complex sound (e.g., quantization noise). These thresholds are determined by the energy within a critical band [Gäss54]. Masking properties and masking thresholds are described next.

5.4 SIMULTANEOUS MASKING, MASKING ASYMMETRY, AND THE SPREAD OF MASKING

Masking refers to a process where one sound is rendered inaudible because of the presence of another sound. Simultaneous masking may occur whenever two

Table 5.1. Idealized critical band filter bank (after [Scha70]). Band edges and center frequencies for a collection of 25 critical bandwidth auditory filters that span the audio spectrum..

Band number	Center frequency (Hz)	Bandwidth (Hz)
1	50	–100
2	150	100–200
3	250	200–300
4	350	300–400
5	450	400–510
6	570	510–630
7	700	630–770
8	840	770–920
9	1000	920–1080
10	1175	1080–1270
11	1370	1270–1480
12	1600	1480–1720
13	1850	1720–2000
14	2150	2000–2320
15	2500	2320–2700
16	2900	2700–3150
17	3400	3150–3700
18	4000	3700–4400
19	4800	4400–5300
20	5800	5300–6400
21	7000	6400–7700
22	8500	7700–9500
23	10,500	9500–12000
24	13,500	12000–15500
25	19,500	15500–

or more stimuli are simultaneously presented to the auditory system. From a frequency-domain point of view, the relative shapes of the masker and maskee magnitude spectra determine to what extent the presence of certain spectral energy will mask the presence of other spectral energy. From a time-domain perspective, phase relationships between stimuli can also affect masking outcomes. A simplified explanation of the mechanism underlying simultaneous masking phenomena is that the presence of a strong noise or tone masker creates an excitation of sufficient strength on the basilar membrane at the critical band location to block effectively detection of a weaker signal. Although arbitrary audio spectra may contain complex simultaneous masking scenarios, for the purposes of shaping coding distortions it is convenient to distinguish between three types of simultaneous masking, namely *noise-masking-tone* (NMT) [Scha70], *tone-masking-noise* (TMN) [Hell72], and *noise-masking-noise* (NMN) [Hall98].

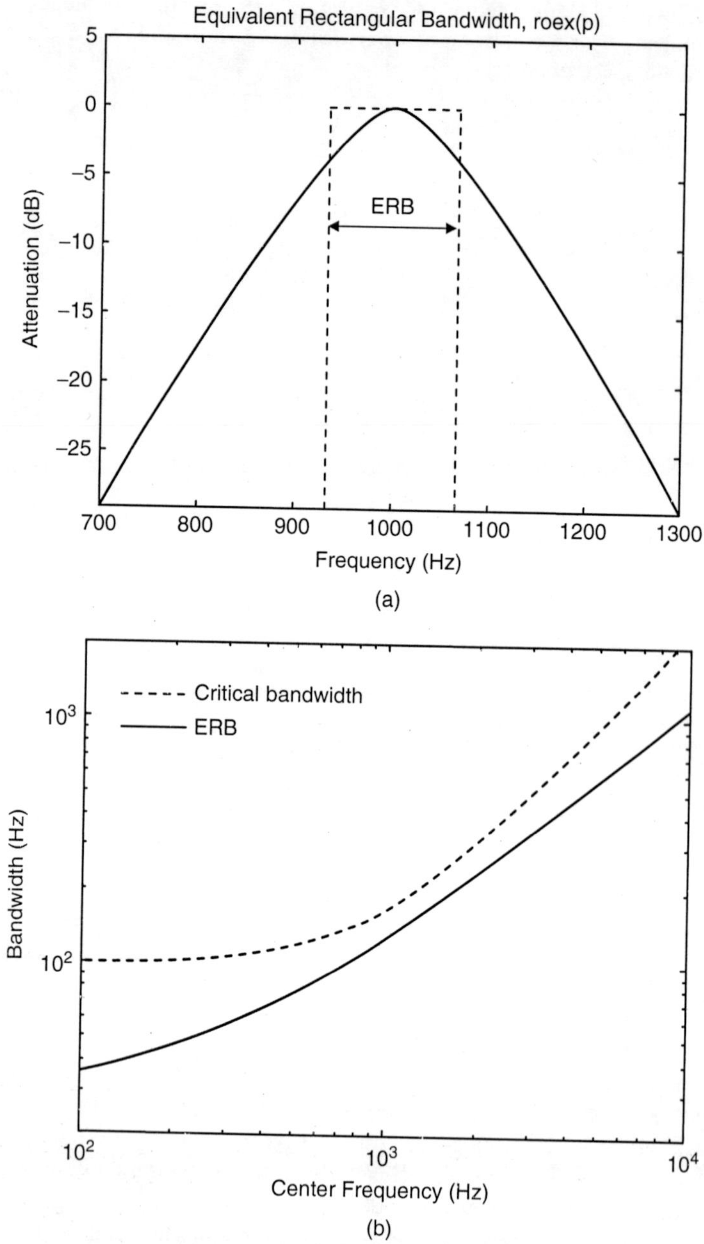

Figure 5.6. Equivalent rectangular bandwidth (ERB). (a) Example ERB for a *roex(p)* single-parameter estimate of the shape of the auditory filter centered at 1 kHz. The solid line represents an estimated spectral weighting function for a single-parameter fit to data from a notched noise masking experiment; the dashed line represents the equivalent rectangular bandwidth. (b) ERB *vs* critical bandwidth – the ERB scale of Eq. (5.6) (solid) *vs* critical bandwidth of Eq. (5.2) (dashed) as a function of center frequency.

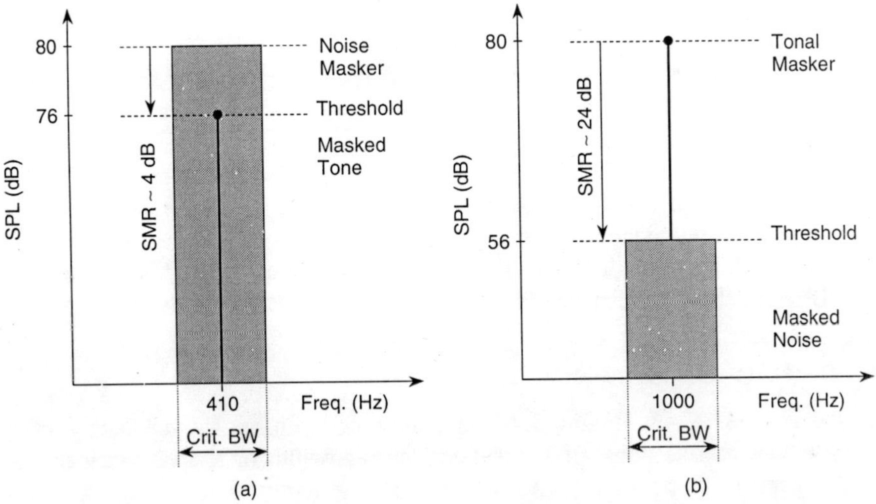

Figure 5.7. Example to illustrate the asymmetry of simultaneous masking: (a) Noise-masking-tone – At the threshold of detection, a 410-Hz pure tone presented at 76-dB SPL is just masked by a critical bandwidth narrowband noise centered at 410 Hz (90 Hz BW) of overall intensity 80 dB SPL. This corresponds to a threshold minimum signal-to-mask (SMR) ratio of 4 dB. The threshold SMR increases as the probe tone is shifted either above or below 410 Hz. (b) Tone-masking-noise – At the threshold of detection, a 1-kHz pure tone presented at 80-dB SPL just masks a critical-band narrowband noise centered at 1 kHz of overall intensity 56-dB SPL. This corresponds to a threshold minimum SMR of 24 dB. As for the NMT experiment, threshold SMR for the TMN increases as the masking tone is shifted either above or below the noise center frequency, 1 kHz. When comparing (a) to (b), it is important to notice the apparent "masking asymmetry," namely that NMT produces a significantly smaller threshold minimum SMR (4 dB) than does TMN (24 dB).

5.4.1 Noise-Masking-Tone

In the NMT scenario (Figure 5.7a), a narrowband noise (e.g., having 1 Bark bandwidth) masks a tone within the same critical band, provided that the intensity of the masked tone is below a predictable threshold directly related to the intensity and, to a lesser extent, center frequency of the masking noise. Numerous studies characterizing NMT for random noise and pure-tone stimuli have appeared since the 1930s (e.g., [Flet37] [Egan50]). At the threshold of detection for the masked tone, the minimum signal-to-mask ratio (SMR), i.e., the smallest difference between the intensity (SPL) of the masking noise ("signal") and the intensity of the masked tone ("mask") occurs when the frequency of the masked tone is close to the masker's center frequency. In most studies, the minimum SMR tends to lie between −5 and +5 dB. For example, a sample threshold SMR result from the NMT investigation [Egan50] is schematically represented in Figure 5.7a. In the figure, a critical band noise masker centered at 410 Hz with an intensity of 80 db SPL masks a 410 Hz tone, and the resulting SMR at the threshold of detection is 4 dB.

Masking power decreases (i.e., SMR increases) for probe tones above and below the frequency of the minimum SMR tone, in accordance with a level- and frequency-dependent spreading function that is described later. We note that temporal factors also affect simultaneous masking. For example, in the NMT scenario, an overshoot effect is possible when the probe tone onset occurs within a short interval immediately following masker onset. Overshoot can boost simultaneous masking (i.e., decrease the threshold minimum SMR) by as much as 10 dB over a brief time span [Zwic90].

5.4.2 Tone-Masking-Noise

In the case of TMN (Figure 5.7b), a pure tone occurring at the center of a critical band masks noise of any subcritical bandwidth or shape, provided the noise spectrum is below a predictable threshold directly related to the strength and, to a lesser extent, the center frequency of the masking tone. In contrast to NMT, relatively few studies have attempted to characterize TMN. At the threshold of detection for a noise band masked by a pure tone, however, it was found in both [Hell72] and [Schr79] that the minimum SMR, i.e., the smallest difference between the intensity of the masking tone ("signal") and the intensity of the masked noise ("mask") occurs when the masker frequency is close to the center frequency of the probe noise, and that the minimum SMR for TMN tends to lie between 21 and 28 dB. A sample result from the TMN study [Schr79] is given in Figure 5.7b. In the figure, a narrowband noise of one Bark bandwidth centered at 1 kHz is masked by a 1 kHz tone of intensity 80 dB SPL. The resulting SMR at the threshold of detection is 24 dB. As with NMT, the TMN masking power decreases for critical bandwidth probe noises centered above and below the minimum SMR probe noise.

5.4.3 Noise-Masking-Noise

The NMN scenario, in which a narrowband noise masks another narrowband noise, is more difficult to characterize than either NMT or TMN because of the confounding influence of phase relationships between the masker and maskee [Hall98]. Essentially, different relative phases between the components of each can lead to different threshold SMRs. The results from one study of intensity difference detection thresholds for wideband noise [Mill47] produced threshold SMRs of nearly 26 dB for NMN [Hall98].

5.4.4 Asymmetry of Masking

The NMT and TMN examples in Figure 5.7 clearly show an asymmetry in masking power between the noise masker and the tone masker. In spite of the fact that both maskers are presented at a level of 80 dB SPL, the associated threshold SMRs differ by 20 dB. This asymmetry motivates our interest in both the

TMN and NMT masking paradigms, as well as NMN. In fact, knowledge of all three is critical to success in the task of shaping coding distortion such that it is undetectable by the human auditory system. For each temporal analysis interval, a codec's perceptual model should identify across the frequency spectrum noise-like and tone-like components within both the audio signal and the coding distortion. Next, the model should apply the appropriate masking relationships in a frequency-specific manner. In conjunction with the spread of masking (below), NMT, NMN, and TMN properties can then be used to construct a global masking threshold. Although several methods for masking threshold estimation have proven effective, we note that a deeper understanding of masking asymmetry may provide opportunities for improved perceptual models. In particular, Hall [Hall97] has shown that masking asymmetry can be explained in terms of relative masker/maskee bandwidths, and not necessarily exclusively in terms of absolute masker properties. Ultimately, this implies that the *de facto* standard energy-based schemes for masking power estimation among perceptual codecs may be valid only so long as the masker bandwidth equals or exceeds maskee (probe) bandwidth. In cases where the probe bandwidth exceeds the masker bandwidth, an envelope-based measure be embedded in the masking calculation [Hall97] [Hall98].

5.4.5 The Spread of Masking

The simultaneous masking effects characterized before by the paradigms NMT, TMN, and NMN are not bandlimited to within the boundaries of a single critical band. Interband masking also occurs, i.e., a masker centered within one critical band has some predictable effect on detection thresholds in other critical bands. This effect, also known as the spread of masking, is often modeled in coding applications by an approximately triangular spreading function that has slopes of +25 and −10 dB per Bark. A convenient analytical expression [Schr79] is given by

$$SF_{dB}(x) = 15.81 + 7.5(x + 0.474) - 17.5\sqrt{1 + (x + 0.474)^2}\text{dB}, \tag{5.7}$$

where x has units of Barks and $SF_{db}(x)$ is expressed in dB. After critical band analysis is done and the spread of masking has been accounted for, masking thresholds in perceptual coders are often established by the [Jaya93] decibel (dB) relations:

$$TH_N = E_T - 14.5 - B \tag{5.8}$$

$$TH_T = E_N - K, \tag{5.9}$$

where TH_N and TH_T, respectively, are the noise-and tone-masking thresholds due to tone-masking-noise and noise-masking-tone; E_N and E_T are the critical band noise and tone masker energy levels; and B is the critical band number. Depending upon the algorithm, the parameter K is typically set between 3 and 5 dB. Of course, the thresholds of Eqs. (5.8) and (5.9) capture only the contributions of individual tone-like or noise-like maskers. In the actual coding scenario, each frame typically contains a collection of both masker types. One can see easily that Eqs. (5.8) and (5.9) capture the masking asymmetry described previously. After they have been identified, these individual masking thresholds are combined to form a global masking threshold.

The global masking threshold comprises an estimate of the level at which quantization noise becomes just noticeable. Consequently, the global masking threshold is sometimes referred to as the level of "just-noticeable distortion," or JND. The standard practice in perceptual coding involves first classifying masking signals as either noise or tone, next computing appropriate thresholds, then using this information to shape the noise spectrum beneath the JND level. Two illustrated examples are given later in Sections 5.6 and 5.7, which address the perceptual entropy and the ISO/IEC MPEG Model 1, respectively. Note that the absolute threshold (T_q) of hearing is also considered when shaping the noise spectra, and that MAX (JND, T_q) is most often used as the permissible distortion threshold. Notions of critical bandwidth and simultaneous masking in the audio coding context give rise to some convenient terminology illustrated in Figure 5.8, where we consider the case of a single masking tone occurring at the center of a critical band. All levels in the figure are given in terms of dB SPL.

A hypothetical masking tone occurs at some masking level. This generates an excitation along the basilar membrane that is modeled by a spreading function and a corresponding *masking threshold*. For the band under consideration, the *minimum masking threshold* denotes the spreading function in-band minimum. Assuming the masker is quantized using an m-bit uniform scalar quantizer, noise might be introduced at the level m. *Signal-to-mask ratio* (SMR) and *noise-to-mask ratio* (NMR) denote the log distances from the minimum masking threshold to the masker and noise levels, respectively.

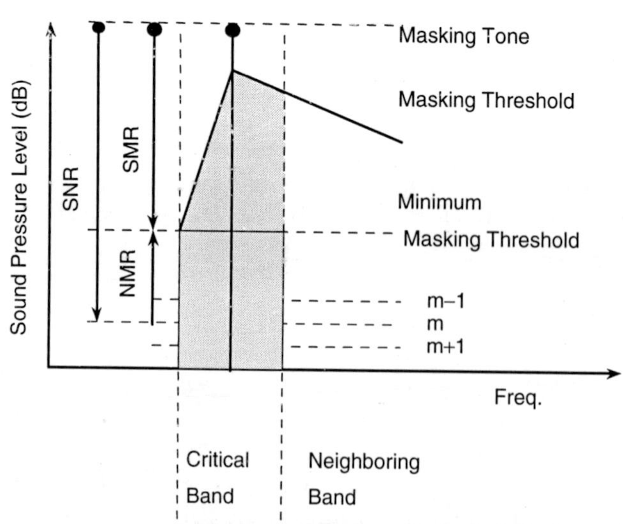

Figure 5.8. Schematic representation of simultaneous masking (after [Noll93]).

5.5 NONSIMULTANEOUS MASKING

As shown in Figure 5.9, masking phenomena extend in time beyond the window of simultaneous stimulus presentation. In other words, for a masker of finite duration, non simultaneous (also sometimes denoted "temporal") masking occurs both prior to masker onset as well as after masker removal. The skirts on both regions are schematically represented in Figure 5.9. Essentially, absolute audibility thresholds for masked sounds are artificially increased prior to, during, and following the occurrence of a masking signal. Whereas significant premasking tends to last only about 1–2 ms, postmasking will extend anywhere from 50 to 300 ms, depending upon the strength and duration of the masker [Zwic90]. Below, we consider key nonsimultaneous masking properties that should be embedded in audio codec perceptual models.

Of the two nonsimultaneous masking modes, forward masking is better understood. For masker and probe of the same frequency, experimental studies have shown that the amount of forward (post-) masking depends in a predictable way on stimulus frequency [Jest82], masker intensity [Jest82], probe delay after masker cessation [Jest82], and masker duration [Moor96]. Forward masking also exhibits frequency-dependent behavior similar to simultaneous masking that can be observed when the masker and probe frequency relationship is varied [Moor78]. Although backward (pre) masking has also been the subject of many studies, it is not well understood [Moor96]. As shown in Figure 5.9, backward masking decays much more rapidly than forward masking. For example, one study at Thomson Consumer Electronics showed that only 2 ms prior to masker onset, the masked threshold was already 25 dB below the threshold of simultaneous masking [Bran98].

We note, however, that the literature lacks consensus over the maximum time persistence of significant backward masking. Despite the inconsistent results across studies, it is generally accepted that the amount of measured backward

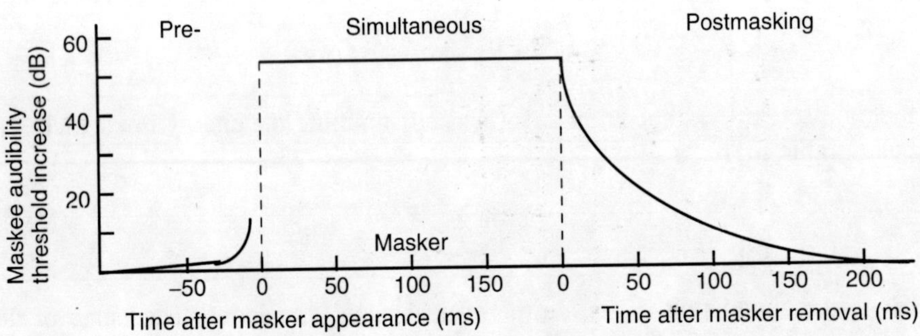

Figure 5.9. Nonsimultaneous masking properties of the human ear. Backward (pre-) masking occurs prior to masker onset and lasts only a few milliseconds; Forward (post-) masking may persist for more than 100 ms after masker removal (after [Zwic90]).

masking depends significantly on the training of the experimental subjects. For the purposes of perceptual coding, abrupt audio signal transients (e.g., the onset of a percussive musical instrument) create pre- and postmasking regions during which a listener will not perceive signals beneath the elevated audibility thresholds produced by a masker. In fact, temporal masking has been used in several audio coding algorithms (e.g., [Bran94a] [Papa95] [ISOI96a] [Fiel96] [Sinh98a]). Premasking in particular has been exploited in conjunction with adaptive block size transform coding to compensate for pre-echo distortions (Chapter 6, Sections 6.9 and 6.10).

5.6 PERCEPTUAL ENTROPY

Johnston [John88a] combined notions of psychoacoustic masking with signal quantization principles to define perceptual entropy (PE), a measure of perceptually relevant information contained in any audio record. Expressed in bits per sample, PE represents a theoretical limit on the compressibility of a particular signal. PE measurements reported in [John88a] and [John88b] suggest that a wide variety of CD-quality audio source material can be transparently compressed at approximately 2.1 bits per sample. The PE estimation process is accomplished as follows. The signal is first windowed and transformed to the frequency domain. A masking threshold is then obtained using perceptual rules. Finally, a determination is made of the number of bits required to quantize the spectrum without injecting perceptible noise. The PE measurement is obtained by constructing a PE histogram over many frames and then choosing a worst-case value as the actual measurement.

The frequency-domain transformation is done with a Hann window followed by a 2048-point fast Fourier transform (FFT). Masking thresholds are obtained by performing critical band analysis (with spreading), making a determination of the noise-like or tone-like nature of the signal, applying thresholding rules for the signal quality, then accounting for the absolute hearing threshold. First, real and imaginary transform components are converted to power spectral components

$$P(\omega) = \mathrm{Re}^2(\omega) + \mathrm{Im}^2(\omega), \tag{5.10}$$

then a discrete Bark spectrum is formed by summing the energy in each critical band (Table 5.1)

$$B_i = \sum_{\omega=bl_i}^{bh_i} P(\omega), \tag{5.11}$$

where the summation limits are the critical band boundaries. The range of the index, i, is sample rate dependent and, in particular, $i \in \{1, 25\}$ for CD-quality signals. A spreading function, Eq. (5.7) is then convolved with the discrete Bark spectrum

$$C_i = B_i * SF_i \tag{5.12}$$

to account the spread of masking. An estimation of the tone-like or noise-like quality for C_i is then obtained using the spectral flatness measure (SFM)

$$SFM = \frac{\mu_g}{\mu_a}, \tag{5.13}$$

where μ_g and μ_a, respectively, correspond to the geometric and arithmetic means of the PSD components for each band. The SFM has the property that it is bounded by 0 and 1. Values close to 1 will occur if the spectrum is flat in a particular band, indicating a decorrelated (noisy) band. Values close to zero will occur if the spectrum in a particular band is narrowband. A coefficient of tonality, α, is next derived from the SFM on a dB scale

$$\alpha = \min\left(\frac{SFM_{db}}{-60}, 1\right) \tag{5.14}$$

and this is used to weight the thresholding rules given by Eqs. (5.8) and (5.9) (with K = 5.5) as follows for each band to form an offset

$$O_i = \alpha(14.5 + i) + (1 - \alpha)5.5(\text{ in dB}). \tag{5.15}$$

A set of JND estimates in the frequency power domain are then formed by subtracting the offsets from the Bark spectral components

$$T_i = 10^{\log_{10}(C_i) - \frac{O_i}{10}}. \tag{5.16}$$

These estimates are scaled by a correction factor to simulate deconvolution of the spreading function, and then each T_i is checked against the absolute threshold of hearing and replaced by $\max(T_i, T_q(i))$. In a manner essentially identical to the SPL calibration procedure that was described in Section 5.2, the PE estimation is calibrated by equating the minimum absolute threshold to the energy in a 4 kHz signal of +/− 1 bit amplitude. In other words, the system assumes that the playback level (volume control) is configured such that the smallest possible signal amplitude will be associated with an SPL equal to the minimum absolute threshold. By applying uniform quantization principles to the signal and associated set of JND estimates, it is possible to estimate a lower bound on the number of bits required to achieve transparent coding. In fact, it can be shown that the perceptual entropy in bits per sample is given by

$$\begin{aligned} PE = \sum_{i=1}^{25} \sum_{\omega=bl_i}^{bh_i} & \log_2\left(2\left|nint\left(\frac{\text{Re}(\omega)}{\sqrt{6T_i/k_i}}\right)\right| + 1\right) \\ & + \log_2\left(2\left|nint\left(\frac{\text{Im}(\omega)}{\sqrt{6T_i/k_i}}\right)\right| + 1\right) \text{(bits/sample)}, \end{aligned} \tag{5.17}$$

where i is the index of critical band, bl_i and bh_i are the lower and upper bounds of band i, k_i is the number of transform components in band i, T_i is the masking threshold in band i, (Eq. (5.16)), and *nint* denotes rounding to the nearest integer. Note that if 0 occurs in the log we assign 0 for the result. The masking thresholds used in the above PE computation also form the basis for a transform coding algorithm described in Chapter 7. In addition, the ISO/IEC MPEG-1 psychoacoustic model 2, which is often used in .MP3 encoders, is closely related to the PE procedure.

We note, however, that there have been evolutionary improvements since the PE estimation scheme first appeared in 1988. For example, the PE calculation in many systems (e.g., [ISOI94]) relies on improved tonality estimates relative to the SFM-based measure of Eq. (5.14). The SFM-based measure is both time- and frequency-constrained. Only one spectral estimate (analysis frame) is examined in time, and in frequency, the measure by definition lumps together multiple spectral lines. In contrast, other tonality estimation schemes, e.g., the "chaos measure" [ISOI94] [Bran98], consider the predictability of individual frequency components across time, in terms of magnitude and phase-tracking properties. A predicted value for each component is compared against its actual value, and the Euclidean distance is mapped to a measure of predictability. Highly predictable spectral components are considered to be tonal, while unpredictable components are treated as noise-like. A tonality coefficient that allows weighting towards one extreme or the other is computed from the chaos measure, just as in Eq. (5.14). Improved performance has been demonstrated in several instances (e.g., [Bran90] [ISOI94] [Bran98]). Nevertheless, the PE measurement as proposed in its original form conveys valuable insight on the application of simultaneous masking asymmetry to a perceptual model in a practical system.

5.7 EXAMPLE CODEC PERCEPTUAL MODEL: ISO/IEC 11172-3 (MPEG - 1) PSYCHOACOUSTIC MODEL 1

It is useful to consider an example of how the psychoacoustic principles described thus far are applied in actual coding algorithms. The ISO/IEC 11172-3 (MPEG-1, layer 1) psychoacoustic model 1 [ISOI92] determines the maximum allowable quantization noise energy in each critical band such that quantization noise remains inaudible. In one of its modes, the model uses a 512-point FFT for high-resolution spectral analysis (86.13 Hz), then estimates for each input frame individual simultaneous masking thresholds due to the presence of tone-like and noise-like maskers in the signal spectrum. A global masking threshold is then estimated for a subset of the original 256 frequency bins by (power) additive combination of the tonal and nontonal individual masking thresholds. The remainder of this section describes the step-by-step model operations. Sample results are given for one frame of CD-quality pop music sampled at 44.1 kHz/16-bits per sample. We note that although this model is suitable for any of the MPEG-1 coding layers I–III, the standard [ISOI92] recommends that model 1 be used with layers I and II, while model 2 is recommended for layer III (MP3). The five steps leading to computation of global masking thresholds are described in the following Sections.

5.7.1 Step 1: Spectral Analysis and SPL Normalization

Spectral analysis and normalization are performed first. The goal of this step is to obtain a high-resolution spectral estimate of the input, with spectral components expressed in terms of sound pressure level (SPL). Much like the PE calculation described previously, this SPL normalization guarantees that a 4 kHz signal of +/−1 bit amplitude will be associated with an SPL near 0 dB (close to an acceptable T_q value for normal listeners at 4 kHz), whereas a full-scale sinusoid will be associated with an SPL near 90 dB. The spectral analysis procedure works as follows. First, incoming audio samples, $s(n)$, are normalized according to the FFT length, N, and the number of bits per sample, b, using the relation

$$x(n) = \frac{s(n)}{N(2^{b-1})}. \tag{5.18}$$

Normalization references the power spectrum to a 0-dB maximum. The normalized input, $x(n)$, is then segmented into 12-ms frames (512 samples) using a 1/16th-overlapped Hann window such that each frame contains 10.9 ms of new data. A power spectral density (PSD) estimate, $P(k)$, is then obtained using a 512-point FFT, i.e.,

$$P(k) = PN + 10\log_{10}\left|\sum_{n=0}^{N-1} w(n)x(n)e^{-j\frac{2\pi kn}{N}}\right|^2, 0 \leqslant k \leqslant \frac{N}{2}, \tag{5.19}$$

where the power normalization term, *PN*, is fixed at 90.302 dB and the Hann window, $w(n)$, is defined as

$$w(n) = \frac{1}{2}\left[1 - \cos\left(\frac{2\pi n}{N}\right)\right]. \tag{5.20}$$

Because playback levels are unknown during psychoacoustic signal analysis, the normalization procedure (Eq. (5.18)) and the parameter *PN* in Eq. (5.19) are used to estimate SPL conservatively from the input signal. For example, a full-scale sinusoid that is precisely resolved by the 512-point FFT in bin k_o will yield a spectral line, $P(k_0)$, having 84 dB SPL. With 16-bit sample resolution, SPL estimates for very-low-amplitude input signals will be at or below the absolute threshold. An example PSD estimate obtained in this manner for a CD-quality pop music selection is given in Figure 5.10(a). The spectrum is shown both on a linear frequency scale (upper plot) and on the Bark scale (lower plot). The dashed line in both plots corresponds to the absolute threshold of hearing approximation used by the model.

5.7.2 Step 2: Identification of Tonal and Noise Maskers

After PSD estimation and SPL normalization, tonal and nontonal masking components are identified. Local maxima in the sample PSD that exceed neighboring

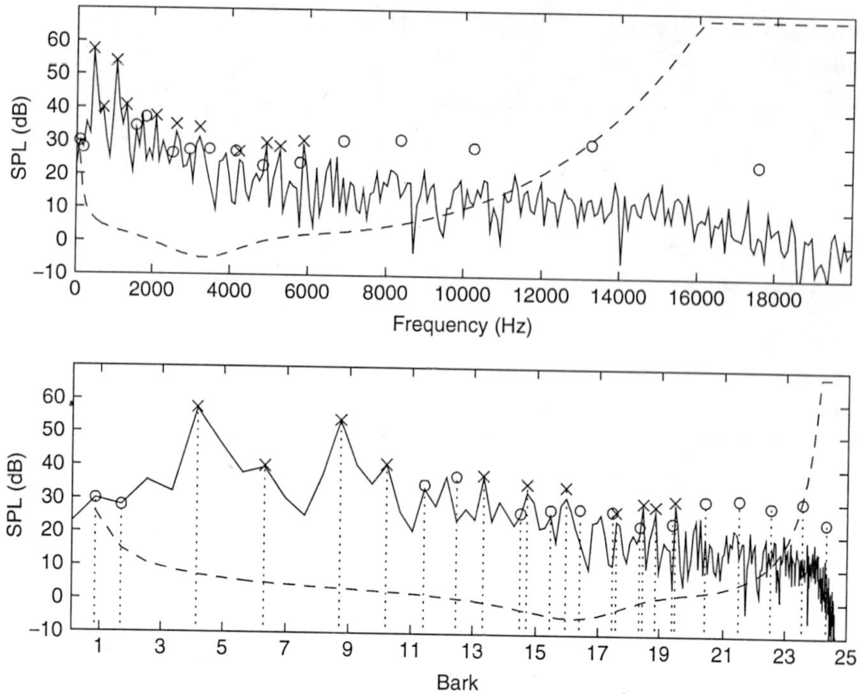

(a)

Figure 5.10a. ISO/IEC MPEG-1 psychoacoustic analysis model 1 for an example pop music selection, steps 1–5 as described in the text: (a) *Step 1:* Obtain PSD, express in dB SPL. Top panel gives linear frequency scale, bottom panel gives Bark frequency scale. Absolute threshold superimposed. *Step 2:* Tonal maskers identified and denoted by 'x' symbol; noise maskers identified and denoted by 'o' symbol. (b) Collection of prototype spreading functions (Eq. (5.31)) shown with level as the parameter. These illustrate the incorporation of excitation pattern level-dependence into the model. Note that the prototype functions are defined to be piecewise linear on the Bark scale. These will be associated with maskers in steps 3 and 4. (c) *Steps 3 and 4:* Spreading functions are associated with each of the individual tonal maskers satisfying the rules outlined in the text. Note that the Signal-to-Mask Ratio (SMR) at the peak is close to the widely accepted tonal value of 14.5 dB. (d) Spreading functions are associated with each of the individual noise maskers that were extracted after the tonal maskers had been eliminated from consideration, as described in the text. Note that the peak SMR is close to the widely accepted noise-masker value of 5 dB. (e) *Step 5:* A global masking threshold is obtained by combining the individual thresholds as described in the text. The maximum of the global threshold and the absolute threshold is taken at each point in frequency to be the final global threshold. The figure clearly shows that some portions of the input spectrum require SNRs better than 20 dB to prevent audible distortion, while other spectral regions require less than 3 dB SNR.

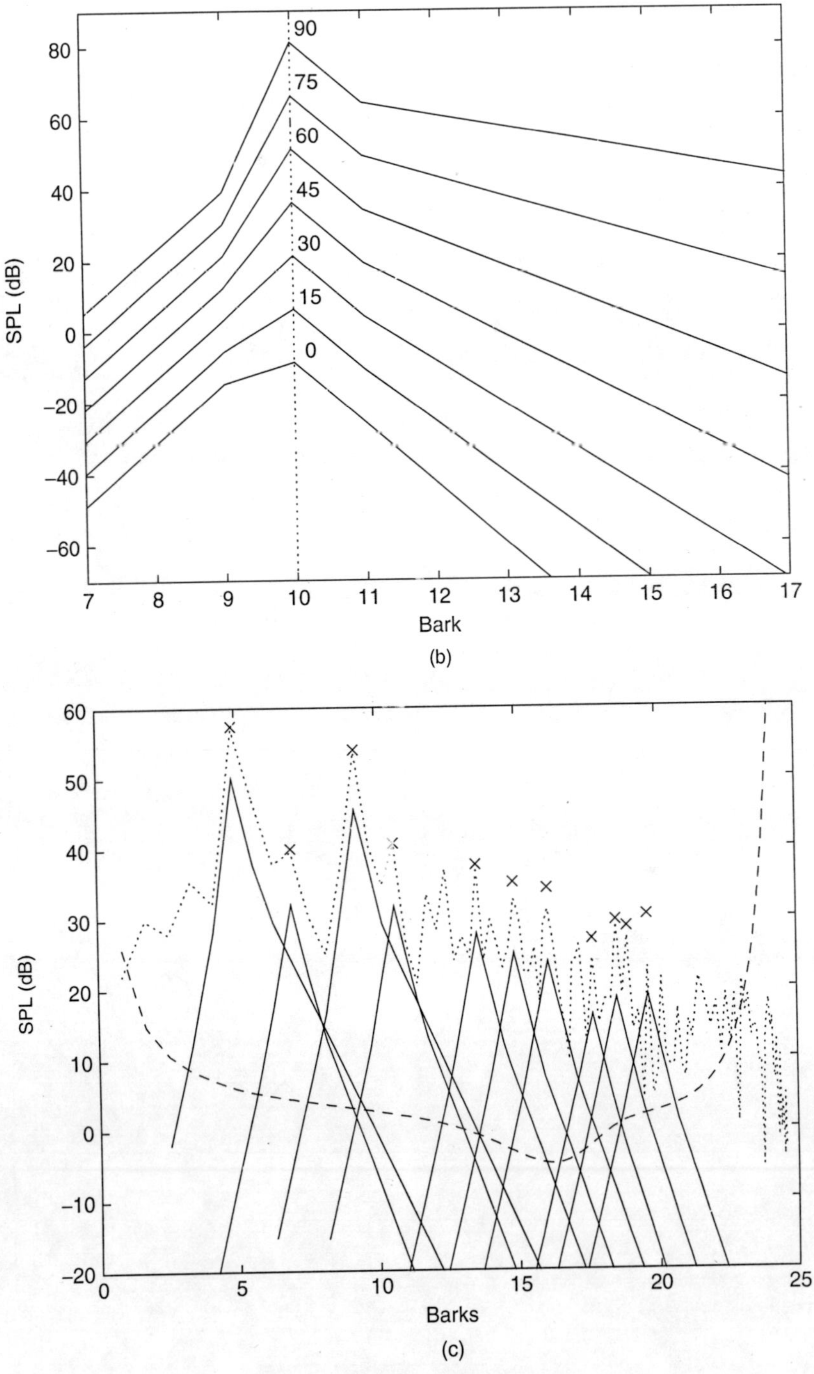

Figure 5.10b, c

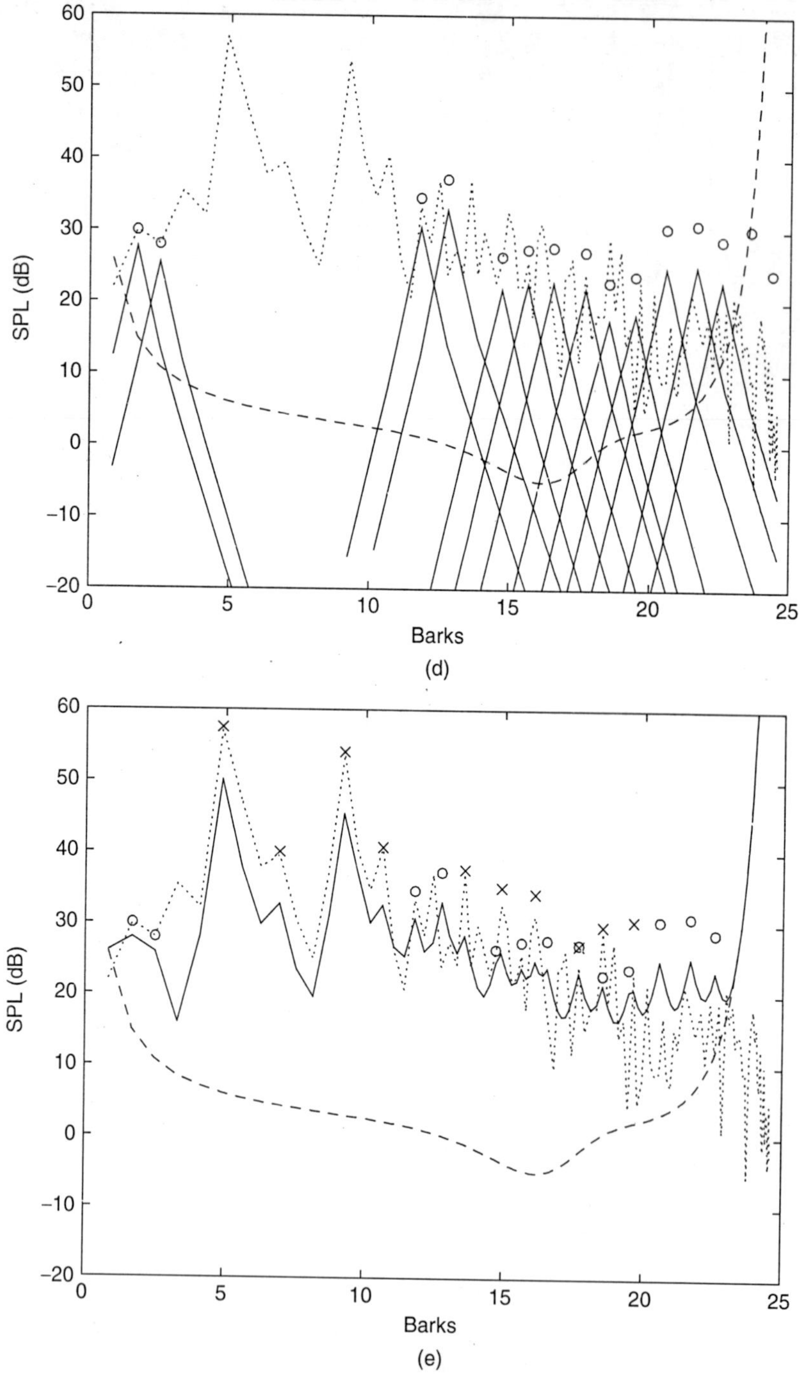

Figure 5.10d, e

components within a certain Bark distance by at least 7 dB are classified as tonal. Specifically, the "tonal" set, S_T, is defined as

$$S_T = \left\{ P(k) \middle| \begin{array}{c} P(k) > P(k \pm 1), \\ P(k) > P(k \pm \Delta_k) + 7dB \end{array} \right\}, \tag{5.21}$$

where

$$\Delta_k \in \begin{cases} 2 & 2 < k < 63 \quad (0.17 - 5.5 \text{ kHz}) \\ [2,3] & 63 \leqslant k < 127 \quad (5.5 - 11 \text{ kHz}) \\ [2,6] & 127 \leqslant k \leqslant 256 \quad (11 - 20 \text{ kHz}). \end{cases} \tag{5.22}$$

Tonal maskers, $P_{TM}(k)$, are computed from the spectral peaks listed in S_T as follows

$$P_{TM}(k) = 10 \log_{10} \sum_{j=-1}^{1} 10^{0.1P(k+j)} (\text{dB}). \tag{5.23}$$

In other words, for each neighborhood maximum, energy from three adjacent spectral components centered at the peak are combined to form a single tonal masker. Tonal maskers extracted from the example pop music selection are identified using 'x' symbols in Figure 5.10(a). A single noise masker for each critical band, $P_{NM}(\overline{k})$, is then computed from (remaining) spectral lines not within the $\pm\Delta_k$ neighborhood of a tonal masker using the sum

$$P_{NM}(\overline{k}) = 10 \log_{10} \sum_{j} 10^{0.1P(j)} (\text{dB}), \forall P(j) \notin \{P_{TM}(k, k \pm 1, k \pm \Delta_k)\}, \tag{5.24}$$

where $\overline{k}$ is defined to be the geometric mean spectral line of the critical band, i.e.,

$$\overline{k} = \left(\prod_{j=l}^{u} j \right)^{1/(l-u+1)}, \tag{5.25}$$

where l and u are the lower and upper spectral line boundaries of the critical band, respectively. The idea behind Eq. (5.24) is that residual spectral energy within a critical bandwidth not associated with a tonal masker must, by default, be associated with a noise masker. Therefore, in each critical band, Eq. (5.24) combines into a single noise masker all of the energy from spectral components that have not contributed to a tonal masker within the same band. Noise maskers are denoted in Figure 5.10 by 'o' symbols. Dashed vertical lines are included in the Bark scale plot to show the associated critical band for each masker.

5.7.3 Step 3: Decimation and Reorganization of Maskers

In this step, the number of maskers is reduced using two criteria. First, any tonal or noise maskers below the absolute threshold are discarded, i.e., only maskers that satisfy

$$P_{TM,NM}(k) \geqslant T_q(k) \tag{5.26}$$

are retained, where $T_q(k)$ is the SPL of the threshold in quiet at spectral line k. In the pop music example, two high-frequency noise maskers identified during step 2 (Figure 5.10(a)) are dropped after application of Eq. (5.26) (Figure 5.10(c–e)). Next, a sliding 0.5 Bark-wide window is used to replace any pair of maskers occurring within a distance of 0.5 Bark by the stronger of the two. In the pop music example, two tonal maskers appear between 19.5 and 20.5 Barks (Figure 5.10(a)). It can be seen that the pair is replaced by the stronger of the two during threshold calculations (Figure 5.10(c–e)). After the sliding window procedure, masker frequency bins are reorganized according to the subsampling scheme

$$P_{TM,NM}(i) = P_{TM,NM}(k) \tag{5.27}$$

$$P_{TM,NM}(k) = 0, \tag{5.28}$$

where

$$i = \begin{cases} k, & 1 \leqslant k \leqslant 48 \\ k + (k \bmod 2), & 49 \leqslant k \leqslant 96 \\ k + 3 - ((k-1) \bmod 4), & 97 \leqslant k \leqslant 232. \end{cases} \tag{5.29}$$

The net effect of Eq. (5.29) is 2:1 decimation of masker bins in critical bands 18–22 and 4:1 decimation of masker bins in critical bands 22–25, with no loss of masking components. This procedure reduces the total number of tone and noise masker frequency bins under consideration from 256 to 106. Tonal and noise maskers shown in Figure 5.10(c–e) have been relocated according to this decimation scheme.

5.7.4 Step 4: Calculation of Individual Masking Thresholds

Using the decimated set of tonal and noise maskers, individual tone and noise masking thresholds are computed next. Each individual threshold represents a masking contribution at frequency bin i due to the tone or noise masker located at bin j (reorganized during step 3). Tonal masker thresholds, $T_{TM}(i, j)$, are given by

$$T_{TM}(i, j) = P_{TM}(j) - 0.275Z_b(j) + SF(i, j) - 6.025(\text{dB SPL}), \tag{5.30}$$

where $P_{TM}(j)$ denotes the SPL of the tonal masker in frequency bin j, $Z_b(j)$ denotes the Bark frequency of bin j (Eq. (5.3)), and the spread of masking from masker bin j to maskee bin i, $SF(i, j)$, is modeled by the expression

$$SF(i, j) = \begin{cases} 17\Delta_{Z_b} - 0.4P_{TM}(j) + 11, & -3 \leqslant \Delta_{Z_b} < -1 \\ (0.4P_{TM}(j) + 6)\Delta_{Z_b}, & -1 \leqslant \Delta_{Z_b} < 0 \\ -17\Delta_{Z_b}, & 0 \leqslant \Delta_{Z_b} < 1 \\ (0.15P_{TM}(j) - 17)\Delta_{Z_b} - 0.15P_{TM}(j), & 1 \leqslant \Delta_{Z_b} < 8 \end{cases} (\text{dB SPL}), \tag{5.31}$$

i.e., as a piecewise linear function of masker level, $P(j)$, and Bark maskee-masker separation, $\Delta_{Z_b} = Z_b(i) - Z_b(j)$. $SF(i, j)$ approximates the basilar spreading (excitation pattern) described in Section 5.4. Prototype individual masking thresholds, $T_{TM}(i, j)$, are shown as a function of masker level in Figure 5.10(b) for an example tonal masker occurring at $Z_b = 10$ Barks. As shown in the figure, the slope of $T_{TM}(i, j)$ decreases with increasing masker level. This is a reflection of psychophysical test results, which have demonstrated [Zwic90] that the ear's frequency selectivity decreases as stimulus levels increase. It is also noted here that the spread of masking in this particular model is constrained to a 10-Bark neighborhood for computational efficiency. This simplifying assumption is reasonable given the very low masking levels that occur in the tails of the excitation patterns modeled by $SF(i, j)$. Figure 5.10(c) shows the individual masking thresholds (Eq. (5.30)) associated with the tonal maskers in Figure 5.10(a) ('x'). It can be seen here that the pair of maskers identified near 19 Barks has been replaced by the stronger of the two during the decimation phase. The plot includes the absolute hearing threshold for reference. Individual noise masker thresholds, $T_{NM}(i, j)$, are given by

$$T_{NM}(i, j) = P_{NM}(j) - 0.175Z_b(j) + SF(i, j) - 2.025(\text{dB SPL}), \qquad (5.32)$$

where $P_{NM}(j)$ denotes the SPL of the noise masker in frequency bin j, $Z_b(j)$ denotes the Bark frequency of bin j (Eq. (5.3)), and $SF(i, j)$ is obtained by replacing $P_{TM}(j)$ with $P_{NM}(j)$ in Eq. (5.31). Figure 5.10(d) shows individual masking thresholds associated with the noise maskers identified in step 2 (Figure 5.10(a) 'o'). It can be seen in Figure 5.10(d) that the two high frequency noise maskers that occur below the absolute threshold have been eliminated.

Before we proceed to step 5 and compute a global masking threshold, it is worthwhile to consider the relationship between Eq. (5.8) and Eq. (5.30), as well as the connection between Eq. (5.9) and Eq. (5.32). Equations (5.8) and (5.30) are related in that both model the TMN paradigm (Section 5.4) in order to generate a masking threshold for quantization noise masked by a tonal signal component. In the case of Eq. (5.8), a Bark-dependent offset that is consistent with experimental TMN data for the threshold minimum SMR is subtracted from the masker intensity, namely, the quantity $14.5 + B$. In a similar manner, Eq. (5.30) estimates for a quantization noise maskee located in bin i the intensity of the masking contribution due the tonal masker located in bin j. Like Eq. (5.8), the psychophysical motivation for Eq. (5.30) is the desire to model the relatively weak masking contributions of a TMN. Unlike Eq. (5.8), however, Eq. (5.30) uses an offset of only $6.025 + 0.275B$, i.e., Eq. (5.30) assumes a smaller minimum SMR at threshold than does Eq. (5.8). The connection between Eqs. (5.9) and (5.32) is analogous. In the case of this equation pair, however, the psychophysical motivation is to model the masking contributions of NMT. Equation (5.9) assumes a Bark-independent minimum SMR of 3–5 dB, depending on the value of the parameter K. Equation (5.32), on the other hand, assumes a Bark-dependent threshold minimum SMR of $2.025 + 0.175B$ dB. Also, whereas the spreading function (SF) terms embedded in Eqs. (5.30) and (5.32) explicitly account for the spread of masking, equations (5.8) and (5.9) assume that the spread of masking was captured during the computation of the terms E_T and E_N, respectively.

5.7.5 Step 5: Calculation of Global Masking Thresholds

In this step, individual masking thresholds are combined to estimate a *global* masking threshold for each frequency bin in the subset given by Eq. (5.29). The model assumes that masking effects are additive. The global masking threshold, $T_g(i)$, is therefore obtained by computing the sum

$$T_g(i) = 10\log_{10}\left(10^{0.1T_q(i)} + \sum_{l=1}^{L} 10^{0.1T_{TM}(i,l)} + \sum_{m=1}^{M} 10^{0.1T_{NM}(i,m)}\right)(\text{dB SPL}), \tag{5.33}$$

where $T_q(i)$ is the absolute hearing threshold for frequency bin i, $T_{TM}(i, l)$ and $T_{NM}(i, m)$ are the individual masking thresholds from step 4, and L and M are the numbers of tonal and noise maskers, respectively, identified during step 3. In other words, the global threshold for each frequency bin represents a signal-dependent, power-additive modification of the absolute threshold due to the basilar spread of all tonal and noise maskers in the signal power spectrum. Figure 5.10(e) shows global masking threshold obtained by adding the power of the individual tonal (Figure 5.10(c)) and noise (Figure 5.10(d)) maskers to the absolute threshold in quiet.

5.8 PERCEPTUAL BIT ALLOCATION

In this section, we will extend the uniform- and optimal-bit allocation algorithms presented in Chapter 3, Section 3.5, with perceptual bit-assignment strategies. In perceptual bit allocation method, the number of bits allocated to different bands is determined based on the *global masking thresholds* obtained from the psychoacoustic model. The steps involved in the computation of the global masking thresholds have been presented in detail in the previous section. The signal-to-mask ratio (SMR) determines the number of bits to be assigned in each band for perceptually transparent coding of the input audio. The noise-to-mask ratios (NMRs) are computed by subtracting the SMR from the SNR in each subband, i.e.,

$$NMR = SNR - SMR(\text{dB}). \tag{5.34}$$

The main objective in a perceptual bit allocation scheme is to keep the quantization noise below a masking threshold. For example, note that the NMR in Figure 5.11(a) is relatively more compared to NMR in Figure 5.11(b). Hence, the (quantization) noise in case of Figure 5.11(a) can be masked relatively easily than in case of Figure 5.11(b). Therefore, it is logical to assign sufficient number of bits to the subband with the lowest NMR. This criterion will be applied to all the subbands and until all the bits are exhausted. Typically, in audio coding standards an iterative procedure is employed that satisfies both the bit rate and global masking threshold requirements.

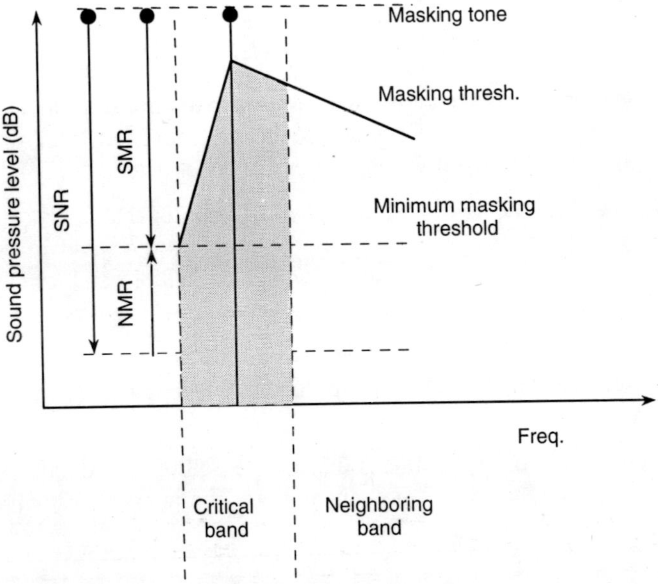

(a)

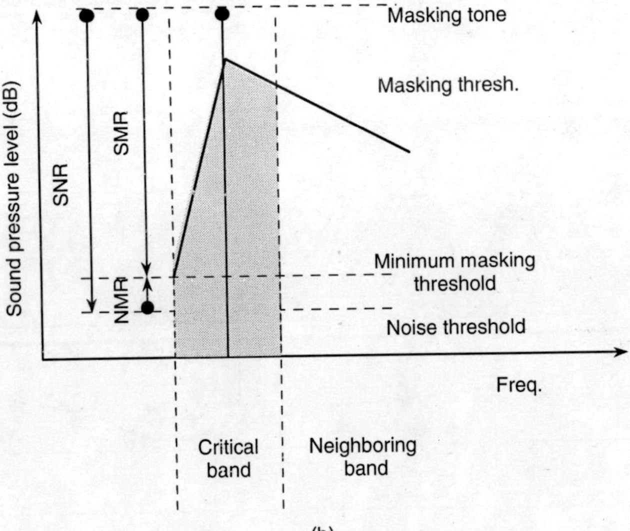

(b)

Figure 5.11. Simultaneous masking depicting relatively large NMR in (a) compared to (b).

5.9 SUMMARY

This chapter dealt with some of the basics of psychoacoustics. We covered the absolute threshold of hearing, the Bark scale, the simultaneous and temporal masking effects, and the perceptual entropy. A step-by-step procedure that describes the ISO/IEC psychoacoustic model 1 was provided.

PROBLEMS

5.1. Describe the difference between a Mel scale and a Bark scale. Give tables and itemize side-by-side the center frequencies and bandwidth for 0–5 kHz. Describe how the two different scales are constructed.

5.2. In Figure 5.12, the solid line indicates the just noticeable distortion (JND) curve and the dotted line indicates the absolute threshold in quiet. State which of the tones A, B, C, or D would be audible and which ones are likely to be masked. Explain.

5.3. In Figure 5.13, state whether tone B would be masked by tone A. Explain. Also indicate whether tone C would mask the narrow-band noise. Give reasons.

5.4. In Figure 5.14, the solid line indicates the JND curve obtained from the psychoacoustic model 1. A broadband noise component is shown that spans

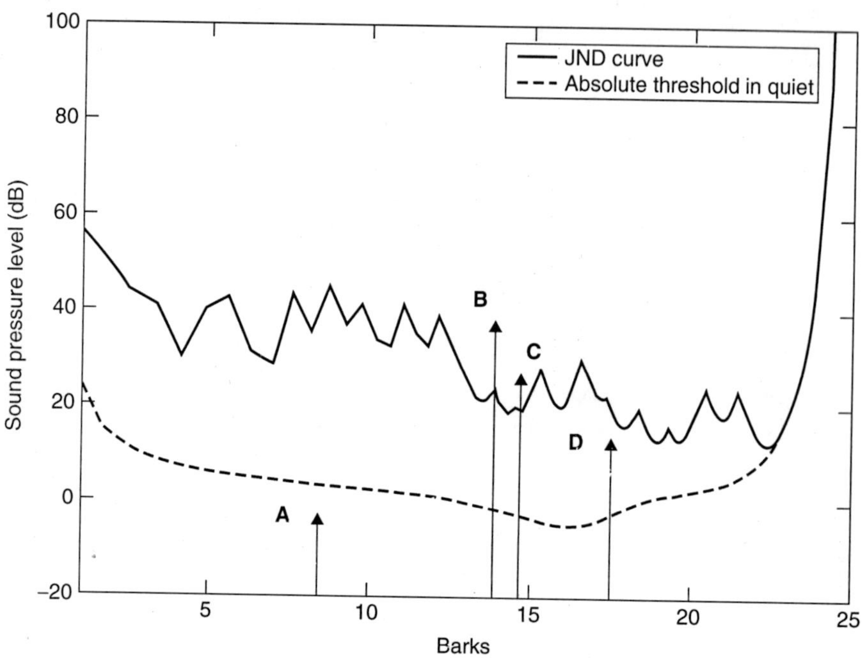

Figure 5.12. JND curve for Problem 5.2.

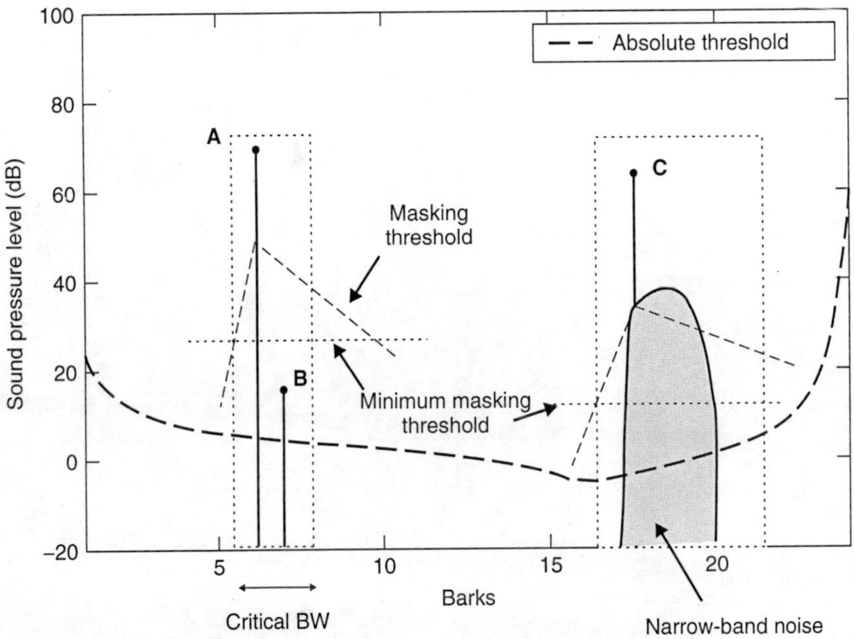

Figure 5.13. Masking experiment, Problem 5.3.

from 3 to 11 Barks and a tone is present at 10 Barks. Sketch the portions of the noise and the tone that could be considered perceptually relevant.

COMPUTER EXERCISES

5.5. Design a 3-band equalizer using the peaking filter equations of Chapter 2. The center frequencies should correspond to the auditory filters (see Table 5.1) at center frequencies 450 Hz, 1000 Hz, and 2500 Hz. Compute the Q-factors associated with each of these filters using, $Q = f_0/BW$, where f_0 is the center frequency and BW is the filter bandwidth (obtain from Table 5.1). Choose $g = 5\text{dB}$ for all the filters. Give the frequency response of the 3-band equalizer in terms of Bark scale.

5.6. Write a program to plot the absolute threshold of hearing in quiet Eq. (5.1). Give a plot in terms of a linear Hz scale.

5.7. Use the program of Problem 5.6 and plot the absolute threshold of hearing in a Bark scale.

5.8. Generate four sinusoids with frequencies, 400 Hz, 1000 Hz, 2500 Hz, and 6000 Hz; $f_s = 44.1$ kHz. Obtain $s(n)$ by adding these individual sinusoids as follows,

$$s(n) = \sum_{i=1}^{4} \sin\left(\frac{2\pi f_i n}{f_s}\right), n = 1, 2, \ldots, 1024.$$

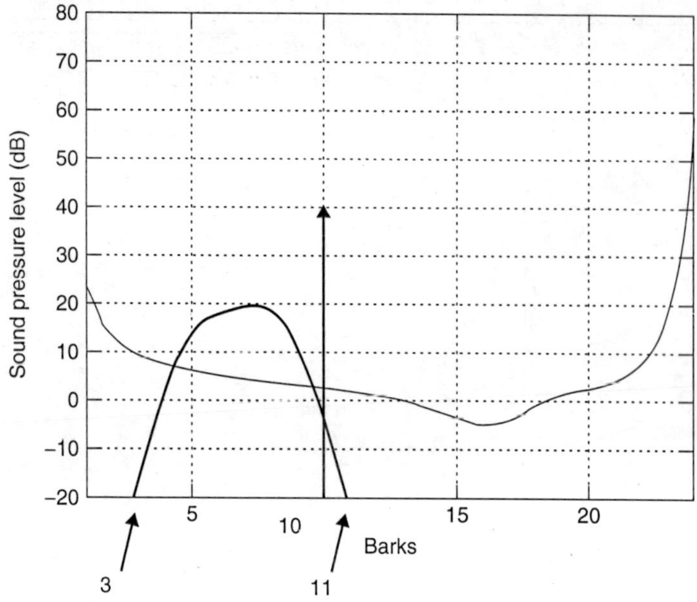

Figure 5.14. Perceptual bit-allocation, Problem 5.4.

Give power spectrum plots of $s(n)$ (in dB) in terms of a Bark scale and in terms of a linear Hz scale. List the Bark-band numbers where the four peaks are located. (Hint: see Table 5.1 for the bark band numbers.)

5.9. Extend the above problem and give the power spectrum plot in dB SPL. See Section 5.7.1 for details. Also include the absolute threshold of hearing in quiet in your plot.

5.10. Write a program to compute the perceptual entropy (in bits/sample) of the following signals:

a. *ch5_malespeech.wav* (8 kHz, 16 bit)

b. *ch5_music.wav* (44.1 kHz, 16 bit)

(Hint: Use equations (5.10)–(5.17) in Chapter 5, Section 5.6.) Choose the frame size as 512 samples. Also, the perceptual entropy (PE) measurement is obtained by constructing a PE histogram over many frames and then choosing a worst-case value as the actual measurement.

5.11. FFT-based perceptual audio synthesis using the MPEG 1 psychoacoustic model 1.

In this computer exercise, we will consider an example to show how the psychoacoustic principles are applied in actual audio coding algorithms. Recall that in Chapter 2, Computer Exercise 2.25, we employed the peak-picking method to select a subset of FFT components for audio synthesis. In this exercise, we will use the just-noticeable-distortion (JND) curve as the "reference" to select the perceptually important FFT components. All the FFT components below the

JND curve are assigned a minimal value (for example, −50 dB SPL), such that these perceptually irrelevant FFT components receive minimum number of bits for encoding. The ISO/IEC MPEG 1 psychoacoustic model 1 (See Section 5.7; Steps 1 through 5) is used to compute the JND curve.

Use the MATLAB software from the Book website to simulate the ISO/IEC MPEG-1 psychoacoustic model 1. The software package consists of three MATLAB files, psymain.m, psychoacoustics.m, audio_synthesis.m, a wave file *ch5_music.wav*, and a hints.doc file. The psymain.m file is the main file that contains the complete flow of your computer exercise. The psychoacoustics.m file contains the steps performed in the psychoacoustic analysis.

Deliverable 1:

a. Using the comments included in the *hints.doc* file, fill in the MATLAB commands in the audio_synthesis.m file to complete the program.

b. Give plots of the input and the synthesized audio. Is the psychoacoustic criterion for picking FFT components equivalent to using a Parseval's related criterion? In the audio synthesis, what happens if the FFT conjugate symmetry was not maintained?

c. How is this FFT-based perceptual audio synthesis method different from a typical peak-picking method for audio synthesis?

Deliverable 2:

d. Provide a subjective evaluation of the synthesized audio in terms of a MOS scale? Did you hear any clicks between the frames in the synthesized audio? If yes, what would you do to eliminate such artifacts? Compute the over-all and segmental SNR values between the input and the synthesized audio.

e. On the average, how many FFT components per frame were selected? If only 30 FFT components out of 512 were to be picked (because of the bit rate considerations), would the application of the psychoacoustic model to select the FFT components yield the best possible SNR?

5.12. In this computer exercise, we will analyze the asymmetry of simultaneous masking. Use the MATLAB software from the Book website. The software package consists of two MATLAB files, asymm_mask.m and psychoacoustics.m. The asymm_mask.m includes steps to generate a pure tone, $s_1(n)$, with $f = 4$ kHz and $f_s = 44.1$ kHz,

$$s_1(n) = \sin\left(\frac{2\pi f n}{f_s}\right), n = 0, 1, 2, \ldots, 44099$$

a. Simulate a broadband noise, $s_2(n)$, by band-pass filtering uniform white noise ($\mu = 0$ and $\sigma^2 = 1$) using a Butterworth filter (of appropriate order, e.g., 8) with center frequency, 4 kHz. Assume that 3-dB cut-off frequencies of the bandpass filter as 3500 Hz and 4500 Hz. Generate a test signal, $s(n) = \alpha s_1(n) + \beta s_2(n)$. Choose $\alpha = 0.025$ and $\beta = 1$. Observe if the broad-band noise completely masks the tone. Experiment for different values of α, β and find out when 1) the broadband noise masks the tone, and 2) the tone masks the broadband noise.

b. Simulate the two cases of masking (i.e., the NMT and the TMN) given in Figure, 5.7, Section, 5.4.

CHAPTER 6

TIME-FREQUENCY ANALYSIS: FILTER BANKS AND TRANSFORMS

6.1 INTRODUCTION

Audio codecs typically use a time-frequency analysis block to extract a set of parameters that is amenable to quantization. The tool most commonly employed for this mapping is a filter bank of bandpass filters. The filter bank divides the signal spectrum into frequency subbands and generates a time-indexed series of coefficients representing the frequency-localized signal power within each band. By providing explicit information about the distribution of signal and hence masking power over the time-frequency plane, the filter bank plays an essential role in the identification of perceptual irrelevancies. Additionally, the time-frequency parameters generated by the filter bank provide a signal mapping that is conveniently manipulated to shape the coding distortion. On the other hand, by decomposing the signal into its constituent frequency components, the filter bank also assists in the reduction of statistical redundancies.

This chapter provides a perspective on filter-bank design and other techniques of particular importance in audio coding. The chapter is organized as follows. Sections 6.2 and 6.3 introduce filter-bank design issues for audio coding. Sections 6.4 through 6.7 review specific filter-bank methodologies found in audio codecs, namely, the two-band quadrature mirror filter (QMF), the M-band tree-structured QMF, the M-band pseudo-QMF bank, and the M-band Modified Discrete Cosine Transform (MDCT). The 'MP3' or MPEG-1, Layer III audio

Audio Signal Processing and Coding, by Andreas Spanias, Ted Painter, and Venkatraman Atti

codec pseudo-QMF and MDCT are discussed in Sections 6.6 and 6.7, respectively. Section 6.8 provides filter-bank interpretations of the discrete Fourier and discrete cosine transforms. Finally, Sections 6.9 and 6.10 examine the time-domain "pre-echo" artifact in conjunction with pre-echo control techniques.

Beyond the references cited in this chapter, the reader is referred to in-depth tutorials on filter banks that have appeared in the literature [Croc83] [Vaid87] [Vaid90] [Malv91] [Vaid93] [Akan96]. The reader may also wish to explore the connection between filter banks and wavelets that has been well documented in [Riou91] [Vett92] and in several texts [Akan92] [Wick94] [Akan96] [Stra96].

6.2 ANALYSIS-SYNTHESIS FRAMEWORK FOR *M*-BAND FILTER BANKS

Filter banks are perhaps most conveniently described in terms of an analysis-synthesis framework (Figure 6.1), in which the input signal, $s(n)$, is processed at the encoder by a parallel bank of $(L-1)$-th order FIR bandpass filters, $H_k(z)$. The bandpass analysis outputs,

$$v_k(n) = h_k(n) * s(n) = \sum_{m=0}^{L-1} s(n-m)h_k(m), \quad k = 0, 1, \ldots, M-1 \tag{6.1}$$

are decimated by a factor of M, yielding the subband sequences

$$y_k(n) = v_k(Mn) = \sum_{m=0}^{L-1} s(nM-m)h_k(m), \quad k = 0, 1, \ldots, M-1, \tag{6.2}$$

which comprise a *critically sampled* or *maximally decimated* signal representation, i.e., the number of subband samples is equal to the number of input samples. Because it is impossible to achieve perfect "brickwall" magnitude responses with finite-order bandpass filters, there is unavoidable aliasing between the decimated subband sequences. Quantization and coding are performed on the subband sequences $y_k(n)$. In the perceptual audio codec, the quantization noise is usually shaped according to a perceptual model. The quantized subband samples, $\hat{y}_k(n)$, are eventually received by the decoder, where they are upsampled by M to form the intermediate sequences

$$w_k(n) = \begin{cases} \hat{y}_k(n/M), & n = 0, M, 2M, 3M, \ldots \\ 0, & \text{otherwise.} \end{cases} \tag{6.3}$$

In order to eliminate the imaging distortions introduced by the upsampling operations, the sequences $w_k(n)$ are processed by a parallel bank of synthesis filters, $G_k(z)$, and then the filter outputs are combined to form the overall output, $\hat{s}(n)$.

The analysis and synthesis filters are carefully designed to cancel aliasing and imaging distortions. It can be shown [Akan96] that

$$\hat{s}(n) = \frac{1}{M} \sum_{m=-\infty}^{\infty} \sum_{l=-\infty}^{\infty} \sum_{k=0}^{M-1} s(m) h_k(lM - m) g_k(l - Mn) \tag{6.4}$$

or, in the frequency domain,

$$\hat{S}(\Omega) = \frac{1}{M} \sum_{k=0}^{M-1} \sum_{l=0}^{M-1} S\left(\Omega + \frac{2\pi l}{M}\right) H_k\left(\Omega + \frac{2\pi l}{M}\right) G_k(\Omega). \tag{6.5}$$

For perfect reconstruction (PR) filter banks, the output, $\hat{s}(n)$, will be identical to the input, $s(n)$, within a delay, i.e., $\hat{s}(n) = s(n - n_0)$, as long as there is no quantization noise introduced, i.e., $y(n) = \hat{y}_k(n)$. This is naturally not the case for a codec, and therefore quantization sensitivity is an important filter bank property, since PR guarantees are lost in the presence of quantization.

Figures 6.2 and 6.3, respectively, give example magnitude responses for banks of uniform and nonuniform bandwidth filters that can be realized within the framework of Figure 6.1. A uniform bandwidth M-channel filter bank is shown in Figure 6.2. The M analysis filters have normalized center frequencies $(2k + 1)\pi/2M$, and are characterized by individual impulse responses $h_k(n)$, as well as frequency responses $H_k(\Omega)$, $0 \leqslant k < M - 1$.

Some of the popular audio codecs contain parallel bandpass filters of uniform bandwidth similar to Figure 6.2. Other coders strive for a "critical band" analysis by relying upon filters of nonuniform bandwidth. The octave-band filter bank,

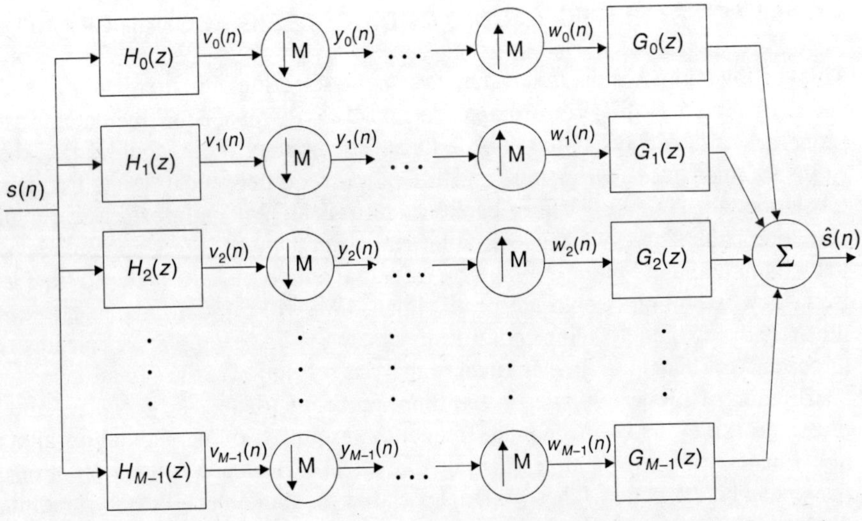

Figure 6.1. Uniform M-band maximally decimated analysis-synthesis filter bank.

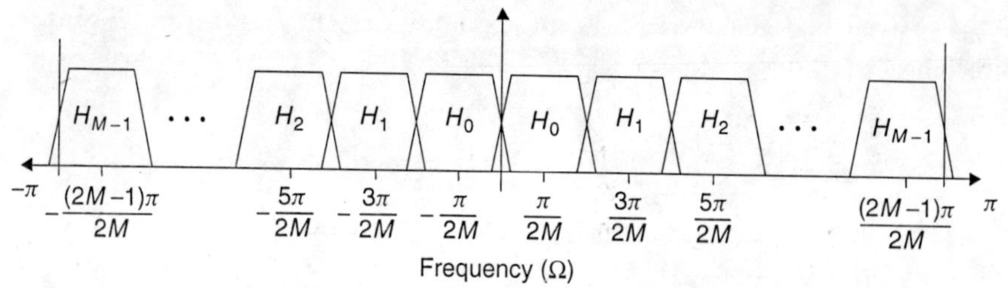

Figure 6.2. Magnitude frequency response for a uniform M-band filter bank (oddly stacked).

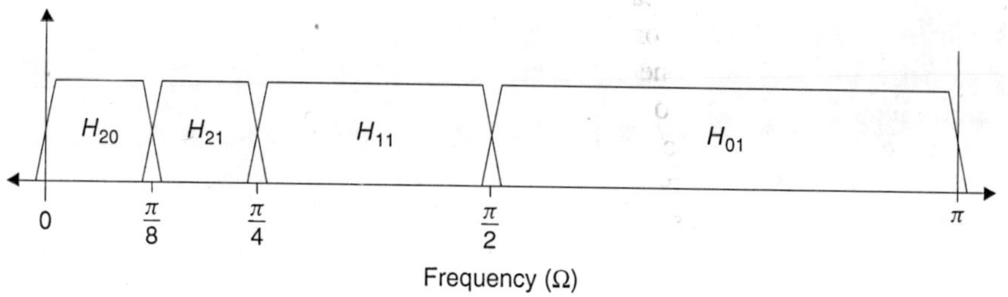

Figure 6.3. Magnitude frequency response for an octave-band filter bank.

for which the four-band case is illustrated in Figure 6.3, is sometimes used as an approximation to the auditory filter bank, albeit a poor one.

As shown in Figure 6.3, octave-band analysis filters have normalized center frequencies and filter bandwidths that are dyadically related. Naturally, much better approximations are possible.

6.3 FILTER BANKS FOR AUDIO CODING: DESIGN CONSIDERATIONS

This section addresses the issues that govern the selection of a filter bank for audio coding. Efficient coding performance depends heavily on adequately matching the properties of the analysis filter bank to the characteristics of the input signal. Algorithm designers face an important and difficult tradeoff between time and frequency resolution when selecting a filter-bank structure [Bran92a]. Failure to choose a suitable filter bank can result in perceptible artifacts in the output (e.g., pre-echoes) or low coding gain and therefore high bit rates. No single tradeoff between time and frequency resolution is optimal for all signals. We will present three examples to illustrate the challenge facing codec designers. In the first example, we consider the importance of matching time-frequency analysis resolution to the signal-dependent distribution of masking power in the time-frequency plane. The second example illustrates the effect of inadequate frequency resolution on perceptual bit allocation. Finally, the third example illustrates the effect of inadequate time resolution on perceptual bit allocation. These examples clarify the fundamental tradeoff required during filter-bank selection for perceptual coding.

6.3.1 The Role of Time-Frequency Resolution in Masking Power Estimation

Through schematic representations of masking thresholds for castanets and piccolo, Figure 6.4(a,b) illustrates the difficulty of selecting a single filter bank to satisfy the diverse time and frequency resolution requirements associated with different classes of audio. In the figures, darker regions correspond to higher masking thresholds. To realize maximum coding gain, the strongly harmonic piccolo signal clearly calls for fine frequency resolution and coarse time resolution, because the masking thresholds are quite localized in frequency. Quite the opposite is true of the castanets. The fast attacks associated with this percussive sound create highly time-localized masking thresholds that are also widely dispersed in frequency. Therefore, adequate time resolution is essential for accurate estimation of the highly time-varying masked threshold. Naturally, similar resolution properties should also be associated with the filter bank used to decompose the signal into a parametric set for quantization and encoding. Using real signals and filter banks, the next two examples illustrate the bit rate impact of adequate and inadequate resolutions in each domain.

6.3.2 The Role of Frequency Resolution in Perceptual Bit Allocation

To demonstrate the importance of matching a filter bank's resolution properties with the noise-shaping requirements imposed by a perceptual model, the next two examples combine high- and low-resolution filter banks with two input extremes, namely those of a sinusoid and an impulse. First, we consider the importance of adequate frequency resolution. The need for high-resolution frequency analysis is most pronounced when the input contains strong sinusoidal components. Given a tonal input, inadequate frequency resolution can produce unreasonably high signal-to-noise ratio (SNR) requirements within individual subbands, resulting in high bit rates. To see this, we compare the results of processing a 2.7-kHz pure tone first with a 32-channel, and then with a 1024-channel MDCT filter bank, as shown in Figure 6.5(a) and Figure 6.5(b), respectively. The vertical line in each figure represents the frequency and level (80 dB SPL) of the input tone. In the figures, the masked threshold associated with the sinusoid is represented by a solid, nearly triangular line. For each filter bank in Figure 6.5(a) and Figure 6.5(b), the band containing most of the signal energy is quantized with sufficient resolution to create an in-band SNR of 15.4 dB. Then, the quantization noise is superimposed on the masked threshold. In the 32-band case (Figure 6.5a), it can be seen that the quantization noise at an SNR of 15.4 dB spreads considerably beyond the masked threshold, implying that significant artifacts will be audible in the reconstructed signal. On the other hand, the improved selectivity of the 1024-channel filter bank restricts the spread of quantization at 15.4 dB SNR to well within the limits of the masked threshold (Figure 6.5b). The figure clearly shows that for a tonal signal, good frequency selectivity is essential for low bit rates. In fact, the 32-channel filter bank for this signal requires greater than a 60 dB SNR to satisfy the masked threshold (Figure 6.5c). This high cost (in terms of bits required) results

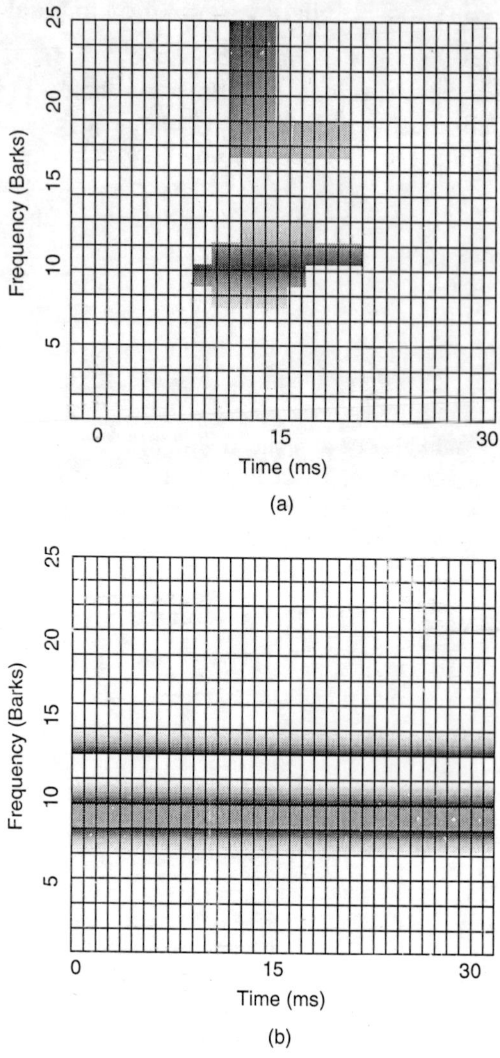

Figure 6.4. Masking thresholds in the time-frequency plane: (a) castanets, (b) piccolo (after [Prin95]).

from the mismatch between the filter bank's poor selectivity and the very limited downward spread of masking in the human ear. As this experiment would imply, high-resolution frequency analysis is usually appropriate for tonal audio signals.

6.3.3 The Role of Time Resolution in Perceptual Bit Allocation

A time-domain dual of the effect observed in the previous example can be used to illustrate the importance of adequate time resolution. Whereas the previous experiment showed that simultaneous masking criteria determine how much frequency

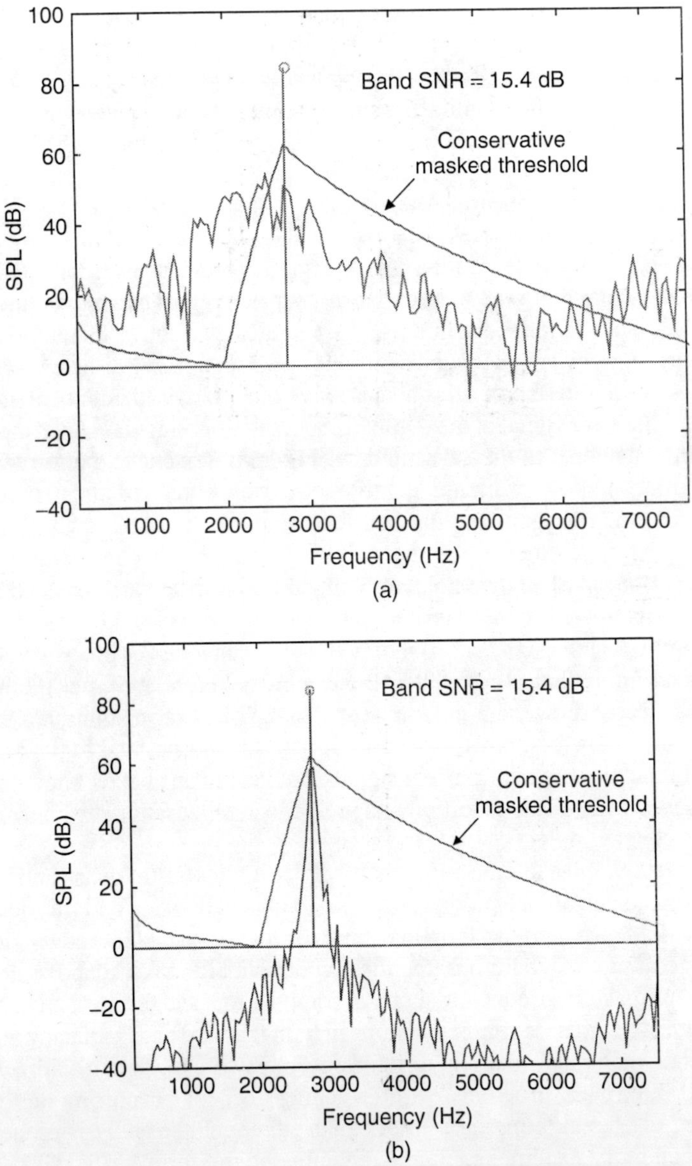

Figure 6.5. The effect of frequency resolution on perceptual SNR requirements for a 2.7 kHz pure tone input. Input tone is represented by the spectral line at 2.7 kHz with 80 dB SPL. Conservative masked threshold due to presence of tone is shown. Quantization noise for a given number of bits per sample is superimposed for several precisions: (a) 32-channel MDCT with 15.4 dB in-band SNR, quantization noise spreads beyond masked threshold; (b) 1024-channel MDCT with 15.4 dB in-band SNR, quantization noise remains below masked threshold; (c) 32-channel MDCT requires 63.7 dB in-band SNR to satisfy masked threshold, i.e., requires larger bit allocation to mask quantization noise.

resolution is necessary for good performance, one can also imagine that temporal masking effects dictate time resolution requirements. In fact, the need for good time resolution is pronounced when the input contains sharp transients. This is best illustrated with an impulse. Unlike the sinusoid, with its highly frequency-localized masking power, the broadband impulse contains broadband masking power that is highly time-localized. Given a transient input, therefore, lacking time resolution can result in a temporal smearing of quantization noise beyond the time window of effective masking. To illustrate this point, consider the results obtained by processing an impulse with the same filter banks as in the previous example. The results are shown in Figure 6.6(a) and Figure 6.6(b), respectively. The vertical line in each figure corresponds to the input impulse. The figures also show the temporal envelope of masking power associated with the impulse as a solid, nearly triangular window. For each filter bank, all subbands were quantized with identical bit allocations. Then, the error signal at the output (quantization noise) was superimposed on the masking envelope. In the 32-band case (Figure 6.6a), one can observe that the time resolution (impulse response length) restricts the spread of quantization noise to well within the limits of the masked threshold. For the 1024-band filter bank (Figure 6.6b), on the other hand, quantization noise spreads considerably beyond the time envelope masked threshold, implying that significant artifacts will be audible in the reconstructed signal. The only remedy in this case would be to "overcode" the transient so that the signal surrounding the transient receives precision adequate to satisfy the limited masking power in the regions before and after the impulse. The figures clearly show that for a transient signal, good time resolution is essential for low bit rates. The combination of long impulse responses and limited masking windows in the presence of transient signals can lead to an artifact known as "pre-echo" distortion. Pre-echo distortion and pre-echo compensation are covered at the end of this chapter in Sections 6.9 and 6.10.

Unfortunately, most audio source material is highly nonstationary and contains significant tonal and atonal energy, as well as both steady-state and transient intervals. As a rule, signal models [John96a] tend to remain constant for long periods and then change abruptly. Therefore, the ideal coder should make adaptive decisions regarding optimal time-frequency signal decomposition, and the ideal analysis filter bank would have time-varying resolutions in both the time and frequency domains. This fact has motivated many algorithm designers to experiment with switched and hybrid filter-bank structures, with switching decisions occurring on the basis of the changing signal properties. Filter banks emulating the analysis properties of the human auditory system, i.e., those containing nonuniform "critical bandwidth" subbands, have proven highly effective in the coding of highly transient signals such as the castanets, glockenspiel, or triangle. For dense, harmonically structured signals such as the harpsichord or pitch pipe, on the other hand, the "critical band" filter banks have been less successful because of their reduced coding gain relative to filter banks with a large number of subbands. In short, several bank characteristics are highly desirable for audio coding:

- Signal-adaptive time-frequency tiling
- Low-resolution, "critical-band" mode (e.g., 32 subbands)

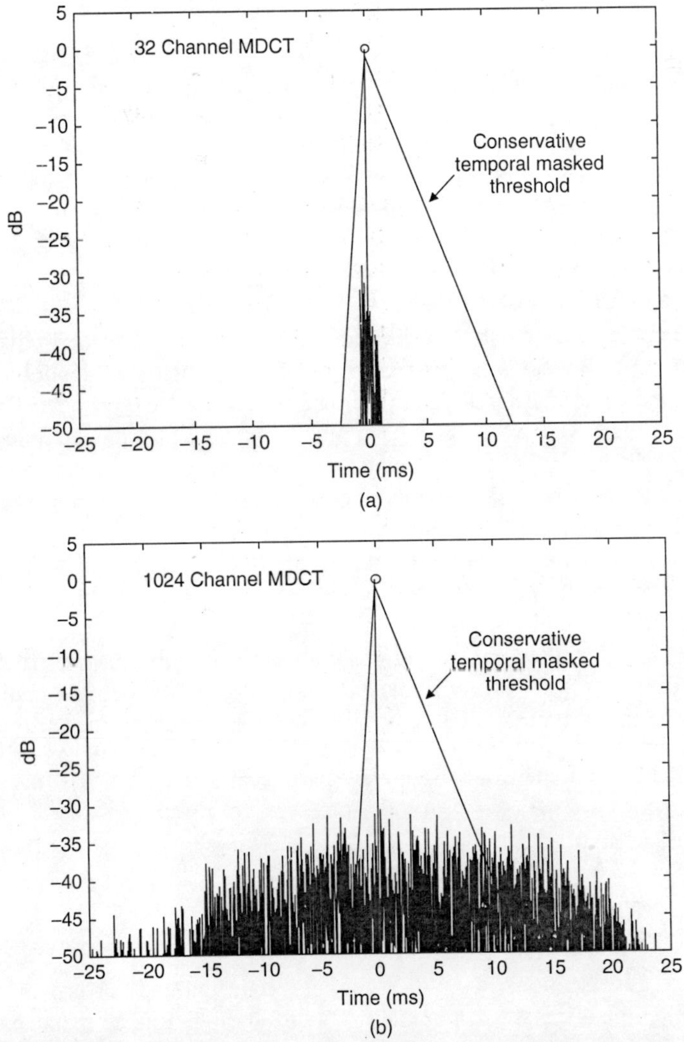

Figure 6.6. The effect of time resolution on perceptual bit allocation for an impulse. Impulse input occurs at time 0. Conservative temporal masked threshold due to the presence of impulse is shown. Quantization noise for a fixed number of bits per sample is superimposed for low- and high-resolution filter banks: (a) 32-channel MDCT; (b) 1024-channel MDCT.

- High-resolution mode, e.g., 4096 subbands
- Efficient resolution switching
- Minimum blocking artifacts
- Good channel separation
- Strong stop-band attenuation

- Perfect reconstruction
- Critical sampling
- Fast implementation.

Good channel separation and stop-band attenuation are particularly desirable for signals containing very little irrelevancy, such as the harpsichord. Maximum redundancy removal is essential for maintaining high quality at low bit rates for these signals. Blocking artifacts in time-varying filter banks can lead to audible distortion in the reconstruction. Beyond filter-bank-specific architectural and performance criteria, system-level considerations may also influence the best choice of filter bank for a codec design. For example, the codec architecture could contain two separate, parallel time-frequency analysis blocks, one for the perceptual model and one for generating the parametric set that is ultimately quantized and encoded. The parallel scenario offers the advantage that each filter bank can be optimized independently. This is possible since the perceptual analysis section does not typically require signal reconstruction, whereas the coefficients for coding must eventually be mapped back to the time-domain. In the interest of computational efficiency, however, many audio codecs have only one time-frequency analysis block that does "double duty," in the sense that the perceptual model obtains information from the same set of coefficients that are ultimately quantized and encoded.

Algorithms for filter-bank design as well as fast algorithms for efficient filter-bank realizations offer many choices to designers of perceptual audio codecs. Among the many types available are those characterized by the following:

- Uniform or nonuniform frequency partitioning
- An arbitrary number of subbands
- Perfect or almost perfect reconstruction
- Critically sampled or oversampled representations
- FIR or IIR constituent filters.

In the next few Sections, we will focus on the design and performance of well-known filter banks that are popular in audio coding. Rather than dealing with efficient implementation structures that are available, we have elected for each filter-bank architecture to describe the individual bandpass filters in terms of impulse and frequency response functions that are easily related to the analysis-synthesis framework of Figure 6.1. These descriptions are intended to provide insight regarding the filter-bank response characteristics, and to allow for comparisons across different methods. The reader should be aware, however, that structures for efficient realizations are almost always used in practice, and because computational efficiency is of paramount importance, most audio coding filter-bank realizations, although functionally equivalent, may or may not resemble the maximally decimated analysis-synthesis structure given in Figure 6.1. In other words, most of the filter banks used in audio coders have equivalent parallel forms and can be conveniently analyzed in terms of this analysis-synthesis framework. The framework provides a useful interpretation for the sets of coefficients generated by the unitary transforms often embedded in audio coders such as the discrete cosine transform (DCT), the discrete Fourier transform (DFT), the discrete wavelet transform (DWT), and the discrete wavelet packet transform (DWPT).

6.4 QUADRATURE MIRROR AND CONJUGATE QUADRATURE FILTERS

The two-band quadrature mirror and conjugate quadrature filter (QMF and CQF) banks are logical starting points for the discussion on filter banks for audio coding. Two-band QMF banks were used in early subband algorithms for speech coding [Croc76], and later for the first standardized 7-kHz wideband audio algorithm, the ITU G.722 [G722]. Also, the strong connection between two-band perfect reconstruction (PR) CQF filter banks and the discrete wavelet transform [Akan96] has played a significant role in the development of high-performance audio coding filter banks. Ultimately, tree-structured cascades of the CQF filters have been used to construct several "critical-band" filter banks in a number of high quality algorithms. The two-channel bank, which can provide a building block for structured M-channel banks, is developed as follows. If the analysis-synthesis filter bank (Figure 6.1) is constrained to two channels, i.e., if $M = 2$, then Eq. (6.5) becomes

$$\hat{S}(\Omega) = \frac{1}{2}S(\Omega)[H_0^2(\Omega) - H_1^2(\Omega)]. \tag{6.6}$$

Esteband and Galand showed [Este77] that aliasing is cancelled between the upper and lower bands if the QMF conditions are satisfied, namely

$$\begin{aligned} H_1(\Omega) &= H_0(\Omega+\pi) \Rightarrow h_1(n) = (-1)^n h_0(n) \\ G_0(\Omega) &= H_0(\Omega) \Rightarrow g_0(n) = h_0(n) \\ G_1(\Omega) &= -H_0(\Omega+\pi) \Rightarrow g_1(n) = -(-1)^n h_0(n). \end{aligned} \tag{6.7}$$

Thus, the two-band filter-bank design task is reduced to the design of a single, lowpass filter, $h_0(n)$, under the constraint that the overall transfer function, Eq. (6.6), be an allpass function with constant group delay (linear phase). Although filter families satisfying the QMF criteria with good stop-band and transition-band characteristics have been designed (e.g., [John80]) to minimize overall distortion, the QMF conditions actually make perfect reconstruction impossible. Smith and Barnwell showed in [Smit86], however, that PR two-band filter banks based on a lowpass prototype are possible if the CQF conditions are satisfied, namely

$$\begin{aligned} h_1(n) &= (-1)^n h_0(L-1-n) \\ g_0(n) &= h_0(L-1-n) \\ g_1(n) &= -(-1)^n h_0(n). \end{aligned} \tag{6.8}$$

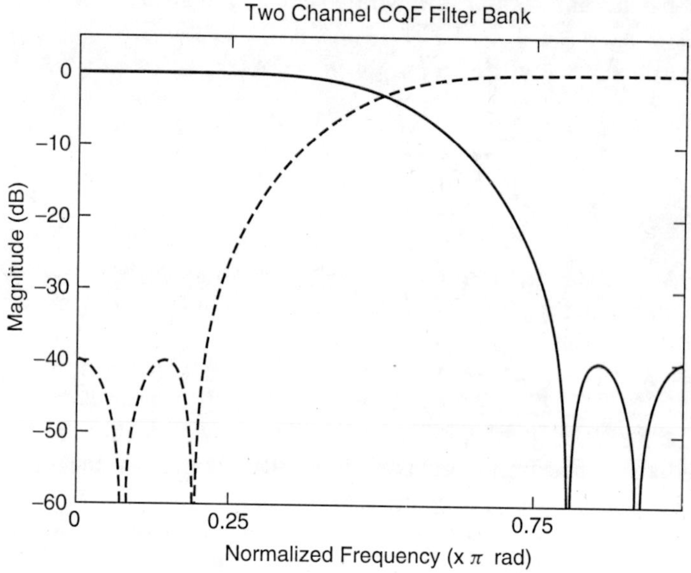

Figure 6.7. Two-band Smith-Barnwell CQF filter bank magnitude frequency response, with $L = 8$ [Smit86].

The magnitude response of an example Smith-Barnwell CQF filter bank from [Smit86] with $L = 8$ is shown in Figure 6.7. The lowpass response is shown as a solid line, while the highpass is dashed. One can observe the significant overlap between channels, as well as monotonic passband and equiripple stop-band characteristics, with minimum stop-band rejection of 40 dB. As mentioned previously, efficiency concerns dictate that filter banks are rarely implemented in the direct form of Eq. (6.1). The QMF banks are most often realized using a polyphase factorization [Bell76], (i.e.,

$$H(z) = \sum_{l=0}^{M-1} z^{-l} E_l(z^M), \tag{6.9}$$

where

$$E_l(z) = \sum_{n=-\infty}^{\infty} h(Mn + l) z^{-n}, \tag{6.10}$$

which yields better than a 2:1 computational load reduction. On the other hand, the CQF filters are incompatible with the polyphase factorization but can be efficiently realized using alternative structures such as the lattice [Vaid88].

6.5 TREE-STRUCTURED QMF AND CQF *M*-BAND BANKS

Clearly, audio coders require better frequency resolution than either the QMF or CQF two-band decompositions can provide in order to realize sufficient coding

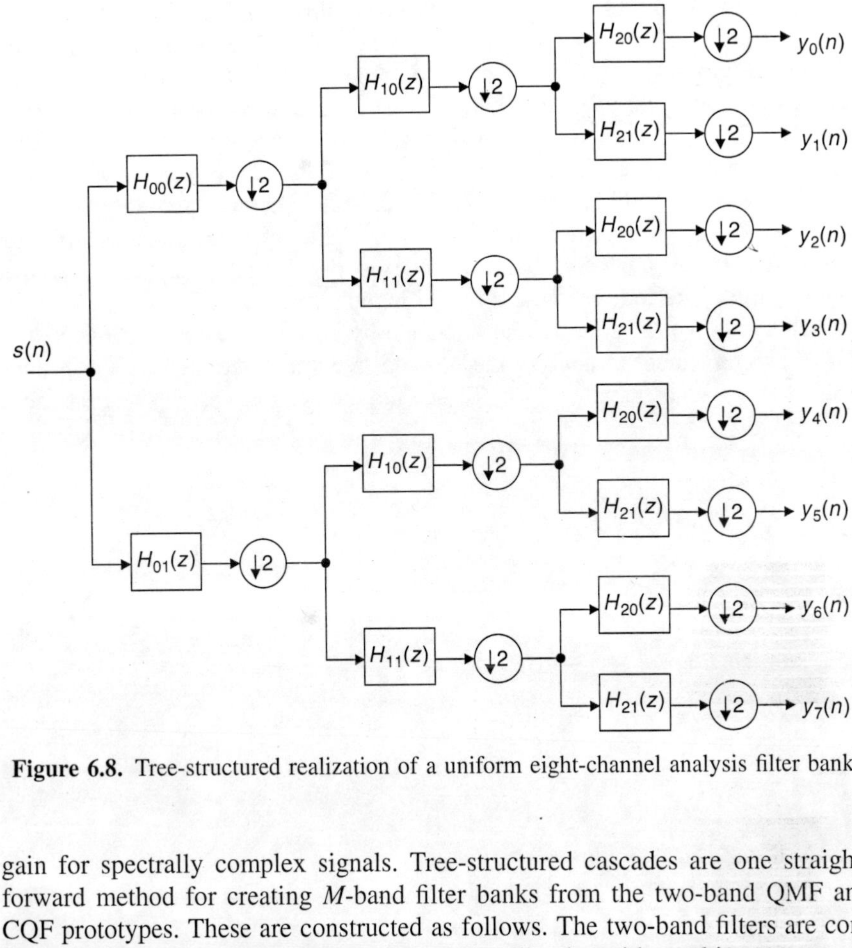

Figure 6.8. Tree-structured realization of a uniform eight-channel analysis filter bank.

gain for spectrally complex signals. Tree-structured cascades are one straightforward method for creating M-band filter banks from the two-band QMF and CQF prototypes. These are constructed as follows. The two-band filters are connected in a cascade that can be represented well using either a binary tree or a pruned binary tree. The root node of the tree is formed by a single two-band QMF or CQF section. Then, each of the root node outputs is connected to a cascaded QMF or CQF bank. The cascade structure may be continued to the depth necessary to achieve the desired magnitude response characteristics. At each node in the tree, a two-channel QMF or CQF bank operates on the output from a higher-level two-channel bank. Thus, frequency subdivision occurs through a series of two-band splits. Tree-structured filter banks have several advantages. First of all, the designer can approximate an arbitrary partitioning of the frequency axis by creating an appropriate cascade. Consider, for example, the uniform subband tree (Figure 6.8) or the octave-band tree (Figure 6.9). The ability to partition the frequency axis in a nonuniform manner also has implications for multi-resolution temporal analysis, or nonuniform tiling of the time-frequency plane. This property can be advantageous if the ultimate objective is to approximate the analysis properties of the human ear, and in fact many algorithms make use of tree-structured filter banks for this very reason [Bran90] [Sinh93b]

[Tsut98]. In addition, the designer has the flexibility to optimize the length and other properties of constituent filters at each node in the tree. This flexibility has been exploited to enhance the performance of several experimental audio codecs [Sinh93b] [Phil95a] [Phil95b]. Tree-structured filter banks are also attractive for their computational efficiency relative to other M-band techniques. One disadvantage of the tree-structured filter bank is that delays accumulate through the cascaded nodes and hence the overall delay can become quite large.

The example tree in Figure 6.8 shows an eight-band uniform analysis filter bank in which the analysis filters are indexed first by level and then by type. Lowpass filters are indexed with a 0, and highpass with a 1. For instance, highpass filters at level 2 in the tree are denoted by $H_{21}(z)$, and lowpass by $H_{20}(z)$. It is often convenient to analyze the M-band tree-structured CQF or QMF bank using an equivalent parallel form. To see the connection between the cascaded

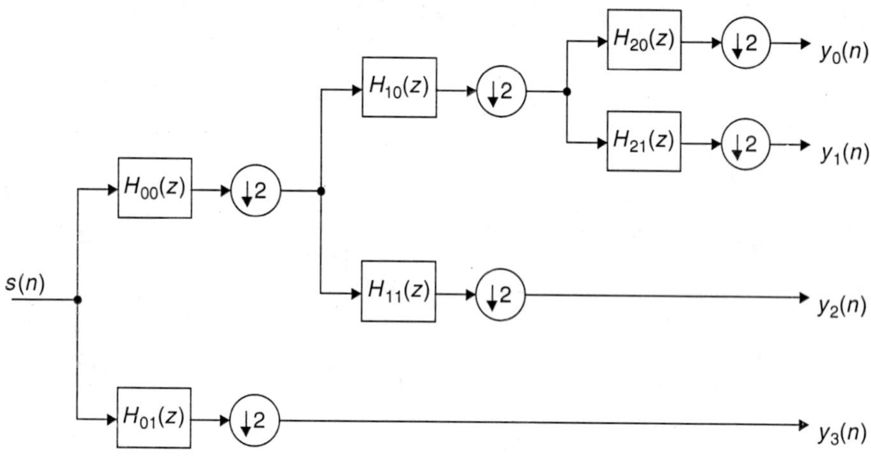

Figure 6.9. Tree-structured realization of an octave-band four-channel analysis filter bank.

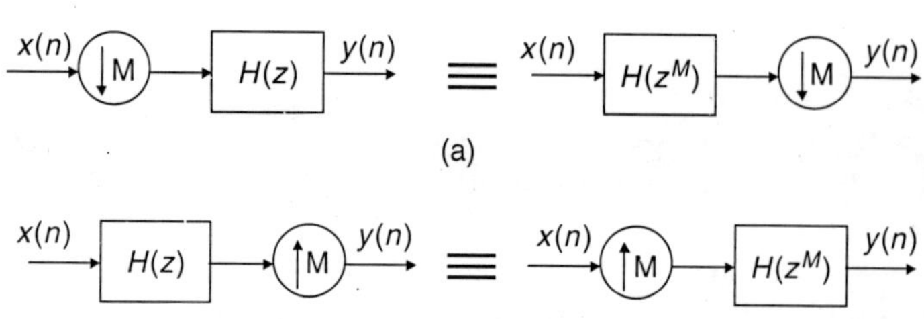

Figure 6.10. The Noble identities. In each picture, the structure on the left is equivalent to the structure on the right: (a) Interchange of a filter and a downsampler. The positions are swapped after the complex variable z is replaced by z^M in the system function, $H(z)$. (b) Interchange of a filter and an upsampler. The positions are swapped after the complex variable z is replaced by z^M in the system function, $H(z)$.

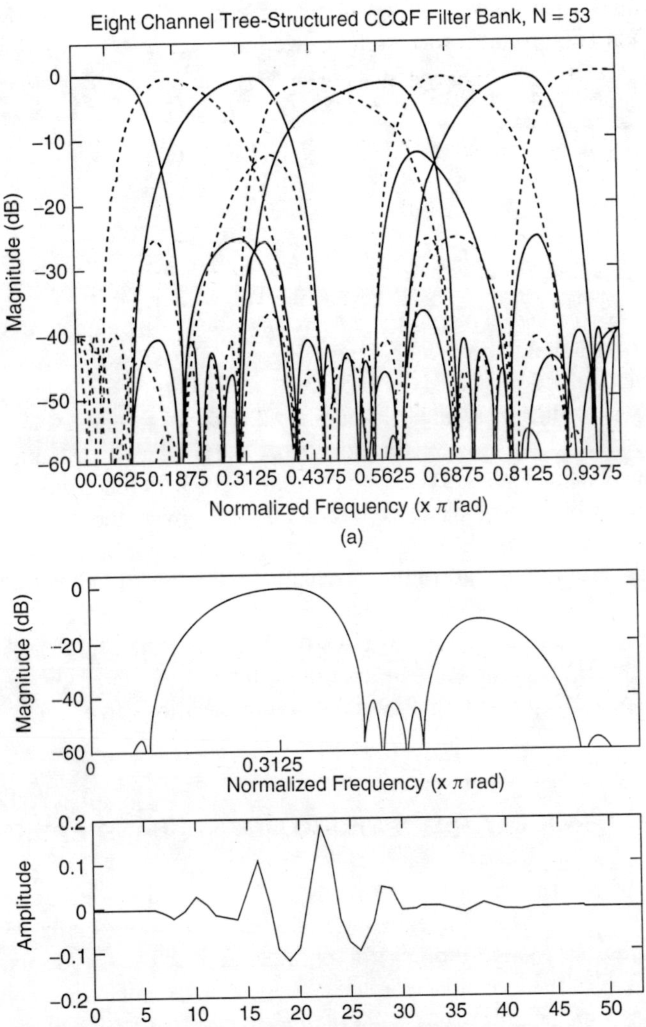

Figure 6.11. Eight-channel cascaded CQF (CCQF) filter bank: (a) Magnitude frequency responses for all eight channels. Odd-numbered channel responses are drawn with dashed lines, even-numbered channel responses are drawn with solid lines. (b) Isolated view of the magnitude frequency response and time-domain impulse response for channel 3. This view highlights the presence of a significant sidelobe in the stop-band In this figure, N is the length of the impulse response.

tree structures (Figures 6.8 and 6.9) and the parallel analysis-synthesis structure (Figure 6.1), one can apply the “noble identities” (Figure 6.10), which allow for the interchange of the down-sampling and filtering operations. In a straightforward manner, this practice collapses the cascaded filter transfer functions into single parallel-form analysis filters for each channel. In the case of the eight-channel bank (Figure 6.8), for example, we have

$$\begin{aligned} H_0(z) &= H_{00}(z)H_{10}(z^2)H_{20}(z^4) \\ H_1(z) &= H_{00}(z)H_{10}(z^2)H_{21}(z^4) \\ &\vdots \\ H_7(z) &= H_{01}(z)H_{11}(z^2)H_{21}(z^4) \end{aligned} \tag{6.11}$$

Figure 6.11(a) shows the magnitude spectrum of an eight-channel, tree-structured filter bank based on a three-level cascade (Figure 6.8) of the same Smith-Barnwell CQF filter examined previously (Figure 6.7). Even-numbered channel responses are drawn with solid lines, and odd-numbered channel responses are drawn with dashed lines. One can observe the effect that the cascaded structure has on the shape of the channel responses. Bands 0 through 3 are each uniquely shaped, and are mirror images of bands 4 through 7. Moreover, the M-band stop-band characteristics are significantly different than the prototype filter, i.e., the equiripple property does not extend to the M-channels. Figure 6.11(b) shows $|H_2(\Omega)|^2$, making it is possible to observe clearly a sidelobe of significant magnitude in the stop-band. The figure also illustrates the impulse response, $h_2(n)$, associated with the filter $H_2(z)$. One can see that the effective length of the parallel-form impulse response represents the cumulative contributions from each of the cascaded filters.

6.6 COSINE MODULATED "PSEUDO QMF" *M*-BAND BANKS

A tree-structured cascade of two-channel prototypes is only one of several well-known methods available for realization of an M-band filter bank. Although the tree structures offer opportunities for optimization at each node and are conceptually simple, the potential for long delay and irregular channel responses is sometimes unappealing. As an alternative to the tree-structured architecture, cosine modulation of a lowpass prototype filter has been used since the early 1980s [Nuss81] [Roth83] [Chu85] [Mass85] [Cox86] to realize parallel M-channel filter banks with nearly perfect reconstruction. Because they do not achieve perfect reconstruction, these filter banks are known collectively as "pseudo QMF," and they are characterized by several attractive properties:

- Constrained design; single FIR prototype filter
- Uniform, linear phase channel responses
- Overall linear phase; hence constant group delay
- Low complexity, i.e., one filter plus modulation
- Amenable to fast block algorithms
- Critical sampling.

In the pseudo QMF (PQMF) bank derivation phase distortion is completely eliminated from the overall transfer function, Eq. (6.5), by forcing the analysis and synthesis filters to satisfy the mirror image condition

$$g_k(n) = h_k(L-1-n) \tag{6.12}$$

Moreover, adjacent channel aliasing is cancelled by establishing precise relationships between the analysis and synthesis filters, $H_k(z)$ and $G_k(z)$, respectively. In the critically sampled analysis-synthesis notation of Figure 6.1, these conditions ultimately yield analysis filters given by

$$h_k(n) = 2w(n)\cos\left[\frac{\pi}{M}(k+0.5)\left(n-\frac{(L-1)}{2}\right)+\Omega_k\right] \tag{6.13}$$

and synthesis filters given by

$$g_k(n) = 2w(n)\cos\left[\frac{\pi}{M}(k+0.5)\left(n-\frac{(L-1)}{2}\right)-\Omega_k\right] \tag{6.14}$$

$$\text{where, } \Omega_k = (-1)^k\frac{\pi}{4} \tag{6.15}$$

and the sequence $w(n)$ corresponds to the L-sample "window," a real-coefficient, linear phase FIR prototype lowpass filter, with normalized cutoff frequency $\pi/2M$. Given that aliasing and phase distortions have been eliminated in this formulation, the filter-bank design procedure is reduced to the design of the window, $w(n)$, such that overall amplitude distortion (Eq. (6.5)) is minimized. One approach [Vaid93] is to minimize a composite objective function, i.e.,

$$C = \alpha c_1 + (1-\alpha)c_2 \tag{6.16}$$

where constraint c_1, of the form

$$c_1 = \int_0^{\pi/M}\left[|W(\Omega)|^2 + \left|W\left(\Omega-\frac{\pi}{M}\right)\right|^2 - 1\right]^2 d\Omega \tag{6.17}$$

minimizes spectral nonflatness in the reconstruction, and constraint c_2, of the form

$$c_2 = \int_{\frac{\pi}{2M}+\varepsilon}^{\pi} |W(\Omega)|^2 d\Omega \tag{6.18}$$

maximizes stop-band attenuation. The parameter ε is related to transition bandwidth, and the parameter α determines which design constraint is more dominant.

The magnitude frequency response of an example eight-channel pseudo QMF bank designed using Eqs. (6.6) and (6.7) is shown in Figure 6.12. In contrast to the previous CCQF example, one can observe that all of the channel magnitude responses are identical, modulated versions of the lowpass prototype, and therefore the passband and stop-band characteristics are uniform. The impulse

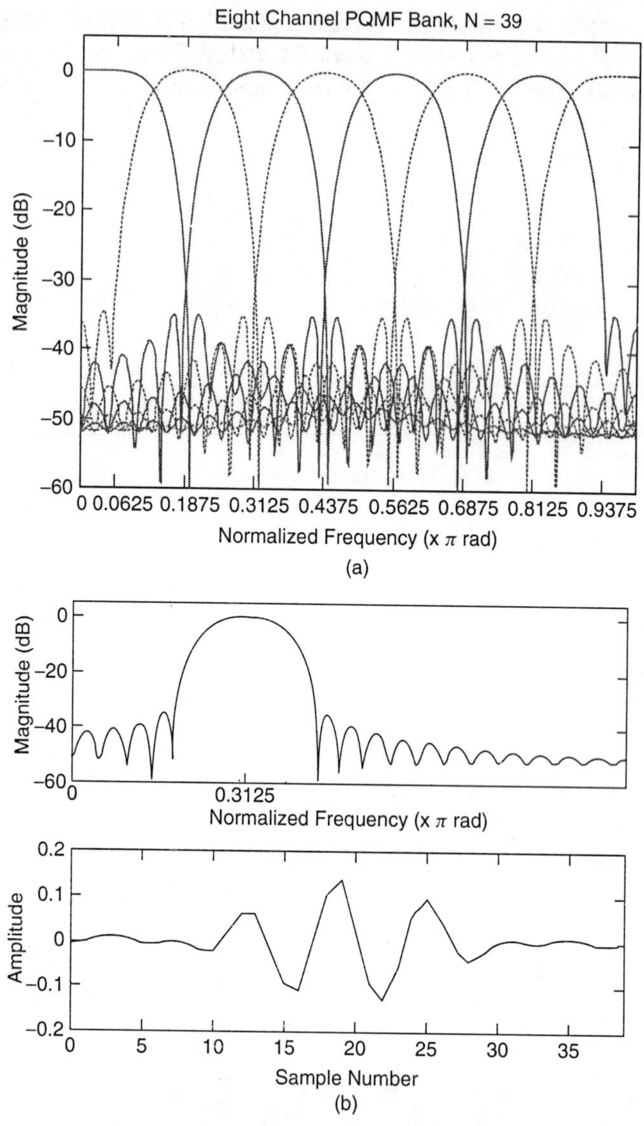

Figure 6.12. Eight-channel PQMF bank: (a) Magnitude frequency responses for all eight channels. Odd-numbered channel responses are drawn with dashed lines, even-numbered channel responses are drawn with solid lines. (b) Isolated view of the magnitude frequency response and time-domain impulse response for channel 3 Here, N is the impulse response length.

response symmetry associated with a linear phase filter is also evident in an examination of Figure 6.12(b).

The PQMF bank plays a significant role in several popular audio coding algorithms. In particular, the IS11172-3 and IS13818-3 algorithms ("MPEG-1"

[ISOI92] and "MPEG-2 BC/LSF" [ISOI94a]) employ a 32-channel PQMF bank for spectral decomposition in both layers I and II. The prototype filter, $w(n)$, contains 512 samples, yielding better than 96-dB sidelobe suppression in the stop-band of each analysis channel. Output ripple (non-PR) is less than 0.07 dB. In addition, the same PQMF is used in conjunction with a PR cosine modulated filter bank (discussed in the next section) in layer III to form a hybrid filter-bank architecture with time-varying properties. The MPEG-1 algorithm has reached a position of prominence with the widespread use of ".MP3" files (MPEG-1, layer 3) on the World Wide Web (WWW) for the exchange of audio recordings, as well as with the deployment of MPEG-1, layer II in direct broadcast satellite (DBS/DSS) and European digital audio broadcast (DBA) initiatives. Because of the availability of common algorithms for pseudo QMF and PR QMF banks, we defer the discussion on generic complexity and efficient implementation strategies until later. In the particular case of MPEG-1, however, note that the 32-band pseudo QMF analysis bank as defined in the standard requires approximately 80 real multiplies and 80 real additions per output sample [ISOI92], although a more efficient implementation based on a fast algorithm for the DCT was also proposed [Pan93] [Kons94].

6.7 COSINE MODULATED PERFECT RECONSTRUCTION (PR) *M*-BAND BANKS AND THE MODIFIED DISCRETE COSINE TRANSFORM (MDCT)

Although PQMF banks have been used quite successfully in perceptual audio coders, the overall system design still must compensate for the inherent distortion induced by the lack of perfect reconstruction to avoid audible artifacts in the codec output. The compensation strategy may be a simple one (e.g., increased prototype filter length), but perfect reconstruction is actually preferable because it constrains the sources of output distortion to the quantization stage. Beginning in the early 1990s, independent work by Malvar [Malv90b], Ramstad [Rams91], and Koilpillai and Vaidyanathan [Koil91] [Koil92] showed that generalized perfect reconstruction (PR) cosine modulated filter banks are possible by appropriately constraining the prototype lowpass filter, $w(n)$, and synthesis filters, $g_k(n)$, for $0 \leqslant k \leqslant M-1$. In particular, perfect reconstruction is guaranteed for a cosine-modulated filter bank with analysis filters, $h_k(n)$, given by Eqs. (6.13) and (6.15) if *four* conditions are satisfied. First, the length, L, of the window, $w(n)$, must be integer multiple of the number of subbands, i.e.,

1.
$$L = 2mM \tag{6.19}$$

where the parameter m is an integer greater than zero. Next, the synthesis filters, $g_k(n)$, must be related to the analysis filters by a time-reversal, such that

2.
$$g_k(n) = h_k(L-1-n) \tag{6.20}$$

In addition, the FIR lowpass prototype must have linear phase, which means that

3.

$$w(n) = w(L - 1 - n) \tag{6.21}$$

and, finally, the polyphase components of $w(n)$ must satisfy the pairwise power complementary requirement, i.e.,

4.

$$\tilde{E}_k(z)E_k(z) + \tilde{E}_{M+k}(z)E_{M+k}(z) = \alpha \tag{6.22}$$

where the constant α is greater than 0, the functions $E_k(z)$ are the $k = 0, 1, 2, \ldots, M - 1$ polyphase components (Eq. (6.9)) of $W(z)$, and the tilde notation denotes the *paraconjugate*, i.e.,

$$\tilde{E}(z) = E * (z^{-1}) \tag{6.23}$$

or, in other words, the coefficients of $E(z)$ are conjugated, and then the complex variable z^{-1} is substituted for the complex variable z.

The generalized PR cosine-modulated filter banks developed in Eqs.(6.19) through (6.23) are of considerable interest in many applications. This Section, however, concentrates on the special case that has become of central importance in the advancement of perceptual audio coding algorithms, namely, the filter bank for which $L = 2M$, i.e., $m = 1$. The PR properties of this special case were first demonstrated by Princen and Bradley [Prin86] using time-domain arguments for the development of the time domain aliasing cancellation (TDAC) filter bank. Later, Malvar [Malv90a] developed the modulated lapped transform (MLT) by restricting attention to a particular prototype filter and formulating the filter bank as a lapped orthogonal block transform. More recently, the consensus name in the audio coding literature for lapped block transform interpretation of this special case filter bank has evolved into the modified discrete cosine transform (MDCT). To avoid confusion, we will denote throughout this book by MDCT the PR cosine-modulated filter bank with $L = 2M$, and we will restrict the window, $w(n)$, in accordance with Eqs. (6.19) and (6.21). In short, the reader should be aware that the different acronyms TDAC, MLT, and MDCT all refer essentially to the same PR cosine modulated filter bank. Only Malvar's MLT label implies a particular choice for $w(n)$, as described below. From the perspective of an analysis-synthesis filter bank (Figure 6.1), the MDCT analysis filter impulse responses are given by

$$h_k(n) = w(n)\sqrt{\frac{2}{M}}\cos\left[\frac{(2n + M + 1)(2k + 1)\pi}{4m}\right] \tag{6.24}$$

and the synthesis filters, to satisfy the overall linear phase constraint, are obtained by a time reversal, i.e.,

$$g_k(n) = h_k(2M - 1 - n) \tag{6.25}$$

This perspective is useful for visualizing individual channel characteristics in terms of their impulse and frequency responses. In practice, however, the MDCT is typically realized as a block transform, usually via a fast algorithm. The remainder of this section treats several MDCT facets that are of importance in audio coding applications, including its forward and inverse transform interpretations, prototype filter (window) design criteria, window design examples, time-varying forms, and fast algorithms.

6.7.1 Forward and Inverse MDCT

The analysis filter bank (Figure 6.13(a)) is realized as a block transform of length $2M$ samples, while using a block advance of only M samples, i.e., with 50% overlap between blocks. Thus, the MDCT basis functions extend across two blocks in time, leading to virtual elimination of the blocking artifacts that plague the reconstruction of nonoverlapped transform coders. Despite the 50% overlap, however, the MDCT is still critically sampled, and only m coefficients are generated by the forward transform for each $2M$-sample input block. Given an input block, $x(n)$, the transform coefficients, $X(k)$, for $0 \leqslant k \leqslant M-1$, are obtained by means of the forward MDCT, defined as

$$X(k) = \sum_{n=0}^{2M-1} x(n)h_k(n). \tag{6.26}$$

Clearly, the forward MDCT performs a series of inner products between the M analysis filter impulse responses, $h_k(n)$, and the input, $x(n)$. On the other hand, the inverse MDCT (Figure 6.13(b)) obtains a reconstruction by computing a sum of the basis vectors weighted by the transform coefficients from two blocks. The first M samples of the k-th basis vector, for $h_k(n)$, $0 \leqslant n \leqslant M-1$, are weighted by k-th coefficient of the current block, $X(k)$. Simultaneously, the second M samples of the k-th basis vector, $h_k(n)$, for $M \leqslant n \leqslant 2M-1$, are weighted by the k-th coefficient of the previous block, $X^P(K)$. Then, the weighted basis vectors are overlapped and added at each time index, n. Note that the extended basis functions require that the inverse transform maintains an M sample memory to retain the previous set of coefficients. Thus, the reconstructed samples $x(n)$, for $0 \leqslant n \leqslant M-1$, are obtained via the inverse MDCT, defined as

$$x(n) = \sum_{k=0}^{M-1} [X(k)h_k(n) + X^P(k)h_k(n+M)], \tag{6.27}$$

where $x^P(k)$ denotes the previous block of transform coefficients. The overlapped analysis and overlap-add synthesis processes are illustrated in Figure 6.13(a) and Figure 6.13(b), respectively.

6.7.2 MDCT Window Design

Given the forward (Eq. (6.26)) and inverse (Eq. (6.27)) transform definitions, one still must design a suitable FIR prototype filter (window), $w(n)$. Several general

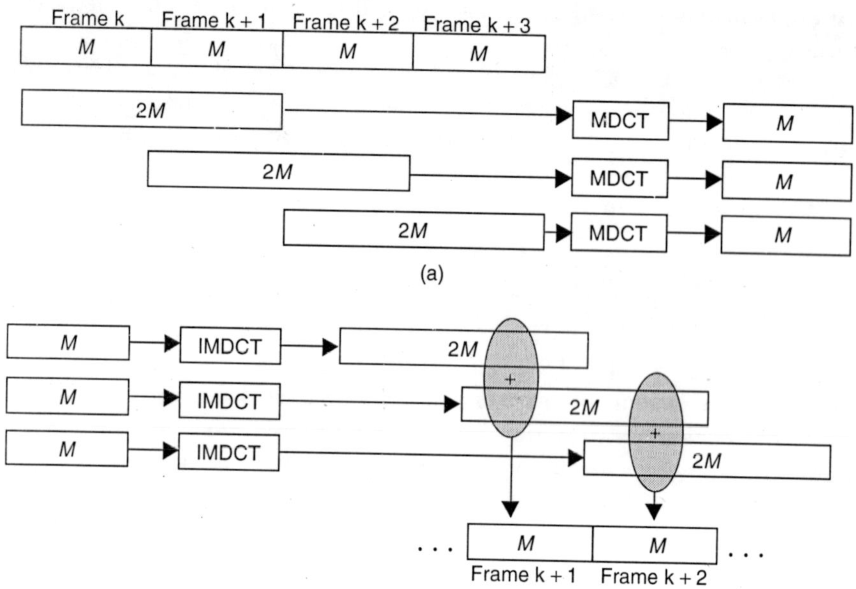

Figure 6.13. Modified discrete cosine transform (MDCT): (a) Lapped forward transform (analysis) – 2M samples are mapped to M spectral components (Eq. (6.26)). Analysis block length is 2M samples, but analysis stride (hop size) and time resolution are M-samples. (b) Inverse transform (synthesis) – M spectral components are mapped to a vector of 2M samples (Eq. (6.27)) that is overlapped by M samples and added to the vector of 2M samples associated with the previous frame.

purpose orthogonal [Prin86] [Prin87] [Malv90a] and biorthogonal [Jawe95] [Smar95] [Matv96] windows that have been proposed, while still other orthogonal [USAT95a] [Ferr96a] [ISOI96a] [Fiel96] and biorthogonal [Cheu95] [Malv98] windows are optimized explicitly for audio coding. In the orthogonal case, the generalized PR conditions [Vaid93] given in Eqs. (6.19)–(6.23) can be reduced to linear phase and Nyquist constraints on the window, namely,

$$w(2M - 1 - n) = w(n) \tag{6.28a}$$

$$w^2(n) + w^2(n + M) = 1 \tag{6.28b}$$

for the sample indices $0 \leqslant n \leqslant M - 1$. These constraints give rise to two considerations. First, unlike the pseudo-QMF bank, linear phase in the MDCT lowpass prototype does not translate into linear phase for the modulated analysis filters on each subband channel. The overall MDCT analysis-synthesis filter bank, however, is characterized by perfect reconstruction and hence linear phase with a constant group delay of $L - 1$ samples. Secondly, although Eqs. (6.28a) and (6.28b) guarantee an orthogonal basis for the MDCT, an orthogonal basis is not required to satisfy the PR constraints in Eqs. (6.19)–(6.23). In fact, it can be shown [Cheu95] that Eq. (6.28b) can be revised, and that perfect reconstruction for the MDCT is still guaranteed as long as it is true that,

$$w_s(n) = \frac{w_a(n)}{w_a^2(n) + w_a^2(n+M)} \tag{6.29}$$

for the sample indices $0 \leqslant n \leqslant M-1$, where $w_s(n)$ denotes the synthesis window, and $w_a(n)$ denotes the analysis window. From the transform perspective, Eqs. (6.28a) and (6.29) guarantee a biorthogonal MDCT basis. Clearly, this relaxation of prototype FIR lowpass filter design requirements increases the degrees of freedom available to the filter bank designer from $M/2$ to M. In effect, it is no longer necessary to use the same analysis and synthesis windows. In any case, whether an orthogonal or biorthogonal basis is used, the MDCT window design problem can be formulated in the same manner as it was for the PQMF bank (Eq. (6.16)), except that the PR property of the MDCT eliminates the spectral flatness constraint (Eq. (6.17)), such that the designer can concentrate solely on minimizing either the stop-band energy or the maximum stop-band magnitude of $W(\Omega)$. Well-known tools are available (e.g., [Pres89]) for minimizing Eq. (6.16), but in many cases one can safely forego the design process and rely instead upon the general purpose orthogonal [Prin86] [Prin87] [Malv90a] or biorthogonal [Jawe95] [Smar95] [Matv96] MDCT windows that have been proposed in the literature. In fact, several existing orthogonal [Ferr96a] [ISOI96a] [USAT95a] and biorthogonal [Cheu95] [Malv98] transform windows were explicitly designed to be in some sense optimal for audio coding.

6.7.3 Example MDCT Windows (Prototype FIR Filters)

It is instructive to consider some example MDCT windows in order to appreciate more fully the characteristics well suited to audio coding, as well as the tradeoffs that are involved in the window selection process.

6.7.3.1 Sine Window Malvar [Malv90a] denotes by MLT the MDCT filter bank that makes use of the sine window, defined as

$$w(n) = \sin\left[\left(n + \frac{1}{2}\right)\frac{\pi}{2M}\right] \tag{6.30}$$

for $0 \leqslant n \leqslant M-1$. This particular window is perhaps the most popular in audio coding. It appears, for example, in the MPEG-1 layer III (MP3) hybrid filter bank [ISOI92], the MPEG-2 AAC/MPEG-4 time-frequency filter bank [ISOI96a], and numerous experimental MDCT-based coders that have appeared in the literature. In fact, this window has become the *de facto* standard in MDCT audio applications, and its properties are typically referenced as performance benchmarks when windows are proposed. The sine window (Figure 6.14) has several unique properties that make it advantageous. First, DC energy is concentrated in a single transform coefficient, because all basis functions except for the first one have infinite attenuation at DC. Secondly, the filter bank channels achieve 24 dB sidelobe attenuation when the sine window (Figure 6.14, dashed line) is used. Finally, the sine window has been shown [Malv90a] to make the MDCT asymptotically optimal in terms of coding gain for a lapped transform. Coding gain is desirable because it quantifies the factor by which the mean square error (MSE) is reduced when using the filter bank relative to using direct pulse code modulated (PCM) quantization of the time-domain signal at the same rate.

6.7.3.2 Parametric Phase-Modulated Sine Window Optimization criteria other than coding gain or DC localization are possible and have also been investigated for the MDCT. Ferreira [Ferr96a] proposed a parametric window for the orthogonal MDCT that offers a controlled tradeoff between reduction of the time-domain ringing artifacts produced by coarse quantization and reduction of stop-band leakage relative to the sine window. The window (Figure 6.14, solid), which is defined in terms of three parameters for any value of M, i.e.,

$$w(n) = \sin\left[\left(n + \frac{1}{2}\right)\frac{\pi}{2M} + \phi_{opt}(n)\right] \tag{6.31}$$

$$\text{where,} \quad \phi_{opt}(n) = \frac{4\pi\beta}{(1-\delta^2)}\left[\left(\frac{4n}{2M-2}\right)^{\alpha} - \delta\right]\left[\left(\frac{4n}{2M-2}\right)^{\alpha} - 1\right] \tag{6.32}$$

was motivated by the observation that explicit simultaneous minimization of time-domain aliasing and stop-band energy resulted in a window well approximated by a nonlinear phase difference with respect to the sine window. Moreover, the parametric solution provided nearly optimal results and was tractable, while the explicit minimization was numerically unstable for long windows. Parameters are given in [Ferr96a] for three windows that offer, respectively, time-domain aliasing/stop-band leakage percentage improvements relative to the sine window of 6.3/10.1%, 8.3/0.7%, and 13.3/−31%. Figure 6.14 compares the latter parametric window ($\beta = 0.03125$, $\alpha = 0.92$, $\delta = 0.0$) in both time and frequency against the sine window. It can be seen that the negative gain in stop-band attenuation is caused by a slight increase in the first sidelobe energy. It is also clear, however, that the stop-band attenuation characteristics improve with increasing frequency. In fact, the Ferreira window has a broader range of better than 110 dB attenuation than does the sine window. This characteristic of improved ultimate stop-band rejection can be beneficial for perceptual gain, particularly for strongly harmonic signals.

6.7.3.3 Separate Analysis and Synthesis Windows – Biorthogonal MDCT Basis Even more dramatic improvements in ultimate stop-band rejection are possible when the orthogonality constraint is removed. Cheung and Lim [Cheu95] derived for the MDCT the biorthogonality window constraint given by Eq. (6.29), and then demonstrated with a Kaiser analysis window, $w_a(n)$, the potential for improved stop-band attenuation. In a similar fashion, Figure 6.15(a) shows the analysis (solid) and synthesis (dashed) windows that result for the biorthogonal MDCT when $w_a(n)$ is a Kaiser window [Oppe99] with $\beta = 11$. The most significant benefit of this arrangement is apparent from the frequency response plot for

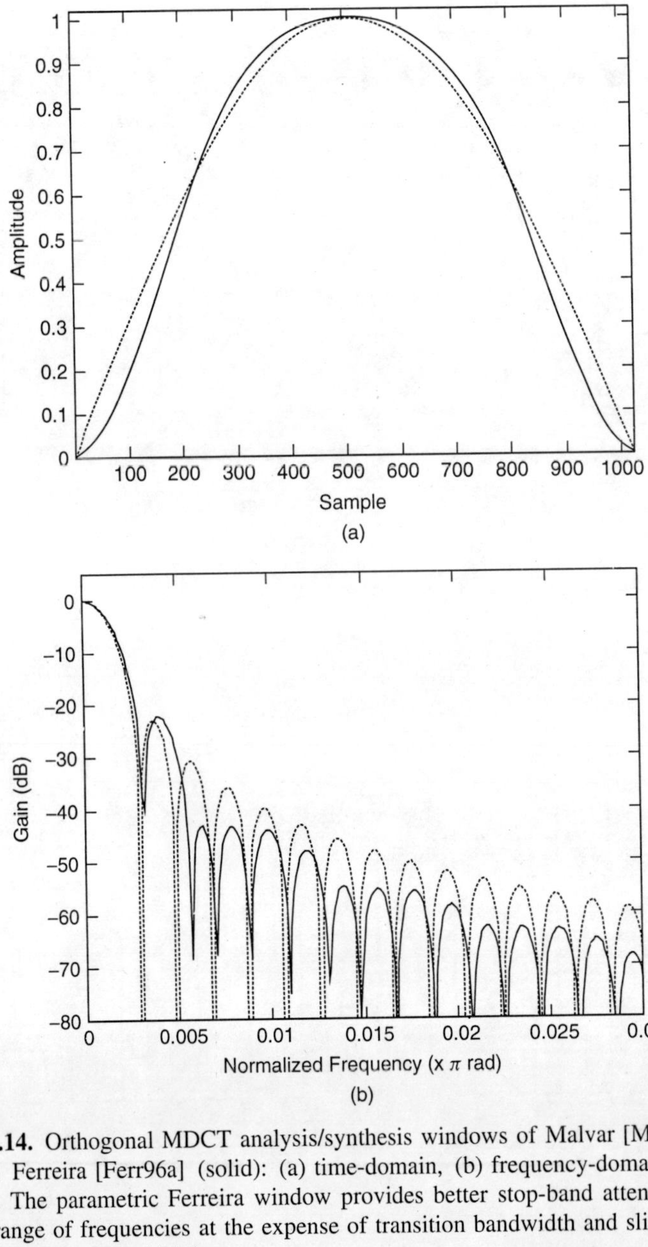

Figure 6.14. Orthogonal MDCT analysis/synthesis windows of Malvar [Malv90a] (dashed) and Ferreira [Ferr96a] (solid): (a) time-domain, (b) frequency-domain magnitude response. The parametric Ferreira window provides better stop-band attenuation over a broader range of frequencies at the expense of transition bandwidth and slightly reduced attenuation of the first sidelobe.

two 256-channel filter banks depicted in Figure 6.15(b). In this figure, the dashed line represents the frequency response associated with channel four of the sine window MDCT, and the lighter solid line corresponds to the frequency response associated with the same channel in the Kaiser window MDCT filter bank. Also

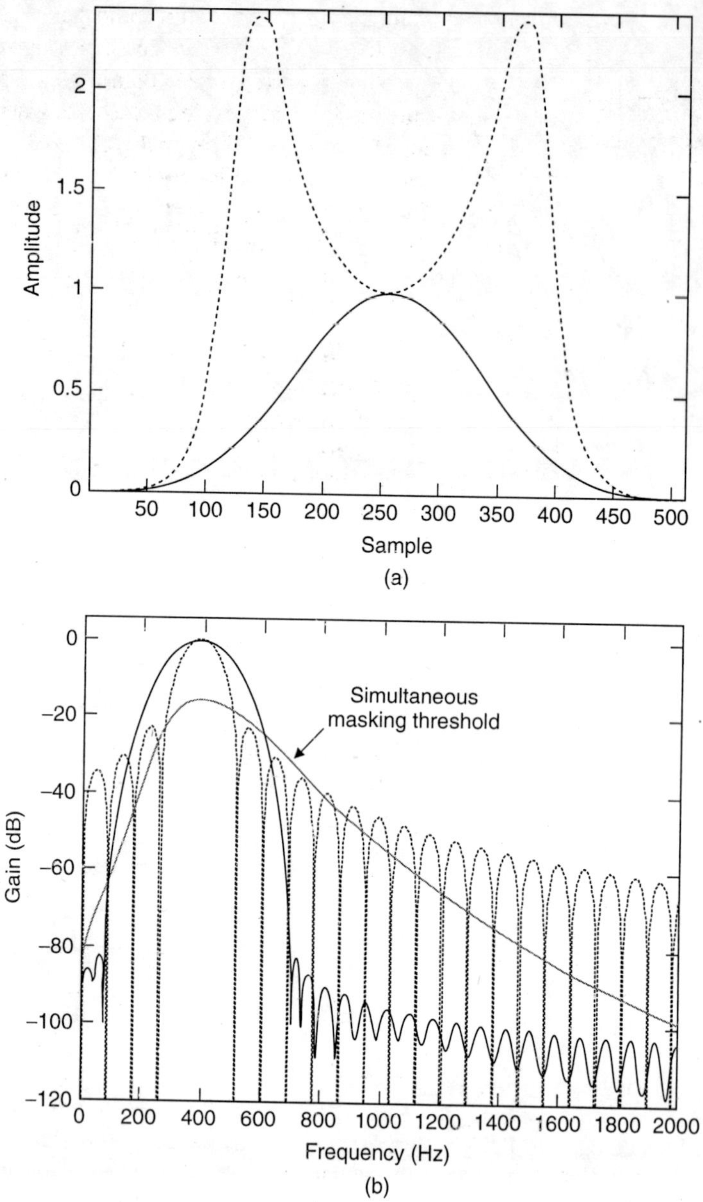

Figure 6.15. Biorthogonal MDCT basis: (a) Time-domain view of the analysis (solid) and synthesis (dashed) windows (Cheung and Lim [Cheu95]) that are associated with a biorthogonal MDCT basis. (b) Frequency-domain magnitude responses associated with MDCT channel four of 256 for the sine (orthogonal basis dashed) and Kaiser (biorthogonal basis solid) windows. Simultaneous masking threshold is superimposed for a pure tone occurring at the channel center frequency, 388 Hz. Picture demonstrates the potential for super-threshold leakage associated with the sine window and the improved stop-band attenuation realized with the Kaiser window.

superimposed on the plot is the simultaneous masking threshold generated by a 388-Hz pure tone occurring at the channel center. It can be seen that although the main lobe for the Kaiser MDCT is somewhat broader than the sine MDCT, the stop-band attenuation is significantly below the masking threshold, whereas the sine window MDCT stop-band leakage has substantial super-threshold energy. The sine window, therefore, has the potential to cause artificially high bit rates because of its greater leakage. This type of artifact motivated the designers of the Dolby AC-2/AC-3 [USAT95a] and MPEG-2 AAC/MPEG-4 T-F [ISOI96a] algorithms to use customized windows rather than the standard sine window in their respective orthogonal MDCT filter banks.

6.7.3.4 The Dolby AC-2/Dolby AC-3/MPEG-2 AAC KBD Window The Kaiser-Bessel Derived (KBD) window was obtained in a procedure devised at Dolby Laboratories. The AC-2 and AC-3 designers showed [Fiel96] that the prototype filter for an M-channel orthogonal MDCT filter bank satisfying the PR conditions (Eqs. (6.28a) and (6.28b)) can be derived from any symmetric kernel window of length $M + 1$ by applying a transformation of the form

$$w_a(n) = w_s(n) = \sqrt{\frac{\sum_{j=0}^{n} v(j)}{\sum_{j=0}^{M} v(j)}}, 0 \leqslant n < M \tag{6.33}$$

where the sequence $v(n)$ represents the symmetric kernel. The resulting identical analysis and synthesis windows, $w_a(n)$ and $w_s(n)$, respectively, are of length $M + 1$ and symmetric, i.e., $w(2M - n - 1) = w(n)$. Note that although a more general form of Eq. (6.33) appeared [Fiel96], we have simplified it here for the particular case of the 50%-overlapped MDCT. During the development of the AC-2 and AC-3 algorithms, novel MDCT prototype filters optimized to satisfy a minimum masking template (e.g., Figure 6.16(a) for AC-3) were designed using Eq. (6.33) with a parametric Kaiser-Bessel kernel, $v(n)$. At the expense of some passband selectivity, the KBD windows achieve considerably better stop-band attenuation (greater than 40 dB improvement) than the sine window (Figure 6.16b). Thus, for a pure tone occurring at the center of a particular MDCT channel, the KBD filter bank concentrates more energy into a single transform coefficient. The remaining dispersed energy tends to generate coefficient magnitudes that lie below the worst-case pure tone excitation pattern ("masking template" (Figure 6.16b)). Particularly for signals with adequately spaced tonal components, the presence of fewer supra-threshold MDCT components reduces the perceptual bit allocation and therefore tends to improve coding gain. In spite of the reduced bit allocation, the filter bank still renders the quantization noise inaudible since the uncoded coefficients have smaller magnitudes than the masked threshold. A KBD filter bank simulation exemplifying this behavior for the MPEG-2 AAC algorithm is given later.

6.7.3.5 Parametric Windows for a Biorthogonal MDCT Basis In another example of biorthogonal window design, Malvar [Malv98] proposed the 'modulated biorthogonal lapped transform (MBLT),' a biorthogonal version of the

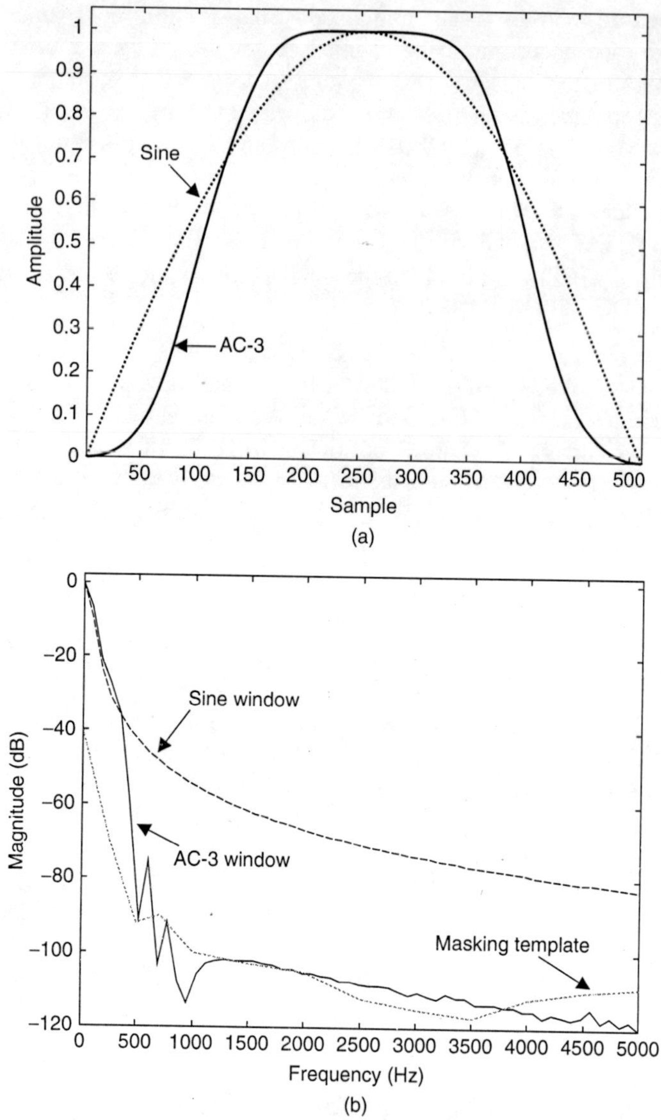

Figure 6.16. Dolby AC-3 (solid) *vs* sine (dashed) MDCT windows: (a) time-domain views, (b) frequency-domain magnitude responses in relation to worst-case masking template. Improved stop-band attenuation of the AC-3 (KBD) window is shown to approximate well the minimum masking template.

MDCT based on a parametric window, defined as

$$w_s(n) = \frac{1 - \cos\left[\left(\frac{n+1}{2M}\right)^{\alpha}\pi\right] + \beta}{2 + \beta} \tag{6.34}$$

for $0 \leqslant n \leqslant M - 1$. Like [Cheu95], Eq. (6.34) was also motivated by a desire to realize improved stop-band attenuation. Additionally, it was used to achieve good characteristics in a novel nonuniform filter bank based on a straightforward manipulation of the MBLT. In this design, the parameter α controls window width, while the parameter β controls its end values.

6.7.3.6 *Summary and Example Eight-Channel Filter Bank (MDCT) Using a Sine Window* The foregoing examples demonstrate that MDCT window designs are predominantly concerned with optimizing in some sense the trade-off between mainlobe width and stopband attenuation, as is true of any FIR filter design. We also note that biorthogonal MDCT extensions are a recent development and consequently most current audio coders incorporate primarily design innovations that have occurred within the orthogonal MDCT framework. To facilitate comparisons with the previously described filter bank methodologies (QMF, CQF, tree-structured QMF, pseudo-QMF, etc.), the analysis filter magnitude responses for an example eight-channel MDCT filter bank using the sine window are shown in Figure 6.17(a). Examination of the channel-3 impulse response in Figure 6.17(b) reveals the asymmetry that precludes linear phase for the analysis filters.

6.7.3.7 *Time-Varying Forms of the MDCT* One final point regarding MDCT window design is of particular relevance for perceptual audio coders. The earlier examples for tone-like and noise-like signals (Chapter 6, Section 6.2) demonstrated clearly that characteristics of the "best" filter bank for audio are signal-specific and therefore time-varying. In practice, it is very common for codecs using the MDCT (e.g., MPEG-1 [ISOI92a], MPEG-2 AAC [ISOI96a], Dolby AC-3 [USAT95a], Sony ATRAC [Tsut96], etc.) to change the number of channels and hence the window length to match the signal properties of the input. Typically, a binary classification scheme identifies the input as either stationary or nonstationary/transient. Then, a long window is used to maximize coding gain and achieve good channel separation during segments identified as stationary, or a short window is used to localize time-domain artifacts when pre-echoes are likely. Although the strategy has proven to be highly effective, it does complicate the codec structure. In particular, because of the time overlap between basis vectors, either boundary filters [Herl95] or special transitional windows [Herl93] are required to preserve perfect reconstruction when window switching occurs. Other schemes are also available to achieve perfect reconstruction with time-varying filter bank properties [Quei93] [Soda94] but for practical reasons these are not typically used. Consequently, window switching has been the method of choice. In this scenario, the transitional window function does not need to be symmetrical. It can be shown that the PR property is preserved as long as the transitional window satisfies the following constraints:

$$w^2(n) + w^2(M - n) = 1, n < M \tag{6.35a}$$

$$w^2(M + n) + w^2(2M - n) = 1, n \geqslant M \tag{6.35b}$$

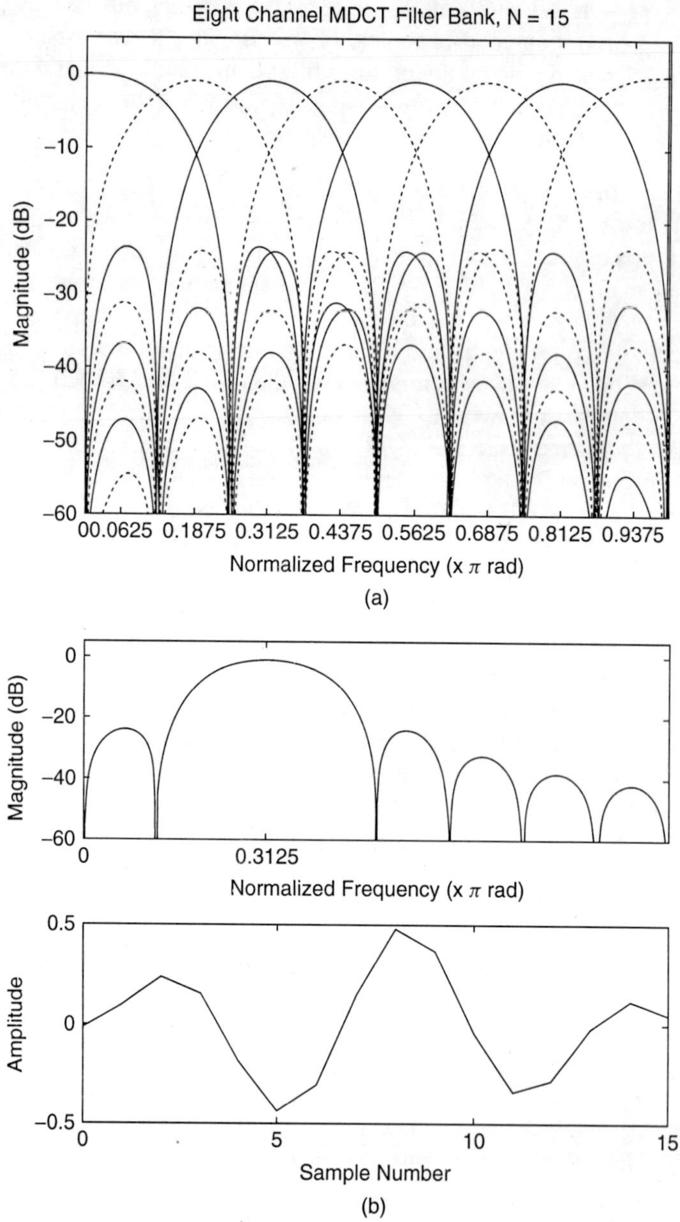

Figure 6.17. Eight-channel MDCT filter bank constructed with the sine window: (a) Magnitude frequency responses for all eight channels of the analysis filter bank. Odd-numbered channel responses are drawn with dashed lines, even-numbered channel responses are drawn with solid lines. (b) Isolated view of the magnitude frequency response and time-domain impulse response for channel 3. Asymmetry is clearly visible in the channel impulse response, precluding the possibility of linear phase, although the overall analysis-synthesis filter bank has linear phase on all channels.

and provided that the relationship between the transitional window and the adjoining, new length window obeys

$$w_1(M+n) = w_2(M-n), \tag{6.36}$$

where $w_1(n)$ and $w_2(n)$ are the left and right window functions, respectively. In spite of the preserved PR property, it should be noted that MDCT transitional windows are highly non ideal in the sense that they seriously impair the channel selectivity and stop-band attenuation of the filter bank.

The Dolby AC-3 algorithm as well as the MPEG MDCT-based coders employ MDCT window switching to maximize filter bank-to-signal matching. The MPEG-1 layer III and MPEG-2 AAC window switching schemes use transitional windows that are described in some detail later (Section 6.10). Unlike the MPEG approach, the AC-3 algorithm maintains perfect reconstruction while avoiding transitional windows. The AC-3 applies high-resolution frequency analysis to stationary signals using an MDCT as defined in Eqs. (6.26) and (6.27), with $M = 256$. During transient segments, a pair of two half-length transforms ($M = 128$), given by

$$X_1(k) = \sum_{n-0}^{2M-1} x(n)h_{k,1}(n) \tag{6.37a}$$

$$X_2(k) = \sum_{n=0}^{2M-1} x(n+2M)h_{k,2}(n+2M) \tag{6.37b}$$

replaces the single long-block transform, and the short block filter impulse responses, $h_{k,1}$, and $h_{k,2}$, are defined as

$$h_{k,1}(n) = w(n)\sqrt{\frac{2}{M}}\cos\left[\frac{(2n+1)(2k+1)\pi}{4M}\right] \tag{6.38a}$$

$$h_{k,2}(n) = w(n)\sqrt{\frac{2}{M}}\cos\left[\frac{(2n+2M+1)(2k+1)\pi}{4M}\right]. \tag{6.38b}$$

The window function, $w(n)$, remains identical for both the long and short transforms. Here, the key to maintaining the PR property is that the different phase shifts in Eqs. (6.38a) and (6.38b) relative to Eq. (6.24) guarantee an orthogonal basis. Also note that the AC-3 window is customized and incorporates into its design some perceptual properties [USAT95a]. The spectral and temporal analysis tradeoffs involved in transitional window designs are well illustrated in [Shli97] for both the MPEG-1 layer III [ISOI92a] and the Dolby AC-3 [USAT95a] filter banks.

6.7.3.8 *Fast Algorithms, Complexity, and Implementation Issues* One of the attractive properties that has contributed to the widespread use of the MDCT, particularly in the standards, is the availability of FFT-based fast algorithms [Duha91] [Sevi94] that make the filter bank viable for real-time applications. A unified fast algorithm [Liu98] is available for the MPEG-1, -2, -4, and AC-3 long block MDCT (Eq. (6.26) and Eq. (6.27)), the AC-3 short block MDCT (Eq. (6.38a) and Eq. (6.38b)), and the MPEG-1 pseudo-QMF bank (Eq. (6.13) and Eq. (6.14)). The computational load of [Liu98] for an $M = 1024$ (2048-point) MDCT (e.g., MPEG-2 AAC, AT&T PAC), is 8,192 multiplies and 13,920 adds. This translates into complexity of $O(M \log_2 M)$ for multiplies and $O(2M \log_2 M)$ for adds. The complexity scales accordingly for other values of M. Both [Duha91] and [Liu98] exploit the fact that the forward MDCT can be decomposed into two cascaded stages (Figure 6.18), namely, a set of $M/2$ butterflies followed by an M-point discrete cosine transform (DCT). The inverse transform is decomposed in the inverse manner, i.e., a DCT followed by a butterfly network. In both cases, the butterflies capture the windowing behavior, while the DCT performs the modulation and filtering. The decompositions are efficient as well-known fast algorithms are available for the various DCT types [Rao90]. The butterflies are of low complexity, typically $O(2M)$ for both multiplies and adds. In addition to the computationally efficient algorithms of [Duha91] and [Liu98], a regressive structure suitable for parallel VLSI implementation of the Eq. (6.26) forward MDCT was proposed in [Chia96] with complexity of $3M$ adds and $2M$ multiplies per output for the forward transform and $3M$ adds and M multiplies for the inverse transform.

As far as other implementation issues are concerned, several researchers have addressed the quantization sensitivity of the MDCT. There are available

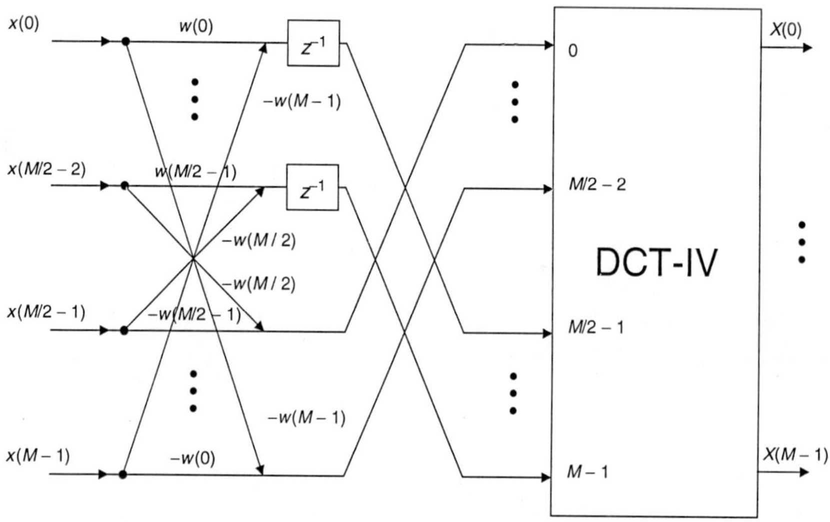

Figure 6.18. A fast algorithm for the $2M$-point (M-channel) forward MDCT (Eq. (6.26)) consists of a butterfly network and memory, followed by a Type IV DCT. The inverse structure can be formed to compute the inverse MDCT (Eq. (6.27)). Efficient FFT-based algorithms are available for the Type IV DCT.

expressions [Jako96] for the reconstruction error of the quantized system in terms of signal-correlated and uncorrelated components that can be used to assist algorithm designers in the identification and optimization of perceptually disturbing reconstruction artifacts induced by quantization noise. A more general treatment of quantization issues for PR cosine modulated filter banks has also appeared [Akan92].

6.7.3.9 Remarks on the MDCT The MDCT has become of central importance in audio coding, and the majority of standardized algorithms make some use of this filter bank. This section has traced the origins of the MDCT, reviewed common terminology and definitions, addressed the major window design issues, examined the strategies for time-varying implementations, and noted the availability fast algorithms for efficient realization. It has also provided numerous examples. The important properties of the MDCT filter bank can be summarized as follows:

- Perfect reconstruction
- Overlapping basis vectors
- Linear overall filter bank phase response
- Extended ringing artifacts due to quantization
- Critical sampling
- Virtual elimination of blocking artifacts
- Constant group delay $= L - 1$
- Nonlinear analysis filter phase responses
- Low complexity; one filter and modulation
- Orthogonal version, $M/2$ degrees of freedom for $w(n)$
- Amenable to time-varying implementations, with some performance sacrifices
- Amenable to fast algorithms
- Constrained design; a single FIR lowpass prototype filter
- Biorthogonal version, M degrees of freedom for $w_a(n)$ or $w_s(n)$

One can see from this synopsis that the MDCT possesses many of the qualities suitable for audio coding (Section 6.3). As a PR cosine-modulated filter bank, it inherits all of the advantages realized for the pseudo-QMF except for phase linearity on individual analysis channels, and it does so at the expense of less than a 5 dB reduction (typically) in stop-band attenuation. Moreover, the MDCT offers the added advantage that the number of parameters to be optimized in design of the lowpass prototype is essentially reduced to $M/2$ in the orthogonal case. If more freedom is desired, however, one can opt for the biorthogonal construction. Finally, we have presented the MDCT as both a filter bank and a block transform. To maintain consistency, we recognize the filter-bank/transform duality of some of the other tools presented in this chapter. Recall that the MDCT is a special case of the PR cosine modulated filter bank for which $L = 2mM$, and $m = 1$. Note, then, that the PQMF bank (Chapter 6, Section 6.6) can also be interpreted as a lapped transform for which it is possible to have $L = 2mM$. In the case of the MPEG-1 filter bank for layers I and II, for example, $L = 512$ and $M = 32$, or in other words, $m = 8$. As the coder architectures described in Chapters 7 through 10 will demonstrate, many filter banks for audio coding are most efficiently realized as block transforms, particularly when fast algorithms are available.

6.8 DISCRETE FOURIER AND DISCRETE COSINE TRANSFORM

This section offers abbreviated filter bank interpretations of the discrete Fourier transform (DFT) and the discrete cosine transform (DCT). These classical block transforms were often used to achieve high-resolution frequency analysis in the early experimental transform-based audio coders (Chapter 7) that preceded the adaptive spectral entropy coding (ASPEC), and ultimately, the MPEG-1 algorithms, layers I–III (Chapter 10). For example, the FFT realization of the DFT plays an important role in layer III of MPEG-1 (MP3). The FFT is embedded in efficient realizations of both MP3 hybrid filter bank stages (pseudo-QMF and MDCT), as well as in the spectral estimation blocks of the psychoacoustic models 1 and 2 recommended in the MPEG-1 standard [ISOI92]. It can be seen that block transforms are a special case of the more general uniform-band analysis-synthesis filter bank of Figure 6.1. For example, consider the unitary DFT and its inverse [Akan92], which can be written as, respectively,

$$X(k) = \frac{1}{\sqrt{2M}} \sum_{n=0}^{2M-1} x(n) W^{-nk}, 0 \leqslant k \leqslant 2M - 1 \tag{6.39a}$$

$$x(n) = \frac{1}{\sqrt{2M}} \sum_{k=0}^{2M-1} X(k) W^{nk}, 0 \leqslant n \leqslant 2M - 1, \tag{6. 39b}$$

where $W = e^{j\pi/M}$. If the analysis filters in Eq. (6.1) all have the same length and $L = 2M$, then the filter bank could be interpreted as taking contiguous L sample blocks of the input and applying to each block the transform in Eq. (6.39a). Although the DFT is usually defined with a block size of N instead of $2M$, Eqs. (6.39a) and (6.39b) are given using notation slightly different from the usual to remain consistent with the convention of this chapter, throughout which the number of filter bank channels is denoted by the parameter M. The DFT has conjugate symmetry for real signals, and thus from the audio filter bank perspective, effectively half as many channels as its block length. Also from the filter bank viewpoint, the impulse response of the k-th-channel analysis filter is given by the k-th DFT basis vector, i.e.,

$$h_k(n) = \frac{1}{\sqrt{2M}} W^{kn}, 0 \leqslant n \leqslant 2M - 1, 0 \leqslant k \leqslant M - 1 \tag{6.40}$$

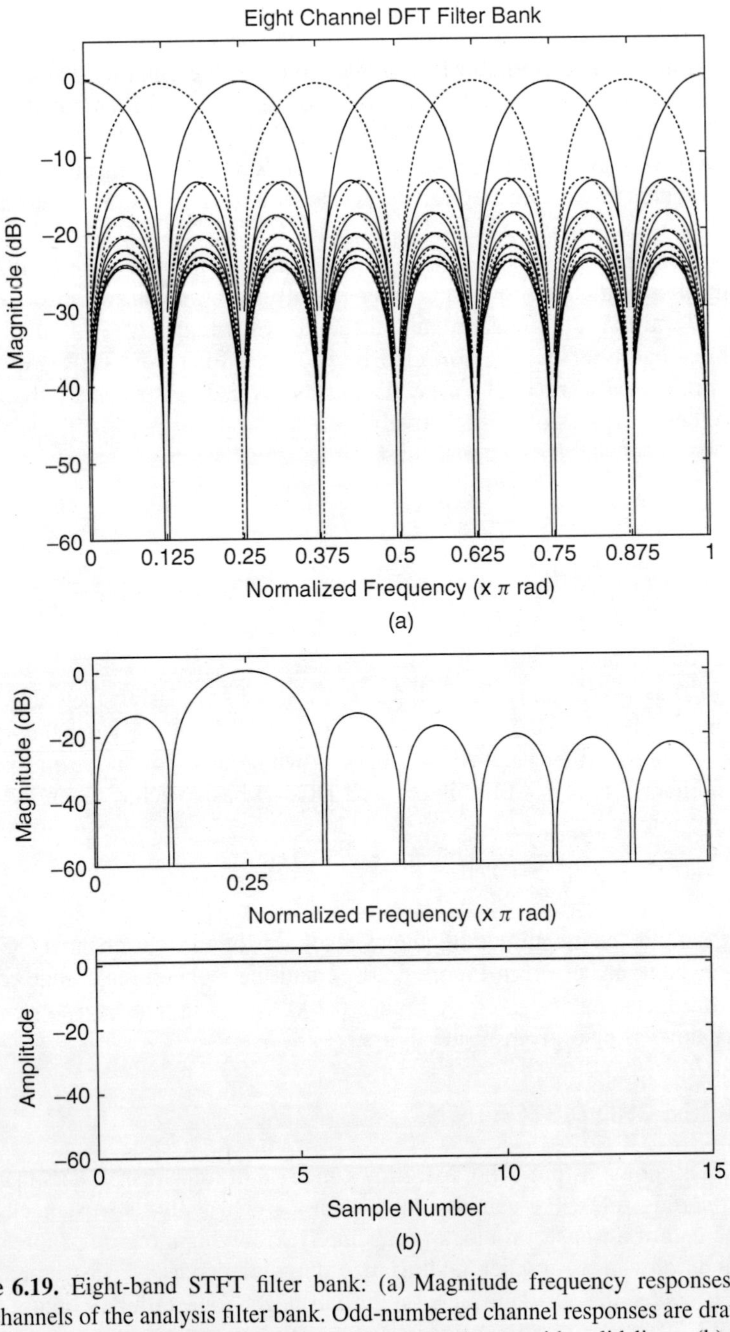

Figure 6.19. Eight-band STFT filter bank: (a) Magnitude frequency responses for all eight channels of the analysis filter bank. Odd-numbered channel responses are drawn with dashed lines, even-numbered channel responses are drawn with solid lines. (b) Isolated view of the magnitude frequency response and time-domain impulse response for channel 3. Note that the impulse response is complex-valued and that only its magnitude is shown in the figure.

An example eight-channel DFT analysis filter bank magnitude frequency response appears in Figure 6.19, with the magnitude response of the third channel magnified in Figure 6.19(b). The magnitude of the complex-valued impulse response for the same channel also appears in Figure 6.19(b). Note that the DFT filter bank is *evenly stacked*, whereas the cosine-modulated filter banks in Sections 6.6 and 6.7 of this chapter were *oddly stacked*. In other words, the center frequencies for the DFT analysis and synthesis filters occur at $k\pi/M$ for $0 \leqslant k \leqslant M-1$, while the center frequencies for the oddly stacked filters occur at $(2k+1)\pi/2M$ for $0 \leqslant k \leqslant M-1$. As is evident from Figure 6.19(a), even stacking means that the low-band filter is only half the bandwidth of the other channels, and that it "wraps-around" the fold-over frequency.

A filter bank perspective can also be provided for to the DCT. As a block transform, the forward DCT (Type II) and its inverse, are given by the analysis and synthesis expressions, respectively,

$$X(k) = c(k)\sqrt{\frac{2}{M}}\sum_{n=0}^{M-1} x(n)\cos\left[\frac{\pi}{M}\left(n+\frac{1}{2}\right)k\right], 0 \leqslant k \leqslant M-1 \quad (6.41a)$$

$$x(n) = \sqrt{\frac{2}{M}}\sum_{k=0}^{M-1} c(k)X(k)\cos\left[\frac{\pi}{M}\left(n+\frac{1}{2}\right)k\right], 0 \leqslant n \leqslant M-1, \quad (6.41b)$$

where $c(0) = 1/\sqrt{2}$, and $c(k) = 1$ for $1 \leqslant k \leqslant M-1$. Using the same duality arguments as for the DFT, one can view the DCT from the perspective of the analysis-synthesis filter bank (Figure 6.1), in which case the impulse response of the k-th-channel analysis filter is the k-th DCT-II basis vector, given by

$$h_k(n) = c(k)\sqrt{\frac{2}{M}}\cos\left[\frac{\pi}{M}\left(n+\frac{1}{2}\right)k\right], 0 \leqslant n, k \leqslant M-1. \quad (6.42)$$

As an example, the magnitude frequency responses of an eight-channel DCT analysis filter bank are given in Figure 6.20(a), and the isolated magnitude response of the third channel is given in Figure 6.20(b). The impulse response for the same channel is also given in the figure.

6.9 PRE-ECHO DISTORTION

An artifact known as pre-echo distortion can arise in transform coders using perceptual coding rules. Pre-echoes occur when a signal with a sharp attack begins near the end of a transform block immediately following a region of low energy. This situation can arise when coding recordings of percussive instruments such as the triangle, the glockenspiel, or the castanets for example (Figure 6.21a). For a block-based algorithm, when quantization and encoding are performed in order to satisfy the masking thresholds associated with the block average spectral estimate, time-frequency uncertainty dictates that the inverse transform will spread quantization distortion evenly in time throughout the reconstructed block

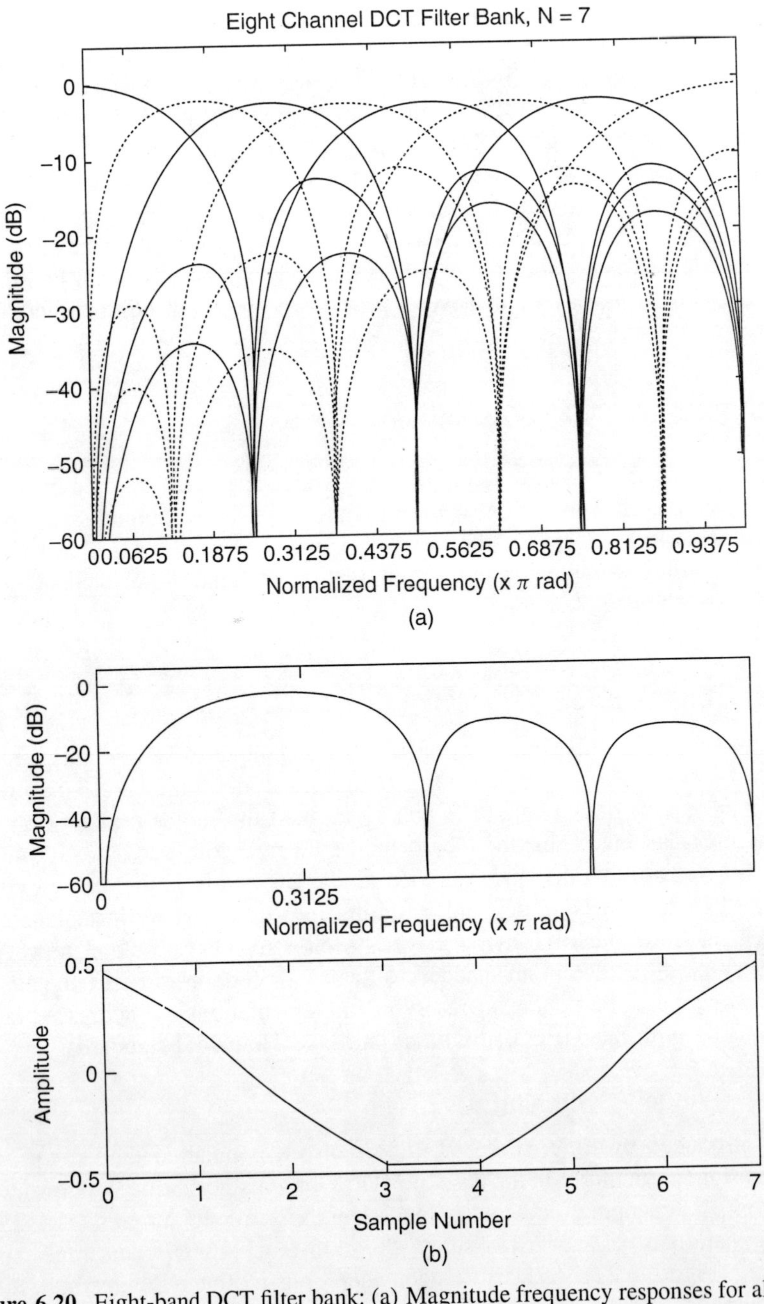

Figure 6.20. Eight-band DCT filter bank: (a) Magnitude frequency responses for all eight channels of the analysis filter bank. Odd-numbered channel responses are drawn with dashed lines, even-numbered channel responses are drawn with solid lines. (b) Isolated view of the magnitude frequency response and time-domain impulse response for channel 3.

(Figure 6.21b). This results in unmasked distortion throughout the low-energy region preceding in time the signal attack at the decoder. Although it has the potential to compensate for pre-echo, temporal premasking is possible only if the transform block size is sufficiently small (minimal coder delay). Percussive sounds are not the only signals likely to produce pre-echoes. Such artifacts also often plague coders when processing "pitched" signals containing nearly impulsive bursts at the beginning of each pitch period, e.g., the "German male speech" recording [Herr96]. For a male speaker with a fundamental frequency of 125 Hz, the interval between impulsive events is only 8 ms, which is much less than the typical analysis block length. Several methods proposed to eliminate pre-echoes are reviewed next.

6.10 PRE-ECHO CONTROL STRATEGIES

Several methodologies have been proposed and successfully applied in the effort to mitigate the pre-echoes that tend to plague block-based coding schemes. This section describes several of the most widespread techniques, including the bit reservoir, window switching, gain modification, switched filter banks, and temporal noise shaping. Advantages and drawbacks associated with each method are also discussed.

6.10.1 Bit Reservoir

Some coders [ISOI92] [John96c] utilize this technique to satisfy the greater bit demand associated with transients. Although most algorithms are fixed rate, the instantaneous bit rates required to satisfy masked thresholds on each frame are in fact time-varying. Thus, the idea behind a bit reservoir is to store surplus bits during periods of low demand, and then to allocate bits from the reservoir during localized periods of peak demand, resulting in a time-varying instantaneous bit rate but at the same time a fixed average bit rate. One problem, however, is that very large reservoirs are needed to deal with certain transient signals, e.g., "pitched signals." Particular bit reservoir implementations are addressed later in conjunction with the MPEG [ISOI92a] and PAC [John96c] standards.

6.10.2 Window Switching

First introduced by Edler [Edle89], this is also a popular method for pre-echo suppression, particularly in the case of MDCT-based algorithms. Window switching works by changing the analysis block length from long duration (e.g., 25 ms) during stationary segments to "short" duration (e.g., 4 ms) when transients are detected (Figure 6.22). At least two considerations motivate this method. First, a short window applied to the frame containing the transient will tend to minimize the temporal spread of quantization noise such that temporal premasking effects might preclude audibility. Secondly, it is desirable to constrain the high bit rates associated with transients to the shortest possible temporal regions. Although window switching has been successful [ISOI92] [John96c] [Tsut98], it also has

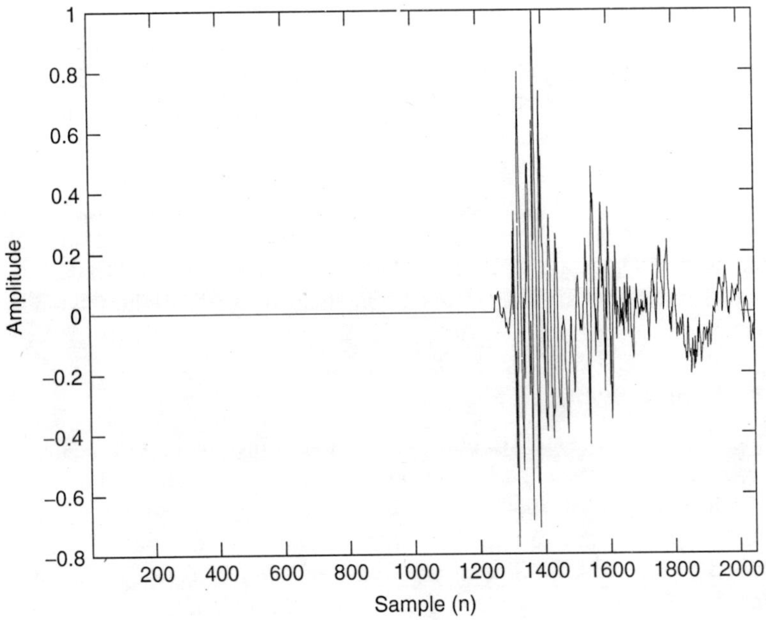

(a)

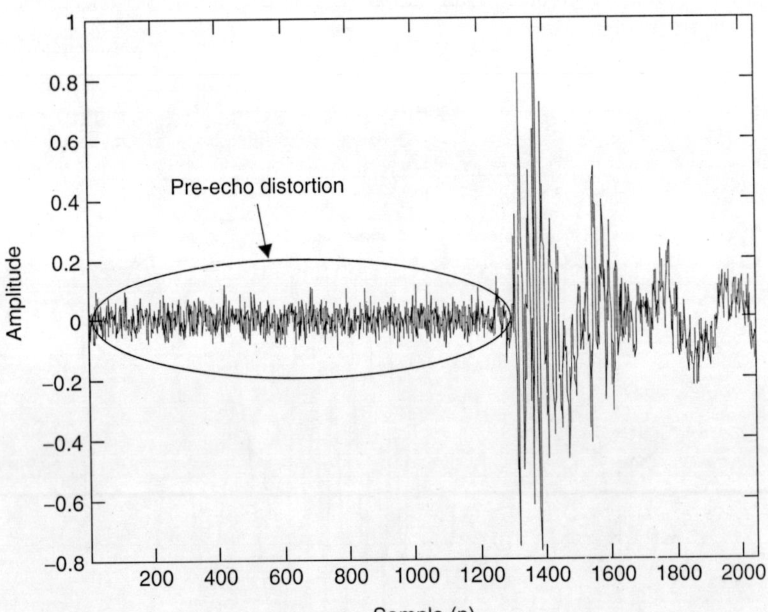

(b)

Figure 6.21. Pre-echo example (time-domain waveforms): (a) Uncoded castanets, (b) transform coded castanets, 2048-point block size. Pre-echo distortion is clearly visible in the first 1300 samples of the reconstructed signal.

significant drawbacks. For one, the perceptual model and lossless coding portions of the coder must support multiple time resolutions. This usually translates into increased complexity. Furthermore, most coders nowadays use lapped transforms such as the MDCT. To satisfy PR constraints, window switching typically requires transition windows between the long and short blocks. Even when suitable transition windows (Figure 6.22) satisfy the PR constraints, they do so at the expense of poor time and frequency localization properties [Shli97], resulting in reduced coding gain. Other difficulties inherent to window switching schemes are increased coder delay, undesirable latency for closely spaced transients (e.g., long-start-short-stop-start-short), and impractical overuse of short windows for "pitched" signals.

6.10.3 Hybrid, Switched Filter Banks

Window switching essentially relies upon a fixed filter bank with adaptive window lengths. In contrast, the hybrid and switched filter-bank architectures rely upon distinct filter bank modes. In hybrid schemes (e.g., [Prin95]), compatible filter-bank elements are cascaded in order to achieve the time-frequency tiling best suited to the current input signal. Switched filter banks (e.g., [Sinh96]), on the other hand, make hard switching decisions on each analysis interval in order to select a single monolithic filter bank tailored to the current input. Examples of these methods are given in later chapters, along with some discussion of their associated tradeoffs.

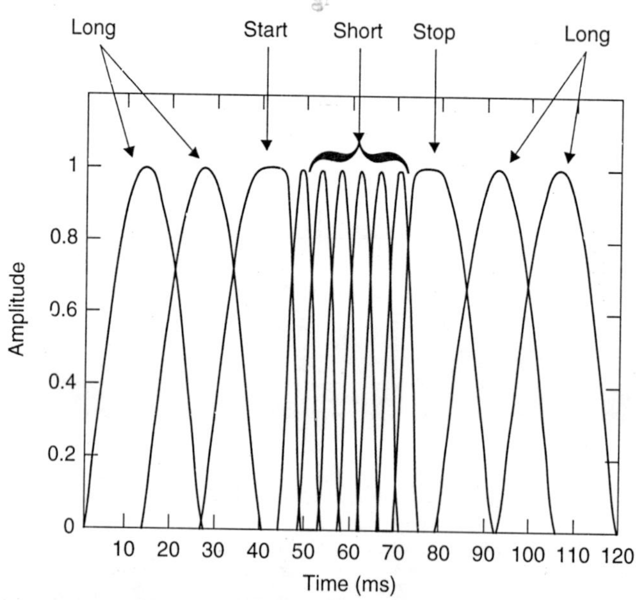

Figure 6.22. Example window switching scheme (MPEG-1, layer III or "MP3"). Transitional start and stop windows are required in between the long and short blocks to preserve the PR properties of the filter bank.

6.10.4 Gain Modification

The gain modification approach (Figure 6.23) has also shown promise in the task of pre-echo control [Vaup91] [Link93]. The gain modification procedure smoothes transient peaks in the time-domain prior to spectral analysis. Then, perceptual coding may proceed as it does for normal, stationary blocks. Quantization noise is shaped to satisfy masking thresholds computed for the equalized long block without compensating for an undesirable temporal spread of quantization noise. A time-varying gain and the modification time interval are transmitted as side information. Inverse operations are performed at the decoder to recover the original signal. Like the other techniques, caveats also apply to this method. For example, gain modification effectively distorts the spectral analysis time window. Depending upon the chosen filter bank, this distortion could have the unintended consequence of broadening the filter-bank responses at low frequencies beyond critical bandwidth. One solution for this problem is to apply independent gain modifications selectively within only frequency bands affected by the transient event. This selective approach, however, requires embedding of the gain blocks within a hybrid filter-bank structure, which increases coder complexity [Akag94].

6.10.5 Temporal Noise Shaping

The final pre-echo control technique considered in this section is temporal noise shaping (TNS). As shown in Figure 6.24, TNS [Herr96] is a frequency-domain technique that operates on the spectral coefficients, $X(k)$, generated by the analysis filter bank. TNS is applied only during input attacks susceptible to pre-echoes.

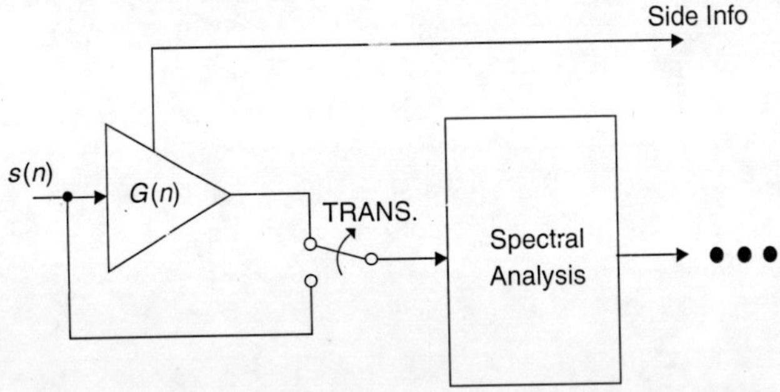

Figure 6.23. Gain modification scheme for pre-echo control.

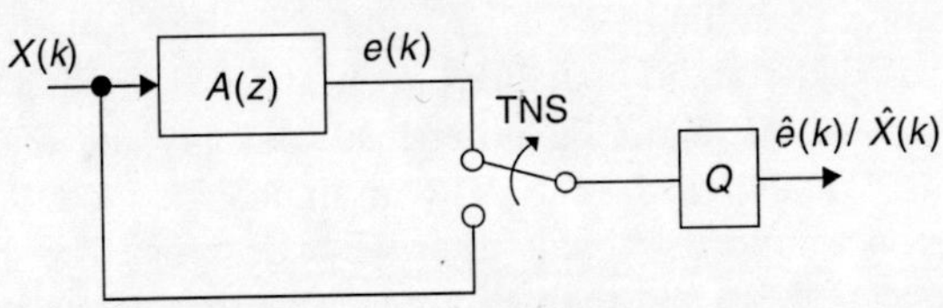

Figure 6.24. Temporal noise shaping scheme (TNS) for pre-echo control.

The idea is to apply linear prediction (LP) across frequency (rather than time), since for an impulsive time signal, frequency-domain coding gain is maximized using prediction techniques. The method works as follows. Parameters of a spectral LP synthesis filter, $A(z)$, are estimated via application of standard minimum MSE estimation methods (e.g., Levinson-Durbin) to the spectral coefficients, $X(k)$. The resulting prediction residual, $e(k)$, is quantized and encoded using standard perceptual coding according to the original masking threshold. Prediction coefficients are transmitted to the receiver as side information to allow recovery of the original signal. The convolution operation associated with spectral domain prediction is associated with multiplication in time. In a manner analogous to the source-system separation realized by time-domain LP analysis for traditional speech codecs TNS effectively separates the time-domain waveform into an envelope and temporally flat "excitation." Then, because quantization noise is added to the flattened residual, the time-domain multiplicative envelope corresponding to $A(z)$ shapes the quantization noise such that it follows the original signal envelope.

Quantization noise for the castanets applied to a DCT-based coder is shown in Figure 6.25(a) and Figure 6.25(b) both without and with TNS active, respectively. TNS clearly shapes the quantization noise to follow the input signal's energy envelope. TNS mitigates pre-echoes since the error energy is now concentrated in the time interval associated with the largest masking threshold. Although they are related as time-frequency dual operations, TNS is advantageous relative to gain shaping because it is easily applied selectively in specific frequency subbands. Moreover, TNS has the advantages of compatibility with most filter-bank structures and manageable complexity. Unlike window switching schemes, for example, TNS does not require modification of the perceptual model or lossless coding stages to a new time-frequency mapping. TNS was reported in [Herr96] to dramatically improve performance on a five-point mean opinion score (MOS) test from 2.64 to 3.54 for a particularly troublesome pitched signal "German Male Speech" for the MPEG-2 nonbackward compatible (NBC) coder [Herr96]. A MOS improvement of 0.3 was also realized for the well-known "Glockenspiel" test signal. This ultimately led to the adoption of TNS in the MPEG NBC scheme [Bosi96a] [ISOI96a].

6.11 SUMMARY

This chapter covered the basics of time-frequency analysis techniques for audio signal processing and coding. We also highlighted the time-frequency tradeoff challenges faced by the audio codec designers when designing a filter bank. We discussed both the QMF and CQF filter-bank designs and their extended tree-structured forms. We also dealt in detail with the cosine modulated pseudo-QMF and perfect reconstruction M-band filter-bank designs. The modified discrete cosine transform (MDCT) and the various window designs were also covered in detail.

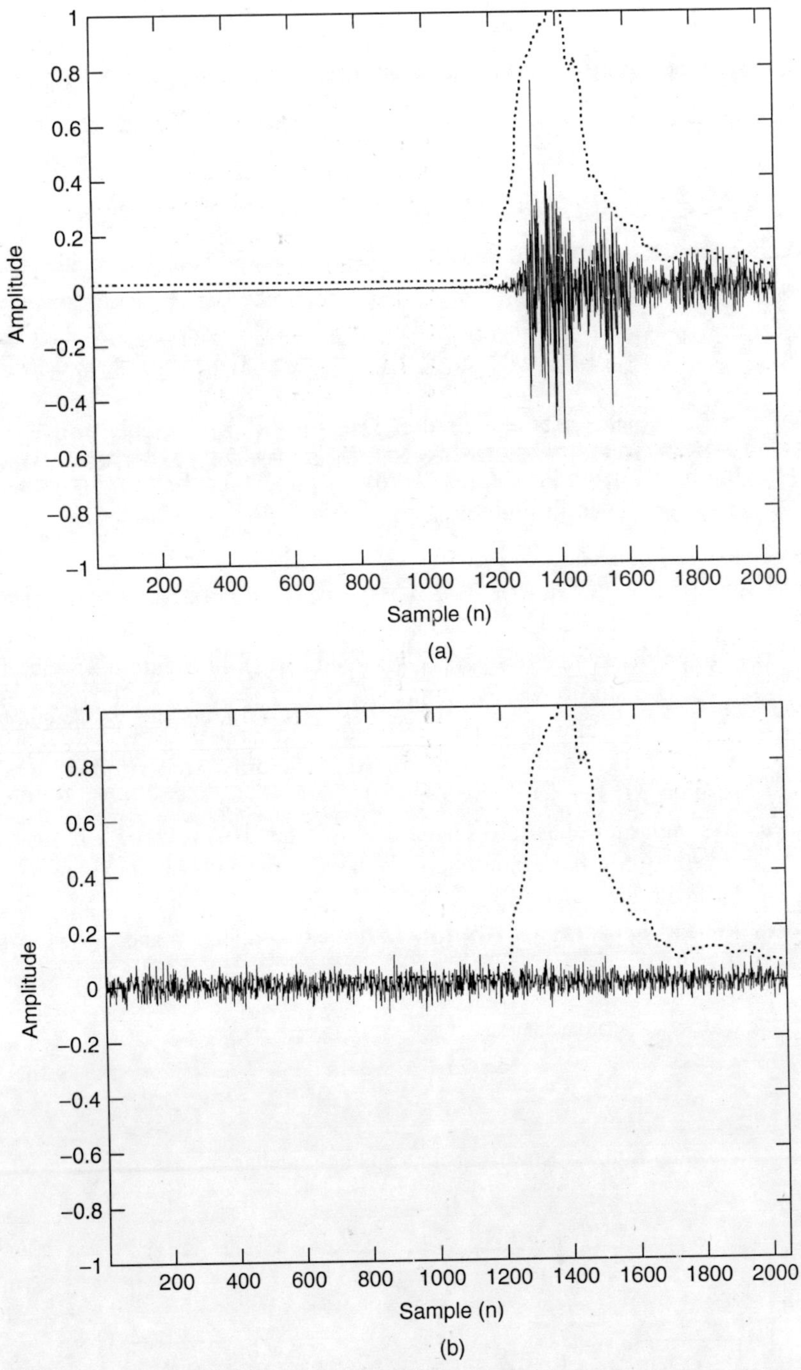

Figure 6.25. Temporal noise shaping example showing quantization noise and the input signal energy envelope for castanets: (a) without TNS, and (b) with TNS

PROBLEMS

6.1. Prove the identities shown in Figure 6.10.

6.2. Consider the analysis-synthesis filter bank shown in Figure 6.1 with input signal, $s(n)$. Show that $\hat{s}(n) = \frac{1}{M}\sum_{m=-\infty}^{\infty}\sum_{l=-\infty}^{\infty}\sum_{k=0}^{M-1} s(m)$ $h_k(lM - m)g_k(l - Mn)$, where M is the number of subbands.

6.3. For the down-sampling and up-sampling processes given in Figure 6.26, show that

$$S_d(\Omega) = \frac{1}{M}\sum_{l-0}^{M-1} S\left(\frac{\Omega + 2\pi l}{M}\right) H\left(\frac{\Omega + 2\pi l}{M}\right)$$

and $S_u(\Omega) = S(\Omega M)G(\Omega)$

6.4. Using results from Problem 6.3, Prove Eq. (6.5) for the analysis-synthesis framework shown in Figure 6.1.

6.5. Consider Figure 6.27,
Given $s(n) = 0.75\sin(\pi n/3) + 0.5\cos(\pi n/6), n = 0, 1, \ldots, 6, H_0(z) = 1 - z^{-1}$, and $H_1(z) = 1 + z^{-1}$

a. Design the synthesis filters, $G_0(z)$ and $G_1(z)$, in Figure 6.27 such that aliasing distortions are minimized.

b. Write the closed-form expression for $v_0(n), v_1(n), y_0(n), y_1(n)$, $w_0(n), w_1(n)$, and the synthesized waveform, $\hat{s}(n)$. In Figure 6.27, assume $y_i(n) = \hat{y}_i(n)$, for $i = 0, 1$.

c. Assuming an alias-free scenario, show that $\hat{s}(n) = \alpha s(n - n_0)$, where α is the QMF bank gain, n_0 is a delay that depends on $H_i(z)$ and $G_i(z)$. Estimate the value of n_0.

d. Repeat steps (a) and (c) for $H_0(z) = 1 - 0.75z^{-1}$ and $H_1(z) = 1 + 0.75z^{-1}$.

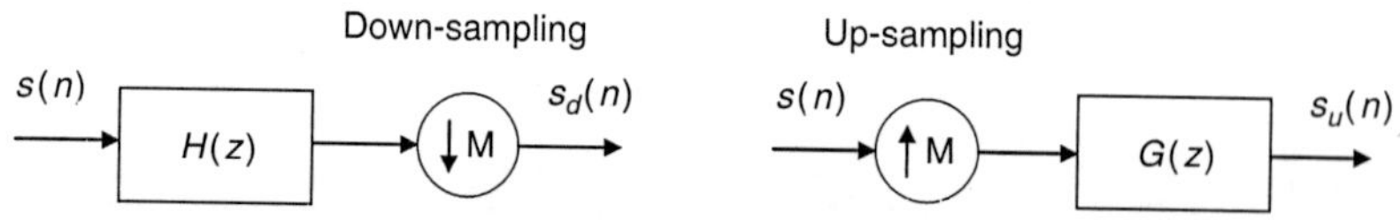

Figure 6.26. Down-sampling and up-sampling processes.

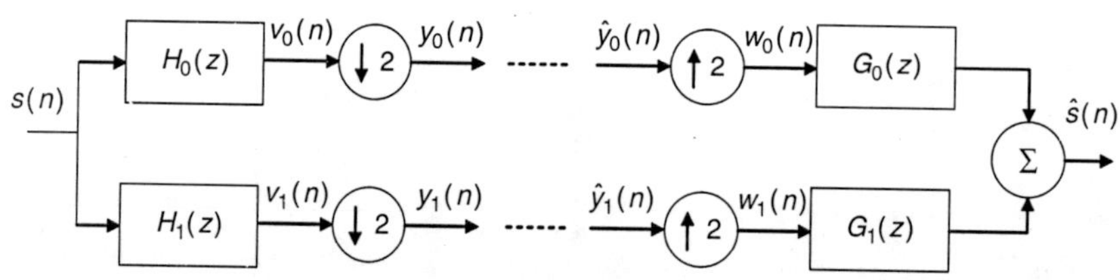

Figure 6.27. A two-band maximally decimated analysis-synthesis filter bank.

6.6. In this problem, we will compare the two-band QMF and CQF designs. Given $H_0(z) = 1 - 0.9z^{-1}$.

a. Design a two-band (i.e., $M = 2$) QMF [Use Eq. (6.7)].

b. Design a two-band CQF for $L = 8$ [Use Eq. (6.8)].

c. Consider the two-band QMF and CQF banks in Figure 6.28 with input signal $s(n) = 0.75\sin(\pi n/3) + 0.5\cos(\pi n/6), n = 0, 1, \ldots, 6$. Compare the designs in (a) and (b) and check for the alias-free reconstruction in case of CQF. Give the delay values d_1 and d_2.

d. Extend the two-band QMF design in part (a) to polyphase factorization [use Equations (6.9) and (6.10)]. What are the advantages of employing polyphase factorization?

6.7. In this problem, we will design and analyze a four-channel uniform tree-structured QMF bank.

a. Given $H_{00}(z) = 1 + 0.1z^{-1}$ and $H_{10}(z) = 1 + 0.9z^{-1}$. Complete the tree-structured QMF bank (use Figure 6.8) for four channels.

b. Using the identities given in Figure 6.10 (or Eq. (6.11)), construct a parallel analysis-synthesis filter bank. The parallel analysis-synthesis filter bank structure must be similar to the one shown in Figure 6.1 with $M = 4$.

c. Plot the frequency response of the resulting parallel filter bank analysis filters, i.e., $H_0(z)$, $H_1(z)$, $H_2(z)$, and $H_3(z)$. Comment on the pass-band and stopband structures of the magnitude responses associated with these filters.

d. Plot the impulse response $h_1(n)$. Is $h_1(n)$ symmetric?

6.8. Repeat Problem 6.7 for a four-channel uniform tree-structured CQF bank with $L = 4$.

6.9. A time-domain plot of an audio signal is shown in Figure 6.29. Given the flexibility to encode the regions A through E with varying frame sizes. Which of the following choices is preferred?

Choice I:

Long frames in regions B and D.

Choice II:

Long frames in regions A, C, and E.

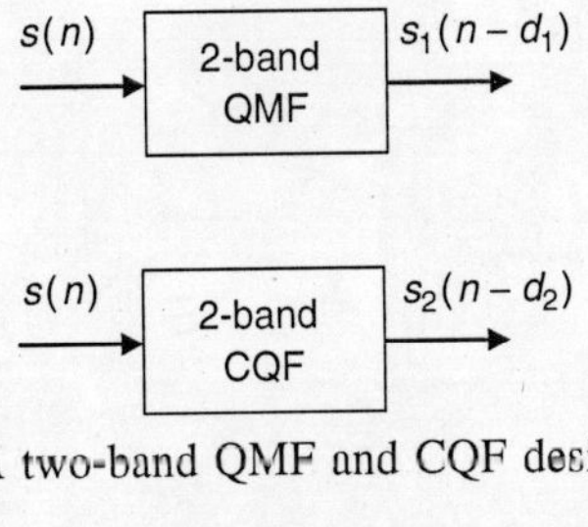

Figure 6.28. A two-band QMF and CQF design comparison.

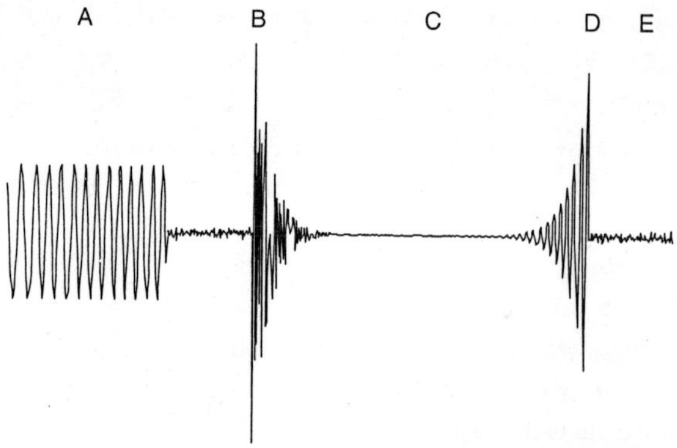

Figure 6.29. An example audio segment with harmonics (region A), a transient (region B), background noise (region C), exponentially weighted harmonics (region D), and background noise (region E) segments.

Choice III:
Short frames in region B only.
Choice IV:
Short frames in regions B and D.
Choice V:
Short frames in region A only.

Explain how would you assign frequency-resolution (high or low) among the regions A, B, C, D, and E.

6.10. A pure tone at f_0 with P_0 dB SPL is encoded such that the quantization noise is masked. Let us assume that a 256-point MDCT produced an in-band signal-to-noise ratio of SNR_A and encodes the tone with b_A bits/sample. And, a 1024-point MDCT yielded SNR_B and encodes the tone with b_B bits/sample.

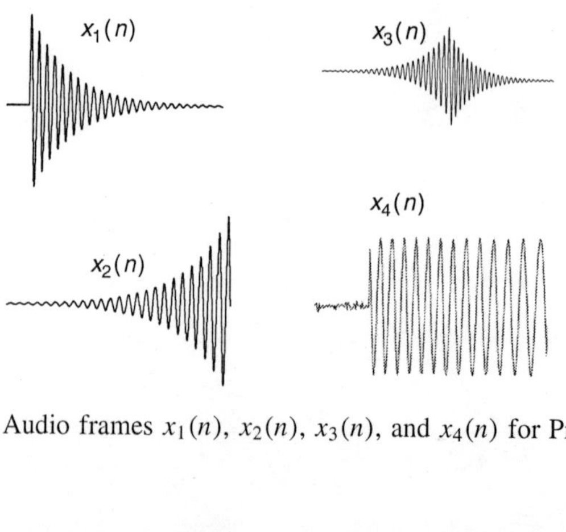

Figure 6.30. Audio frames $x_1(n)$, $x_2(n)$, $x_3(n)$, and $x_4(n)$ for Problem 6.11.

In which of the two cases we will require the most bits/sample (state if $b_A > b_B$ or $b_A < b_B$) to mask the quantization noise.

6.11. Given the signals, $x_1(n)$, $x_2(n)$, $x_3(n)$, and $x_4(n)$ as shown in Figure 6.30. Let all the signals be of length 1024 samples. When transform coded using a 512-point MDCT, which of the signals will result in pre-echo distortion?

COMPUTER EXERCISES

6.12. In this problem, we will study the filter banks that are based on the DFT. Use Eq. (6.39a) to implement a $2M$-point DFT of $x(n)$ given in Figure 6.31. Assume $M = 8$.

a. Give the plots of $|X(k)|$.

b. Plot the frequency response of the second- and third-channel analysis filters that are associated with the basis vectors $h_1(n)$ and $h_2(n)$.

c. State whether the DFT filter bank is evenly stacked or oddly stacked.

6.13. In this problem, we will study the filter banks that are based on the DCT. Use Eq. (6.41a) to implement a M-point DCT of $x(n)$ given in Figure 6.31. Assume $M = 8$.

a. Give the plots of $|X(k)|$.

b. Also plot the frequency response of the second and third channel analysis filters that are associated with the basis vectors $h_1(n)$ and $h_2(n)$.

c. Plot the impulse response of $h_1(n)$ and see if it is symmetric.

d. Is the DCT filter bank evenly stacked or oddly stacked?

6.14. In this problem, we will study the filter banks based on the MDCT.

a. First, design a sine window, $w(n) = \sin[(2n + 1)\pi/4M]$ with $M = 8$.

b. Check if the sine window satisfies the generalized perfect reconstruction conditions, i.e., Eqs. (6.28a) (6.28b).

c. Next, design a MDCT analysis filter bank, $h_k(n)$, for $0 < k < 7$.

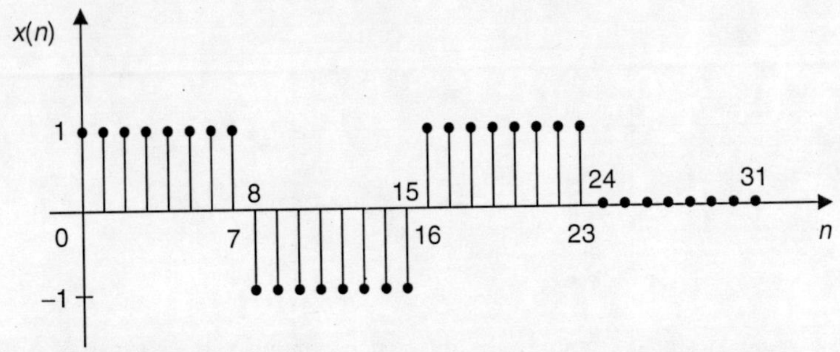

Figure 6.31. Input signal, $x(n)$, for Problems 6.12, 6.13, and 6.14.

d. Plot both the impulse response and the frequency response of the analysis filters, $h_1(n)$ and $h_2(n)$.

e. Compute the MDCT coefficients, $X(k)$, of the input signal, $x(n)$, shown in Figure 6.31.

f. Is the impulse response of the analysis filter, $h_1(n)$, symmetric?

6.15. Show analytically that a DCT can be implemented using FFTs. Also, use $x(n)$ given in Figure 6.32 as your test signal and verify your software implementation.

6.16. Give expressions for DCT-I, DCT-II, DCT-III, and DCT-IV orthonormal transforms (e.g., see [Rao90]). Use the signals, $x_1(n)$ and $x_2(n)$, shown in Figure 6.33 to study the differences in the 4-point DCT coefficients obtained from different types of DCT. Describe, in general, whether choosing a particular type of DCT affects the energy compaction of a signal.

6.17. In this problem, we will design a two-band ($M = 2$) cosine-modulated PQMF bank with $L = 8$.

a. First, design a linear phase FIR prototype lowpass filter (i.e., $w(n)$), with normalized cutoff frequency $\pi/4$. Plot the frequency response of this window. Use fir2 command in MATLAB to design the lowpass filter.

b. Use Eq. (6.13) and (6.14) to design the PQMF analysis and synthesis filters, respectively.

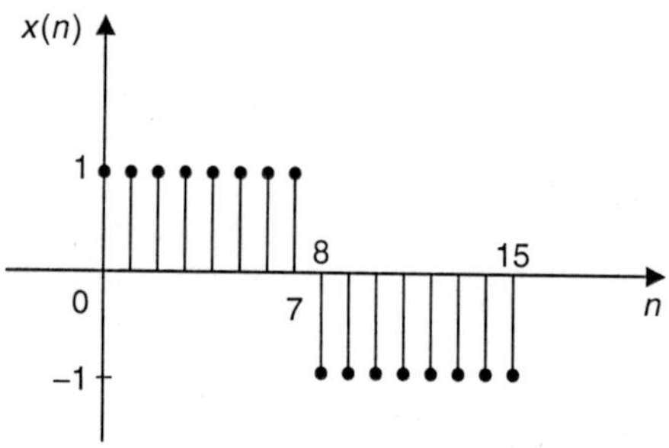

Figure 6.32. Input signal, $x(n)$ for Problem 6.15.

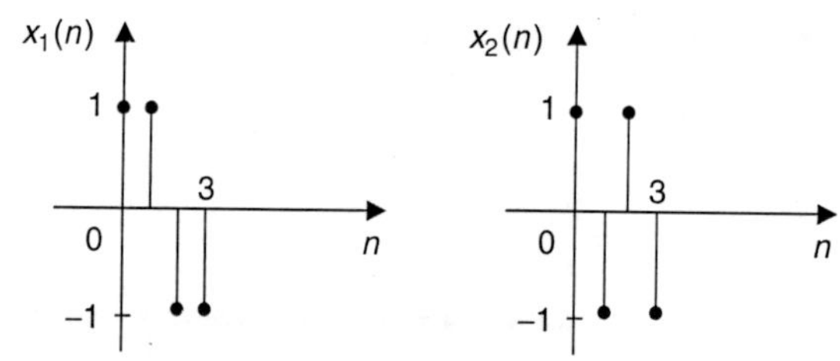

Figure 6.33. Test signals to study the differences among various types of orthonormal DCT transforms.

Figure 6.34. A two-band PQMF design.

c. In Figure 6.34, use $s(n) = 0.75\sin(\pi n/3) + 0.5\cos(\pi n/6), n = 0, 1, \ldots, 6$, and generate $s_3(n)$. Compare $s(n)$ and $s_3(n)$ and comment on any type of distortion that you may observe.

d. What are the advantages of employing a cosine modulated filter bank over the two-band QMF and two-band CQF.

e. List some of the key differences between the CQF and the PQMF in terms of 1) the analysis filter bank frequency responses, 2) phase distortion, 3) impulse response symmetries.

f. Use sine window, $w(n) = \sin[(2n + 1)\pi/4M]$ with $M = 8$, and repeat steps (b) and (c).

CHAPTER 7

TRANSFORM CODERS

7.1 INTRODUCTION

Transform coders make use of unitary transforms (e.g., DFT, DCT, etc.) for the time/frequency analysis section of the audio coder shown in Figure 1.1. Many transform coding schemes for wideband and high-fidelity audio have been proposed, starting with some of the earliest perceptual audio codecs. For example, in the mid-1980s, Krahe applied psychoacoustic bit allocation principles to a transform coding scheme [Krah85] [Krah88]. Schroeder [Schr86] later extended these ideas into multiple adaptive spectral audio coding (MSC). The MSC utilizes a 1024-point DFT, then groups coefficients into 26 subbands, inspired by the critical bands of the ear. This chapter gives overview of algorithms that were proposed for transform coding of high-fidelity audio following the early work of Schroeder [Schr86].

The Chapter is organized as follows. Sections 7.2 through 7.5 describe in some detail the transform coding algorithms proposed by Brandenburg, Johnston, and Mahieux [Bran87b] [John88a] [Mahi89] [Bran90]. Most of this research became connected with the MPEG standardization, and the ISO/IEC eventually clustered these algorithms into a single candidate algorithm called adaptive spectral entropy coding (ASPEC) [Bran91] of high quality music signals. The ASPEC algorithm (Section 7.6) has become part of the ISO/IEC MPEG-1 [ISOI92] and the MPEG-2/BC-LSF [ISOI94a] audio coding standards. Sections 7.7 and 7.8 are concerned with two transform coefficient substitution schemes, namely the differential perceptual audio coder (DPAC), and the DFT noise substitution algorithm. Finally,

Audio Signal Processing and Coding, by Andreas Spanias, Ted Painter, and Venkatraman Atti

Sections 7.9 and 7.10 address several early applications of vector quantization (VQ) to transform coding of high-fidelity audio.

The algorithms described in the Chapter that make use of modulated filter banks (e.g., ASPEC, DPAC, TwinVQ) can also be characterized as high-resolution subband coders. Typically, transform coders perform high-resolution frequency analysis and subband coders rely on a coarse division of the frequency spectrum. In many ways, the transform and subband coder categories overlap, and in some cases it is hard to categorize a coder in a definite manner. The source of this overlapping of transform/subband categories come from the fact that block transform realizations are used for cosine modulated filter banks.

7.2 OPTIMUM CODING IN THE FREQUENCY DOMAIN

Brandenburg in 1987 proposed a 132 kb/s algorithm known as optimum coding in the frequency domain (OCF) [Bran87b], which is in some respects an extension of the well-known adaptive transform coder (ATC) for speech. The OCF was refined several times over the years, with two enhanced versions appearing after the original algorithm. The OCF is of interest because of its influence on current standards.

The original OCF (Figure 7.1) works as follows. The input signal is first buffered in 512 sample blocks and transformed to the frequency domain using the DCT. Next, transform components are quantized and entropy coded. A single quantizer is used for all transform components. Adaptive quantization and entropy coding work together in an iterative procedure to achieve a fixed bit rate. The initial quantizer step size is derived from the spectral flatness measure (Eq. (5.13)).

In the inner loop of Figure 7.1, the quantizer step size is iteratively increased and a new entropy-coded bit stream is formed at each update until the desired bit

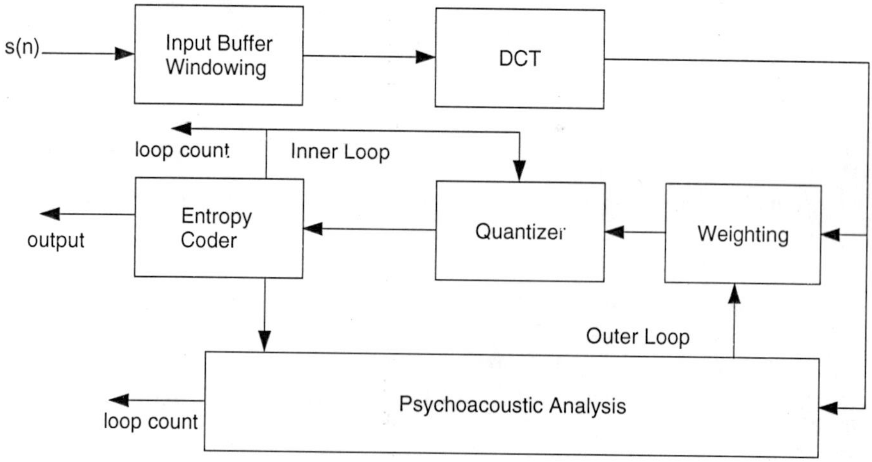

Figure 7.1. OCF encoder (after [Bran88b]).

rate is achieved. Increasing the step size at each update produces fewer levels, which in turn reduces the bit rate. Using a second iterative procedure, a perceptual analysis is introduced after the inner loop is done. First, critical band analysis is applied. Then, a masking function is applied which combines a flat −6 dB masking threshold with an interband masking threshold, leading to an estimate of JND for each critical band. If after inner-loop quantization and entropy encoding the measured distortion exceeds JND in at least one critical band, then quantization step sizes are adjusted only in the out-of-tolerance critical bands. The outer loop repeats until JND criteria are satisfied or a maximum loop count is reached. Entropy coded transform components are then transmitted to the receiver, along with side information, which includes the log encoded SFM, the number of quantizer updates during the inner loop, and the number of step size reductions that occurred for each critical band in the outer loop. This side information is sufficient to decode the transform components and perform reconstruction at the receiver.

Brandenburg in 1988 reported an enhanced OCF (OCF-2), which achieved subjective quality improvements at a reduced bit rate of only 110 kb/s [Bran88a]. The improvements were realized by replacing the DCT with the MDCT and adding a pre-echo detection/compensation scheme. Reconstruction quality is improved due to the effective time resolution increase (i.e., 50% time overlap) associated with the MDCT. OCF-2 quality is also improved for difficult signals such as triangle and castanets due to a simple pre-echo detection/compensation scheme. The encoder detects pre-echoes using analysis-by-synthesis. Pre-echoes are detected when noise energy in a reconstructed segment (16 samples = 0.36 ms @ 44.1 kHz) exceeds signal energy. The encoder then determines the frequency below which 90% of signal energy is contained and transmits this cutoff to the decoder. Given pre-echo detection at the encoder (1 bit) and a cutoff frequency, the decoder discards frequency components above the cutoff, in effect low-pass filtering pre-echoes. Due to these enhancements, the OCF-2 was reported to achieve transparency over a wide variety of source material.

Later in 1988, Brandenburg reported further OCF enhancements (OCF-3) in which better quality was realized at a lower bit rate (64 kb/s) with reduced complexity [Bran88b]. This was achieved through differential coding of spectral components to exploit correlation between adjacent samples, an enhanced psychoacoustic model modified to account for temporal masking, and an improved rate-distortion loop.

7.3 PERCEPTUAL TRANSFORM CODER

While Brandenburg developed the OCF algorithm, similar work was simultaneously underway at AT&T Bell Labs. Johnston developed several DFT-based transform coders [John88a] [John89] for audio during the late 1980s that became an integral part of the ASPEC proposal. Johnston's work in perceptual entropy [John88b] forms the basis for a transform coder reported in 1988 [John88a] that

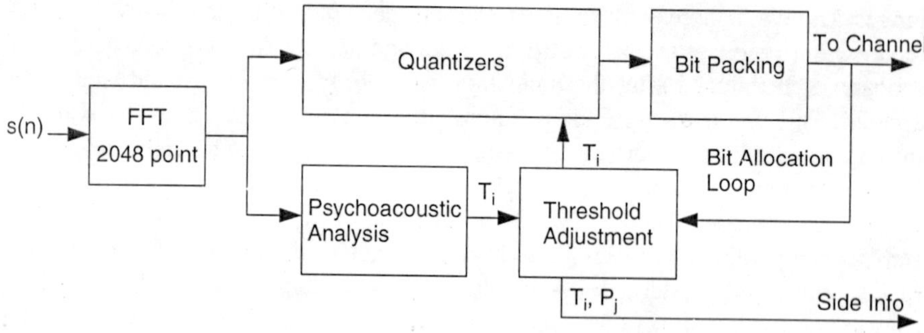

Figure 7.2. PXFM encoder (after [John88a]).

achieves transparent coding of FM-quality monaural audio signals (Figure 7.2). A stereophonic coder based on similar principles was developed later.

7.3.1 PXFM

A monaural algorithm, the perceptual transform coder (PXFM), was developed first. The idea behind the PXFM is to estimate the amount of quantization noise that can be inaudibly injected into each transform domain subband using PE estimates. The coder works as follows. The signal is first windowed into overlapping (1/16) segments and transformed using a 2048-point FFT. Next, the PE procedure described in Section 5.6, is used to estimate JND thresholds for each critical band. Then, an iterative quantization loop adapts a set of 128 subband quantizers to satisfy the JND thresholds until the fixed bit rate is achieved. Finally, quantization and bit packing are performed. Quantized transform components are transmitted to the receiver along with appropriate side information. Quantization subbands consist of 8-sample blocks of complex-valued transform components. The quantizer adaptation loop first initializes the $j \in [1, 128]$ subband quantizers (1024 unique FFT components/8 components per subband) with k_j levels and step sizes of T_i as follows:

$$k_j = 2\, nint\left(\frac{P_j}{T_i}\right) + 1, \tag{7.1}$$

where T_i are the quantized critical band JND thresholds, P_j is the quantized magnitude of the largest real or imaginary transform component in the j-th subband, and *nint*() is the nearest integer rounding function. The adaptation process involves repeated application of two steps. First, bit packing is attempted using the current quantizer set. Although many bit packing techniques are possible, one simple scenario involves sorting quantizers in k_j order, then filling 64-bit words with encoded transform components according to the sorted results. After bit packing, T_i are adjusted by a carefully controlled scale factor, and the adaptation cycle repeats. Quantizer adaptation halts as soon as the packed data length satisfies the desired bit rate. Both P_j and the modified T_i are quantized on a dB scale using 8-bit uniform quantizers with a 170 dB dynamic range. These parameters are transmitted as side information and used at the receiver to recover quantization levels (and thus implicit bit allocations) for each subband, which are in turn used to decode quantized transform components. The DC FFT component is quantized with 16 bits and is also transmitted as side information.

7.3.2 SEPXFM

In 1989, Johnston extended the PXFM coder to handle stereophonic signals (SEPXFM) and attained transparent coding of a CD-quality stereophonic channel at 192 kb/s, or 2.2 bits/sample. SEPXFM [John89] realizes performance improvements over PXFM by exploiting inherent stereo cross-channel redundancy and by assuming that both channels are presented to a single listener rather than being used as separate signal sources. The SEPXFM structure is similar to that of PXFM, with variable radix bit packing replaced by adaptive entropy coding. Side information is therefore reduced to include only adjusted JND thresholds (step sizes) and pointers to the entropy codebooks used in each transform domain subband. The coder works in the following manner. First, sum $(L + R)$ and difference $(L - R)$ signals are extracted from the left (L) and right (R) channels to exploit left/right redundancy. Next, the sum and difference signals are windowed and transformed using the FFT. Then, a single JND threshold for each critical band is established via the PE method using the summed power spectra from the $L + R$ and $L - R$ signals. A single combined JND threshold is applied to quantization noise shaping for both signals ($L + R$ and $L - R$), based upon the assumption that a listener is more than one "critical distance" [Jetz79] away from the stereo speakers.

Like PXFM, a fixed bit rate is achieved by applying an iterative threshold adjustment procedure after the initial determination of JND levels. The adaptation process, analogous to PXFM bit rate adjustment and bit packing, consists of several steps. First, transform components from both $(L + R)$ and $(L - R)$ are split into subband blocks, each averaging 8 real/imaginary samples. Then, one of six entropy codebooks is selected for each subband based on the average component magnitude within that subband. Next, transform components are quantized given the JND levels and encoded using the selected codebook. Subband codebook selections are themselves entropy encoded and transmitted as side information. After encoding, JND thresholds are scaled by an estimator and the quantizer adaptation process repeats. Threshold adaptation stops when the combined bitstream of quantized JND levels, Huffman-encoded $(L + R)$ components, Huffman-encoded $(L - R)$ components, and Huffman-encoded average magnitudes achieves the desired bit rate. The Huffman codebooks are developed using a large music and speech database. They are optimized for difficult signals at the expense of mean compression rate. It is also interesting to note that headphone listeners reported no noticeable acoustic mixing, despite the critical distance assumption and single combined JND level estimate for both channels, $(L + R)$ and $(L - R)$.

7.4 BRANDENBURG-JOHNSTON HYBRID CODER

Johnston and Brandenburg [Bran90] collaborated in 1990 to produce a hybrid coder that, strictly speaking, is both a subband and transform coding algorithm. The idea behind the hybrid coder is to improve time and frequency resolution relative to OCF and PXFM by constructing a filter bank that more closely resembles the auditory filter bank. This is accomplished at the encoder by first splitting the input signal into four octave-width subbands using a QMF filter bank.

The decimated output sequence from each subband is then followed by one or more transforms to achieve the desired time/frequency resolution, Figure 7.3(a). Both the DFT and the MDCT were investigated. Given the tiling of the time-frequency plane shown in Figure 7.3(b), frequency resolution at low frequencies (23.4 Hz) is well matched to the ear, while the time resolution at high frequencies (2.7 ms) is sufficient for pre-echo control.

The quantization and coding schemes of the hybrid coder combine elements from both PXFM and OCF. Masking thresholds are estimated using the PXFM

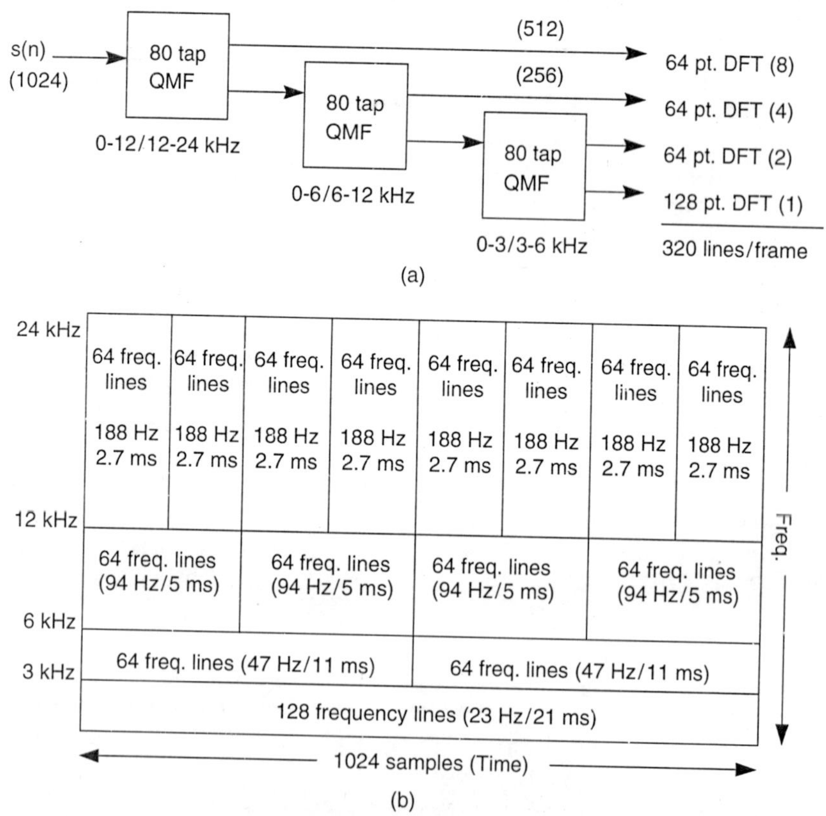

Figure 7.3. Brandenburg-Johnston coder: (a) filter bank structure, (b) time/freq tiling (after [Bran90]).

approach for eight time slices in each frequency subband. A more sophisticated tonality estimate was defined to replace the SFM (Eq. (5.13)) used in PXFM, however, such that tonality is estimated in the hybrid coder as a local characteristic of each individual spectral line. Predictability of magnitude and phase spectral components across time is used to evaluate tonality instead of just global spectral shape within a single frame. High temporal predictability of magnitudes and phases is associated with the presence of a tonal signal. In contrast, low predictability implies the presence of a noise-like signal. The hybrid coder employs a quantization and coding scheme borrowed from OCF. As far as quality, the hybrid coder without any explicit pre-echo control mechanism was reported to achieve quality better than or equal to OCF-3 at 64 kb/s [Bran90]. The only disadvantage noted by the authors was increased complexity. A similar hybrid structure was eventually adopted in MPEG-1 and -2, layer III.

7.5 CNET CODERS

Research at the Centre National d'Etudes des Telecommunications (CNET) resulted in several transform coders based on the DFT and the MDCT.

7.5.1 CNET DFT Coder

In 1989, Mahieux, Petit, *et al.* proposed a DFT-based audio coding system that introduced a novel scheme to exploit DFT interblock redundancy. Nearly transparent quality was reported for 15-kHz (FM-grade) audio at 96 kb/s [Mahi89], except for some highly harmonic signals. The encoder applies first-order backward-adaptive predictors (across time) to DFT magnitude and differential phase components, then quantizes separately the prediction residuals. Magnitude and differential phase residuals are quantized using an adaptive nonuniform pdf-optimized quantizer designed for a Laplacian distribution and an adaptive uniform quantizer, respectively. The backward-adaptive quantizers are reinitialized during transients. Bits are allocated during step-size adaptation to shape quantization noise such that a psychoacoustic noise threshold is satisfied for each block. The perceptual model used is similar to Johnston's model that was described earlier in Section 5.6. The use of linear prediction is justified because it exploits magnitude and differential phase time redundancy, which tends to be large during periods when the audio signal is quasi-stationary, especially for signal harmonics. Quasi-stationarity might occur, for example, during a sustained note. A similar technique was eventually embedded in the MPEG-2 AAC algorithm.

7.5.2 CNET MDCT Coder 1

In 1990, Mahieux and Petit reported on the development of a similar MDCT-based transform coder for which they reported transparent CD-quality at 64 kb/s [Mahi90]. This algorithm introduced a novel spectrum descriptor scheme for representing the power spectral envelope. The algorithm first segments input audio into frames of 1024 samples, corresponding to 12 ms of new data per frame, given 50% MDCT time overlap. Then, a bit allocation is computed at the encoder using a set of "spectrum descriptors." Spectrum descriptors consist of quantized sample variances for MDCT coefficients grouped into 35 nonuniform frequency subbands. Like their DFT coder, this algorithm exploits either interblock or intrablock redundancy by differentially encoding the spectrum descriptors with respect to time or frequency and transmitting them to the receiver as side information. A decision whether to code with respect to time or frequency is made on the basis of which method requires fewer bits; the binary decision requires only 1 bit. Either way, spectral descriptor encoding is done using log DPCM with a first-order predictor and a 16-level uniform quantizer with a step size of 5 dB. Huffman coding of the spectral descriptor codewords results in less than 2 bits/descriptor. A global masking threshold is estimated by convolving the spectral descriptors with a basilar spreading function on a bark scale, somewhat like the approach taken by Johnston's PXFM. Bit allocations for quantization of normalized transform coefficients are obtained from the masking threshold estimate. As usual, bits are allocated such that quantization noise is below the masking threshold at every spectral line. Transform coefficients are normalized by the appropriate spectral descriptor, then quantized and coded, with one exception. Masked transform coefficients, which have lower energy than the global masking threshold, are treated differently. The authors found that masked coefficient bins tend to be clustered, therefore, they can be compactly represented using run length encoding (RLE). RLE codewords are Huffman coded for maximum coding gain. The first CNET MDCT coder was reported to perform well for broadband signals with many harmonics but had some problems in the case of spectrally flat signals.

7.5.3 CNET MDCT Coder 2

Mahieux and Petit enhanced their 64 kb/s algorithm by incorporating a sophisticated pre-echo detection and postfiltering scheme, as well as by incorporating a novel quantization scheme for two-coefficient (low-frequency) spectral descriptor bands [Mahi94]. For improved quantization performance, two-component spectral descriptors are efficiently vector encoded in terms of polar coordinates. Pre-echoes are detected at the encoder and flagged using 1 bit. The idea behind the pre-echo compensation is to temporarily activate a postfilter at the decoder in the corrupted quiet region prior to the signal attack, and therefore a stopping index must also be transmitted. The second-order IIR postfilter difference equation is given by,

$$\hat{s}_{pf}(n) = b_0\hat{s}(n) + a_1\hat{s}_{pf}(n-1) + a_2\hat{s}_{pf}(n-2), \tag{7.2}$$

where $\hat{s}(n)$ is the nonpostfiltered output signal that is corrupted by pre-echo distortion, $\hat{s}_{pf}(n)$ is the postfiltered output signal, and a_i are related to the parameters α_i by,

$$a_1 = \alpha_1\left[1 - \left(\frac{p(0,0)}{p(0,0) + \sigma_b^2}\right)\right], \tag{7.3a}$$

$$a_2 = \alpha_2\left[1 - \left(\frac{p(1,0)}{p(0,0) + \sigma_b^2}\right)\right], \tag{7.3b}$$

where α_i are the parameters of a second-order autoregressive (AR-2) spectral estimate of the output audio, $\hat{s}(n)$, during the previous nonpostfiltered frame. The AR-2 estimate, $\dot{s}(n)$, can be expressed in the time domain as

$$\dot{s}(n) = w(n) + \alpha_1 \dot{s}(n-1) + \alpha_2 \dot{s}(n-2), \tag{7.4}$$

where $w(n)$ represents Gaussian white noise. The prediction error is then defined as

$$e(n) = \hat{s}(n) - \dot{s}(n). \tag{7.5}$$

The parameters $p(i,j)$ in Eqs. (7.3a) and (7.3b) are elements of the prediction error covariance matrix, **P**, and the parameter σ_b^2 is the pre-echo distortion variance, which is derived from side information. Pre-echo postfiltering and improved quantization schemes resulted in a subjective score of 3.65 for two-channel stereo coding at 64 kb/s per channel on the 5-point CCIR 5-grade impairment scale (described in Section 12.3), over a wide range of listening material. The CCIR J.41 reference audio codec (MPEG-1, layer II) achieved a score of 3.84 at 384 kb/s/channel over the same set of tests.

7.6 ADAPTIVE SPECTRAL ENTROPY CODING

The MSC, OCF, PXFM, Brandenburg-Johnston hybrid, and CNET transform coders were eventually clustered into a single proposal by the ISO/IEC JTC1/SC2 WG11 committee. As a result, Schroeder, Brandenburg, Johnston, Herre, and Mahieux collaborated in 1991 to propose for acceptance as the new MPEG audio compression standard a flexible coding algorithm, ASPEC, which incorporated the best features of each coder in the group. ASPEC [Bran91] was claimed to produce better quality than any of the individual coders at 64 kb/s.

The structure of ASPEC combines elements from all of its predecessors. Like OCF and the later CNET coders, ASPEC uses the MDCT for time-frequency mapping. The masking model is similar to that used in PXFM and the Brandenburg-Johnston hybrid coders, including the sophisticated tonality estimation scheme at lower bit rates. The quantization and coding procedures use the pair of nested loops proposed for OCF, as well as the block differential coding scheme developed at CNET. Moreover, long runs of masked coefficients are run-length and Huffman encoded. Quantized scale factors and transform coefficients are Huffman coded also. Pre-echoes are controlled using a dynamic window switching mechanism, like the Thomson coder [Edle89]. ASPEC offers several modes for different quality levels, ranging from 64 to 192 kb/s per channel. A real-time ASPEC implementation for coding one channel at 64 kb/s was realized on a pair of 33-MHz Motorola DSP56001 devices. ASPEC ultimately formed the basis for layer III of the MPEG-1 and MPEG-2/BC-LSF standards. We note that similar contributions were made in the area of transform coding for audio outside of the ASPEC cluster. For example, Iwadare, *et al.* reported on DCT-based [Sugi90] and MDCT-based [Iwad92] perceptual adaptive transform coders that control pre-echo distortion using an adaptive window size.

7.7 DIFFERENTIAL PERCEPTUAL AUDIO CODER

Other investigators have also developed promising schemes for transform coding of audio. Paraskevas and Mourjopoulos [Para95] reported on a differential perceptual audio coder (DPAC), which makes use of a novel scheme for exploiting long-term correlations. DPAC works as follows. Input audio is transformed using the MDCT. A two-state classifier then labels each new frame of transform coefficients as either a "reference" frame or a "simple" frame. The classifier labels as "reference" frames that contain significant audible differences from the previous frame. The classifier labels nonreference frames as "simple." Reference frames are quantized and encoded using scalar quantization and psychoacoustic bit allocation strategies similar to Johnston's PXFM. Simple frames, however, are subjected to coefficient substitution. Coefficients whose magnitude differences with respect to the previous reference frame are below an experimentally optimized threshold are replaced at the decoder by the corresponding reference frame coefficients. The encoder, then, replaces subthreshold coefficients with zeros, thus saving transmission bits. Unlike the interframe predictive coding schemes of Mahieux and Petit, the DPAC coefficient substitution system is advantageous in that it guarantees that the "simple" frame bit allocation will always be less than or equal to the bit allocation that would be required if the frame was coded as a "reference" frame. Suprathreshold "simple" frame coefficients are coded in the same way as reference frame coefficients. DPAC performance was evaluated for frame classifiers that utilized three different selection criteria:

1. *Euclidean distance*: Under the Euclidean criterion, test frames satisfying the inequality

$$\left[\frac{\mathbf{s}_d{}^T\mathbf{s}_d}{\mathbf{s}_r{}^T\mathbf{s}_r}\right]^{\frac{1}{2}} \leqslant \lambda \tag{7.6}$$

   are classified as simple, where the vectors $\mathbf{s}_r$ and, $\mathbf{s}_t$, respectively, contain reference and test frame time-domain samples, and the difference vector, $\mathbf{s}_d$, is defined as

$$\mathbf{s}_d = \mathbf{s}_r - \mathbf{s}_t. \tag{7.7}$$

2. *Perceptual entropy*: Under the PE criterion (Eq. 5.17), a test frame is labeled as "simple" if it satisfies the inequality

$$\frac{PE_S}{PE_R} \leqslant \lambda, \tag{7.8}$$

   where PE_S corresponds to the PE of the "simple" (coefficient-substituted) version of the test frame, and PE_R corresponds to the PE of the unmodified test frame.

3. *Spectral flatness measure*: Finally, under the SFM criterion (Eq. 5.13), a test frame is labeled as "simple" if it satisfies the inequality

$$abs\left(10\log_{10}\frac{SFM_T}{SFM_R}\right) \leqslant \lambda, \tag{7.9}$$

where SFM_T corresponds to the test frame SFM, and SFM_R corresponds to the SFM of the previous reference frame. The decision threshold, λ, was experimentally optimized for all three criteria. Best performance was obtained while encoding source material using a PE criterion. As far as overall performance is concerned, noise-to-mask ratio (NMR) measurements were compared between DPAC and Johnston's PXFM algorithm at 64, 88, and 128 kb/s. Despite an average drop of 30–35% in PE measured at the DPAC coefficient substitution stage output relative to the coefficient substitution input, comparative NMR studies indicated that DPAC outperforms PXFM only below 88 kb/s and then only for certain types of source material such as pop or jazz music. The desirable PE reduction led to an undesirable drop in reconstruction quality. The authors concluded that DPAC may be preferable to algorithms such as PXFM for low-bit-rate, non transparent applications.

7.8 DFT NOISE SUBSTITUTION

Whereas DPAC exploits temporal correlation, a substitution technique that exploits decorrelation was devised for coding efficiently noise-like portions of the spectrum. In a noise substitution procedure [Schu96], Schulz parameterizes transform coefficients corresponding to noise-like portions of the spectrum in terms of average power, frequency range, and temporal evolution, resulting in an increased coding efficiency of 15% on average. A temporal envelope for each parametric noise band is required because transform block sizes for most codecs are much longer (e.g., 30 ms) than the human auditory system's temporal resolution (e.g., 2 ms). In this method, noise-like spectral regions are identified in the following way. First, least-mean-square (LMS) adaptive linear predictors (LP) are applied to the output channels of a multi-band QMF analysis filter bank that has as input the original audio, $s(n)$. A predicted signal, $\hat{s}(n)$, is obtained by passing the LP output sequences through the QMF synthesis filter bank. Prediction is done in subbands rather than over the entire spectrum to prevent classification errors that could result if high-energy noise subbands are allowed to dominate predictor adaptation, resulting in misinterpretation of low-energy tonal subbands as noisy. Next, the DFT is used to obtain magnitude ($S(k)$,$\hat{S}(k)$) and phase components ($\theta(k)$,$\hat{\theta}(k)$), of the input, $s(n)$, and prediction, $\hat{s}(n)$, respectively. Then, tonality, $T(k)$, is estimated as a function of the magnitude and phase predictability, i.e.,

$$T(k) = \alpha\left|\frac{S(k)-\hat{S}(k)}{S(k)}\right| + \beta\left|\frac{\theta(k)-\hat{\theta}(k)}{\theta(k)}\right|, \tag{7.10}$$

where α and β are experimentally determined constants. Noise substitution is applied to contiguous blocks of transform coefficient bins for which $T(k)$ is very small. The 15% average bit savings realized using this method in conjunction with transform coding is offset to a large extent by a significant complexity increase due to the additions of the adaptive linear predictors and a multi-band analysis-synthesis QMF filter bank. As a result, the author focused his attention on the application of noise substitution to QMF-based subband coding algorithms. A modified version of this scheme was adopted as part of the MPEG-2 AAC time-frequency coder within the MPEG-4 reference model [Herr98].

7.9 DCT WITH VECTOR QUANTIZATION

For the most part, the algorithms described thus far rely upon scalar quantization of transform coefficients. This is not unreasonable, since scalar quantization in combination with entropy coding can achieve very good performance. As one might expect, however, vector quantization (VQ) has also been applied to transform coding of audio, although on a much more limited scale. Gersho and Chan investigated VQ schemes for coding DCT coefficients subject to a constraint of minimum perceptual distortion. They reported on a variable rate coder [Chan90] that achieves high quality in the range of 55–106 kb/s for audio sequences bandlimited to 15 kHz (32 kHz sample rate). After computing the DCT on 512 sample blocks, the algorithm utilizes a novel multi-stage tree-structured VQ (MSTVQ) scheme for quantization of normalized vectors, with each vector containing four DCT components. Bit allocation and vector normalization are derived at both the encoder and decoder from a sampled power spectral envelope which consists of 29 groups of transform coefficients. A simplified masking model assumes that each sample of the power envelope represents a single masker. Masking is assumed to be additive, as in the ASPEC algorithms. Thresholds are computed as a fixed offset from the masking level. The authors observed a strong correlation between the SFM and the amount of offset required to achieve high quality. Two-segment scalar quantizers that are piecewise linear on a dB scale are used to encode the power spectral envelope. Quadratic interpolation is used to restore full resolution to the subsampled envelope.

Gersho and Chan later enhanced [Chan91b] their algorithm by improving the power envelope and transform coefficient quantization schemes. In the new approach to quantization of transform coefficients, constrained-storage VQ [Chan91a] techniques are combined with the MSTVQ from the original coder, allowing the new coder to handle peak noise-to-mask ratio (NMR) requirements without impractical codebook storage requirements. In fact, CS-MSTVQ enabled quantization of 127 four-coefficient vectors using only four unique quantizers. Power spectral envelope quantization is enhanced by extending its resolution to 127 samples. The power envelope samples are encoded using a two-stage process. The first stage applies nonlinear interpolative VQ (NLIVQ), a dimensionality reduction process which represents the 127-element power spectral envelope vector using only a 12-dimensional "feature power envelope." Unstructured VQ is applied to the feature power envelope. Then, a full-resolution quantized envelope is obtained from the unstructured VQ index into a corresponding interpolation codebook. In the second stage, segments of a power envelope residual are encoded using 8-, 9-, and 10-element TSVQ. Relative to their first VQ/DCT coder, the authors reported savings of 10–20 kb/s with no reduction in quality due to the CS-VQ and NLIVQ schemes. Although VQ schemes with this level of sophistication typically have not been seen in the audio coding literature since [Chan90] and [Chan91b] first appeared, there have been successful applications of less-sophisticated VQ in some of the standards (e.g., [Sree98a] [Sree98b]).

7.10 MDCT WITH VECTOR QUANTIZATION

Iwakami *et al.* developed transform-domain weighted interleave vector quantization (TWIN-VQ), an MDCT-based coder that also involves transform coefficient VQ [Iwak95]. This algorithm exploits LPC analysis, spectral interframe redundancy, and interleaved VQ.

At the encoder (Figure 7.4.), each frame of MDCT coefficients is first divided by the corresponding elements of the LPC spectral envelope, resulting in a spectrally flattened quotient (residual) sequence. This procedure flattens the MDCT envelope but does not affect the fine structure. The next step, therefore, divides the first step residual by a predicted fine structure envelope. This predicted fine structure envelope is computed as a weighted sum of three previous quantized fine structure envelopes, i.e., using backward prediction. Interleaved VQ is applied to the normalized second step residual. The interleaved VQ vectors are structured in the following way. Each N-sample normalized second step residual vector is split into K subvectors, each containing N/K coefficients. Second step residuals from the N-sample vector are interleaved in the K subvectors such that the i-th subvector contains elements $i + nK$, where $n = 0, 1, \ldots, (N/K) - 1$. Perceptual weighting is also incorporated by weighting each subvector by a nonlinearly transformed version of its corresponding LPC envelope component prior to the codebook search. VQ indices are transmitted to the receiver. Side information

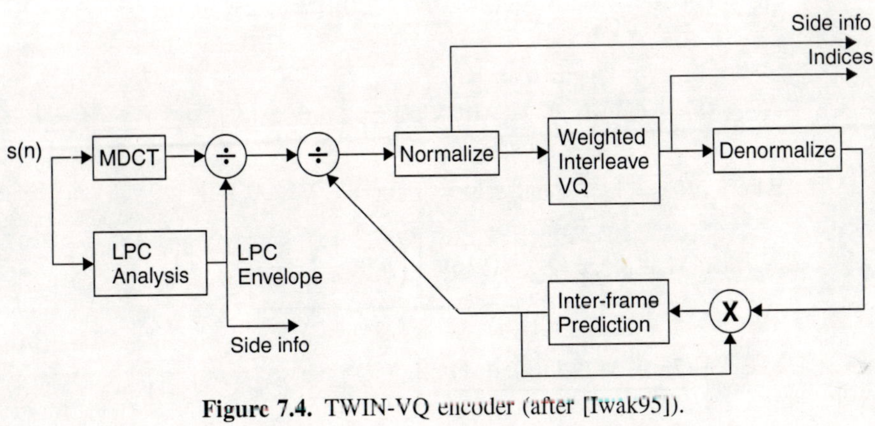

Figure 7.4. TWIN-VQ encoder (after [Iwak95]).

consists of VQ normalization coefficients and the LPC envelope encoded in terms of LSPs. The authors claimed higher subjective quality than MPEG-1 layer II at 64 kb/s for 48 kHz CD-quality audio, as well as higher quality than MPEG-1 layer II for 32 kHz audio at 32 kb/s.

TWIN-VQ performance at lower bit rates has also been investigated. At least three trends were identified during ISO-sponsored comparative tests [ISOI98] of TWIN-VQ and MPEG-2 AAC. First, AAC outperformed TWIN-VQ for bit rates above 16 kb/s. Secondly, TWIN-VQ and AAC achieved similar performance at 16 kb/s, with AAC having a slight edge. Finally, the performance of TWIN-VQ exceeded that of AAC at a rate of 8 kb/s. These results ultimately motivated a combined AAC/TWIN-VQ architecture for inclusion in MPEG-4 [Herre98]. Enhancements to the weighted interleaving scheme and LPC envelope representation [Mori96] enabled real-time implementation of stereo decoders on Pentium-I and PowerPC platforms. Channel error robustness issues are addressed in [Iked95]. A later version of the TWIN-VQ scheme is embedded in the set of tools for MPEG-4 audio.

7.11 SUMMARY

Transform coders for high-fidelity audio were described in this Chapter. The transform coding algorithms presented include

- the OCF algorithm
- the monaural and stereophonic perceptual transform coders (PXFM and SEPXFM)
- the CNET DFT and MDCT coders
- the ASPEC
- the differential PAC
- the TWIN-VQ algorithm.

PROBLEMS

7.1. Given the expressions for the DFT, the DCT, and the MDCT,

$$X_{DFT}(k) = \frac{1}{\sqrt{2M}} \sum_{n=0}^{2M-1} x(n) e^{-j\pi nk/M}, 0 \leqslant k \leqslant 2M-1$$

$$X_{DCT}(k) = c(k)\sqrt{\frac{2}{M}} \sum_{n=0}^{M-1} x(n) \cos\left[\frac{\pi}{M}\left(n+\frac{1}{2}\right)k\right], 0 \leqslant k \leqslant M-1$$

where $c(0) = 1\sqrt{2}$, and $c(k) = 1$ for $1 \leqslant k \leqslant M-1$

$$X_{MDCT}(k) = \sqrt{\frac{2}{M}} \sum_{n=0}^{2M-1} x(n) \underbrace{\sin\left[\left(n+\frac{1}{2}\right)\frac{\pi}{2M}\right]}_{w(n)}$$

$$\cos\left[\frac{(2n+M+1)(2k+1)\pi}{4M}\right], \quad \text{for} \quad 0 \leqslant k \leqslant M-1$$

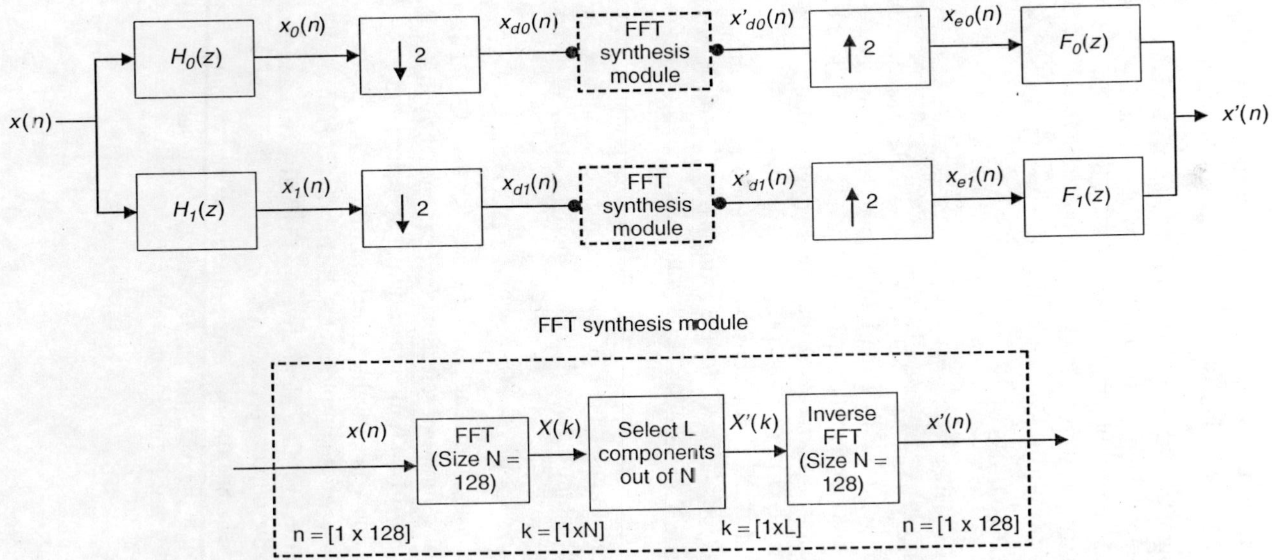

Figure 7.5. FFT analysis/synthesis within the two bands of QMF bank.

Write the three transforms in matrix form as follows

$$\mathbf{X}_{\mathrm{T}} = \mathbf{H}\mathbf{x},$$

where $\mathbf{H}$ is the transform matrix, and $\mathbf{x}$ and $\mathbf{X}_{\mathrm{T}}$ denote the input and transformed vector, respectively. Note the structure in the transform matrices.

7.2. Give the signal flowgraph of the FFT butterfly structure for an 8-point DFT, an 8-point DCT, and an 8-point MDCT. Specify clearly the values on the nodes and the branches. [Hint: See Problem 6.16 and Figure 6.18 in Chapter 6.]

COMPUTER EXERCISES

7.3. In this problem, we will study the energy compaction of the DFT and the DCT. Use $x(n) = e^{-0.5n}\sin(0.4\pi n), n = 0, 1, \ldots, 15$. Plot the 16-point DFT and 16-point DCT of the input signal, $x(n)$. See how the energy of the sequence is concentrated. Now pick two peaks of the DFT vector and the DCT vector and synthesize the input signal, $x(n)$. Let the synthesized signals be, $\hat{x}_{DFT}(n)$ and $\hat{x}_{DCT}(n)$. Compute the MSE values between the input signal and the two reconstructed signals. Repeat this for four peaks, six peaks, and eight peaks. Plot the estimated MSE values across the number of peaks selected and comment on your result.

7.4. This computer exercise is a combination of Problems 2.24 and 2.25 in Chapter 2. In particular, the FFT analysis/synthesis module, in Problem 2.25, will be used within the two bands of the QMF bank. The configuration is shown in Figure 7.5.

a. Given, $H_0(z) = 1 - z^{-1}$, $H_1(z) = 1 + z^{-1}$. Choose $F_0(z)$ and $F_1(z)$ such that the aliasing term can be cancelled. Use $L = 32$ and the peak-picking method for component selection. Perform speech synthesis and give time-domain plots of both input and output speech records.

b. Use the same voiced frame selected in Problem 2.24. Give time-domain and frequency-domain plots of $x'_{d0}(n)$ and $x'_{d1}(n)$ in Figure7.5.

c. Compute the overall SNR (between $x(n)$ and $x'(n)$) and estimate a MOS score for the output speech.

d. Describe whether the perceptual quality of the output speech improves if the FFT analysis/synthesis module is employed within the subbands instead of using it for the entire band.

CHAPTER 8

SUBBAND CODERS

8.1 INTRODUCTION

Similar to the transform coders described in the previous chapter, subband coders also exploit signal redundancy and psychoacoustic irrelevancy in the frequency domain. The audible frequency spectrum (20 Hz–20 kHz) is divided into frequency subbands using a bank of bandpass filters. The output of each filter is then sampled and encoded. At the receiver, the signals are demultiplexed, decoded, demodulated, and then summed to reconstruct the signal. Audio subband coders realize coding gains by efficiently quantizing decimated output sequences from perfect reconstruction filter banks. Efficient quantization methods usually rely upon psychoacoustically controlled dynamic bit allocation rules that allocate bits to subbands in such a way that the reconstructed output signal is free of audible quantization noise or other artifacts. In a generic subband audio coder, the input signal is first split into several uniform or nonuniform subbands using some critically sampled, perfect reconstruction (or nearly perfect reconstruction) filter bank. Nonideal reconstruction properties in the presence of quantization noise are compensated for by utilizing subband filters that have good sidelobe attenuation. Then, decimated output sequences from the filter bank are normalized and quantized over short, 2–10 ms blocks. Psychoacoustic signal analysis is used to allocate an appropriate number of bits for the quantization of each subband. The usual approach is to allocate an adequate number of bits to mask quantization noise in each block while simultaneously satisfying some bit rate constraint. Since masking thresholds and hence bit allocation requirements are time-varying, buffering is often introduced to match the coder output to a fixed rate. The encoder

Audio Signal Processing and Coding, by Andreas Spanias, Ted Painter, and Venkatraman Atti

sends to the decoder quantized subband output samples, normalization scale factors for each block of samples, and bit allocation side information. Bit allocation may be transmitted as explicit side information, or it may be implicitly represented by some parameter such as the scale factor magnitudes. The decoder uses side information and scale factors in conjunction with an inverse filter bank to reconstruct a coded version of the original input.

The purpose of this chapter is to expose the reader to subband coding algorithms for high-fidelity audio. This chapter is organized much like Chapter 7. The first portion of this chapter is concerned with early subband algorithms that not only contributed to the MPEG-1 standardization, but also had an impact on later developments in the field. The remainder of the chapter examines a variety of recent experimental subband algorithms that make use of discrete wavelet transforms (DWT), discrete wavelet packet transforms (DWPT), and hybrid filter banks. The chapter is organized as follows. Section 8.1.1 concentrates upon the early subband coding algorithms for high-fidelity audio, including the Masking Pattern Adapted Universal Subband Integrated Coding and Multiplexing (MUSICAM). Section 8.2 presents the filter-bank interpretations of the DWT and the DWPT. Section 8.3 addresses subband audio coding algorithms in which time-invariant and time-varying, signal adaptive filter banks are constructed from the DWT and the DWPT. Section 8.4 examines the use of nonuniform filter banks related the DWPT. Sections 8.5 and 8.6 are concerned with hybrid subband architectures involving sinusoidal modeling and code-excited linear prediction (CELP). Finally, Section 8.7 addresses subband audio coding with IIR filter banks.

8.1.1 Subband Algorithms

This section is concerned with early subband algorithms proposed by researchers from the Institut fur Rundfunktechnik (IRT) [Thei87] [Stoll88], Philips Research Laboratories [Veld89], and CCETT. Much of this work was motivated by standardization activities for the European Eureka-147 digital broadcast audio (DBA) system. The ISO/IEC eventually clustered the IRT, Philips, and CCETT proposals into the MUSICAM algorithm [Wies90] [Dehe91], which was adopted as part of the ISO/IEC MPEG-1 and MPEG-2 BC-LSF audio coding standards.

8.1.1.1 Masking Pattern Adapted Subband Coding (MASCAM) The MUSICAM algorithm is derived from coders developed at IRT, Philips, and CNET. At IRT, Theile, Stoll, and Link developed Masking Pattern Adapted Subband Coding (MASCAM), a subband audio coder [Thei87] based upon a tree-structured quadrature mirror filter (QMF) filter bank that was designed to mimic the critical band structure of the auditory filter bank. The coder has 24 nonuniform subbands, with bandwidths of 125 Hz below 1 kHz, 250 Hz in the range 1–2 kHz, 500 Hz in the range 2–4 kHz, 1 kHz in the range 4–8 kHz, and 2 kHz from 8 kHz to 16 kHz. The prototype QMF has 64 taps. Subband output sequences are processed in 2-ms blocks. A normalization scale factor is quantized and transmitted for each block from each subband. Subband bit allocations are derived from a simplified psychoacoustic analysis. The original coder reported in [Thei87] considered only in-band simultaneous masking. Later, as described in [Stol88], interband simultaneous masking and temporal masking were added to the bit rate calculation. Temporal postmasking is exploited by updating scale factors less frequently during periods of signal decay. The MASCAM coder was reported to achieve high-quality results for 15 kHz bandwidth input signals at bit rates between 80 and 100 kb/s per channel. A similar subband coder was developed at Philips during this same period. As described by Velhuis *et al.* in [Veld89], the Philips group investigated subband schemes based on 20- and 26-band nonuniform filter banks. Like the original MASCAM system, the Philips coder relies upon a highly simplified masking model that considers only the upward spread of simultaneous masking. Thresholds are derived from a prototype basilar excitation function under worst-case assumptions regarding the frequency separation of masker and maskee. Within each subband, signal energy levels are treated as single maskers. Given SNR targets due to the masking model, uniform ADPCM is applied to the normalized output of each subband. The Philips coder was claimed to deliver high-quality coding of CD-quality signals at 110 kb/s for the 26-band version and 180 kb/s for the 20-band version.

8.1.1.2 Masking Pattern Adapted Universal Subband Integrated Coding and Multiplexing (MUSICAM) Based primarily upon coders developed at IRT and Philips, the MUSICAM algorithm [Wies90] [Dehe91] was successful in the 1990 ISO/IEC competition [SBC90] for a new audio coding standard. It eventually formed the basis for MPEG-1 and MPEG-2 audio layers I and II. Relative to its predecessors, MUSICAM (Figure 8.1) makes several practical tradeoffs between complexity, delay, and quality. By utilizing a uniform bandwidth, 32-band pseudo-QMF bank (aka "polyphase" filter bank) instead of a tree-structured QMF bank, both complexity and delay are greatly reduced relative to the IRT and Phillips coders. Delay and complexity are 10.66 ms and 5 MFLOPS, respectively. These improvements are realized at the expense of using a sub-optimal

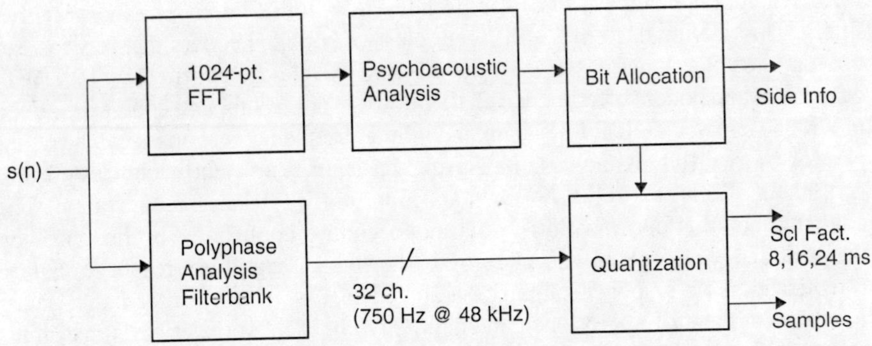

Figure 8.1. MUSICAM encoder (after [Wies90]).

filter bank, however, in the sense that filter bandwidths (constant 750 Hz for 48 kHz sample rate) no longer correspond to the critical band rate. Despite these excessive filter bandwidths at low frequencies, high-quality coding is still possible with MUSICAM due to its enhanced psychoacoustic analysis. High-resolution spectral estimates (46 Hz/line at 48 kHz sample rate) are obtained through the use of a 1024-point FFT in parallel with the PQMF bank. This parallel structure allows for improved estimation of masking thresholds and hence determination of more accurate minimum signal-to-mask ratios (SMRs) required within each subband.

The MUSICAM psychoacoustic analysis procedure is essentially the same as the MPEG-1 psychoacoustic model 1. The remainder of MUSICAM works as follows. Subband output sequences are processed in 8-ms blocks (12 samples at 48 kHz), which is close to the temporal resolution of the auditory system (4–6 ms). Scale factors are extracted from each block and encoded using 6 bits over a 120-dB dynamic range. Occasionally, temporal redundancy is exploited by repetition over 2 or 3 blocks (16 or 24 ms) of slowly changing scale factors within a single subband. Repetition is avoided during transient periods such as sharp attacks. Subband samples are quantized and coded in accordance with SMR requirements for each subband as determined by the psychoacoustic analysis. Bit allocations for each subband are transmitted as side information. On the CCIR five-grade impairment scale, MUSICAM scored 4.6 (std. dev. 0.7) at 128 kb/s, and 4.3 (std. dev. 1.1) at 96 kb/s per monaural channel, compared to 4.7 (std. dev. 0.6) on the same scale for the uncoded original. Quality was reported to suffer somewhat at 96 kb/s for critical signals which contained sharp attacks (e.g., triangle, castanets), and this was reflected in a relatively high standard deviation of 1.1. MUSICAM was selected by ISO/IEC for MPEG-1 audio due to its desirable combination of high quality, reasonable complexity, and manageable delay. Also, bit error robustness was found to be very good (errors nearly imperceptible) up to a bit error rate of 10^{-3}.

8.2 DWT AND DISCRETE WAVELET PACKET TRANSFORM (DWPT)

The previous section described subband coding algorithms that utilize banks of fixed resolution bandpass QMF or pseudo-QMF finite impulse response (FIR) filters. This section describes a different class of subband coders that rely instead upon a filter-bank interpretation of the discrete wavelet transform (DWT). DWT-based subband coders offer increased flexibility over the subband coders described previously since identical filter-bank magnitude frequency responses can be obtained for many different choices of a wavelet basis, or equivalently, choices of filter coefficients. This flexibility presents an opportunity for basis optimization. The advantage of this optimization in the audio coding application is illustrated by the following example. First, a desired filter-bank magnitude response can be established. This response might be matched to the auditory filter bank. Then, for each segment of audio, one can adaptively choose a wavelet basis that minimizes the rate for some target distortion level. Given a psychoacoustically derived distortion target, the encoding remains perceptually transparent.

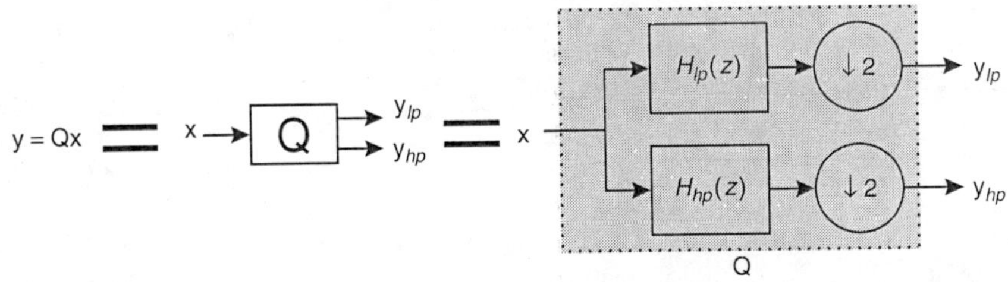

Figure 8.2. Filter-bank interpretation of the DWT.

A detailed discussion of specific technical conditions associated with the various wavelet families is beyond the scope of this book, and this chapter therefore concentrates upon high-level coder architectures. In-depth treatment of wavelets is available from many sources, e.g., [Daub92]. Before describing the wavelet-based coders, however, it is useful to summarize some basic wavelet characteristics. Wavelets are a family of basis functions for the space of square integrable signals. A finite energy signal can be represented as a weighted sum of the translates and dilates of a single wavelet. Continuous-time wavelet signal analysis can be extended to discrete-time and square summable sequences. Under certain assumptions, the DWT acts as an orthonormal linear transform $T: R^N \rightarrow R^N$. For a compact (finite) support wavelet of length K, the associated transformation matrix, $\mathbf{Q}$, is fully determined by a set of coefficients $\{c_k\}$ for $0 \leqslant k \leqslant K-1$. As shown in Figure 8.2, this transformation matrix has an associated filter-bank interpretation. One application of the transform matrix, $\mathbf{Q}$, to an $N \times 1$ signal vector, $\mathbf{x}$, generates an $N \times 1$ vector of wavelet-domain transform coefficients, $\mathbf{y}$. The $N \times 1$ vector $\mathbf{y}$ can be separated into two $\frac{N}{2} \times 1$ vectors of approximation and detail coefficients, $\mathbf{y}_{lp}$ and $\mathbf{y}_{hp}$, respectively. The spectral content of the signal $\mathbf{x}$ captured in $\mathbf{y}_{lp}$ and $\mathbf{y}_{hp}$ corresponds to the frequency subbands realized in the 2:1 decimated output sequences from a QMF bank (Section 6.4), which obeys the "power complimentary condition", i.e.,

$$|H_{lp}(\Omega)|^2 + |H_{lp}(\Omega + \pi)|^2 = 1, \tag{8.1}$$

where $H_{lp}(\Omega)$ is the frequency response of the lowpass filter. Therefore, recursive DWT applications effectively pass input data through a tree-structured cascade of lowpass (LP) and highpass (HP) filters followed by 2:1 decimation at every node. The forward/inverse transform matrices of a particular wavelet are associated with a corresponding QMF analysis/synthesis filter bank. The usual wavelet decomposition implements an octave-band filter bank structure as shown in Figure 8.3. In the figure, frequency subbands associated with the coefficients from each stage are schematically represented for an audio signal sampled at 44.1 kHz.

Wavelet packet (WP) or discrete wavelet packet transform (DWPT) representations, on the other hand, decompose both the detail and approximation coefficients at each stage of the tree, as shown in Figure 8.4. In the figure, frequency subbands

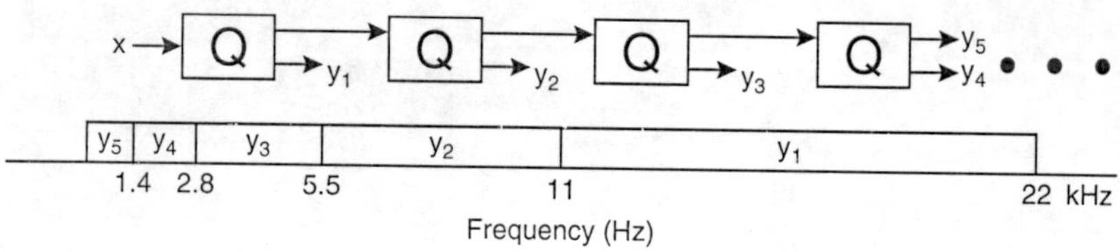

Figure 8.3. Octave-band subband decomposition associated with a discrete wavelet transform ("DWT").

associated with the coefficients from each stage are schematically represented for a 44.1-kHz sample rate.

A filter-bank interpretation of wavelet transforms is attractive in the context of audio coding algorithms. Wavelet or wavelet packet decompositions can be tree structured as necessary (unbalanced trees are possible) to decompose input audio into a set of frequency subbands tailored to some application. It is possible, for example, to approximate the critical band auditory filter bank utilizing a wavelet packet approach. Moreover, many K-coefficient finite support wavelets are associated with a single magnitude frequency response QMF pair, and therefore a specific subband decomposition can be realized while retaining the freedom to choose a wavelet basis which is in some sense "optimal." These considerations have motivated the development of several experimental wavelet-based subband coders in recent years. The basic idea behind DWT and DWPT-based subband coders is to quantize and encode efficiently the coefficient sequences associated with each stage of the wavelet decomposition tree using the same noise shaping techniques as the previously described perceptual subband coders.

The next few sections of this chapter, Sections 8.3 through 8.5, expose the reader to several WP-based subband coders developed in the early 1990s by Sinha, Tewfik, *et al.* [Sinh93a] [Sinh93b] [Tewf93], as well as more recently proposed hybrid sinusoidal/WPT algorithms developed by Hamdy and Tewfik [Hamd96], Boland and Deriche [Bola97], and Pena *et al.* [Pena96] [Prel96a] [Prel96b] [Pena97a]. The core of least one experimental WP audio coder [Sinh96] has been embedded in a commercial standard, namely the AT&T Perceptual Audio Coder (PAC) [Sinh98]. Although not addressed in this chapter, we note that other studies of DWT and DWPT-based audio coding schemes have appeared. For example, experimental coder architectures for low-complexity, low-delay, combined wavelet/multipulse LPC coding, and combined scalar/vector quantization of transform coefficients were reported, respectively, by Black and Zeytinoglu [Blac95], Kudumakis and Sandler [Kudu95a] [Kudu95b] [Kudu96], and Boland and Deriche [Bola95][Bola96]. Several bit rate scalable DWPT-based schemes have also been investigated recently. For example, a fixed-tree DWPT coding scheme capable of nearly transparent quality with scalable bitrates below 100 kb/s was proposed by Dobson *et al.* and implemented in real-time on a 75 MHz Pentium-class platform [Dobs97]. Additionally, Lu and Pearlman investigated a rate-scalable DWPT-based coder that applies set partitioning in hierarchical trees (SPIHT) to generate an embedded bitstream. Nearly transparent quality was reported at bit rates between 55 and 66 kb/s [Lu98].

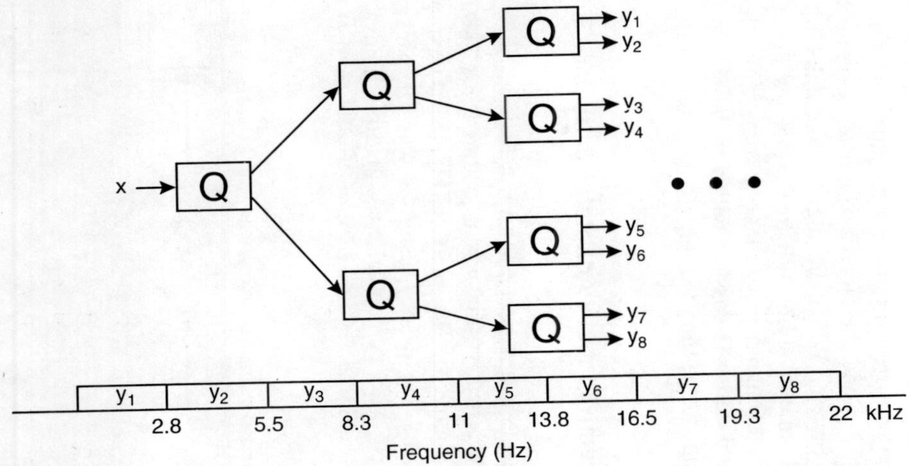

Figure 8.4. Subband decomposition associated with a particular wavelet packet transform ("WPT" or "WP"). Although the picture illustrates a balanced binary tree and the associated uniform bandwidth subbands, nodes could be pruned in order to achieve nonuniform frequency subdivision.

8.3 ADAPTED WP ALGORITHMS

The "best basis" methodologies [Coif92] [Wick94] for adapting the WP *tree structure* to signal properties are typically formulated in terms of Shannon entropy [Shan48] and other perceptually blind statistical measures. For a given WP tree, related research directed towards *optimal filter selection* [Hedg97] [Hedg98a] [Hedg98b] has also emphasized optimization of statistical rather than perceptual properties. The questions of perceptually motivated filter selection and tree construction are central to successful application of WP analysis in audio coding algorithms. The WP tree structure determines the time and frequency resolution of the transform and therefore also creates a particular tiling of the time-frequency plane. Several WP audio algorithms [Sinh93b] [Dobs97] have successfully employed time-invariant WP tree structures that mimic the ear's critical band frequency resolution properties. In some cases, however, a more efficient perceptual bit allocation is possible with a signal-specific time-frequency tiling that tracks the shape of the time-varying masking threshold. Some examples are described next.

8.3.1 DWPT Coder with Globally Adapted Daubechies Analysis Wavelet

Sinha and Tewfik developed a variable-rate wavelet-based coding scheme for which they reported nearly transparent coding of CD-quality audio at 48–64 kb/s [Sinh93a] [Sinh93b]. The encoder (Figure 8.5) exploits redundancy using a VQ scheme and irrelevancy using a wavelet packet (WP) signal decomposition combined with perceptual masking thresholds. The algorithm works as follows. Input audio is segmented into $N \times 1$ vectors, which are then

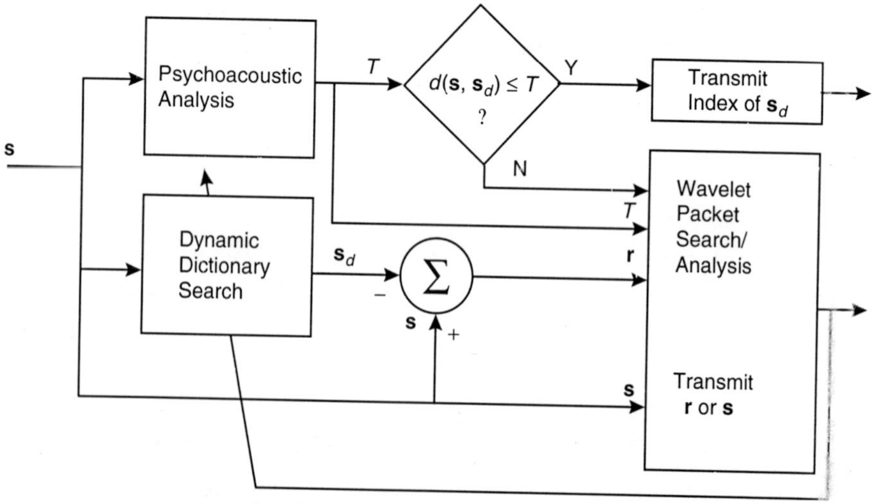

Figure 8.5. Dynamic dictionary/optimal wavelet packet encoder (after [Sinh93a]).

windowed using a 1/16-th overlap square root Hann window. The dynamic dictionary (DD), which is essentially an adaptive VQ subsystem, then eliminates signal redundancy. A dictionary of $N \times 1$ codewords is searched for the vector perceptually closest to the input vector. The effective size of the dictionary is made larger than its actual size by a novel correlation lag search/time-warping procedure that identifies two $N/2$-sample codewords for each N-sample input vector. At both the transmitter and receiver, the dictionary is systematically updated with N-sample reconstructed output audio vectors according to a perceptual distance criterion and last-used-first-out rule. After the DD procedure has been completed, an optimized WP decomposition is applied to the original signal as well as the DD residual. The decomposition tree is structured such that its 29 frequency subbands roughly correspond to the critical bands of the auditory filter bank. A masking threshold, obtained as in [Veld89], is assumed constant within each subband and then used to compute a perceptual bit allocation.

The encoder transmits the particular combination of DD and WP information that minimizes the bit rate while maintaining perceptual quality. Three combinations are possible. In one scenario, the DD index and time-warping factor are transmitted alone if the DD residual energy is below the masking threshold at all frequencies. Alternatively, if the DD residual has audible noise energy, then WP coefficients of the DD residual are also quantized, encoded, and transmitted. In some cases, however, WP coefficients corresponding to the original signal are more compactly represented than the combination of the DD plus WP residual information. In this case, the DD information is discarded and only quantized and encoded WP coefficients are transmitted. In the latter two cases, the encoder also transmits subband scale factors, bit allocations, and energy normalization side information.

This algorithm is unique in that it contains the first reported application of adapted WP analysis to perceptual subband coding of high-fidelity, CD-quality audio. During each frame, the WP basis selection procedure applies an optimality criterion of minimum bit rate for a given distortion level. The adaptation is "global" in the sense that the same analysis wavelet is applied to the entire decomposition. The authors reached several conclusions regarding the optimal compact support (K-coefficient) wavelet basis when selecting from among the Daubechies orthogonal wavelet bases ([Daub88]).

First, optimization produced average bit rate savings dependent on filter length of up to 15%. Average bit rate savings were 3, 6.5, 8.75, and 15% for wavelets selected from the sets associated with coefficient sequences of lengths 10, 20, 40, and 60, respectively. In an extreme case, a savings of 1.7 bits/sample is realized for transparent coding of a difficult castanets sequence when using best-case rather than worst-case wavelets (0.8 *vs* 2.5 bits/sample for $K = 40$). The second conclusion reached by the researchers was that it is not necessary to search exhaustively the space of all wavelets for a particular value of K. The search can be constrained to wavelets with $K/2$ vanishing moments (the maximum possible number) with minimal impact on bit rate. The frequency responses of the filters associated with a p-th-order vanishing moment wavelet have p-th-order zeros at the foldover frequency, i.e., $\Omega = \pi$. Only a 3.1% bitrate reduction was realized for an exhaustive search versus a maximal vanishing moment constrained search. Third, the authors found that larger K, i.e., more taps, and deeper decomposition trees tended to yield better results. Given identical distortion criteria for a castanets sequence, bit rates of 2.1 bits/sample for $K = 4$ wavelets were realized versus 0.8 bits/sample for $K = 40$ wavelets.

As far as quality is concerned, subjective tests showed that the algorithm produced transparent quality for certain test material including drums, pop, violin with orchestra, and clarinet. Subjects detected differences, however, for the castanets and piano sequences. These difficulties arise, respectively, because of inadequate pre-echo control, and inefficient modeling of steady sinusoids. The coder utilizes only an adaptive window scheme which switches between 1024 and 2048-sample windows. Shorter windows ($N = 1024$ or 23 ms) are used for signals that are likely to produce pre-echoes. The piano sequence contained long segments of nearly steady or slowly decaying sinusoids. The wavelet coder does not handle steady sinusoids as well as other signals. With the exception of these troublesome signals in a comparative test, one additional expert listener also found that the WP coder outperformed MPEG-1, layer II at 64 kb/s.

Tewfik and Ali later enhanced the WP coder to improve pre-echo control and increase coding efficiency. After elimination of the dynamic dictionary, they reported improved quality in the range of 55 to 63 kb/s, as well as a real-time implementation of a simplified 64 to 78 kb/s coder on two TMS320C31 devices [Tewf93]. Other improvements included exploitation of auditory temporal masking for pre-echo control, more efficient quantization and encoding of scale-factors, and run-length coding of long zero sequences. The improved WP coder also upgraded its psychoacoustic analysis section with a more sophisticated model similar to Johnston's PXFM coder [John88a]. The most notable improvement occurred in the area of pre-echo control. This was accomplished in the following manner. First, input frames likely to produce pre-echoes are identified using a normalized energy measure criterion. These frames are parsed into 5-ms time slots (256 samples). Then, WP coefficients from all scales within each time slot are combined to estimate subframe energies. Masking thresholds computed over the global 1024-sample frame are assumed only to apply during high-energy time slots. Masking thresholds are reduced across all subbands for low-energy time slots utilizing weighting factors proportional to the energy ratio between high- and low-energy time-slots. The remaining enhancements of improved scale factor coding efficiency and run-length coding of zero sequences more than compensated for removal of the dynamic dictionary.

8.3.2 Scalable DWPT Coder with Adaptive Tree Structure

Srinivasan and Jamieson proposed a WP-based audio coding scheme [Srin97] [Srin98] in which a signal-specific perceptual best basis is constructed by adapting the WP tree structure on each frame such that perceptual entropy and, ultimately, the bit rate are minimized. While the tree structure is signal-adaptive, the analysis

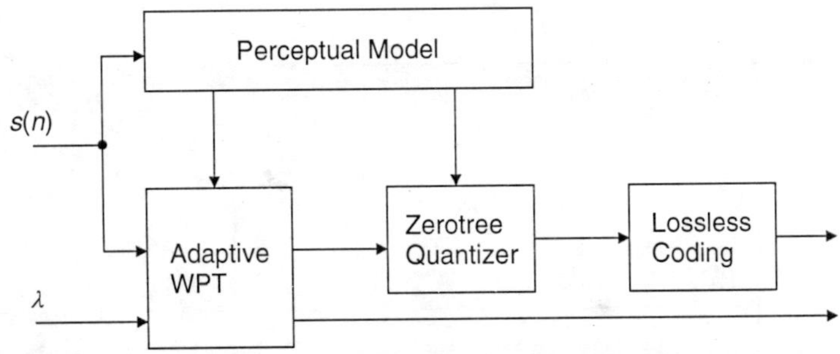

Figure 8.6. Masking-threshold adapted WP audio coder [Srin98]. On each frame, the WP tree structure is adapted in order to minimize a perceptually motivated rate constraint.

filters are time-invariant and obtained from the family of spline-based biorthogonal wavelets [Daub92]. The algorithm (Figure 8.6) is also unique in the sense that it incorporates mechanisms for both bit rate and complexity scaling. Before the tree adaptation process can commence for a given frame, a set of 63 masking thresholds corresponding to a set of threshold frequency partitions roughly 1/3 Bark wide is obtained from the ISO/IEC MPEG-1 psychoacoustic model recommendation 2 [ISOI92]. Of course, depending upon the WP tree, the subbands may or may not align with the threshold partitions. For any particular WP tree, the associated bit rate (cost) is computed by extracting the minimum masking thresholds from each subband and then allocating sufficient bits to guarantee that the quantization noise in each band does not exceed the minimum threshold.

The objective of the tree adaptation process, therefore, is to construct a minimum cost subband decomposition by maximizing the minimum masking threshold in every subband. Figure 8.7a shows a possible subband structure in which subband 0 contains five threshold partitions. This choice of bandsplitting is clearly undesirable since the minimum masking threshold for partition 1 is far below partition 4. Bit allocation for subband 0 will be forced to satisfy partition 1 with a resulting overallocation for partitions 2 through 5.

It can be seen that subdividing the band (Figure 8.7b) relaxes the minimum masking threshold in band 1 to the level of partition 5. Naturally, the ideal bandsplitting would in this case ultimately match the subband boundaries to the threshold partition boundaries. On each frame, therefore, the tree adaptation process performs the following top-down, iterative "growing" procedure. During any iteration, the existing subbands are assigned individual costs based on the bit allocation required for transparent coding. Then, a decision on whether or not to subdivide further at a node is made on the basis of cost reduction. Subbands are examined for potential splitting in order of decreasing cost, and the search is "breadth-first," meaning that each level is completely decomposed before proceeding to the next level. Subdivision occurs only if the associated bit rate improvement exceeds a threshold. The tree adaptation is also constrained by a complexity scaling mechanism. Top-down tree growth is halted by the complexity scaling constraint, λ, when the estimated total cost of computing the

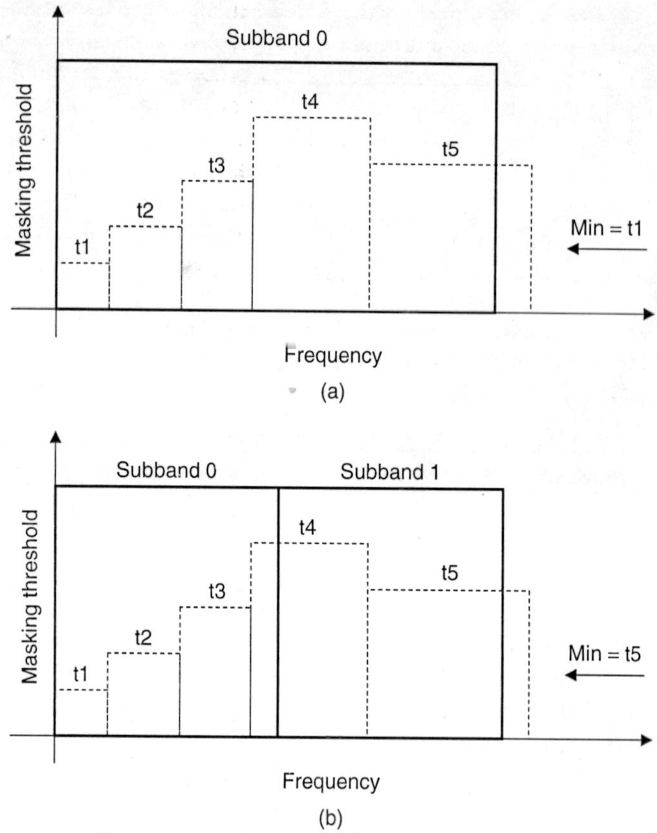

Figure 8.7. Example masking-threshold adapted WP filter bank: (a) initial condition, (b) after one iteration. Threshold partitions are denoted by dashed lines and labeled by t_k. Idealized subband boundaries are denoted by heavy black lines. Under the initial condition, with only one subband, the minimum masking threshold is given by t_1, and therefore the bit allocation will be relatively large in order to satisfy a small threshold. After one band splitting, however, the minimum threshold in subband 1 increases from t_1 to t_5, thereby reducing the perceptual bit allocation. Hence, the cost function is reduced in part (b) relative to part (a).

DWPT reaches a predetermined limit. With this feature, it is envisioned that in a real-time environment the WP adaptation process could respond to changing CPU resources by controlling the cost of the analysis and synthesis filter banks.

In [Srin98], a complexity-constrained tree adaptation procedure is shown to yield a basis requiring the fewest bits for perceptually transparent coding for a given complexity and temporal resolution. After the WP tree adaptation procedure has been completed, Shapiro's zerotree algorithm [Shap93] is applied iteratively to quantize the coefficients and exploit remaining temporal correlation until the perceptual rate-distortion criteria are satisfied, i.e., until sufficient bits have been allocated to satisfy the perceptually transparent bit rate associated with the given

WP tree. The zerotree technique has the added benefit of generating an embedded bitstream, making this coder amenable to progressive transmission. In scalable applications, the embedded bitstream has the property that it can be partially decoded and still guarantee the best possible quality of reconstruction given the number of bits decoded. The complete bitstream consists of the encoded tree structure, the number of zerotree iterations, and a block of zerotree encoded data. These elements are coded in a lossless fashion (e.g., Huffman, arithmetic, etc.) to remove any remaining redundancies and transmitted to the decoder. For informal listening tests over coded program material that included violin, violin/viola, flute, sitar, vocals/orchestra, and sax the coded outputs at rates in the vicinity of 45 kb/s were reported to be indistinguishable from the originals with the exceptions of the flute and sax.

8.3.3 DWPT Coder with Globally Adapted General Analysis Wavelet

Srinivasan and Jamieson [Srin98] demonstrated the advantages of a masking threshold adapted WP tree with a time-invariant analysis wavelet. On the other hand, Sinha and Tewfik [Sinh93b] used a time-invariant WP tree but a globally adapted analysis wavelet to demonstrate that there exists a signal-specific "best" wavelet basis in terms of perceptual coding gain for a particular number of filter taps. The basis optimization in [Sinh93b], however, was restricted to Daubechies' wavelets. Recent research has attempted to identify which wavelet properties portend an optimal basis, as well as to consider basis optimization over a broader class of wavelets. In an effort to identify those wavelet properties that could be associated with the "best" filter, Philippe *et al.* measured the impact on perceptual coding gain of wavelet regularity, AR(1) coding gain, and filter bank frequency selectivity [Phil95a] [Phil95b]. The study compared performance between orthogonal Rioul [Riou94], orthogonal Onno [Onno93], and the biorthogonal wavelets of [More95] in a WP coding scheme that had essentially the same time-invariant critical band WP decomposition tree as [Sinh93b]. Using filters of lengths varying between 4 and 120 taps, minimum bit rates required for transparent coding in accordance with the usual perceptual subband bit allocations were measured for each wavelet. For a given filter length, the results suggested that neither regularity nor frequency selectivity mattered significantly. On the other hand, the minimum bit rate required for transparent coding was shown to decrease with increasing analysis filter AR(1) coding gain, leading the authors to conclude that AR(1) coding gain is a legitimate criterion for WP filter selection in perceptual coding schemes.

8.3.4 DWPT Coder with Adaptive Tree Structure and Locally Adapted Analysis Wavelet

Phillipe *et al.* [Phil96] measured the perceptual coding gain associated with optimization of the WP analysis filters at every node in the tree, as well as optimization of the tree structure. In the first experiment, the WP tree structure was fixed, and then optimal filters were selected for each tree node (local adaptation) such that the bit rate required for transparent coding was minimized. Simulated annealing [Kirk83] was used to solve the discrete optimization problem posed by a search space containing 300 filters of varying lengths from the Daubechies [Daub92], Onno [Onno93], Smith-Barnwell [Smit86], Rioul [Riou94], and Akansu-Caglar [Cagl91] families. Then, the filters selected by simulated annealing were used in a second set of experiments on tree structure optimization. The best WP decomposition tree was constructed by means of a growing procedure starting from a single cell and progressively subdividing. Further splitting at each node occurred only if it significantly reduced the perceptually transparent bit rate. As in [Phil95b], these filter and tree adaptation experiments estimated bit rates required for perceptually transparent coding of 48-kHz sampled source material using statistical signal properties. For a fixed tree, the filter adaptation experiments yielded several noteworthy results. First, a nominal bit rate reduction of 3% was realized for Onno's filters (66.5 kb/s) relative to Daubechies' filters (68 kb/s) when the same filter family was applied in all tree nodes and filter length was the only free parameter. Secondly, simulated annealing over the search space of 300 filters yielded a nominal 1% bit rate reduction (66 kb/s) relative to the Onno-only case. Finally, longer filter bank delay, i.e., longer analysis filters and hence better frequency selectivity, yielded lower bitrates. For low-delay applications, however, a sevenfold delay reduction from 700 down to only 100 samples is realized at the cost of only a 10% increase in bit rate. The tree adaptation experiments showed that a 16-band decomposition yielded the best bit rate when tree description overhead was accounted for. In light of these results and the wavelet adaptation results of [Sinh93b], one might conclude that WP filter and WP tree optimization are warranted if less than a 10% bit rate improvement justifies the added complexity.

8.3.5 DWPT Coder with Perceptually Optimized Synthesis Wavelets

The wavelet-based audio coding schemes as well as WP tree and filter adaptation experiments described in the foregoing sections (e.g., [Sinh93b] [Phil95a] [Phil95b] [Phil96]) seek to maximize perceptual coding efficiency by matching subband bandwidths (i.e., the time-frequency tiling) and/or individual filter magnitude and phase characteristics to incoming signal properties. All of these techniques make use of perfect reconstruction ("PR") DWT or WP filter banks that are designed to split a signal into frequency subbands in the analysis filter bank, and then later recombine the subband signals in the synthesis filter bank to reproduce exactly the original input signal. The PR property only holds, however, so long as distortion is not injected into the subband sequences, i.e., in the absence of quantization. This is an important point to consider in the context of coding. The quantization noise introduced into the subbands during bit allocation leads to filter bank-induced reconstruction artifacts because the synthesis filter bank has carefully controlled spectral leakage properties specifically designed to cancel the aliasing and imaging distortions introduced by the critically sampled analysis-synthesis process. Whether using classical or perceptual bit allocation rules, most subband coders do not account explicitly for the filter bank distortion artifacts introduced by quantization. Using explicit knowledge of the analysis filters and the quantization noise, however, recent research has shown that reconstruction distortion can be minimized in the mean square sense (MMSE) by relaxing PR constraints and tuning the synthesis filters [Chen95] [Hadd95] [Kova95] [Delo96] [Goss97b]. Naturally, mean square error minimization is of limited value for subband audio coders. As a result, Gosse *et al.* [Goss95] [Goss97] extended the MMSE synthesis filter tuning procedure [Goss96] to minimize a mean perceptual error (MMPE) rather than MMSE. Experiments were conducted to determine whether or not tuned synthesis filters outperform the unmodified PR synthesis filters, and, if so, whether or not MMPE filters outperform MMSE filters in subjective listening tests. A WP audio coding scheme configured for 128 kb/s operation and having a time-invariant filter-bank structure (Figure 8.8) formed the basis for the experiments.

The tree and filter selections were derived from the minimum-rate filter and tree adaptation investigation reported in [Phil96]. In the figure, each of the 16 subbands is labeled with its upper cutoff frequency (kHz). The experiments involved first a design phase and then an evaluation phase. During the design phase, optimized synthesis filter coefficients were obtained as follows. For the MMPE filters, coding simulations were run using the unmodified PR synthesis filter bank with psychoacoustically derived bit allocations for each subband on each frame. A mean perceptual error (MPE) was evaluated at the PR filter bank output in terms of a unique JND measure [Duro96]. Then, the filter tuning algorithm [Goss96] was applied to minimize the reconstruction error. Since the bit allocation was perceptually motivated, the tuning and reconstruction error minimization procedure yielded MMPE filter coefficients. For the MMSE filters, coefficients were also obtained using [Goss96] without the benefit of a perceptual bit allocation step.

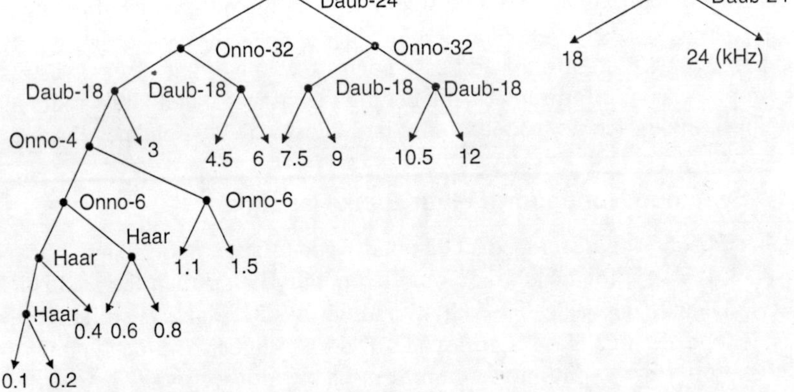

Figure 8.8. Wavelet packet analysis filter-bank optimized for minimum bitrate, used in MMPE experiments.

During the evaluation phase of the experiments, three 128 kb/s coding simulations with psychoacoustic bit allocations were run with,

- PR synthesis filters,
- MMSE-tuned synthesis filters, and
- MMPE-tuned synthesis filters.

Performance was evaluated in terms of a perceptual objective measure (POM) [Colo95], an estimate of the probability that an expert listener can distinguish between the original and coded signal. The POM results were 44% distinguishability for the PR case *versus* only 16% for both the MMSE and MMPE cases. The authors concluded that synthesis filter tuning is worthwhile since some performance enhancement exists over the PR case. They also concluded that MMPE filters failed to outperform MMSE filters because they were designed to minimize the perceptual error over a long period rather than a time-localized basis. Since perceptual signal properties are strongly time-variant, it is possible that time-variant MMPE tuning will realize some performance gain relative to MMSE tuning. The perceptual synthesis filter tuning ideas explored in this work have shown promise, but further investigation is required to better characterize its costs and benefits.

8.4 ADAPTED NONUNIFORM FILTER BANKS

The most popular method for realizing nonuniform frequency subbands is to cascade uniform filters in an unbalanced tree structure, as with, for example, the DWPT. For a given impulse response length, however, cascade structures in general produce poor channel isolation. Recent advances in modulated filter bank design methodologies (e.g., [Prin94]) have made tractable direct form near perfect reconstruction nonuniform designs that are critically sampled. This section is concerned with subband coders that employ signal-adaptive nonuniform modulated filter banks to approximate the time-frequency analysis properties of the auditory system more effectively than the other subband coders. Two examples are given. Beyond the pair of algorithms addressed below, we note that other investigators have proposed nonuniform filter bank coding techniques that address redundancy reduction utilizing lattice [Mont94] and bidimensional VQ schemes [Main96].

8.4.1 Switched Nonuniform Filter Bank Cascade

Princen and Johnston developed a CD-quality coder based upon a signal-adaptive filter bank [Prin95] for which they reported quality better than the sophisticated MPEG-1 layer III algorithm at both 48 and 64 kb/s. The analysis filter bank for this coder consists of a two-stage cascade. The first stage is a 48-band nonuniform modulated filter bank split into four uniform-bandwidth sections. There are 8 uniform subbands from 0 to 750 Hz, 4 uniform subbands from 750 to 1500 Hz, 12 uniform subbands from 1.5 to 6 kHz, and 24 uniform subbands from 6 to 24 kHz.

The second stage in the cascade optionally decomposes nonuniform bank outputs with on/off switchable banks of finer resolution uniform subbands. During filter bank adaptation, a suitable overall time-frequency resolution is attained by selectively enabling or disabling the second stage filters for each of the four uniform bandwidth sections. The low-resolution mode for this architecture corresponds to slightly better than auditory filter-bank frequency resolution. On the other hand, the high-resolution mode corresponds roughly to 512 uniform subband decomposition. Adaptation decisions are made independently for each of the four cascaded sections based on a criterion of minimum perceptual entropy (PE). The second stage filters in each section are enabled only if a reduction in PE (hence bit rate) is realized. Uniform PCM is applied to subband samples under the constraint of perceptually masked quantization noise. Masking thresholds are transmitted as side information. Further redundancy reduction is achieved by Huffman coding of both quantized subband sequences and masking thresholds.

8.4.2 Frequency-Varying Modulated Lapped Transforms

Purat and Noll [Pura96] also developed a CD-quality audio coding scheme based on a signal-adaptive, nonuniform, tree-structured wavelet packet decomposition. This coder is unique in two ways. First of all, it makes use of a novel wavelet packet decomposition [Pura95]. Secondly, the algorithm adapts to the signal the wavelet packet tree decomposition depth and breadth (branching structure) based on a minimum bit rate criterion, subject to the constraint of inaudible distortions. In informal subjective tests, the algorithm achieved excellent quality at a bit rate of 55 kb/s.

8.5 HYBRID WP AND ADAPTED WP/SINUSOIDAL ALGORITHMS

This section examines audio coding algorithms that make use of a hybrid wavelet packet/sinusoidal signal analysis. Hybrid coder architectures often improve coder robustness to diverse program material. In this case, the wavelet portion of a coder might be better suited to certain signal classes (e.g., transient), while the harmonic portion might be better suited to other classes of input signal (e.g., tonal or steady-state). In an effort to improve coder overall performance (e.g., better output quality for a given bit rate), several of the signal-adaptive wavelet and wavelet packet subband coding schemes presented in the previous section have been embedded in experimental hybrid coding schemes that seek to adapt the analysis properties of the coding algorithm to the signal content. Several examples are considered in this section.

Although the WP coder improvements reported in [Tewf93] addressed pre-echo control problems evident in [Sinh93b], they did not rectify the coder's inadequate performance for harmonic signals such as the piano test sequence. This is in part because the low-order FIR analysis filters typically employed in a WP decomposition are characterized by poor frequency selectivity, and therefore wavelet bases tend not to provide compact representations for strongly sinusoidal signals.

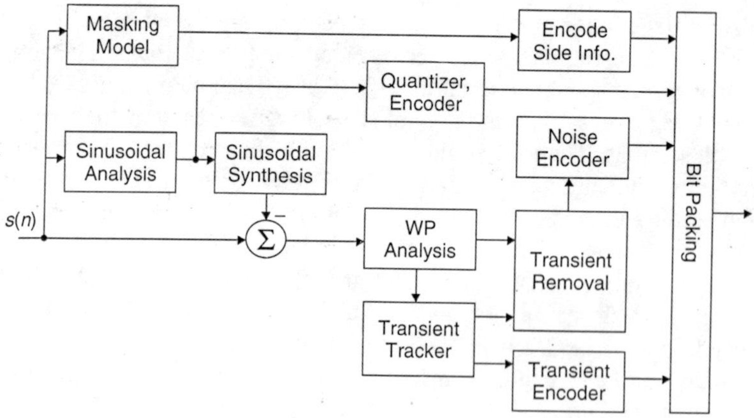

Figure 8.9. Hybrid sinusoidal/wavelet encoder (after [Hamd96]).

On the other hand, wavelet decompositions provide some control over time resolution properties, leading to efficient representations of transient signals. These considerations have inspired several researchers to investigate hybrid coders.

8.5.1 Hybrid Sinusoidal/Classical DWPT Coder

Hamdy *et al.* developed a hybrid coder [Hamd96] designed to exploit the efficiencies of both harmonic and wavelet signal representations. For each frame, the encoder (Figure 8.9) chooses a compact signal representation from combined sinusoidal and wavelet bases. This algorithm is based on the notion that short-time audio signals can be decomposed into tonal, transient, and noise components. It assumes that tonal components are most compactly represented in terms of sinusoidal basis functions, while transient and noise components are most efficiently represented in terms of wavelet bases. The encoder works as follows. First, Thomson's analysis model [Thom82] is applied to extract sinusoidal parameters (frequencies, amplitudes, and phases) for each input frame. Harmonic synthesis using the McAulay and Quatieri reconstruction algorithm [McAu86] for phase and amplitude interpolation is next applied to obtain a residual sequence. Then, the residual is decomposed into WP subbands.

The overall WP analysis tree approximates an auditory filter bank. Edge-detection processing identifies and removes transients in low-frequency subbands. Without transients, the residual WP coefficients at each scale become largely decorrelated. In fact, the authors determined that the sequences are well approximated by white Gaussian noise (WGN) sources having exponential decay envelopes. As far as quantization and encoding are concerned, sinusoidal frequencies are quantized with sufficient precision to satisfy just-noticeable-differences in frequency (JNDF), which requires 8-bit absolute coding for a new frequency track, and then 5-bit differential coding for the duration of the lifetime of the track. The sinusoidal amplitudes are quantized and encoded in a similar absolute/differential manner using simultaneous masking thresholds for shaping of quantization noise. This may require up to 8 bits per component. Sinusoidal phases are uniformly quantized on the interval $[-\pi, \pi]$ and encoded using 6 bits. As for quantization and encoding of WP parameters, all coefficients below 11 kHz are encoded as in [Sinh93b]. Above 11 kHz, however, parametric representations are utilized. Transients are represented in terms of a binary edge mask that can be run length encoded, while the Gaussian noise components are represented in terms of means, variances, and exponential decay constants. The hybrid harmonic-wavelet coder was reported to achieve nearly transparent coding over a wide range of CD-quality source material at bit rates in the vicinity of 44 kb/s [Ali96].

8.5.2 Hybrid Sinusoidal/*M*-band DWPT Coder

During the late 1990s, other researchers continued to explore the potential of hybrid sinusoidal-wavelet signal analysis schemes for audio coding. Boland and Deriche [Bola97] reported on an experimental sinusoidal-wavelet hybrid audio codec with high-level architecture very similar to [Hamd96] but with low-level differences in the sinusoidal and wavelet analysis blocks. In particular, for harmonic analysis the proposed algorithm replaces Thomson's method used in [Hamd96] with a combination of total least squares linear prediction (TLS-LP) and Prony's method. Then, in the harmonic residual wavelet decomposition block, the proposed method replaces the usual DWT cascade of two-band QMF sections with a cascade of four-band QMF sections. The algorithm works as follows. First, harmonic analysis operates on nonoverlapping 12-ms blocks of rectangularly windowed input audio (512 samples @ 44.1 kHz). For each block, sinusoidal frequencies, f_k, are extracted using TLS-LP spectral estimation [Rahm87], a procedure that is formulated to deal with closely spaced sinusoids in low SNR environments. Given the set of TLS-LP frequencies, a classical Prony algorithm [Marp87] next determines the corresponding amplitudes, A_k, and phases, ϕ_k. Masking thresholds for the tonal sequence are calculated in a manner similar to the ISO/IEC MPEG-1 psychoacoustic recommendation 2 [ISOI92]. After masked tones are discarded, the parameters of the remaining sinusoids are uniformly quantized and encoded in a procedure similar to [Hamd96]. Frequencies are encoded according to JNDFs (3 nonuniform bands, 8 bits per component in each band), phases are allocated 6 bits across all frequencies, and amplitudes are block companded with 5 bits for the gain and 6 bits per normalized amplitude. Unlike [Hamd96], however, amplitude bit allocations are fixed rather than signal adaptive. Quantized sinusoidal components are used to synthesize a tonal sequence, $\hat{s}_{tonal}(n)$, as follows:

$$\hat{s}_{tonal}(n) = \sum_{k=1}^{p} A_k e^{j(\Omega_k + \phi_k)}, \tag{8.2}$$

where the parameters $\Omega_k = 2\pi f_k / f_s$ are the normalized radian frequencies and only $p/2$ frequency components are independent since the complex exponentials are organized into conjugate symmetric pairs. As in [Hamd96], the synthetic

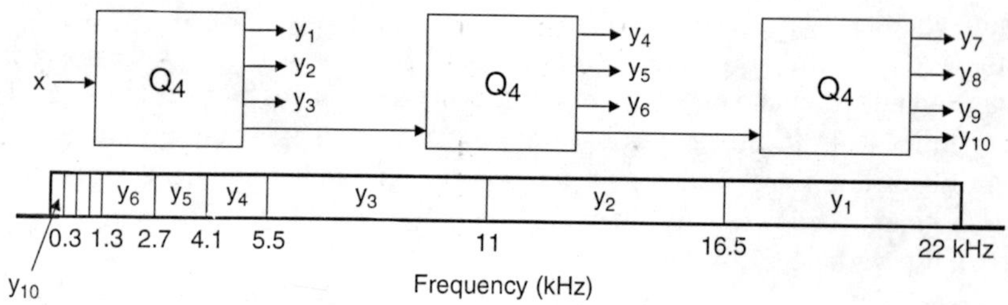

Figure 8.10. Subband decomposition associated with cascaded M-band DWT in [Bola97].

tonal sequence, $\hat{s}_{tonal}(n)$, is subtracted from the input sequence, $s(n)$, to form a spectrally flattened residual, $r(n)$.

In the wavelet analysis section, the harmonic residual, $r(n)$, is decomposed such that critical bandwidths are roughly approximated using a three-level cascade (Figure 8.10) of 4-band analysis filters (i.e., 10 subbands) designed according to the M-band technique in [Alki95]. Compared to the usual DWT cascade of 2-band QMF sections, the M-band cascade offers the advantages of reduced complexity, reduced delay, and linear phase. The DWT coefficients are uniformly quantized and encoded in a block companding scheme with 5 bits per subband gain and a dynamic bit allocation according to a perceptual noise model for the normalized coefficients. A Huffman coding section removes remaining statistical redundancies from the quantized harmonic and DWT coefficient sets. In subjective listening comparisons between the proposed scheme at 60–70 kb/s and MPEG-1, layer III at 64 kb/s on 12 SQAM CD [SQAM88] source items, the authors reported indistinguishable quality for "acoustic guitar," "Eddie Rabbit," "castanets," and "female speech." Slight impairments relative to MPEG-1, layer III were reported for the remaining eight items. No comparisons were reported in terms of delay or complexity.

8.5.3 Hybrid Sinusoidal/DWPT Coder with WP Tree Structure Adaptation (ARCO)

Other researchers have also developed hybrid algorithms that represent audio using a combination of sinusoidal and wavelet packet bases. Pena *et al.* [Pena96] have reported on the Adaptive Resolution COdec (ARCO). This algorithm employs a two-stage hybrid tonal-WP analysis section architecturally similar to both [Hamd96] and [Bola97]. The experimental ARCO algorithm has introduced several novelties in the segmentation, psychoacoustic analysis, tonal analysis, bit allocation, and WP analysis blocks. In addition, recent work on this project has produced a unique MDCT-based filter bank. The remainder of this subsection gives some details on these developments.

8.5.3.1 ARCO Segmentation, Perceptual Model, and Sinusoidal Analysis-by-Synthesis In an effort to match the time-frequency analysis resolution to the signal properties, ARCO includes a segmentation scheme that makes use of both time and frequency block clustering to determine optimal analysis frame lengths [Pena97b]. Similar blocks are assumed to contain stationary signals and are therefore combined into larger frames. Dissimilar blocks, on the other hand, are assumed to contain nonstationarities that are best analyzed using individual short segments. The ARCO psychoacoustic model resembles ISO/IEC MPEG-1 model recommendation 1 [ISOI92], with some enhancements. Unlike [ISOI92], tonality labeling is based on [Terh82], and noise maskers are segregated into narrowband and wideband subclasses. Then, frequency-dependent excitation patterns are associated with the wideband noise maskers. ARCO quantizes tonal signal components in a perceptually motivated analysis-by-synthesis. Using an iterative procedure, bits are allocated on each analysis frame until the synthetic tonal signal's excitation pattern matches the original signal's excitation pattern to within some tolerance.

8.5.3.2 ARCO WP Decomposition The ARCO WP decomposition procedure optimizes both the tree structure, as in [Srin98], and filter selections, as in [Sinh93b] and [Phil96]. For the purposes of WP tree adaptation [Prel96a], ARCO defines for the k-th band a cost, ε_k, as

$$\varepsilon_k = \frac{\int_{f_k-B_k/2}^{f_k+B_k/2}(U(f) - A_k)df}{\int_{f_k-B_k/2}^{f_k+B_k/2} U(f)df}, \tag{8.3}$$

where $U(f)$ is the masking threshold expressed as a continuous function, the parameter f represents frequency, f_k is the center frequency for the k-th subband, B_k is the k-th subband bandwidth, and A_k is the minimum masking threshold in the k-th band. Then, the total cost, C, to be minimized over all M subbands is given by

$$C = \sum_{k=1}^{M} \varepsilon_k. \tag{8.4}$$

By minimizing Eq. (8.4) on each frame, ARCO essentially arranges the subbands such that the corresponding set of idealized brickwall rectangular filters having amplitude equal to the height of the minimum masking threshold in the each band matches as closely as possible the shape of the masking threshold. Then, bits are allocated in each subband to satisfy the minimum masking threshold, A_k. Therefore, uniform quantization in each subband with sufficient bits affects a noise shaping that satisfies perceptual requirements without wasting bits. The method was found to be effective without accounting explicitly for the spectral leakage associated with the filter bank sidelobes [Prel96b]. As far as filter selection is concerned, ARCO employs signal-adaptive filters during steady-state segments and time-invariant filters during transients. Some of the filter selection strategies were reported to have been inspired by Agerkvist's auditory modeling work [Ager94] [Ager96]. In [Pena97a], it was found that the "symmetrization" technique [Bamb94] [Kiya94] was effective for minimizing the boundary distortions associated with the time-varying WP analysis.

8.5.3.3 ARCO Bit Allocation Unlike most other algorithms, ARCO encodes and transmits the masking threshold to the decoder. This has the advantage of efficiently representing both the adapted WP tree and the subband bit allocations with a single piece of information. The disadvantage, however, is that the decoder is no longer decoupled from the details of perceptual bit allocation as is typically the case with other algorithms. The ARCO bit allocation strategy [Sera97] achieves fast convergence to a desired bit rate by shifting the masking threshold up or down using a novel noise scaling procedure. The technique essentially uses a Newton algorithm to converge in only a few iterations to the noise scaling level that achieves the desired bit rate. The technique takes into account bit allocations from previous frames and allocates bits to all subbands simultaneously. Convergence speed and accuracy are controlled by a single parameter, and the procedure is amenable to subband weighting of the threshold to create unique noise profiles. In one set of experiments, convergence to a target rate with perceptual noise shaping was achieved in between two and seven iterations of the low complexity technique. Another unique property of ARCO is its set of high-level "cognitive rules" that seek to minimize the objectionable distortion when insufficient bits are available to guarantee transparent coding [Pena95]. These rules monitor the evolution of coding distortion over many frames and make fine noise-shaping adjustments on individual frames in order to avoid perceptually annoying noise patterns that could not otherwise be detected on a short-time basis.

8.5.3.4 ARCO Developments It is interesting to note that the researchers developing ARCO recently replaced the hybrid sinusoidal-WP analysis filter bank with a novel multiresolution MDCT-based filter bank. In [Casa98], Casal *et al.* developed a "multi-transform" (MT) that retains the lapped properties of the MDCT but creates a nonuniform time-frequency tiling by transforming back into time the high-frequency MDCT components in L-sample blocks. The proposed MT is characterized by high resolution in frequency for the low subbands and high resolution in time for the high frequencies. Like the MDCT upon which it is based, the MT maintains critical sampling and perfect reconstruction in the absence of quantization. Preliminary results for application of the MT in the TARCO (Tonal Adaptive Resolution COdec) are given in [Casa98]. As far as bit rates, reconstruction quality, and complexity are concerned, details on ARCO/TARCO have not yet appeared in the literature.

We conclude this section with the observation that hybrid DWT-sinusoidal and DWPT-sinusoidal architectures such as those advocated by Hamdy [Hamd96], Boland [Bola97], and Pena [Pena96], have been motivated by the notion that a source-robust audio coder must represent radically different signal types with uniform efficiency. The idea behind the hybrid structure is that providing two extreme basis possibilities might yield opportunities for maximally efficient signal adaptive basis selection. By offering superior frequency resolution with inherently narrowband basis elements, sinusoidal signal models are ideally suited for strongly tonal signals, while DWT and WPT filter banks, on the other hand, sacrifice some frequency resolution but offer greater time resolution flexibility, making these bases inherently more efficient for representing transient signals. As this section has demonstrated, the combination of the both signal models within a single codec can provide compact representations for a wide range of input signals. The next section of this chapter examines a different type of hybrid audio coding architecture in which code excited linear prediction (CELP) is embedded within subband coding schemes.

8.6 SUBBAND CODING WITH HYBRID FILTER BANK/CELP ALGORITHMS

While hybrid sinusoidal-DWT and sinusoidal-DWPT signal models seek to maximize robustness and basis flexibility, other hybrid signal models have been motivated by low-delay and low-complexity concerns. In this section, we consider, in particular, algorithms that combine a filter bank front end with subband-specific code-excited linear prediction (CELP) blocks for quantization and coding of the decimated subband sequences. The goal of these experimental hybrid coders is to achieve very low delay and/or low-complexity perceptual coding with reconstruction quality comparable to any state-of-the-art audio codec. Before considering these algorithms, however, we first define what is meant by "code-excited linear prediction."

In the coding literature, the acronym "CELP" denotes an entire class of efficient, analysis-by-synthesis source coding techniques developed primarily for speech applications in which the analyzed signal is treated as the output of a source-system mechanism such as the human vocal apparatus. In the CELP scheme, excitation vectors corresponding to the lower vocal tract "source" contribution drive a slowly time-varying LP synthesis filter that corresponds to the upper vocal tract "system." Parameters of the LP synthesis filter are usually estimated on a block basis, typically every 20 ms, while the excitation vectors are usually updated more frequently, typically every 5 ms. The LP parameters are most often estimated in an open-loop procedure by solving a set of normal equations that have been formulated to minimize the mean square prediction error. In contrast, the excitation vectors are optimized in a closed-loop, analysis-by-synthesis procedure such that the reconstruction error is minimized, most often in the perceptually weighted mean square sense. Given a vector of input speech, the analysis-by-synthesis process essentially reduces to a search during which the encoder must identify within a vector codebook that candidate excitation that generates the best synthetic output speech when processed by the LP synthesis filter. The set of encoded parameters is therefore a set of filter parameters and one (or more) vector indices, depending upon the codebook structure. Since its introduction in the mid-1980s [Schr85], CELP and its derivatives have received considerable attention in the literature. As a result, numerous high-quality, highly efficient algorithms have been proposed and adopted as international standards in speech coding. Although a detailed discussion of CELP is beyond the scope of this book, we refer the reader to the comprehensive tutorial in [Span94] for further details as well as a complete perspective on the CELP research and standards. The remainder of this section assumes that the reader has a basic understanding of CELP coding principles. Several examples of experimental subband/CELP algorithms are examined next.

8.6.1 Hybrid Subband/CELP Algorithm for Low-Delay Applications

One example of a hybrid filter bank/CELP low-delay audio codec was developed jointly by Hay and Saoudi at ENST and Mainard at CCETT. They devised a system for generic audio signals sampled at 32 kHz based on the four-band polyphase quadrature filter bank (pseudo-QMF) borrowed from the ISO/IEC MPEG-2 AAC scalable sample rate profile [Akai95] and a bank of modified ITU G.728 [ITUR92] low-delay CELP speech codecs (Figure 8.11). The primary objective of this system is to achieve transparent coding of the high-fidelity input with very low delay. The coder was first reported in [Hay96], and then enhanced in [Hay97]. The enhanced algorithm works as follows. First, the filter bank decomposes the input into four equal width subbands. Then, each of the decimated subband sequences is quantized and encoded in five-sample blocks (0.625 ms) using modified G.728 codecs (low-delay CELP) for each subband. The backward adaptive G.728 algorithm [ITUR92] generates as output a single vector index for each block of input samples, and therefore a set of four codebook indices, $\{i_1, i_2, i_3, i_4\}$, comprises the complete bitstream for the hybrid audio codec. Algorithmic delay consists of the 3-ms filter bank delay (96-tap filters) plus the additional 2-ms delay contributed by the G.728 stages, resulting in an total delay of only 5 ms. Bit allocation targets for each subband are computed by means of a modified ISO/IEC MPEG-1 psychoacoustic model-1 that computes masking thresholds, signal-to-mask ratios, and ultimately the number of bits required for transparent coding by analyzing the quantized outputs of the i-th band, $\hat{S}_i$, from a 4-ms-old block of data.

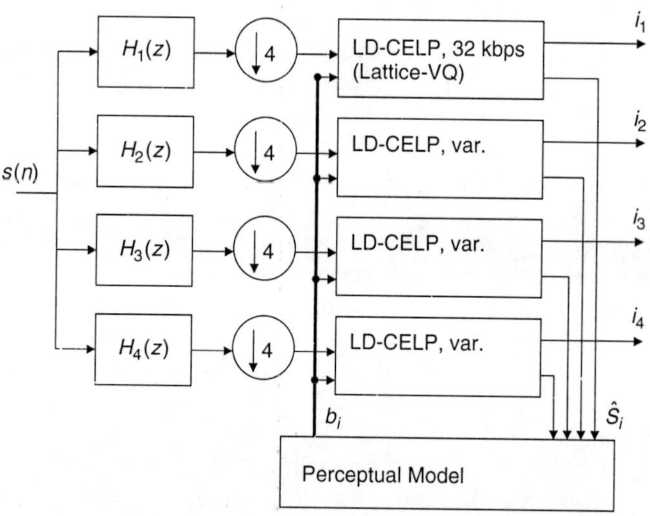

Figure 8.11. Low-delay hybrid filter-bank/LD-CELP algorithm [Hay97].

The perceptual model utilizes an alias-cancelled DFT [Tang95] to compensate for the analysis filter bank's aliasing distortion. Bit allocations are derived at both the transmitter and receiver from the same set of quantized data, making it unnecessary to transmit explicitly any bit allocation information. Average bit allocations on a subset of the standard ISO test material were 31, 18, 12, and 3 kb/s, respectively, for subbands 1 through 4. Given that the G.728 codec is intended to operate at a fixed rate, the primary challenge facing the algorithm designers was implementing dynamic subband bit allocations. Computationally efficient, variable rate versions of G.728 were constructed for bands 2 through 4 by structuring standard LBG (K-means) [Lind80] codebooks to deal with multiple rates (variable precision codebook indices). Unfortunately, the first (low frequency) subband requires an average bit rate of 32 kb/s for perceptual transparency, which translates to an impractical codebook size of 2^{20} vectors. To solve this problem, the authors implemented a highly efficient D_5 lattice VQ scheme [Conw88], which dramatically reduced the search complexity for each input vector by constraining the search space to a 50-vector neighborhood. Lattice vector shapes were assigned 16 bits and gains 4 bits. The lattice scheme was shown to perform nearly as well as an exhaustive search over a codebook containing more than 50,000 vectors. Neither objective nor subjective quality measures were reported for this hybrid system.

8.6.2 Hybrid Subband/CELP Algorithm for Low-Complexity Applications

Intended for achieving CD quality in low-complexity decoder applications, a second example of a hybrid filter bank/CELP algorithm appeared in [Vand98]. Like [Hay97], the proposed algorithm follows a critically sampled filter bank with a quantization and encoding stage of parallel, variable-rate CELP coders, one per subband (Figure 8.12). Unlike [Hay97], however, this algorithm makes use of a higher resolution, longer delay filter bank. Thus, channel separation is gained at the expense of delay. At the same time, this algorithm utilizes relatively low-order LP synthesis filters, which significantly reduce decoder complexity. In contrast, [Hay97] captures significant spectral detail in the high-order (50-th order) predictors that are embedded in the G.728 blocks. The proposed algorithm closely resembles ISO/IEC MPEG-1, layer 1 in its filter bank and psychoacoustic modeling sections. In particular, the filter bank is identical to the 32-band, 512-tap PQMF bank of [ISOI92]. Also like [ISOI92], the subband sequences are processed in 12-sample blocks, corresponding to 384 input samples.

The proposed algorithm, however, replaces the block companding of [ISOI92] with the CELP quantization and encoding for all 32 subbands. For every block of 12 subband samples, bits are allocated to the subbands on the basis of masking thresholds delivered by the perceptual model. This practice establishes minimum SNRs required in each subband to achieve perceptually transparent coding. Then, parallel noise scaling is applied to the target SNRs to adjust the bit rate to a scalable target. Finally, CELP blocks quantize and encode each subband using the number of bits allocated by the perceptual model. The particulars of the 32

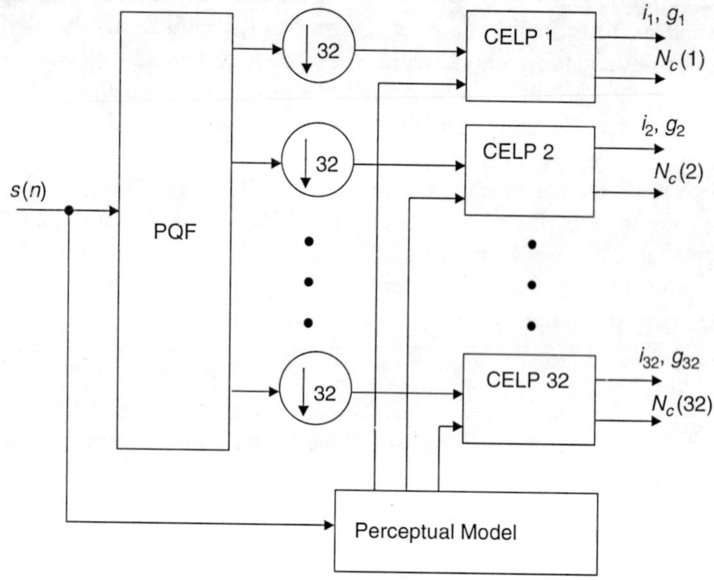

Figure 8.12. Low-complexity hybrid filter-bank/CELP algorithm [Vand98].

identical CELP stages are as follows. In order to maintain low complexity, the backward-adaptive LP synthesis filters are second order. The codebook, which is identical for all stages, contains 12-element stochastic excitation vectors that are structured for gain-shape quantization, with 6 bits allocated to the gains and 8 bits allocated to the shapes for each of the 256 codewords. Because bits are allocated dynamically for each subband in accordance with a masking threshold, the CELP blocks are configured for variable rate operation. Each CELP coder will combine excitation contributions from up to 4 codebooks, meaning that available rates for each subband are 0, 1.67, 2.33, 3.5, and 4.67 bits per sample. The closed-loop analysis-by-synthesis excitation search procedure relies upon a standard MSE minimization codebook search. The total bit budget, R, is given by

$$R = 2N_b + \sum_{i=1}^{N_b} 8N_c(i) + \sum_{i=1}^{N_b} 6N_c(i), \tag{8.5}$$

where N_b is the number of bands (32), $N_c(i)$ is the number of codebooks required in the i-th band to achieve the SNR demanded by the perceptual model. From left to right, the terms in Eq. (8.5) represent the bits required to specify the number of codebooks being used in each subband, the bits required for the shape codewords, and the bits required for the gain codewords. In informal subjective tests over a set of unspecified test material, the algorithm was reported to produce quality "near transparency" at 62 kb/s, "good quality" at 50 and 37 kb/s, and quality that was "weak" at 30 kb/s.

8.7 SUBBAND CODING WITH IIR FILTER BANKS

Although the majority of subband and wavelet audio coding algorithms found in the literature employ banks of perfect reconstruction FIR filters, this does not preclude the possibility of using infinite impulse response (IIR) filter banks for the same purpose. Compared to FIR filters, IIR filters are able to achieve similar magnitude response characteristics with reduced filter orders, and hence with reduced complexity. In the multiband case, IIR filter banks also offer complexity advantages over FIR filter banks. Enhanced performance, however, comes at the expense of an increased sensitivity and implementation cost for IIR filter banks. Creusere and Mitra constructed a template subband audio coding system modeled after [Lokh92] to compare performance and to study the tradeoffs involved when choosing between FIR and IIR filter banks for the audio coding application [Creu96]. In the study, two IIR and two FIR coding schemes were constructed from the template using a structured all-pass filter bank, a parallel all-pass filter bank, a tree-structured QMF bank, and a PQMF bank. Beyond this study, IIR filter banks have not been widely used for audio coding. The application of IIR filter banks to subband audio coding remains a subject that is largely unexplored.

PROBLEMS

8.1. In this problem, we will show that STFT can be interpreted as a bank of subband filters. Given the STFT, $X(n, \Omega_k)$, of the input signal, $x(n)$,

$$X(n, \Omega_k) = \sum_{m=-\infty}^{\infty} x(m)w(n-m)e^{-j\Omega_k m} = w(n) * x(n)e^{-j\Omega_k n},$$

where $w(n)$ is the sliding analysis window. Give a filter-bank realization of the STFT for a discrete frequency variable $\Omega_k = k(\Delta\Omega)$, $k = 0, 1, \ldots, 7$ (i.e., 8 bands). Choose $\Delta\Omega$ such that the speech band (20–4000 Hz) is covered. Assume that the frequencies, Ω_k, are uniformly spaced.

8.2. The mother wavelet function, $\xi(t)$, is given in Figure 8.13. Determine and sketch carefully the wavelet basis functions, $\xi_{\upsilon,\tau}(t)$, for $\upsilon = 0, 1, 2$ and $\tau = 0, 1, 2$ associated with $\xi(t)$,

$$\xi_{\upsilon,\tau}(t) \triangleq 2^{-\upsilon/2}\xi(2^{-\upsilon}t - \tau), \tag{8.6}$$

where υ and τ denote the dilation (frequency scaling) and translation (time shift) indices, respectively.

8.3. Let $h_0(n) = [1/\sqrt{2}, 1/\sqrt{2}]$ and $h_1(n) = [1/\sqrt{2}, -1/\sqrt{2}]$. Compute the scaling and wavelet functions, $\phi(t)$ and $\xi(t)$. Using $\xi(t)$ as the mother wavelet and generate the wavelet basis functions, $\xi_{0,0}(t)$, $\xi_{0,1}(t)$, $\xi_{1,0}(t)$, and $\xi_{1,1}(t)$.

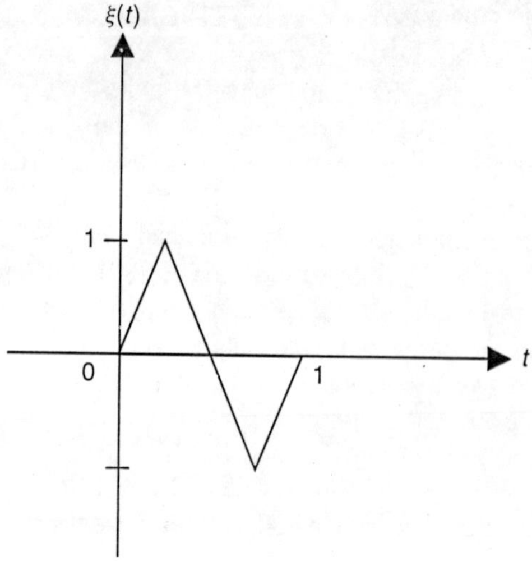

Figure 8.13. An example wavelet function.

Hint: From the DWT theory, the Fourier transforms of $\phi(t)$ and $\xi(t)$ are given by,

$$\Phi(\Omega) = \frac{1}{\sqrt{2}} H_0(e^{j\Omega/2}) \prod_{p=2}^{\infty} H_0(e^{j\Omega/2^p}) \tag{8.7}$$

$$\xi(\Omega) = \frac{1}{\sqrt{2}} H_1(e^{j\Omega/2}) \prod_{p=2}^{\infty} H_0(e^{j\Omega/2^p}), \tag{8.8}$$

where $H_0(e^{j\Omega})$ and $H_1(e^{j\Omega})$ are the DTFTs of the causal FIR filters, $h_0(n)$ and $h_1(n)$, respectively. For convenience, we assumed $T_s = 1$ in $\Omega = \omega T_s$; and $\phi(t) \overset{\text{CFT}}{\longleftrightarrow} \Phi(\omega) \equiv \Phi(\Omega)$, $h_0(n) \overset{\text{DTFT}}{\longleftrightarrow} H_0(e^{j\Omega})$.

8.4. Let $H_0(e^{j\Omega})$ and $H_1(e^{j\Omega})$ be ideal lowpass and highpass filters with cutoff frequency, $\pi/2$, as shown in Figure 8.14. Sketch $\Phi(\Omega)$, $\xi(\Omega)$, and the wavelet basis functions, $\xi_{0,0}(t)$, $\xi_{0,1}(t)$, $\xi_{1,0}(t)$, and $\xi_{1,1}(t)$.

8.5. Show that if both $H_0(e^{j\Omega})$ and $H_1(e^{j\Omega})$ are causal FIR filters of order N, then the wavelet basis functions, $\xi_{\upsilon,\tau}(t)$, will have finite duration of $(N+1)2^{\upsilon}$.

8.6. Using equations (8.7) and (8.8), prove the following: 1) $\Phi(\Omega/2) = \prod_{p=2}^{\infty} H_0(e^{j\Omega/2^p})$, and 2) $|\Phi(\Omega)|^2 + |\xi(\Omega)|^2 = |\Phi(\Omega/2)|^2$.

8.7. From problem 8.6 we have, $\Phi(\Omega) = \frac{1}{\sqrt{2}} H_0(e^{j\Omega/2})\Phi(\Omega/2)$ and $\xi(\Omega) = \frac{1}{\sqrt{2}} H_1(e^{j\Omega/2})\Phi(\Omega/2)$. Show that $\phi(t)$ and $\xi(t)$ can be obtained

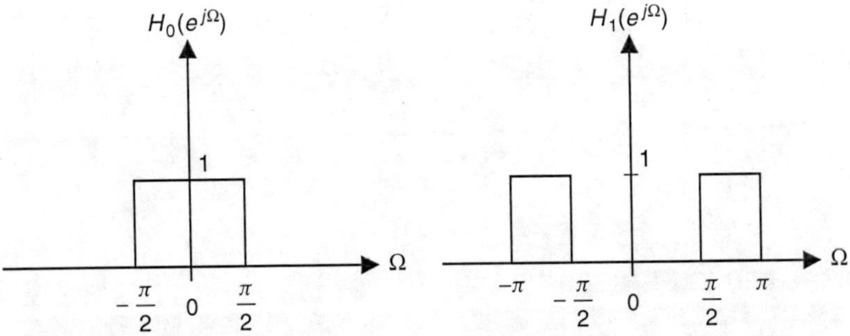

Figure 8.14. Ideal lowpass and highpass filters with cutoff frequency, $\pi/2$.

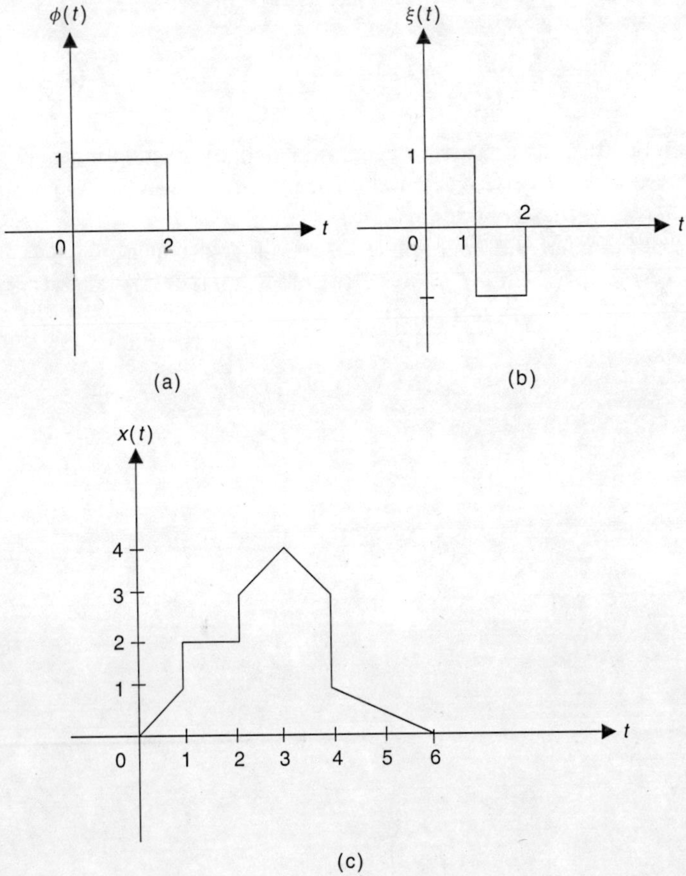

Figure 8.15. (a) The scaling function, $\phi(t)$, (b) the mother wavelet function, $\xi(t)$, and (c) input signal, $x(t)$.

recursively as,

$$\phi(t) = \sqrt{2}\sum_{n} h_0(n)\phi(2t - n) \tag{8.9}$$

$$\xi(t) = \sqrt{2}\sum_{n} h_1(n)\phi(2t - n) \tag{8.10}$$

COMPUTER EXERCISE

8.8. Let the scaling function, $\phi(t)$, and the mother wavelet function, $\xi(t)$, be as shown in Figure 8.15(a) and Figure 8.15(b), respectively. Assume that the input signal, $x(t)$, is as shown in Figure 8.15(c). Given the wavelet series expansion,

$$x(t) = \sum_{\tau=-\infty}^{\infty} \alpha(\tau)\phi(t - \tau) + \sum_{\upsilon=0}^{\infty}\sum_{\tau=-\infty}^{\infty} \beta(\upsilon, \tau)\xi_{\upsilon,\tau}(t), \tag{8.11}$$

where both υ and τ are integers and denote the dilation and translation indices, respectively, $\alpha(\tau)$ and $\beta(\upsilon, \tau)$ are the wavelet expansion coefficients. Solve for $\alpha(\tau)$ and $\beta(\upsilon, \tau)$.
[Hint: Compute the coefficients using inner products, $\alpha(\tau) \triangleq \langle x(t)\phi(t - \tau)\rangle = \int x(t)\phi(t - \tau)dt$ and $\beta(\upsilon, \tau) \triangleq \langle x(t)\xi_{\upsilon,\tau}(t)\rangle = \int x(t)\xi_{\upsilon,\tau}(t)dt = \int x(t)2^{-\upsilon/2}\xi(2^{-\upsilon}t - \tau)dt$.]

CHAPTER 9

SINUSOIDAL CODERS

9.1 INTRODUCTION

This chapter addresses perceptual coding algorithms based on sinusoidal models. Although sinusoidal signal models have been applied successfully since the 1980s in speech coding [Hede81] [Alme83] [McAu86] [Geor87] [Geor92] and music synthesis [Serr90], perceptual properties were not introduced in sinusoidal modeling until later [Edle96c] [Pena96] [Levin98a] [Pain01]. The advent of MPEG-4 standardization established new research goals for high-quality coding of general audio signals at bit rates in the range of 6–24 kb/s. In experiments reported as part of the MPEG-4 standardization effort, it was determined that sinusoidal coding is capable of achieving good quality at low rates without being constrained by a restrictive source model. Furthermore, unlike CELP and other classical low rate speech coding models, the parametric sinusoidal coding is amenable to pitch and time-scale modification at the decoder. Additionally, the emergence of Internet-based streaming audio has motivated considerable research on the application of sinusoidal signal models to high-quality audio coding at low bit rates. For example, Levine and Smith developed a hybrid sinusoidal-filter-bank coding scheme that achieves very high quality at rates around 32 kb/s [Levin98a] [Levi99].

This chapter describes some of the sinusoidal algorithms for low rate audio coding that exploit perceptual properties. In Section 9.2, we review the classical sinusoidal model. Section 9.3 presents the analysis/synthesis audio codec (ASAC), which was eventually considered for MPEG-4 standardization. Section 9.4 describes an enhanced version of ASAC, the harmonic and individual lines plus noise (HILN) algorithm. The HILN algorithm has been adopted

---

Audio Signal Processing and Coding, by Andreas Spanias, Ted Painter, and Venkatraman Atti

as part of the MPEG-4 standard. Section 9.5 examines the use of FM synthesis operators in sinusoidal audio coding. In Section 9.6, we investigate the sines + transients + noise (STN) model. Finally, Section 9.7 is concerned with algorithms that combine sinusoidal modeling with other well-known techniques in various hybrid architectures to achieve efficient low-rate audio coding.

9.2 THE SINUSOIDAL MODEL

This section describes the sinusoidal model that forms the basis for the parametric audio coding and the extended hybrid model given in the latter portions of this chapter. In particular, standard methodologies are presented for sinusoidal analysis, tracking, interpolation, and synthesis. The classical sinusoidal model comprises an analysis-synthesis framework ([McAu86] [Serr90] [Quat02]) that represents a signal, $s(n)$, as the sum of a collection of K sinusoids ("*partials*") with time-varying frequencies, phases, and amplitudes, i.e.,

$$s(n) \approx \hat{s}(n) = \sum_{k=1}^{K} A_k \cos(\omega_k(n)n + \phi_k(n)), \tag{9.1}$$

where A_k represents the amplitude, $\omega_k(n)$ represents the instantaneous frequency, and $\phi_k(n)$ represents the instantaneous phase of the k-th sinusoid. It is assumed that the amplitude, frequency, and phase functions evolve on a time scale substantially longer than a signal period. Analysis for this model amounts to estimating the amplitudes, phases, and frequencies of the constituent partials. Although this estimation is typically accomplished by peak picking in the short-time Fourier domain [McAu86] [Span91] [Serr90], analysis-by-synthesis estimation techniques that minimize explicitly a mean square error in terms of the sinusoidal parameters have also been proposed [Geor87] [Geor90] [Geor92]. Sinusoidal analysis-by-synthesis has also been presented within the more generalized framework of matching pursuits using overcomplete signal dictionaries [Good97] [Verm99]. Whether classical short-time Fourier transform (STFT) peak picking or analysis-by-synthesis is used for parameter estimation, the analysis yields partial parameters on each frame, and the data rate of the parameterization is given by the analysis stride and the order of the model. In the synthesis stage, the frame-rate model parameters are connected from frame to frame by a line tracking process and then interpolated using low-order polynomial models to derive sample-rate control functions for a bank of oscillators. Interpolation is carried out based on synthesis frames, which are implicitly established by the analysis stride. Although the bank of synthesis oscillators can be realized through additive combination of cosines, computationally efficient alternatives are available based on the FFT (e.g., [McAu88] [Rode92]).

9.2.1 Sinusoidal Analysis and Parameter Tracking

The STFT-based analysis scheme [McAu86] [Serr89] that estimates the sinusoidal model parameters is presented here, Figure 9.1. First, the input is segmented into

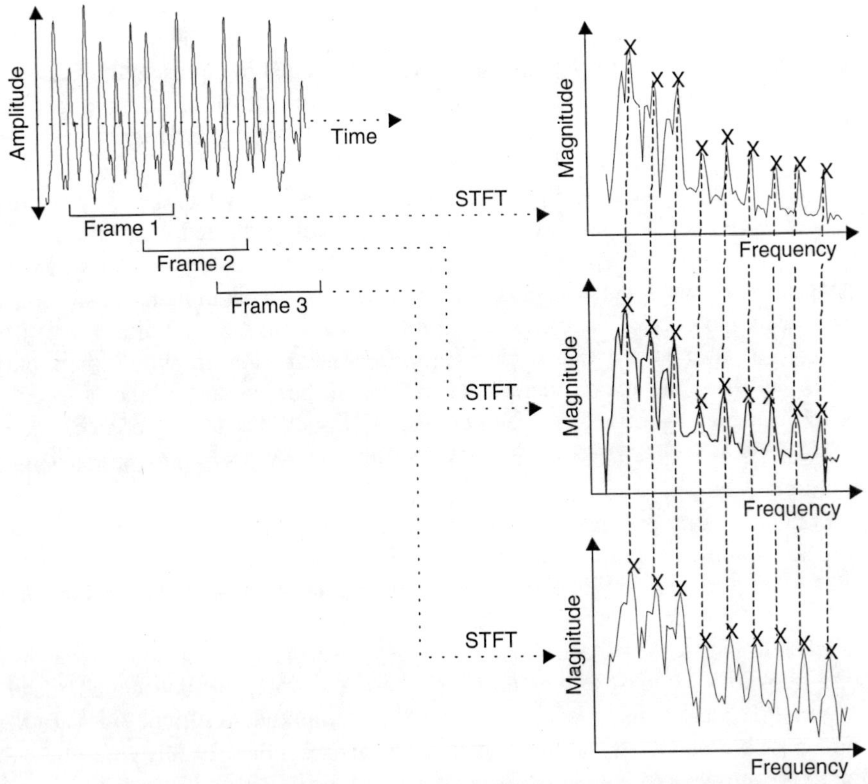

Figure 9.1. Sinusoidal model analysis. The time-domain signal is segmented into overlapping frames that are transformed to the frequency domain using STFT analysis. Local magnitude spectral maxima are identified. It is assumed that each peak is associated with a pure tone (partial) component of the input. For each of the peaks, a parameter triad containing frequency, amplitude, and phase is extracted. Finally, a tracking algorithm forms time trajectories for the sinusoids by matching the amplitude and/or frequency parameters across time.

overlapping frames. In the hybrid signal model, analysis frame lengths are signal adaptive. Frames are typically overlapped by half of their length. After segmentation, the frames are analyzed with the STFT, which yields magnitude and phase spectra. The sinusoidal analysis scheme assumes that magnitude spectral peaks are associated with underlying pure tones in the input. Therefore, spectral peaks are identified by a peak detector and then passed to a tracking algorithm that forms time trajectories by associating peaks from frame to frame. For a time-domain input, $s(n)$, let $S_l(k)$ denote the complex-valued STFT of the signal $s(n)$ on the l-th frame. A spectral peak is defined as a local maximum in the magnitude spectrum $|S_l(k)|$, i.e., an STFT magnitude peak on bin k_0 that satisfies the inequality

$$|S_l(k_0 - 1)| \leqslant |S_l(k_0)| \geqslant |S_l(k_0 + 1)|. \tag{9.2}$$

Following peak identification on frames l and $l + 1$, a tracking procedure forms time trajectories by matching across frames those spectral peaks which satisfy certain matching criteria. The resulting trajectories are intended to represent the smoothly time-varying frequencies, amplitudes, and phases of the sinusoidal partials that comprise the signal under analysis. Several trajectory tracking algorithms have been demonstrated to perform well [McAu86] [Serr89].

The tracking procedure (Figure 9.2) works in the following way. First, denote by ω_i^l the frequencies associated with the sinusoids identified on frame l, with $1 \leqslant i \leqslant p$. Similarly, denote by ω_j^{l+1} the frequencies associated with the sinusoids identified on frame $l + 1$, with $1 \leqslant j \leqslant r$. Given two sets of unmatched sinusoids, the tracking objective is to identify for the i-th sinusoid on frame l the j-th sinusoid on frame $l + 1$ that is closest in frequency and/or amplitude (here only frequency matching is considered). Therefore, in the first step of the procedure, an initial match is formed between ω_i^l and ω_j^{l+1} such that the difference, $\Delta\omega = |\omega_i^l - \omega_j^{l+1}|$ is minimized and such that the distance $\Delta\omega$ is less than a specified maximum, $\Delta\omega_{\max}$.

Following an initial match, three outcomes are possible. First, the trajectory will be continued (Figure 9.2a) if a match is found and there are no match conflicts to be resolved. In this case, the frequency, amplitude, and phase parameters are interpolated from frame l to frame $l + 1$. On the other hand, if no initial match is found during the first step, it is assumed that the trajectory associated with frequency ω_i^l must terminate. In this case, the trajectory is declared "dead" (Figure 9.2c) and is matched to itself with zero amplitude on frame $l + 1$. In the third possible outcome, the initial match creates a conflict. In this case, the i-th trajectory attempts to match with a peak that has already been claimed by another

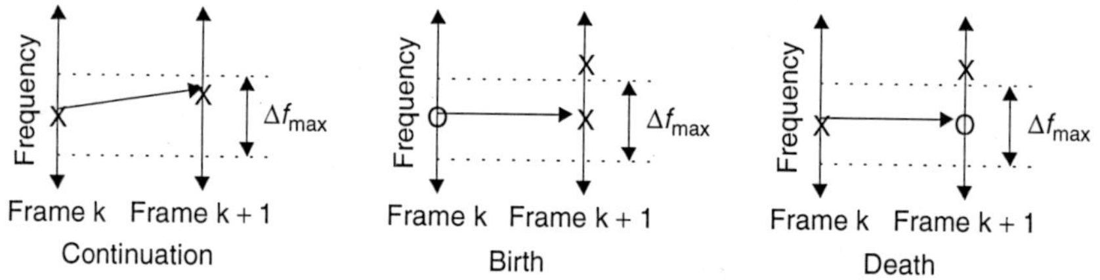

Figure 9.2. Sinusoidal trajectory formation. In the figure, an 'x' denotes the presence of a sinusoid at the specified frequency, while an 'o' denotes the absence of a sinusoid at the specified frequency. In part (a), a sinusoid on frame k is matched to a sinusoid on frame $k + 1$ because the two sinusoids are sufficiently close in frequency and because there are no conflicts. During synthesis, the frequency, amplitude, and phase parameters are interpolated from frame k to frame $k + 1$. In part (b), a sinusoid on frame $k + 1$ is declared "born" because a sufficiently close matching sinusoid does not exist on frame k. In this case, frequency is held constant, but amplitude is interpolated from zero on frame k to the measured amplitude on frame $k + 1$. In part (c), a sinusoid on frame k is declared "dead" because a sufficiently close matching sinusoid does not exist on frame $k + 1$. In this case, frequency is held constant, but amplitude is interpolated from the measured amplitude on frame k to zero on frame $k + 1$.

trajectory. The conflict is resolved in favor of the closest frequency match. If the current trajectory loses, it picks the next best available match that satisfies the difference criterion outlined above. If the pre-existing match loses the conflict, the current trajectory claims the peak and the pre-existing match is returned to the pool of available trajectories. This process is repeated until all trajectories are either matched or declared "dead." At the conclusion of the matching procedure, any unclaimed sinusoids on frame $l+1$ are declared "born." As shown in Figure 9.2(b), trajectories at "birth" are backwards matched to themselves on frame l, with the amplitude interpolated from zero on frame l to the measured amplitude on frame $l+1$.

9.2.2 Sinusoidal Synthesis and Parameter Interpolation

The sinusoidal trajectories of frequency, amplitude, and phase triads are updated at a rate of once per frame. The synthesis portion of the sinusoidal model uses the frame-rate parameters that were extracted during the analysis procedure to generate a sample-rate output sequence, $\hat{s}(n)$ by appropriately controlling the output of a bank of oscillators. One method for generating the model output is as follows. On the l-th frame, let output sample on index $m+lH$ represent the sum of the contributions of the K partials that were estimated on the l-th frame i.e.,

$$\hat{s}(m+lH)=\sum_{k=1}^{K} A_k^l \cos(\omega_k m+\phi_k) \quad 0 \leqslant m < H, \tag{9.3}$$

where the parameter triad $\{\omega_k^l, A_k^l, \phi_k^l\}$ represents the frequency, amplitude, and phase, respectively, of the k-th sinusoid, and the parameter H corresponds to the synthesis hop size (equal to analysis hop size unless time-scale modification is required). The problem with this approach is that the sinusoidal parameters are not interpolated between frames, and therefore the sequence $\hat{s}(n)$ will in general contain jump discontinuities at the frame boundaries. In order to avoid discontinuities and the associated artifacts, a better approach is to use oscillator control functions that interpolate the trajectory parameters from one frame to the next. If the k-th trajectory parameters on frames l and $l+1$ are given by $\{\omega_k^l, A_k^l, \phi_k^l\}$ and $\{\omega_k^{l+1}, A_k^{l+1}, \phi_k^{l+1}\}$, respectively, then the instantaneous amplitude, $\tilde{A}_k^l(m)$, can be linearly interpolated between the measured amplitudes A_k^l and A_k^{l+1} using the relation,

$$\tilde{A}_k^l(m)=A_k^l+\frac{A_k^{l+1}-A_k^l}{H}m \quad 0 \leqslant m < H. \tag{9.4}$$

Measured values for frequency and phase are interpolated next. For clarity, the subscript index k has been dropped throughout the remainder of this discussion, and the frame index l is used in its place. Frequency and phase interpolation are less straightforward than amplitude interpolation because of the fact that frequency is the phase derivative. Before defining a phase interpolation function, it is important to note that the instantaneous phase, $\tilde{\theta}(m)$, is defined as

$$\tilde{\theta}(m)=m\tilde{\omega}+\tilde{\phi} \quad 0 \leqslant m < H, \tag{9.5}$$

where $\tilde{\omega}$ and $\tilde{\phi}$ are the measured frequency and measured phase, respectively. For smooth interpolation between frames, therefore, it is necessary that the instantaneous phase be equal to the measured phases at the frame boundaries and, simultaneously, it is also necessary that the instantaneous phase derivatives be equal to the measured frequencies at the frame boundaries. To accomplish this, a cubic phase interpolation polynomial was proposed [McAu86] of the form,

$$\tilde{\theta}_l(m) = \gamma + \kappa m + \alpha m^2 + \beta \mathrm{m}^3 \quad 0 \leqslant m < H. \tag{9.6}$$

After some manipulation, it can be shown [McAu86] that the instantaneous phase is given by

$$\tilde{\theta}_l(m) = \phi_l + \omega_l m + \alpha(M^*)m^2 + \beta(M^*)\mathrm{m}^3 \quad 0 \leqslant m < H \tag{9.7}$$

and that the parameters α and β are obtained as follows:

$$\begin{bmatrix} \alpha(M) \\ \beta(M) \end{bmatrix} = \begin{bmatrix} \dfrac{3}{H^2} & -\dfrac{1}{H} \\ -\dfrac{2}{H^3} & \dfrac{1}{H^2} \end{bmatrix} \begin{bmatrix} \phi_{l+1} - \phi_l - \omega_l H + 2\pi M \\ \omega_{l+1} - \omega_l \end{bmatrix}. \tag{9.8}$$

Although many values for the parameter M will allow $\tilde{\theta}(m)$ to satisfy the frame boundary conditions, it was shown in [McAu86] that the smoothest function (in the sense that the integral of the square of the second derivative of the function $\tilde{\theta}(m)$ is minimized) is obtained for the value $M = M^*$, which is given by,

$$M^* = round\left(\frac{1}{2\pi}\left[(\phi_l + \omega_l H - \phi_{l+1}) + (\omega_{l+1} - \omega_l)\frac{H}{2}\right]\right), \tag{9.9}$$

where the *round*() operation denotes taking the integer closest to the function argument. We now restore the notational conventions used earlier, i.e., that the subscript corresponds to the parameter index, and that the superscript represents a frame index. Given the interpolated amplitude and instantaneous phase functions $\tilde{A}_k^l(m)$ and $\tilde{\theta}_k^l(m)$, respectively, the interpolated sinusoidal model synthesis expression becomes

$$\hat{s}(m + lH) = \sum_{k=1}^{K} \tilde{A}_k^l(m)\cos(\tilde{\theta}_k^l(m)) \quad 0 \leqslant m < H. \tag{9.10}$$

Unless otherwise specified, this is the standard synthesis expression used in the hybrid sinusoidal model presented throughout the remainder of this chapter. This section has provided the essential details of the basic sinusoidal model. In Section 9.6, the basic model is extended to enhance its robustness over a wider class of signals by including explicit model extensions for transient and noise-like signal energy. In Sections 9.3 through 9.5, we apply the sinusoidal

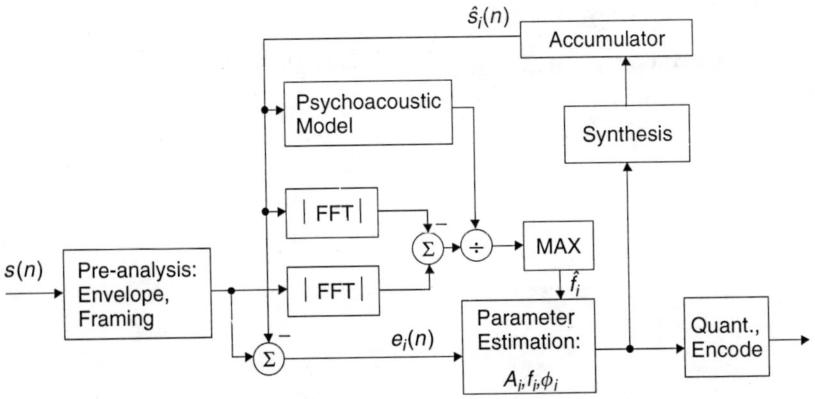

Figure 9.3. ASAC encoder (after [Edle96c]).

model presented here in the context of parametric audio codecs, i.e., the ASAC, the HILN, and the FM.

9.3 ANALYSIS/SYNTHESIS AUDIO CODEC (ASAC)

The sinusoidal analysis/synthesis audio codec (ASAC) for robust coding of general audio signals at rates between 6 and 24 kb/s was developed by Edler *et al.* at the University of Hannover and proposed for MPEG-4 standardization [Edle95] in 1995. An enhanced ASAC proposal later appeared in [Edle96a]. Initially, ASAC segments input audio into analysis frames over which the signal is assumed to be nearly stationary. Sinusoidal synthesis parameters are then extracted according to perceptual criteria, quantized, encoded, and transmitted to the decoder for synthesis. The algorithm distributes synthesis parameters across basic and enhanced bit streams to allow scalable output quality at bitrates of 6 and 24 kb/s. Architecturally, the ASAC scheme (Figure 9.3) consists of a pre-analysis block for window selection and envelope extraction, a sinusoidal analysis-by-synthesis parameter estimation block, a perceptual model, and a quantization and coding block. Although it bears similarities to sinusoidal speech coding [McAu86] [Geor92] [Geor97] and music synthesis [Serr90] algorithms that have been available for some time, the ASAC coder also incorporates some new techniques. In particular, whereas previous speech-specific sinusoidal coders emphasized waveform matching by minimizing reconstruction error norms such as the mean square error (MSE), ASAC disregards classical error minimization criteria and instead selects sinusoids in decreasing order of *perceptual* importance by means of an iterative analysis-by-synthesis loop. The perceptual significance of each component sinusoid is judged with respect to the masking power of the synthesis signal, which is determined by a simplified version of the psychoacoustic model from [Baum95], but without the temporal masking considerations. The remainder of the section presents details of the ASAC encoder [Edle96c].

9.3.1 ASAC Segmentation

Before analysis-by-synthesis begins, a pre-analysis segmentation adapts the analysis window length to match characteristics of the input signal, $s(n)$. A short rectangular window with tapered edges is used during transient events, or a long Hanning window twice the synthesis window length is used during stationary segments. The pre-analysis process also extracts via weighted linear regression a parametric amplitude envelope for each frame to deal with rapid changes in the input signal. During transients, envelopes are parameterized in terms of a peak time, as well as attack and decay slopes. Non-attack envelopes, on the other hand, are parameterized with only a single attack (+) or decay (−) slope.

9.3.2 ASAC Sinusoidal Analysis-by-Synthesis

The iterative analysis-by-synthesis block [Edle96c] estimates one at a time the parameters of the i-th individual constituent sinusoid or partial, and every iteration identifies the most perceptually significant sinusoid remaining in the synthesis residual, $e_i(n) = s(n) - \hat{s}_i(n)$, and adds it to the synthetic output, $\hat{s}_i(n)$. Perceptual significance is assessed by comparing the synthesis residual against the masked threshold associated with the current synthetic output and choosing the residual sinusoid with the largest suprathreshold margin. This perceptual selection criterion assumes that the signal component with the most unmasked energy will be the most perceptually important. After the most important peak frequency, $\hat{f}_i$, has been identified, a high-resolution spectral analysis technique inspired by [Kay89] refines the frequency estimate. Given refined frequency estimates for the current partial at frame start, f_i^{start}, and end, f_i^{end}, the amplitude, A_i, and phase, ϕ_i, parameters are extracted from a complex correlation coefficient obtained from an inner product between the synthesis residual and a complex exponential of linearly time-varying frequency. End-point frequencies for the complex exponential correspond to the frame start and frame end frequencies for the partial of interest. For the synthesis phase, both unscaled and envelope-scaled versions of the current partial are generated. The analysis-by-synthesis loop selects the version that minimizes the residual variance, subtracts the new partial from the current residual, and then repeats the analysis-by-synthesis parameter extraction loop with the updated residual. The loop repeats until the bit budget is exhausted. We note that in contrast to the time-varying global gain used in [Geor92], a unique feature of the ASAC coder is that the temporal envelope may be selectively applied to individual synthesis partials.

9.3.3 ASAC Bit Allocation, Quantization, Encoding, and Scalability

The ASAC quantization and coding block provides bit rate and output quality scaling by generating simultaneously basic and enhanced bitstreams for each frame. The basic bitstream contains the three temporal envelope parameters, as well as parameters representing frequencies, amplitudes, envelope enable bits, and continuation bits for each partial. A tracking algorithm classifies each partial as either new or continued by matching partials from frame to frame on the basis of amplitude and frequency similarities [Edle96c]. Track disputes (multiple matches for a given partial) are resolved in favor of the candidate that maximizes a weighted amplitude and frequency closeness measure. ASAC partial tracking and dispute resolution differs from [McAu86] in that amplitude matching is considered explicitly in addition to frequency proximity. For noncontinued partials, both frequencies and amplitudes are log quantized. For continued partials, on the other hand, frequency differences are uniformly quantized, while amplitude ratios are quantized on a log scale. The enhancement bitstream contains parameters to improve reconstructed signal fidelity over the basic bitstream. Enhanced bitstream parameters include finer quantization bits for the envelope parameters, phases for each partial, and finer frequency quantization bits for noncontinued partials above a threshold frequency. The phases are uniformly quantized. Because of the fixed-rate ASAC architecture, between 10 and 17 spectral lines may fit within the 6 kb/s basic bitstream for a given 32-ms frame. Like the time-frequency MPEG family algorithms [Bran94a] [Bosi96] (Chapter 10) the ASAC quantization and coding block maintains a bit reservoir to smooth local maxima in bit demand. These maxima tend to occur during transients when many noncontinued partials arise and inflate the bit rate. Bits are deposited into the reservoir whenever the number of partials reaches a threshold, and before exhausting the bit budget. Conversely, reservoir bits are withdrawn whenever the number of partials is below a threshold while the bit budget has already been exhausted.

9.3.3.1 ASAC Performance When compared to standard speech codecs at similar bit rates, the first version of ASAC [Edle95] reportedly offered improved quality for nonharmonic tonal signals such as spectrally complex music, similar quality for single instruments, and impaired quality for clean speech [ISOI96b]. The later ASAC [Edle96a] was improved for certain signals [ISOI96c].

9.4 HARMONIC AND INDIVIDUAL LINES PLUS NOISE CODER (HILN)

The ASAC algorithm outperformed speech-specific algorithms at the same bit rate in subjective tests for certain test signals, particularly spectrally complex music characterized by large numbers of nonharmonically related sinusoids. The original ASAC, however, failed to match speech codec performance for other test signals such as clean speech. As a result, the ASAC core was embedded in an enhanced algorithm [Purn98] intended to better match the coder's signal model with diverse input signal characteristics. In research proposed as part of an MPEG-4 "core experiment" [Purn97], Purnhagen *et al.* developed an "object-based" algorithm. In this approach, harmonic sinusoid, individual sinusoid, and colored noise objects could be combined in a hybrid source model to create a parametric signal representation. The enhanced algorithm, known as the "harmonic and individual lines plus noise" (HILN) is architecturally very similar to the original ASAC, with some modifications.

Like ASAC, the HILN coder (Figure 9.4) segments the input signal into 32-ms overlapping analysis frames and extracts a parametric temporal envelope during preanalysis. Unlike ASAC, however, the iterative analysis-synthesis block is extended to include a cascade of analysis stages for each of the available object types. In the enhanced analysis-synthesis system of HILN, harmonic analysis is applied first, followed by individual spectral line analysis, followed by shaped noise modeling of the two-stage residual. The remainder of the section presents some unique details of the HILN encoder, including descriptions of the HILN schemes for sinusoidal analysis-by-synthesis, bit allocation, and quantization and encoding.

9.4.1 HILN Sinusoidal Analysis-by-Synthesis

The HILN sinusoidal analysis-by-synthesis occurs in a cascade of three stages. First, the harmonic section estimates a fundamental frequency and also quantifies the harmonic partial amplitudes using essentially the same method as ASAC. Cepstral analysis is used to estimate the fundamental. Harmonics may occur on either integer or noninteger multiples of the fundamental. A stretching [Flet91] parameter is available to account for the noninteger harmonic phenomena induced by, e.g., stiff-stringed instruments. In the second stage, a partial discriminator separates those partials belonging to the harmonic object from those partials belonging to the individual sinusoid object. In the last stage, the noise extraction module uses a cosine series expansion to estimate the spectral envelope of the noise residual left after all partials have been extracted, including those partials not already extracted by the harmonic and individual stages.

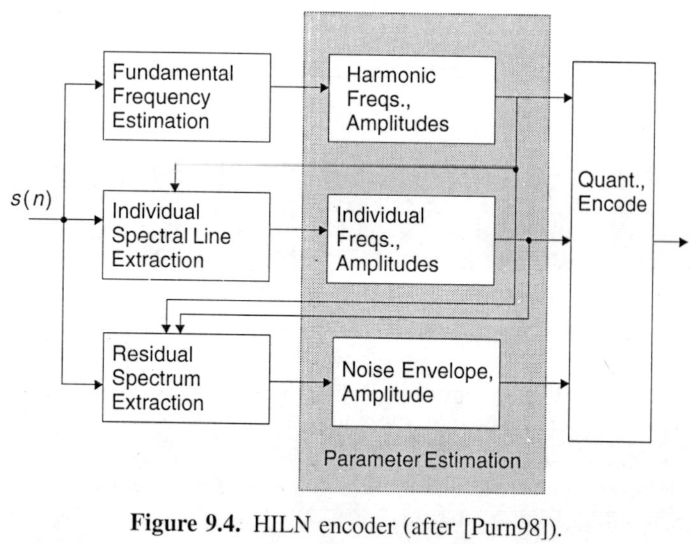

Figure 9.4. HILN encoder (after [Purn98]).

9.4.2 HILN Bit Allocation, Quantization, Encoding, and Decoding

Quantization and encoding is applied to the various object parameters in the following way. For individual sinusoids, quantization, and coding is essentially the same as in ASAC. For the harmonic sinusoids, the fundamental frequency, and partial amplitudes are log quantized and encoded. Since spectrally complex signals may contain in excess of 100 partials, the harmonic partial amplitudes beyond the tenth are grouped and represented by an average amplitude instead of by the individual amplitudes. This approach saves bits by exploiting the reduced frequency resolution of the auditory filter bank at higher frequencies. For the colored noise object, a gain is log quantized and transmitted along with the quantized parameters of the noise envelope. Bits are distributed between the three object types according to perceptual relevance. Unlike ASAC, none of the HILN sinusoidal coding objects include phase information. Start-up phases for new sinusoids are randomized at the decoder, and then continued smoothly for later frames. The decoder also employs smooth fade in and fade out of partial tracks to prevent jump discontinuities in the output. The output noise object is synthesized using the parametric spectral envelope and randomized phase in conjunction with an inverse FFT. As is true of the ASAC algorithm, the parametric nature of HILN means that speed and pitch changes are realized in a straightforward manner.

9.4.2.1 HILN Performance Results from subjective listening tests at 6 kb/s showed significant improvements for HILN over ASAC, particularly for the most critical test items that had previously generated the most objectionable ASAC artifacts [Purn97]. Compared to HILN, the CELP speech codecs are still able to represent more efficiently clean speech at low rates, and time-frequency codecs are able to encode more efficiently general audio at rates above 32 kb/s. Nevertheless, the HILN improvements relative to ASAC inspired the MPEG-4 committee to incorporate HILN into the MPEG-4 committee draft as the recommended low rate parametric audio coder [ISOI97a]. As far as applications are concerned, the HILN algorithm was deployed in a scalable, low-rate Internet streaming audio scheme [Feit98].

9.5 FM SYNTHESIS

The HILN algorithm seeks to optimize coding efficiency by making combined use of three distinct source models. Although the HILN harmonic sinusoid object has been shown to facilitate increased coding gain for certain signals, it is possible that other object types may offer opportunities for greater efficiency when representing spectrally complex harmonic signals. This notion motivated a recent investigation into the use of frequency modulation (FM) synthesis techniques [Chow73] in low-rate sinusoidal audio coding for harmonically structured single instrument sounds [Wind98]. In this section, we first review the basic principles of FM synthesis, and then present an experimental audio codec that makes use of FM synthesis operators for low rate coding.

9.5.1 Principles of FM Synthesis

FM synthesis offers advantages over other harmonic coding methods (e.g., [Ferr96b] [Edle96c]) because of its ability to model with relatively few parameters harmonic signals that have many partials. In the simplest FM synthesis, for example, the frequency of a sine wave (carrier) is modulated by another sine wave (modulator) to generate a complex waveform with spectral characteristics that depend on a modulation index and the parameters of the two sine waves. In continuous time, the FM signal is given by

$$s(t) = A\sin[2\pi f_c t + I\sin(2\pi f_m t)], \tag{9.11}$$

where A is the amplitude, f_c is the carrier frequency, f_m is the modulation frequency, I is the modulation index, and t is the time index. The associated Fourier series representation is

$$s(t) = \sum_{k=-\infty}^{\infty} J_k(I)\sin(2\pi f_c t + 2\pi k f_m t), \tag{9.12}$$

where $J_k(I)$ is the Bessel function of the first kind.

It can be seen from Eq. (9.12) that a large number of harmonic partials can be generated (Figure 9.5) by controlling only three parameters per FM operator. One can observe that the fundamental and harmonic frequencies are determined by f_c and f_m, and that the harmonic partial amplitudes are controlled by the modulation index, I. The Bessel envelope, moreover, essentially determines the FM spectral bandwidth. Example harmonic FM spectra for a unit amplitude 200 Hz carrier are given in Figure 9.5 for modulation indices of 1 (Figure 9.5a) and 15 (Figure 9.5b). While both examples have identical harmonic structure, the amplitude envelopes and bandwidths differ markedly as a function of the index, I. Clearly, the central issue in making effective use of the FM technique for signal modeling is parameter estimation accuracy.

9.5.2 Perceptual Audio Coding Using an FM Synthesis Model

Winduratna [Wind98] proposed an FM synthesis audio coding scheme in which the outputs of parallel FM "operators" are combined to model a single instrument sound. The algorithm (Figure 9.6) works as follows. First, the preanalysis block segments the input into frames, and then extracts parameters for a set of individual spectral lines, as in [Edle96c]. Next, the preanalysis identifies a harmonic structure by maximizing an objective function [Wind98]. Given a fundamental frequency estimate from the pre-analysis, f_0, the iterative parameter extraction loop estimates the parameters of individual FM operators and accumulates their contributions until the composite spectrum from the multiple, parallel FM operators closely resembles the original. Perceptual closeness is judged to be adequate when the magnitude of the original minus synthetic harmonic (difference) spectrum is below the masked

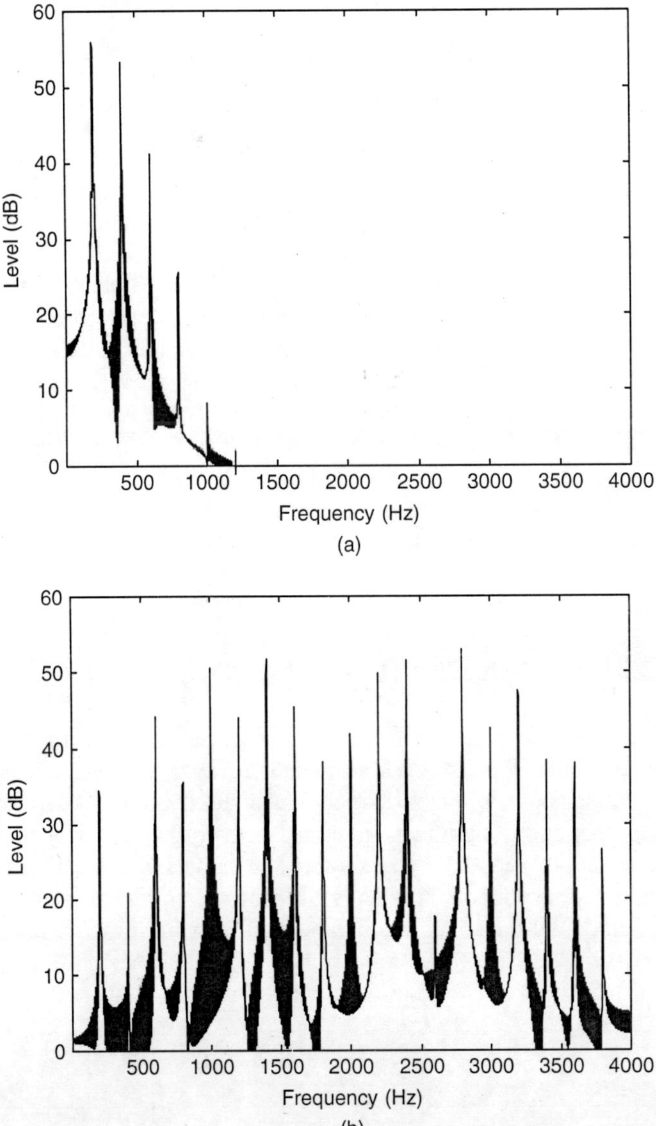

Figure 9.5. Examples of harmonic FM spectra for $f_c = f_m = 200$ Hz with (a) $I = 1$, and (b) $I = 15$.

threshold [Baum95]. During each loop iteration, error minimizing values for the current operator are determined by means of an exhaustive search. Then, the contribution of the current operator is added to the existing output. Finally, the harmonic residual spectrum is checked against the masked threshold. The loop repeats and additional operators are synthesized until the error spectrum is below the masked threshold. Spectral line amplitude prediction is used for adjacent frames having the same fundamental frequency.

In the quantization and encoding block, the parameters f_0, f_c, f_m, I, and A are encoded with 12, 7, 7, 5, and 7 bits, respectively, for each operator. The fundamental frequency and amplitude are log quantized. The carrier and modulation frequencies are encoded as integer multiples of the fundamental, while the amplitude prediction weight, p, is quantized and encoded with 8 bits for adjacent frames having identical fundamental frequencies. At the expense of high complexity, the FM coding scheme was shown to efficiently represent single instrument sounds at bit rates between 2.1 and 4.8 kb/s. Performance of the scheme was investigated not only for audio, but also for speech. Significant performance gains were realized relative to previous sinusoidal schemes. Using a 24-ms analysis window, for example, one critical male speech item was encoded at 21.2 kb/s using FM synthesis compared to 45 kb/s for ASAC [Wind98], with similar output quality. Despite estimation difficulties for signals with more than one fundamental frequency, e.g., polyphonic music, the high efficiency of the FM synthesis technique for parametric representations of complex harmonic spectra makes it a likely candidate for future inclusion in object-based algorithms such as the HILN coder.

9.6 THE SINES + TRANSIENTS + NOISE (STN) MODEL

Although the basic sinusoidal model (Eq. (9.1)) can achieve very efficient representations of some signals (e.g., sustained vowels via harmonically related components), extensions to the basic model have also been proposed to deal with other signals containing significant nontonal energy that can lead to modeling inefficiencies. A viable model for many musical signals is one of a deterministic plus a stochastic component. The deterministic part corresponds to the pitched part of the sound, and the stochastic part accounts for intrinsically random musical characteristics such as breath noise or bow noise. The spectral modeling and

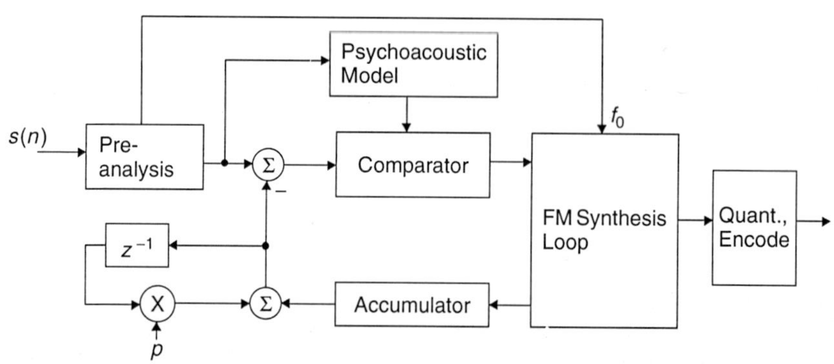

Figure 9.6. FM synthesis coding scheme (after [Wind98]).

synthesis system (SMS) [Serr89] [Serr90] treats audio as the sum of a collection of K sinusoids ("*partials*") with time-varying parameters, (Eq. (9.1)), with the addition of a stochastic component, $e(n)$, i.e.,

$$s(n) \approx \hat{s}(n) = \sum_{k=1}^{K} A_k \cos(\omega_k(n)n + \phi_k(n)) + e(n), \tag{9.13}$$

where A_k represents the amplitude, $\omega_k(n)$ represents the instantaneous frequency, and $\phi_k(n)$ represents the instantaneous phase of the k-th sinusoid. Different parametric models for the stochastic component have been proposed, such as parametric noise spectral envelopes [Serr89] or parametric Bark-band noise [Good96]. Other investigators have constructed hybrid signal models in which the residual of the sinusoidal analysis-synthesis, $s(n) - \hat{s}(n)$, is decomposed using a discrete wavelet packet transform (DWPT) [Hamd96], and then represented in terms of carefully selected principal components. Although the sines + noise signal model has demonstrated improved performance relative to the basic model, one further extension has gained popularity, namely, the addition of a separate model for transient components or, in other words, the construction of a three-part model consisting of sines + transients + noise [Verm98a] [Verm98b].

9.7 HYBRID SINUSOIDAL CODERS

This section examines two experimental audio coders that attempt to enhance source robustness and output quality beyond that of the purely sinusoidal algorithms at low bit rates by embedding additional signal models in the coder architecture. The motivation for this work is as follows. Whereas the waveform-preserving perceptual transform (Chapter 7) and subband (Chapter 8) coders tend to target transparent quality at bitrates between 32 and 128 kb/s per channel, the sinusoidal coders proposed thus far in the literature have concentrated on low rate applications between 2 and 16 kb/s. Rather than transparent quality, these algorithms have emphasized source robustness, i.e., the ability to deal with general audio at low rates without constraining source model dependence. Low rate sinusoidal algorithms (ASAC, HILN, etc.) represent the perceptually significant portions of the magnitude spectrum from the original signal without explicitly treating the phase spectrum. As a result, perceptually transparent coding is typically not achieved with these algorithms. It is generally agreed that different classes of state-of-the-art coding techniques perform most efficiently in terms of output quality achieved for a given bit rate. In particular, CELP speech algorithms offer the best performance for clean speech below 16 kb/s, parametric sinusoidal techniques perform best for general audio between 16 and 32 kb/s, and so-called "time-frequency" audio codecs tend to offer the best performance at rates above 32 kb/s. Designers of comprehensive bit rate scalable coding systems, therefore, must decide whether to cascade multiple stages of fundamentally different coder architectures, with each stage operating on a residual signal from the previous stage, or alternatively to simulcast independent bitstreams from different coder architectures and then select an appropriate decoder at the receiver. In fact, some experimental work performed in the context of MPEG-4 standardization has demonstrated that a cascaded, hybrid sinusoidal/time-frequency coder can not only meet but in some cases even exceed the output quality achieved by the time-frequency (transform) coder alone at the same bit rate for certain critical test signals [Edlc96b]. Several hybrid coder architectures that exploit these ideas by combining sinusoidal modeling with other well-known techniques have been proposed. Two examples are considered next.

9.7.1 Hybrid Sinusoidal-MDCT Algorithm

In one experimental hybrid scheme (Figure 9.7), for example, a locally decoded sinusoidal output signal (generated by, e.g., ASAC) is transformed using an independent MDCT that is configured identically to the second stage coder's internal MDCT. Then, a frequency selective switch (scalability tool) in the second stage coder determines whether quantization and encoding will be more efficient for MDCT coefficients from the original signal or for MDCT coefficient differences between the original and first stage decoded output. Clearly, the coefficient differencing scheme is only beneficial if the difference magnitudes are smaller than the original spectral magnitudes. Several of the issues critical to cascading successfully a parametric sinusoidal coder with a transform-based time-frequency coder are addressed in [Edle98]. For the scheme depicted in Figure 9.7, phase information is essential to minimization of the MDCT coefficient differences. In addition, frequency quantization also influences significantly the maximum residual amplitude. Although log frequency quantization is well suited for stand-alone operation, uniform frequency quantization better controls residual amplitudes in the hybrid scheme. These considerations motivated the inclusion of phase parameters and additional high-frequency spectral line quantization bits in the ASAC enhancement bitstream. In the case of harmonically structured signals, however,

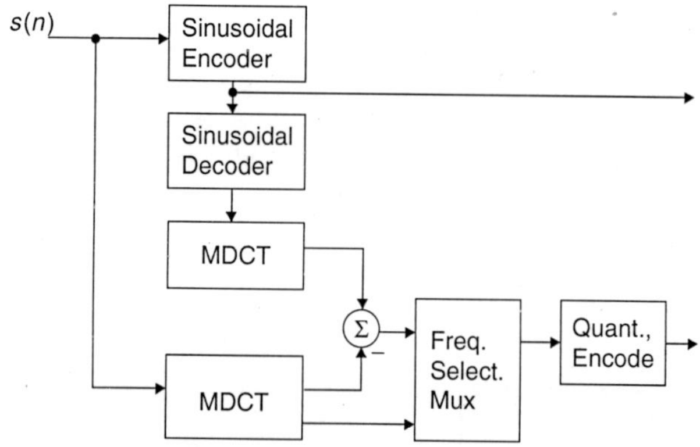

Figure 9.7. Hybrid sinusoidal/T-F (MDCT) encoder (after [Edle98]).

the presence of many partials could overwhelm the enhancement bit allocation if hundreds of partial phases were required. One solution is to limit the enhancement processing to the first ten or so partials. Moreover, for a first stage coder such as the object-based HILN, the noise object would increase rather than decrease the MDCT residuals and should therefore be disabled in a cascade configuration.

9.7.2 Hybrid Sinusoidal-Vocoder Algorithm

It was earlier noted that CELP speech algorithms typically outperform the parametric sinusoidal coders for clean speech inputs at rates below 16 kb/s. There is some uncertainty, however, as to which class of algorithm is best suited when both speech and music are present. A hybrid scheme intended to outperform CELP/parametric "simulcast" for speech/music mixtures was proposed in [Edle98]. The hybrid scheme (Figure 9.8) works as follows. A first-stage parametric coder extracts the dominant harmonic tones, forms an internal residual, and then extracts the remaining harmonic as well as individual tones. Then, an external residual is formed between the original signal and a parametrically synthesized signal containing the contributions of only the *second* set of harmonic and individual tones. This system implicitly assumes that the speech signal is dominant. The idea is that the second stage cascaded vocoder will receive primarily the speech portions of the signal since the first stage parametric coder removes the secondary harmonics and individual tones associated with the musical signal components.

This hybrid structure was reported to outperform simulcast configurations only when the voice signal was dominant [Edle98]. Quality degradations were reported for mixtures containing dominant musical signals. In the future, hybrid structures of this type will benefit from emerging techniques in speech/music discrimination (e.g., [Saun96] [Sche98b]). As observed by Edler, on the other hand, future audio coding research is also quite likely to focus on automatic decomposition of complex input signals into components for which individual coding is more efficient than direct coding of the mixture [Edle97] using hybrid structures. Advances in

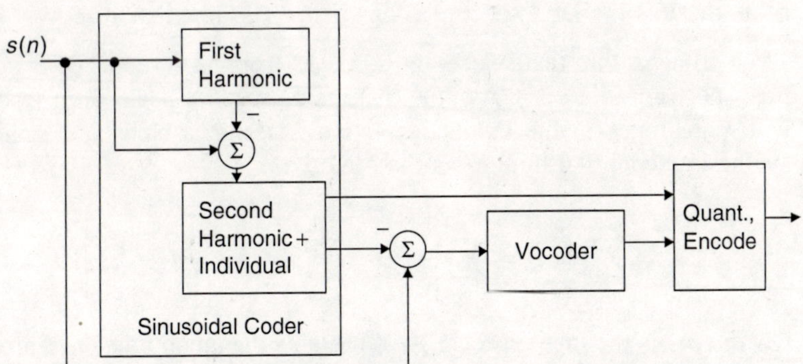

Figure 9.8. Hybrid sinusoidal/vocoder (after [Edle98]).

sound separation and auditory scene analysis [Breg90] [Elli96] techniques will eventually make the automated decomposition process viable.

9.8 SUMMARY

In this Chapter, we reviewed sinusoidal signal models for low-rate audio coding. Some of the key topics we studied include the ASAC, the HILN algorithm, hybrid sinusoidal coders, and the STN model. The parametric representation of sinusoidal coders has the potential for scalable audio compression and delivery of high-quality audio over the Internet. We will discuss the scalable and parametric audio coding tools integrated in the ISO/IEC MPEG 4 audio standard in the next chapter.

PROBLEMS

9.1. Modify the DFT such that the frequency components are computed from $n = -\frac{N}{2}, \ldots, 0, \ldots, \frac{N}{2} - 1$. Why is this approach useful in sinusoidal analysis-synthesis?

9.2. In sinusoidal analysis-synthesis, we require the analysis window to be normalized, i.e., $\sum_{n=-N/2}^{N/2} w(n) = 1$. Explain with mathematical arguments the reason for the use of this normalization.

9.3. Explain with diagrams, text, and equations how the sinusoidal birth-death frequency tracker works.

9.4. Let $s(n)$ be the input audio frame of size N samples. Let $S(k)$ be the N-point DFT.

$$S(k) = \frac{1}{N}\sum_{n=0}^{N-1} s(n)e^{-j2\pi kn/N}, \quad k = 0, 1, 2, \ldots, N-1$$

Obtain a p-th order frequency-domain AR model using least squares algorithm that fits the DFT spectral components. Assume that $p < N - 1$.

9.5. Write an algorithm that selects the K peaks from an N-point DFT magnitude spectrum ($K < N$). Assume that the K spectral peaks are associated with pure tones in the input signal, $s(n)$. Design a sinusoidal analysis-synthesis model that minimizes the MSE, ε,

$$\varepsilon = \sum_n e^2(n) = \sum_n \left[s(n) - \sum_{k=1}^{K} A_k \cos(\omega_k n + \phi_k) \right]^2,$$

where $s(n)$ is the input speech, A_k represents the amplitude, ω_k represents the frequency, and ϕ_k denotes the phase of the k-th sinusoid. Hint: Refer to [McAu86].

9.6. In this problem, we will design a sinusoidal analysis-by-synthesis (A-by-S) algorithm. Use the sinusoidal model, $s(n) \approx \hat{s}(n) = \sum_{k=1}^{K} A_k \cos(\omega_k n + \phi_k)$, where A_k is the amplitude, ω_k represents the frequency, and ϕ_k denotes the phase of the k-th sinusoid. Attempting to minimize $E = \sum_n e^2(n) = \sum_n \left[s(n) - \sum_{k=1}^{K} A_k \cos(\omega_k n + \phi_k)\right]^2$ with respect to A_k, ω_k, and ϕ_k simultaneously would lead to a nonlinear optimization problem. Hence, design a suboptimal A-by-S algorithm to solve for the parameters separately. Hint: Start your design by assuming that the $(K - 1)$ sinusoidal parameters are already determined and the frequencies, ω_k, are known. The MSE will then become,

$$\varepsilon = \sum_n \left[\underbrace{s(n) - \sum_{k=1}^{K-1} A_k \cos(\omega_k n + \phi_k)}_{e_{K-1}(n)} - A_K \cos(\omega_K n + \phi_K) \right]^2,$$

where $e_{K-1}(n)$ is the error after $K - 1$ sinusoidal components were determined. See [Geor87] and [Geor90] to obtain additional information.

9.7. Assuming that a cubic phase interpolation polynomial is used in a sinusoidal analysis-synthesis model, derive Eq. (9.7) using results from [McAu86].

COMPUTER EXERCISES

9.8. Write a MATLAB program to implement the sinusoidal analysis-by-synthesis algorithm that you derived in Problem 9.6. Use the audio file *ch9aud1.wav*. Consider a frame size of $N = 512$ samples, no overlap between frames, and $K = 20$ sinusoids. In your deliverables, provide the MATLAB program; give the plots of the input audio, $s(n)$, the synthesized signal, $\hat{s}(n)$, and the error, $e(n) = s(n) - \hat{s}(n)$ for the entire audio record.

9.9. Use the MATLAB program from the previous problem and estimate the sample mean square error, $E_i = \mathrm{E}[e_i^2(n)]$ for each frame, i, where E[.] is the expectation operator.

a. Calculate the average MSE, $E_{\text{Avg}} = \dfrac{1}{N_f} \sum_{i=1}^{N_f} E_i$, where N_f is the number of frames.

b. Repeat step (a) for a different number of sinusoids, e.g., $K = 5$, 15, 30, 50, and 60. Plot E_{Avg} in dB across the number of sinusoids, K.

c. How does the MSE behave with respect to the number of sinusoids selected?

9.10. In this problem, we will study the significance of phase information in the sinusoidal coding of audio. Use the audio file *ch9aud1.wav*. Consider a frame size of $N = 512$ samples. Let $S_i(k)$ be the 512-point FFT of the

i-th audio frame, and $S_i(k) = |S_i(k)|e^{j\Phi_i(k)}$ where $|S_i(k)|$ is the magnitude spectrum and $\Phi_i(k)$ is the phase at the k-th FFT bin. Write a MATLAB program to pick 30 spectral peaks from the 512-point FFT magnitude spectrum. Set the rest of the spectral component magnitudes to a very small value, e.g., 0.000001. Call the resulting spectral magnitude as $|\hat{S}_i(k)|$.

a. Set the phase $\Phi_i(k) = 0$ and reconstruct the audio signal, $\hat{s}_i(n) = \text{IFFT}_{512}[\hat{S}_i(k)]$ for all the frames, $i = 1, 2, \ldots, N_f$.

b. Calculate the sample MSE, $E_i = \text{E}[e_i^2(n)]$ for each frame, i. Compute the average MSE, $E_{\text{Avg}} = \frac{1}{N_f}\sum_{i=1}^{N_f} E_i$, where N_f is the number of frames.

c. Set the phase $\mathbf{\Phi}_i = \pi(2\text{rand}(1, 512) - 1))$, where $\mathbf{\Phi}_i$ is the $[1 \times 512]$ uniform random phase vector that varies between $-\pi$ and π. Reconstruct the audio signal, $\hat{s}_i(n) = \text{IFFT}_{512}[\hat{S}_i(k)e^{j\mathbf{\Phi}_i}]$ for all the frames. Note that you must avoid using the same set of random phase components in all the frames. Repeat step (b).

d. Set the phase $\Phi_i(k) = \arg[S_i(k)]$, i.e., we are using the input signal phase. Reconstruct the audio, $\hat{s}_i(n) = \text{IFFT}_{512}[\hat{S}_i(k)e^{j\Phi_i(k)}]$ for all the frames. Repeat step (b).

e. Perform an experiment where the low-frequency sinusoids (<1.5 kHz) use the input signal phase and the high-frequency components use random phase as in part (c). Compute E_i and E_{Avg} using step (b).

f. Give a dB plot of E_i obtained from steps (a) and (b) across the number of frames, $i = 1, 2, \ldots, N_f$. Now superimpose the E_i's obtained in steps (c), (d), and (e). Which case performed better in terms of the MSE measure?

g. Evaluate to the synthesized audio obtained from steps (a), (c), (d), and (e). Which case results in better perceptual quality?

9.11. In this problem, we will become familiar with the sinusoidal trajectories. Use the audio file *ch9aud2.wav*. Consider a frame size of $N = 512$ samples, no overlap between frames, and $K = 10$ sinusoids. Write a MATLAB program to implement the sinusoidal A-by-S algorithm [Geor90]. For each frame, i, compute the amplitudes A_i^k, the frequencies ω_i^k, and the phases, ϕ_i^k of the k-th sinusoid.

a. Give a contour plot of the frequencies, ω_i^k versus the frame number, $i = 1, 2, \ldots, N_f$, associated with each of the 10 sinusoids.

b. Now assume a 50% overlap and repeat the above step. Comment on the smoothness and continuity of the frequency trajectories when frames are overlapped 50%.

9.12. Extend the Computer Exercise 9.8 when a 50% overlap is allowed between the frames. Use a Bartlett window or a Hanning window for overlap. See [Geor97] for hints on overlap-add A-by-S implementation. Do you observe any improvements in terms of audio quality when overlapping is employed?

9.13. In this problem, we will design a hybrid subband/sinusoidal algorithm. Use the audio file *ch9aud3_24k.wav*. Consider a frame size of $N = 512$ samples, no overlap between frames, and $K = 20$ sinusoids. Note that the sampling frequency of the given audio file is 24 kHz.

a. Write a MATLAB program to design a sinusoidal A-by-S algorithm to pick 20 sinusoids in each frame. Compute the MSE, $\hat{E} = \frac{1}{N_f}\sum_{i=1}^{N_f} E_i$, where N_f is the number of frames and $E_i = \mathrm{E}[e_i^2(n)]$ is the MSE at the i-the frame.

b. Modify the program in (a) to design a sinusoidal A-by-S algorithm to pick 15 sinusoids between 20 Hz and 6 kHz; and 5 sinusoids between 6 kHz and 12 kHz in each of the frames. Did you observe any performance improvements in terms of computational efficiency and/or perceptual quality?

CHAPTER 10

AUDIO CODING STANDARDS AND ALGORITHMS

10.1 INTRODUCTION

Despite the several advances, research towards developing lower rate coders for *stereophonic* and *multichannel surround sound* systems is strong in many industry and university labs. Multimedia applications such as online radio, web jukeboxes, and teleconferencing created a demand for audio coding algorithms that can deliver real-time wireless audio content. This will in turn require audio compression algorithms to deliver high-quality audio at *low bit-rates* with *resilience/robustness* to bit errors. Motivated by the need for audio compression *algorithms for streaming audio*, researchers pursue techniques such as combined speech/audio architectures, as well as joint source-channel coding algorithms that are optimized for the packet switched Internet [Ben99] [Liu99] [Gril02], Bluetooth [Joha01] [Chen04] [BWEB], and in some cases wideband cellular network [Ji02] [Toh03]. Also the need for transparent reproduction quality coding algorithms in *storage media* such as the super audio CD (SACD) and the DVD-audio provided designers with new challenges. There is in fact an ongoing debate over the *quality* limitations associated with lossy compression. Some experts believe that *uncompressed* digital CD-quality audio (44.1 kHz/16 bit) is inferior to the analog original. They contend that sample rates above 55 kHz and word lengths greater than 20 bits are necessary to achieve transparency in the absence of any compression.

As a result, several standards have been developed [ISOI92] [ISOI94a] [Davi94] [Fiel96] [Wyl96b] [ISOI97b], particularly in the last five years [Gerz99] [ISOI99]

[ISOI00] [ISOI01b] [ISOI02a] [Jans03] [Kuzu03], and several are now being deployed commercially. This chapter and the next address some of the important audio coding algorithms and standards deployed during the last decade. In particular, we describe the lossy audio compression (LAC) algorithms in this chapter, and the lossless audio coding (L^2AC) schemes in Chapter 11.

Some of the LAC schemes (Figure 10.1) described in this chapter include the ISO/MPEG codec series, the Sony ATRAC, the Lucent Technologies PAC/EPAC/MPAC, the Dolby AC-2/AC-3, the APT-x 100, and the DTS-coherent acoustics.

The rest of the chapter is organized as follows. Section 10.2 reviews the MIDI standard. Section 10.3 serves as an introduction to the multichannel surround sound format. Section 10.4 is dedicated to MPEG audio standards. In particular, Sections 10.4.1 through 10.4.6, respectively, describe the MPEG-1, MPEG-2 BC/LSF, MPEG-2 AAC, MPEG-4, MPEG-7, and MPEG-21 audio standards. Section 10.5 presents the adaptive transform acoustic coding (ATRAC) algorithm, the MiniDisc and the Sony dynamic digital sound (SDDS) systems. Section 10.6 reviews the Lucent Technologies perceptual audio coder (PAC), the enhanced PAC (EPAC), and the multichannel PAC (MPAC) coders. Section 10.7 describes the Dolby AC-2 and the AC-3/Dolby Digital algorithms. Section 10.8 is devoted to the Audio Processing Technology – APTx-100 system. Finally, in Section 10.9, we examine the principles of coherent acoustics in coding, that are embedded in the Digital Theater Systems–Coherent Acoustics (DTS-CA).

10.2 MIDI *VERSUS* DIGITAL AUDIO

The musical instrument digital interface (MIDI) encoding is an efficient way of extracting and representing semantic features from audio signals [Lehr93] [Penn95] [Hube98] [Whit00]. MIDI synthesizers, originally established in 1983, are widely used for musical transcriptions. Currently, the MIDI standards are governed by the MIDI Manufacturers Association (MMA) in collaboration with the Japanese Association of Musical Electronics Industry (AMEI).

The digital audio representation contains the actual sampled audio data, while a MIDI synthesizer represents only the instructions that are required to play the sounds. Therefore, the MIDI data files are extremely small when compared to the digital audio data files. Despite being able to represent high-quality stereo data at 10–30 kb/s, there are certain limitations with MIDI formats. In particular, the MIDI protocol uses a slow serial interface for data streaming at 31.25 kb/s [Foss95]. Moreover, MIDI is hardware dependent. Despite such limitations, musicians prefer the MIDI standard because of its simplicity and high-quality sound synthesis capability.

10.2.1 MIDI Synthesizer

A simple MIDI system (Figure 10.2) consists of a MIDI controller, a sequencer, and a MIDI sound module. The keyboard is an example of a MIDI controller

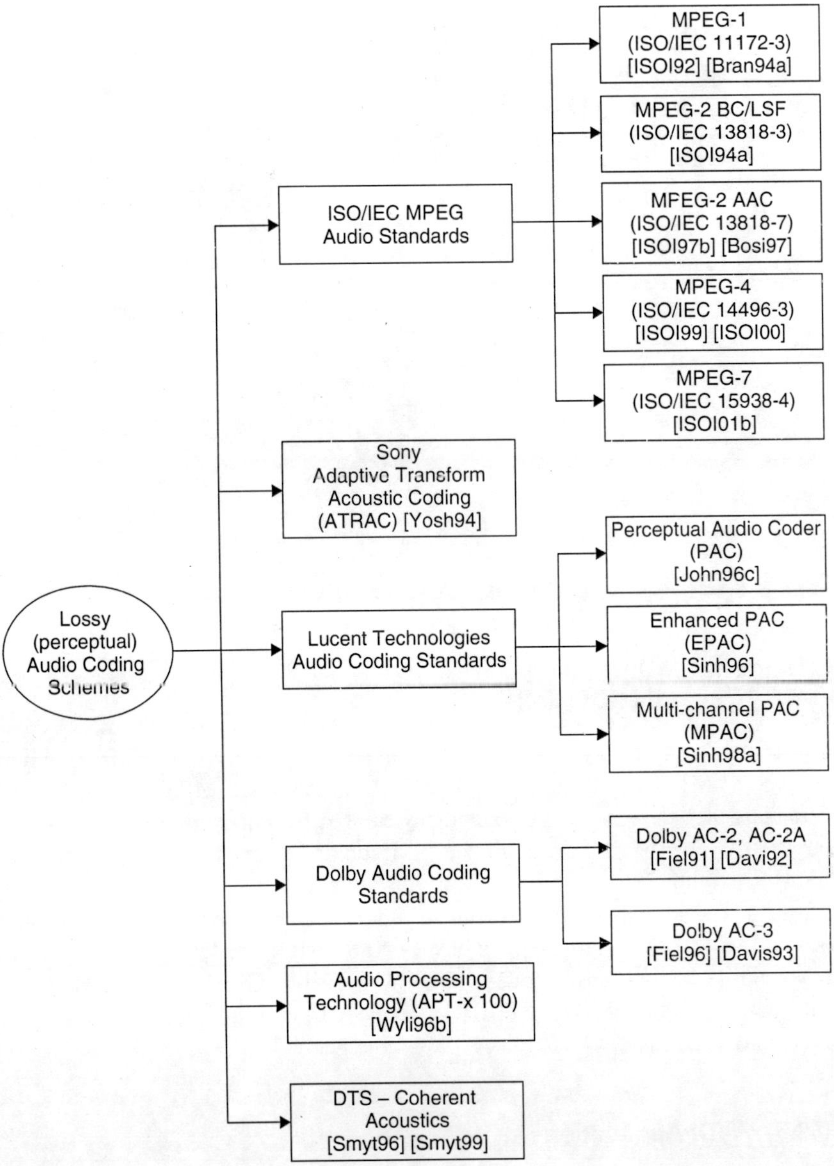

Figure 10.1. A list of some of the lossy audio coding algorithms.

that translates the music notes into a real-time MIDI data stream. The MIDI data stream includes a start bit, 8 data bits, and one stop bit. A MIDI sequencer captures the MIDI data sequence, and allows for various manipulations (e.g., editing, morphing, combining, etc.). On the other hand a MIDI sound module acts as a sound player.

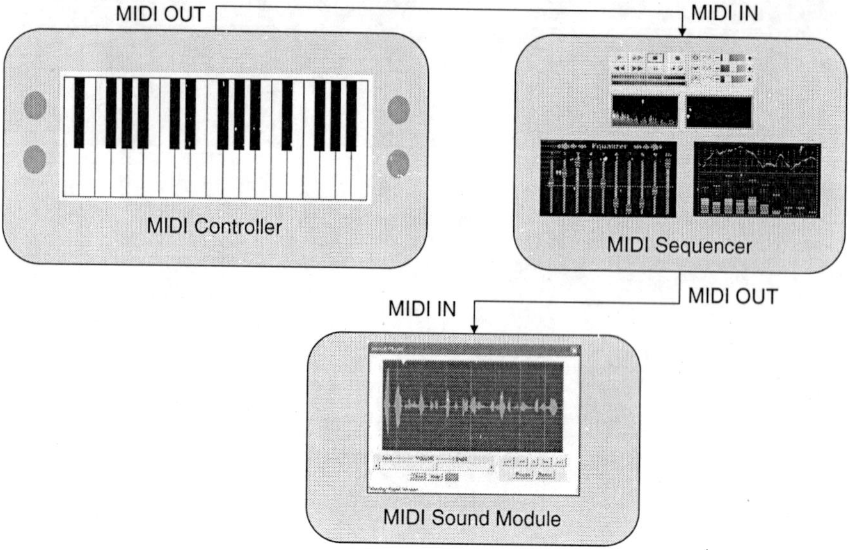

Figure 10.2. A simple MIDI system.

10.2.2 General MIDI (GM)

In order to facilitate a greater degree of file compatibility, the MMA developed the general MIDI (GM) standard. The GM constitutes a MIDI synthesizer with a standard set of voices (16 categories of 8 different sounds = $16 \times 8 = 128$ sounds) that are fixed. Although the GM standard does not describe the sound quality of synthesizer outputs, it provides details on the MIDI compatibility, i.e., the MIDI sounds composed on one sequencer can be reproduced or played back on any other system with reduced or no distortion. Different GM versions are available in the market today, i.e., GM Level-1, GM Level-2, GM lite, and scalable polyphonic MIDI (SPMIDI). Table 10.1 summarizes the various GM levels and versions.

10.2.3 MIDI Applications

MIDI has been successful in a wide range of applications including music-retrieval and classification [Mana02], music databases search [Kost96], musical instrument control [MID03], MIDI karaoke players [MIDI], real-time object-based coding [Bros03], automatic recognition of musical phrases [Kost96], audio authoring [Mode98], waveform-editing [MIDI], singing voice synthesis [Maco97], loudspeaker design [Bald96], and feature extraction [Kost95]. The MPEG-4 structured audio tool incorporates many MIDI-like features. Other applications of MIDI are attributed to MIDI GM Level-2 [Mode00], XMIDI [LKur96], and PCM to MIDI transportation [Mart02] [MID03] [MIDI].

Table 10.1. General MIDI (GM) formats and versions.

GM specifications	GM L-1	GM L-2	GM Lite	SPMIDI
Number of MIDI channels	16	16	16	16
Percussion (drumming) channel	10	10, 11	10	–
Polyphony (voices)	24 voices	32 voices	Limited	–

Other related information [MID03]:
GM L-2: This is the latest standard introduced with capabilities of registered parameter controllers, MIDI tuning, universal system exclusive messages. GM L-2 is backwards compatible with GM L-1.
GM Lite: As the name implies, this is a light version of GM L-1 and is intended for devices with limited polyphony.
SPMIDI: Intended for mobile devices, SPMIDI, functions based on the fundamentals of GM Lite and scalable polyphony. This GM standard has been adopted by the Third-Generation Partnership Project (3GPP) for the multimedia messaging applications in cellular phones.

10.3 MULTICHANNEL SURROUND SOUND

Surround sound tracks (or channels) were included in motion pictures, in the early 1950s, in order to provide a more realistic cinema experience. Later, the popularity of surround sound resulted in its migration from cinema halls to home theaters equipped with matrixed multichannel sound (e.g., Dolby ProLogic™). This can be attributed to the multichannel surround sound format [Bosi93] [Holm99] [DOLBY] and subsequent improvements in the audio compression technology.

Until the early 1990s, almost all surround sound formats were based on matrixing, i.e., the information from all the channels (front and surround) was encoded as a two-channel stereo as shown in Figure 10.3. In the mid-1990s, discrete encoding, i.e., 5.1 separate channels of audio, was introduced by Dolby Laboratories and Digital Theater Systems (DTS).

10.3.1 The Evolution of Surround Sound

Table 10.2 lists some of the milestones in the history of multichannel surround sound systems. In the early 1950s, the first commercial multichannel sound format was developed for cinema applications. "Quad" (Quadraphonic) was the first home-multichannel format, promoted in the early 1970s. But, due to some incompatibility issues in the encoding/decoding techniques, the Quad was not successful. In the mid-1970s, Dolby overcame the incompatibility issues associated with the optical sound tracks and introduced a new format, called the Dolby stereo, a special encoding technique that later became very popular. With the advent of *compact discs* (CDs) in the early 1980s, high-performance stereo systems became quite common. With the emergence of *digital versatile discs* (DVDs) in 1995–1996, content creators began to distribute multichannel music in digital format. Dolby laboratories, in 1992, introduced another coding algorithm (Dolby AC-3, Section 10.7), called the Dolby Digital that offers a high-quality multichannel (5.1-channel) surround sound experience. The Dolby Digital was later chosen as the primary audio coding technique for DVDs and for digital audio broadcasting (DAB). The following year, Digital Theater Systems Inc. (DTS) announced a new format based on the Coherent Acoustics encoding principle (DTS-CA). The same year, Sony proposed the Sony Dynamic Digital Sound (SDDS) system that employs the Adaptive Transform Acoustic Coding (ATRAC) algorithm. Lucent Technologies' Multichannel Perceptual Audio Coder (MPAC) also has a five-channel surround sound configuration. Moreover, the development of two new audio recording technologies, namely, the Meridian Lossless Packing (MLP) and the Direct Stream Digital (DSD), for use in the DVD-Audio [DVD01] and SACD [SACD02] formats, respectively, offer audiophiles listening experiences that promise to be more realistic.

10.3.2 The Mono, the Stereo, and the Surround Sound Formats

Figure 10.4 shows the three most common sound formats, i.e., mono, stereo, and surround. Mono is a simple method of recording sound onto a single channel that is typically played back on one speaker. In stereo encoding, a two-channel recording is employed. Stereo provides a sound field in front, while the multichannel surround sound provides multi-dimensional sound experience. The surround sound systems typically employ a 5.1-channel configuration, i.e., sound tracks are recorded using five main channels: left (L), center (C), right (R), left surround (LS), and right surround (RS). In addition to these five channels, a sixth channel called the low-frequency-effects (LFE) channel is used for the subwoofer. Since the LFE channel covers only a fraction (less than 150 Hz) of the total frequency range, it is referred as the .1-channel.

10.3.3 The ITU-R BS.775 5.1-Channel Configuration

In an effort to evaluate and standardize the so-called 5.1- or 3/2-channel configuration, several technical documents appeared [Bosi93] [ITUR94c] [EBU99] [Holm99] [SMPTE99] [AES00] [Bosi00] [SMPTE02]. Various international standardization bodies became involved in multichannel algorithm adoption/evaluation process. These include: the Audio Engineering Society

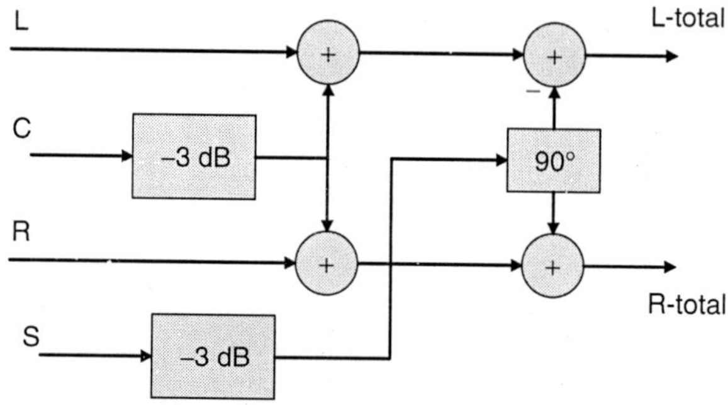

Figure 10.3. Multichannel surround sound matrixing.

Table 10.2. Milestones in multichannel surround sound.

Year	Description
1941	*Fantasia* (Walt-Disney Productions) was the first motion picture to be released in the multichannel format
1955	Introduction of the first 35/70-mm magnetic stripe capable of providing 4/6 channels
1972	Video cassette consumer format – mono (1 channel)
1976	Dolby's stereo in optical format
1978	Videocassette – stereo (2 channels)
1979	Dolby's first stereo surround, called the *split-surround sound format*, offered 3 screen channels, 2 surround channels, and a subwoofer (3/2/0.1)
1982	Dolby surround format implemented on a compact disc (2 channel)
1992	Dolby digital optical (5.1 channel)
1993	Digital Theater Systems (DTS)
1993–94	Sony Dynamic Digital Sound (SDDS) based on ATRAC
1994	The ISO/IEC 13818-3 MPEG-2 Backward compatible audio standard
1995	Dolby digital chosen for DVD (5.1 channel)
1997	DVD video released in market (5.1 channel)
1998	Dolby digital selected for digital audio broadcasting (DAB) in U.S.
1999–	Super Audio CD and DVD-Audio storage formats
2000–	Direct Stream Digital (DSD) and Meridian Lossless Packing (MLP) Technologies

(AES), the European Broadcasting Union (EBU), the Society of Motion Picture and Television Engineers group (SMPTE), the ISO/IEC MPEG, and the ITU-Radio communication sector (ITU-R).

Figure 10.5 shows a 5.1-channel configuration described in the ITU-R BS.775-1 standard [ITUR94c]. Ideally, five full-bandwidth (150 Hz–20 kHz) loudspeakers, i.e., L, R, C, LS, and RS are placed on the circumference of a circle in the following manner: the left (L) and right (R) front loudspeakers are placed at the extremities of an arc subtending, $2\theta = 60^\circ$, at the reference listening position (see Figure 10.5), and the center (C) loudspeaker must be placed at 0° from the listener's axis. This enables the compatibility with the listening arrangement for a conventional two-channel system. The two surround speakers, i.e., LS and RS are usually placed at $\phi = 110^\circ$ to 120° from the listener's axis. In order to achieve synchronization, the front and surround speakers must be equidistant, λ, (usually 2–4 m) from the reference listening point, with their acoustic centers in the horizontal plane as shown in the figure. The sixth channel, i.e., the LFE channel delivers bass-only omnidirectional information (20–150 Hz). This is because low frequencies imply longer-wavelengths where the ears are not sensitive to localization. The subwoofer placement receives less attention in the ITU-R standard; however, we note that the subwoofers are typically placed in a front corner (see Figure 10.4). In [Ohma97], Ingvar discusses the various problems associated with the subwoofer placement. Moreover, [SMPTE02] provides of information on the

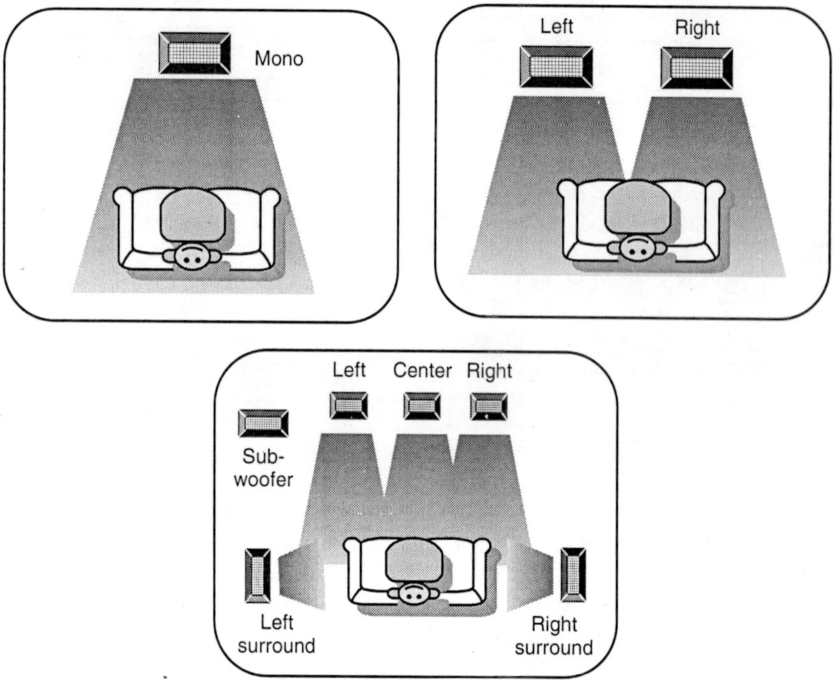

Figure 10.4. Mono, stereo, and surround sound systems.

loudspeaker placement for audio monitoring. The [SMPTE99] specifies the audio channel assignment and their relative levels for audio program recordings (3–6 audio channels) onto storage media for television sound.

10.4 MPEG AUDIO STANDARDS

MPEG is the acronym for *Moving Pictures Experts Group* that forms a work group (WG-11) of ISO/IEC JTC-1 subcommittee (SC-29). The main functions of MPEG are: a) to publish technical results and reports related to audio/video compression techniques; b) to define means to multiplex (combine) video, audio, and information bitstreams into a single bitstream, and c) to provide descriptions and syntax for low bit rate audio/video coding tools for Internet and bandwidth-restricted communications applications. MPEG standards do not characterize or provide any rigid encoder specifications, but rather standardizes the type of information that an encoder has to produce as well as the way in which the decoder has to decompress this information. The MPEG workgroup has its own official webpage that can be accessed at [MPEG]. The MPEG video aspect of the standard is beyond the scope of this book, however, we include some tutorial references and relevant standards [LeGal92] [Scha95] [ISO-V96] [Hask97] [Mitc97] [Siko97a] [Siko97b].

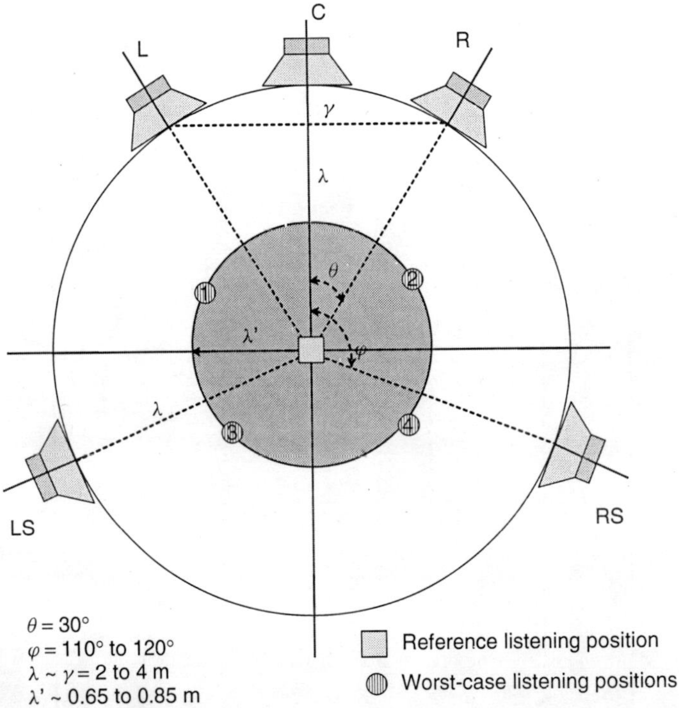

Figure 10.5. A typical 3/2-channel configuration described in the ITU-R BS.775-1 standard [ITUR94c]. L: left, C: center, R: right, LS: left surround, and RS: right surround loud speakers. Note that the above figure is not according to a scale.

MPEG Audio – Background. MPEG has come a long way since the first ISO/IEC MPEG standard was published in 1992. With the emergence of the Internet, MPEG is now also addressing content-based multimedia descriptions and database search. There are five different MPEG audio standards published, i.e., MPEG-1, MPEG-2 BC, MPEG-2 NBC/AAC, MPEG-4, and MPEG-7. MPEG-21 is being formed.

Before proceeding with the details of the MPEG audio standards, however, it is necessary to discuss terminology and notation. The *phases* correspond to the MPEG audio standards type and to a lesser extent to their relative release time, e.g., MPEG-1, MPEG-2, MPEG-4, etc. The *layers* represent a family of coding algorithms within the MPEG standards. Only MPEG-1 and -2 are provided with layers, i.e., MPEG-1 layer-I, -II, and -III; MPEG-2 layer-I, -II, and -III. The *versions* denote the various stages in the audio coding standardization *phase*. MPEG-4 was standardized in two stages (*version*-1 and -2) with new functionality being added to the older version. The newer versions are backward compatible to the older versions. Table 10.3 itemizes the various MPEG audio standards and their specifications. A brief overview of the MPEG standards follows.

Table 10.3. An overview of the MPEG audio standards.

Standard	Standardization details	Bit rates (kb/s)	Sampling rates (kHz)	Channels	Related references	Related information
MPEG-1	ISO/IEC 11172-3 1992	32–448 (Layer I) 32–384 (Layer II) 32–320 (Layer III)	32, 44.1,48	Mono (1), stereo (2)	[ISOI92] [Bran94a] [Shli94] [Pan95] [Noll93] [Noll95] [Noll97] [John99]	A generic compression standard that targets primarily multimedia storage and retrieval.
MPEG-2 BC/LSF	ISO/IEC 13818-3 1994	32–256 (Layer I) 8–160 (Layers II, III)	16, 22.05, 24	Multichannel Surround sound (5.1)	[ISOI94a] [Stol93a] [Gril94] [Noll93] [Noll95] [John99]	First digital television standard that enables lower frequencies and multichannel audio coding.
MPEG-2 NBC/AAC	ISO/IEC 13818-7 1997	8–160	8–96	Multichannel	[ISOI97b] [Bosi96] [Bosi97] [John99] [ISOI99] [Koen99] [Quac98b] [Gril99]	Advanced audio coding scheme that incorporates new coding methodologies (e.g., prediction, noise shaping, etc.).

Table 10.3. (*continued*)

Standard	Standardization details	Bit rates (kb/s)	Sampling rates (kHz)	Channels	Related references	Related information
MPEG-4 (Version 1)	ISO/IEC 14496-3 Oct, 1998	0.2–384	8–96	Multichannel	[Gril97] [Park97] [Edle99] [Purn00a] [Sche98a] [Vaan00] [Sche01]	The first content-based multimedia standard, allowing universality/interactivity and a combination of natural and synthetic material, coded in the form of objects.
MPEG-4 (Version 2)	ISO/IEC 14496-3/AMD-1, Dec, 1999	0.2–384 (finer levels of increment possible)	8–96	Multichannel	[ISOI00] [Kim01] [Herr00a] [Purn99b] [Alla99] [Hilp00] [Sper00] [Sper01]	
MPEG-7	ISO/IEC 15938-4 Sept, 2001	–	–	–	[ISOI01b] [Nack99a] [Nack99b] [Quac01] [Lind99] [Lind00] [Lind01] [Speng01] [ISOI02a]	A normative metadata standard that provides a Multimedia Content Description Interface.
MPEG-21	ISO/IEC-21000	–	–	–	[ISOI03a] [Borm03] [Burn03]	A multimedia framework that provides interoperability in content-access and distribution.

MPEG-1. After four years of extensive collaborative research by audio coding experts worldwide, the first ISO/MPEG audio coding standard, MPEG-1 [ISOI92], for VHS-stereo-CD-quality was adopted in 1992. The MPEG-1 supports video bit rates up to about 1.5 Mb/s providing a Video Home System (VHS) quality, and stereo audio at 192 kb/s. Applications of MPEG-1 range from storing video and audio on CD-ROMs to Internet streaming through the popular MPEG-1 layer III (MP3) format.

MPEG-2 Backward Compatible (BC). In order to extend the capabilities offered by MPEG-1 to support the so-called 3/2 (or 5.1) channel format and to facilitate higher bit rates for video, MPEG-2 [ISOI94a] was published in 1994. The MPEG-2 standard supports digital video transmission in the range of 2–15 Mb/s over cable, satellite, and other broadcast channels; audio coding is defined at the bit rates of 64–192 kb/s/channel. Multichannel MPEG-2 is backward compatible with MPEG-1, hence, the acronym MPEG-2 BC. The MPEG-2 BC standard is used in the high definition Television (HDTV) [ISOI94a] and produces the video quality required in digital television applications.

MPEG-2 Nonbackward Compatible/Advanced Audio Coding (AAC). The backward compatibility constraints imposed on the MPEG-2 BC/LSF algorithm made it impractical to code five channels at rates below 640 kb/s. As a result, MPEG began standardization activities for a nonbackward compatible advanced coding system targeting "indistinguishable" quality at a rate of 384 kb/s for five full bandwidth channels. In less than three years, this effort led to the adoption of the MPEG-2 nonbackward compatible/advanced audio coding (NBC/AAC) algorithm [ISOI97b], a system that exceeded design goals and produced the desired quality at only 320 kb/s for five full bandwidth channels.

MPEG-4. MPEG-4 was established in December 1998 after many proposed algorithms were tested for compliance with the program objectives established by the MPEG committee. MPEG-4 video supports bit rates up to about 1 Gb/s. The MPEG-4 audio [ISOI99] [ISOI00] was released in several steps, resulting in versions 1 and 2. MPEG-4 comprises an integrated family of algorithms with wide-ranging provisions for scalable, object-based speech and audio coding at bit rates from as low as 200 b/s up to 60 kb/s per channel. The distinguishing features of MPEG-4 relative to its predecessors are extensive scalability, object-based representations, user interactivity/object manipulation, and a comprehensive set of coding tools available to accommodate trade-offs between bit rate, complexity, and quality. Very low rates are achieved through the use of structured representations for synthetic speech and music, such as text-to-speech and MIDI. The standard also provides integrated coding tools that make use of different signal models depending upon the desired bit rate, bandwidth, complexity, and quality.

MPEG-7. The MPEG-7 audio committee activities started in 1996. In less than four years, a committee draft was finalized and the first audio standard addressing "multimedia content description interface" was published in September 2001. MPEG-7 [ISOI01b] targets the content-based multimedia applications. In particular, the MPEG-7 audio supports a broad range of applications – multimedia digital libraries, broadcast media selection, multimedia editing and searching, multimedia indexing/searching. Moreover, it provides ways for efficient audio file retrieval and supports both the text-based and context-based queries.

MPEG-21. The MPEG-21 ISO/IEC-21000 standard [MPEG] [ISOI02a] [ISOI03a] defines interoperable and highly automated tools that enable content distribution across different terminals and networks in a programmed manner. This structure enables end-users to have capabilities for universal multimedia access.

10.4.1 MPEG-1 Audio (ISO/IEC 11172-3)

The MPEG-1 audio standard (ISO/IEC 11172-3) [ISOI92] comprises a flexible hybrid coding technique that incorporates several methods including subband decomposition, filter-bank analysis, transform coding, entropy coding, dynamic bit allocation, nonuniform quantization, adaptive segmentation, and psychoacoustic analysis. MPEG-1 audio codec operates on 16-bit PCM input data at sample rates of 32, 44.1, and 48 kHz. Moreover, MPEG-1 offers separate modes for mono, stereo, dual independent mono, and joint stereo. Available bit rates are 32–192 kb/s for mono and 64–384 kb/s for stereo. Several tutorials on the MPEG-1 standards [Noll93] [Bran94a] [Shli94] [Bran95] [Herr95] [Noll95] [Pan95] [Noll97] [John99] have appeared. Chapter 5, Section 5.7, presents step-by-step procedure involved in the ISO/IEC 11172-3 (MPEG-1, layer I) psychoacoustic model 1 [ISOI92] simulation. We summarize these steps in the context of MPEG-1 audio standard.

The MPEG-1 architecture contains three layers of increasing complexity, delay, and output quality. Each higher layer incorporates functional blocks from the lower layers. Figure 10.6 shows the MPEG-1 layer I/II encoder block diagram. The input signal is first decomposed into 32 critically subsampled subbands using a polyphase realization of a pseudo-QMF (PQMF) bank (see also Chapter 6). The channels are equally spaced such that a 48-kHz input signal is split into 750-Hz subbands, with the subbands decimated 32:1. A 511th-order prototype filter was chosen such that the inherent overall PQMF distortion remains below the threshold of audibility. Moreover, the prototype filter was designed

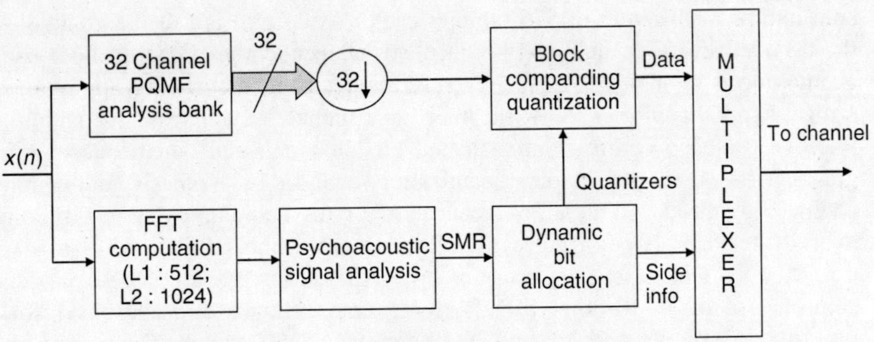

Figure 10.6. ISO/MPEG-1 layer I/II encoder.

for a high sidelobe attenuation (96 dB) to insure that intraband aliasing remains negligible. For the purposes of psychoacoustic analysis and determination of just noticeable distortion (JND) thresholds, a 512 (layer I) or 1024 (layer II) point FFT is computed in parallel with the subband decomposition for each decimated block of 12 input samples (8 ms at 48 kHz). Next, the subbands are block companded (normalized by a scale factor) such that the maximum sample amplitude in each block is unity, then an iterative bit allocation procedure applies the JND thresholds to select an optimal quantizer from a predetermined set for each subband. Quantizers are selected such that both the masking and bit rate requirements are simultaneously satisfied. In each subband, scale factors are quantized using 6 bits and quantizer selections are encoded using 4 bits.

10.4.1.1 Layers I and II For layer I encoding, decimated subband sequences are quantized and transmitted to the receiver in conjunction with side information, including quantized scale factors and quantizer selections. Layer II improves three portions of layer I in order to realize enhanced output quality and reduce bit rates at the expense of greater complexity and increased delay. First, the layer II perceptual model relies upon a higher-resolution FFT (1024 points) than does layer I (512 points). Second, the maximum subband quantizer resolution is increased from 15 to 16 bits. Despite this increase, a lower overall bit rate is achieved by decreasing the number of available quantizers with increasing subband index. Finally, scale factor side information is reduced while exploiting temporal masking by considering properties of three adjacent 12-sample blocks and optionally transmitting one, two, or three scale factors plus a 2-bit side parameter to indicate the scale factor mode. Average mean opinion scores (MOS) of 4.7 and 4.8 were reported [Noll93] for monaural layer I and layer II codecs operating at 192 and 128 kb/s, respectively. Averages were computed over a range of test material.

10.4.1.2 Layer III The layer III MPEG architecture (Figure 10.7) achieves performance improvements by adding several important mechanisms on top of the layer I/II foundation. The MPEG layer-III algorithm operates on consecutive frames of data. Each frame consists of 1152 audio samples; a frame is further split into two subframes of 576 samples each. A subframe is called a granule. At the decoder, every granule can be decoded independently. A hybrid filter bank is introduced to increase frequency resolution and thereby better approximate critical band behavior. The hybrid filter bank includes adaptive segmentation to improve pre-echo control. Sophisticated bit allocation and quantization strategies that rely upon nonuniform quantization, analysis-by-synthesis, and entropy coding are introduced to allow reduced bit rates and improved quality. The hybrid filter bank is constructed by following each subband filter with an adaptive MDCT. This practice allows for higher-frequency resolution and pre-echo control. Use of an 18-point MDCT, for example, improves frequency resolution to 41.67 Hz per spectral line. The adaptive MDCT switches between 6 and 18 points to allow improved pre-echo control. Shorter blocks (4 ms) provide

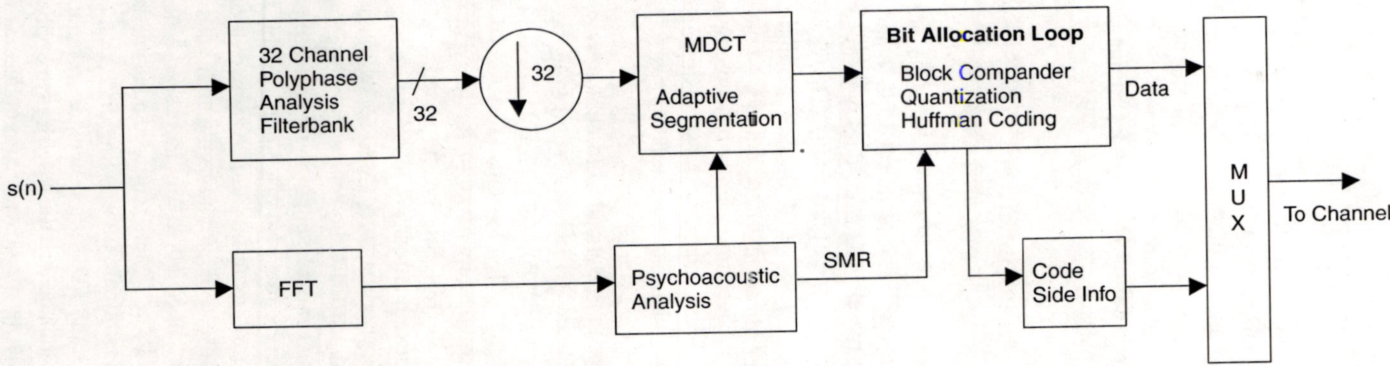

Figure 10.7. ISO/MPEG-1 layer III encoder.

for temporal pre-masking of pre-echoes during transients; longer blocks during steady-state periods improve coding gain by reducing side information and hence bit rates.

Bit allocation and quantization of the spectral lines are realized in a nested loop procedure that uses both nonuniform quantization and Huffman coding. The inner loop adjusts the nonuniform quantizer step sizes for each block until the number of bits required to encode the transform components falls within the bit budget. The outer loop evaluates the quality of the coded signal (analysis-by-synthesis) in terms of quantization noise relative to the JND thresholds. Average MOS of 3.1 and 3.7 were reported [Noll93] for monaural layer II and layer III codecs operating at 64 kb/s.

10.4.1.3 Applications MPEG-1 has been successful in numerous applications. For example, MPEG-1 layer III has become the standard for transmission and storage of compressed audio for both World Wide Web (WWW) and handheld media applications (e.g., IPod™). In these applications, the "MP3" label denotes MPEG-1, layer III. Note that MPEG-1 audio coding has steadily gained acceptance and ultimately has been deployed in several other large scale systems, including the European digital radio (DBA) or Eureka [Jurg96], the direct broadcast satellite or "DBS" [Prit90], and the digital compact cassette or "DCC" [Lokh92]. In particular, the Philips Digital Compact Cassette (DCC) is an example of a consumer product that essentially implements the 384 kb/s stereo mode of MPEG-1 layer I. A discussion of the precision adaptive subband coding (PASC) algorithm and other elements of the DCC system are given in [Lokh92] and [Hoog94].

The collaborative European Advanced Communications Technologies and Services (ACTS) program adopted MPEG audio and video as the core compression technology for the Advanced Television at Low Bit rates And Networked Transmission over Integrated Communication systems (ATLANTIC) project. ATLANTIC is a system intended to provide functionality for television program production and distribution [Stor97] [Gilc98]. This system posed new challenges for MPEG deployment such as seamless bitstream (source) switching [Laub98] and robust transcoding (tandem coding). Bitstream (source) switching becomes nontrivial when different bit rates and/or MPEG layers are associated with different program sources. Robust transcoding is also essential in the video production environment. Editing tasks inevitably require retrieval of compressed bit streams from archival storage, processing of program material in uncompressed form, and then replacement of the recoded compressed bit stream to the archival system. Unfortunately, transcoding is neither guaranteed nor likely to preserve perceptual noise masking [Rits96]. The ATLANTIC designers proposed a buried data "MOLE" signal to mitigate and in some cases eliminate transcoding distortion for cascaded MPEG-1 layer II codecs [Flet98], ideally allowing downstream tandem stages to preserve the original bit stream. The idea behind the MOLE is to apply the same set of quantizers to the same set of data in the downstream codecs as in the original codec. The output bit stream will then be identical to the original bit stream, provided that numerical precision in the analysis filter banks does not bias the data [tenK96b]. It is possible in a cascade of MPEG-1 layer II codecs to regenerate the same set of decimated subband sequences in the downstream codec filter banks as in the original codec filter bank if the full-band PCM signal is properly time aligned at the input to each cascaded stage. Essentially, delays at the filter-bank input must correspond to integer delays at the subband level [tenK96b], and the analysis frames must contain the same block of data in each analysis filter bank. The MOLE signal, therefore, provides downstream codecs with timing synchronization, bit allocation, and scale-factor information for the MPEG bit stream on each frame. The MOLE is buried in the PCM samples between tandem stages and remains inaudible by occupying the LSB of each 20-bit PCM word. Although optimal time-alignment between codecs is possible even without the MOLE [tenK96b], there is unfortunately no easy way to force selection of the same set of quantizers and thus preserve the bit stream.

The widespread use and maturity of MPEG-1 relative to the more recent standards provided several concrete examples for the above discussion of MPEG-1 audio applications. Various real-time implementation schemes of MPEG-1 layers-I, II, and III codecs were proposed [Gbur96] [Hans96] [Main96] [Wang01]. We will next consider the MPEG-2 BC/LSF, MPEG-2 AAC, the MPEG-4, and the MPEG-7 algorithms. The discussion will focus primarily upon architectural novelties and differences with respect to MPEG-1.

10.4.2 MPEG-2 BC/LSF (ISO/IEC-13818-3)

MPEG-2 BC/LSF Audio [Stol93a] [Gril94] [ISOI94a] [Stol96] extends the capabilities offered by MPEG-1 to support the so-called *3/2-channel format* with left (L), right (R), center (C), and left and right surround (LS and RS) channels. The MPEG-2 BC/LSF audio standard is *backward compatible* with MPEG-1, which means that the 3/2 channel information transmitted by an MPEG-2 encoder can be appropriately decoded for 2-channel presentation by an MPEG-1 receiver. Another important feature that was implemented in MPEG-2 BC/LSF is the *multilingual compatibility*. The acronym BC corresponds to the backward compatibility of MPEG-2 towards MPEG-1, and the extension of sampling frequencies to lower ranges (16, 22.05, and 24 kHz) is denoted by LSF. Several tutorials on MPEG-2 [Noll93] [Noll95] [John99] have appeared. Meares and Theile studied the potential application of matrixed surround sound [Mear97] in MPEG audio algorithms.

10.4.2.1 The Backward Compatibility Feature Depending on the bit-demand constraints, interchannel dependencies, and the complexity allowed at the decoder, different methods can be employed to realize compatibility between the 3/2- and 2-channel formats. These methods include mid/side (MS), intensity coding, simulcast, and matrixing. The MS and intensity coding techniques are particularly handy when bit demand imposed by multiple independent channels exceeds the bit budget. The MS scheme is carefully controlled [Davi98] to maintain compatibility among the mono, stereo, and the surround sound formats. Intensity coding, also known as channel coupling, is a multichannel irrelevancy reduction coding technique that exploits properties of spatial hearing. The idea behind intensity coding is to transmit only one envelope with some side information instead of two or more from independent channels. The side information consists of a set of coefficients that is used to recover individual spectra from the intensity channel. The simulcast encoding involves transmission of both stereo and multichannel bitstreams. Two separate bitstreams, i.e., one for 2-channel stereo and another one for the multichannel audio are transmitted, resulting in reduced coding efficiency.

MPEG-2 BC/LSF employs matrixing techniques [tenK92] [tenK94] [Mear97] to down-mix the 3/2 channel format to the 2-channel format. Down-mixing capability is essential for the 5.1-channel system since many of the playback systems are stereophonic or even monaural. Figure 10.8 depicts the matrixing technique employed in the MPEG-2 BC/LSF and can be mathematically expressed as follows

$$L\_total = x(L + yC + zL_s) \tag{10.1}$$

$$R\_total = x(R + yC + zR_s), \tag{10.2}$$

where x, y, and z are constants specified by the IS-13818-3 MPEG-2 standard [ISOI94a]. In Eqs. (10.1) and (10.2), L, C, R, L_s, and R_s represent the 3/2-channel configuration and the parameters *L_total* and *R_total* correspond to the 2-channel format.

Three different choices are provided in the MPEG-2 audio standard [ISOI94a] for choosing the values of x, y, and z to perform the 3/2-channel to 2-channel down-mixing. These include:

$$\text{Choice1}: \quad x = \frac{1}{1+\sqrt{2}},\ y = \frac{1}{\sqrt{2}},\ \text{and } z = \frac{1}{\sqrt{2}} \tag{10.3}$$

$$\text{Choice2}: \quad x = \frac{2}{3+\sqrt{2}};\ y = \frac{1}{\sqrt{2}};\ \text{and } z = \frac{1}{2} \tag{10.4}$$

$$\text{Choice3}: \quad x = 1;\ y = 1;\ \text{and } z = 1 \tag{10.5}$$

The selection of the down-mixing parameters is encoder dependent. The availability of the basic stereo format channels, i.e., *L_total* and *R_total* and the surround sound extension channels, i.e., C,L_s, and R_s at the decoder helps to decode both 3/2-channel and 2-channel bitstreams. This insures the backwards compatibility in the MPEG-2 BC/LSF audio coding standard.

10.4.2.2 MPEG-2 BC/LSF Encoder The steps involved in the reduction of the objective redundancies and the removal of the perceptual irrelevancies in MPEG-2 BC/LSF encoding are the same as in MPEG-1 audio standard. However, the differences arise from employing multichannel and multilingual bitstream

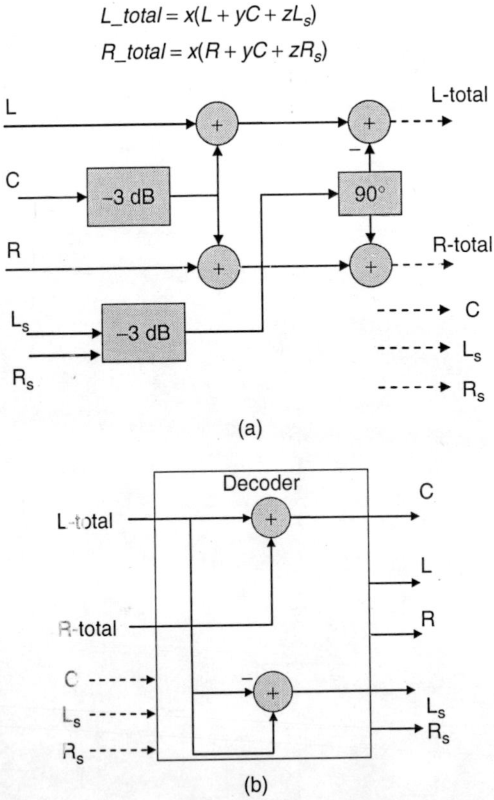

Figure 10.8. Multichannel surround sound matrixing: (a) encoder (b) decoder.

format in the MPEG-2 audio. A *matrixing* module is used for this purpose. Another important feature employed in the MPEG-2 BC/LSF is the "dynamic cross-talk," a multichannel irrelevancy reduction technique. This feature exploits properties of spatial hearing and encodes only one envelope instead of two or more together with some side information (i.e., scale factors). Note that this technique is in some sense similar to the intensity coding that we discussed earlier. In summary *matrixing* enables backwards compatibility between the MPEG-2 and MPEG-1 bitstreams, and *dynamic cross-talk* reduces the interchannel redundancies.

In Figure 10.9, first the segmented audio frames are decomposed into 32 critically subsampled subbands using a polyphase realization of a *pseudo QMF* (PQMF) bank. Next, a *matrixing module* is employed for down-mixing purposes. Matrixing results in two stereo-format channels, i.e., *L_total* and *R_total* and three extension channels, i.e., C, L_s, and R_s. In order to remove statistical redundancies associated with these channels a second-order linear *predictor* is employed [Fuch93] [ISOI94a]. The predictor coefficients are updated on each subband using a backward adaptive LMS algorithm [Widr85]. The resulting

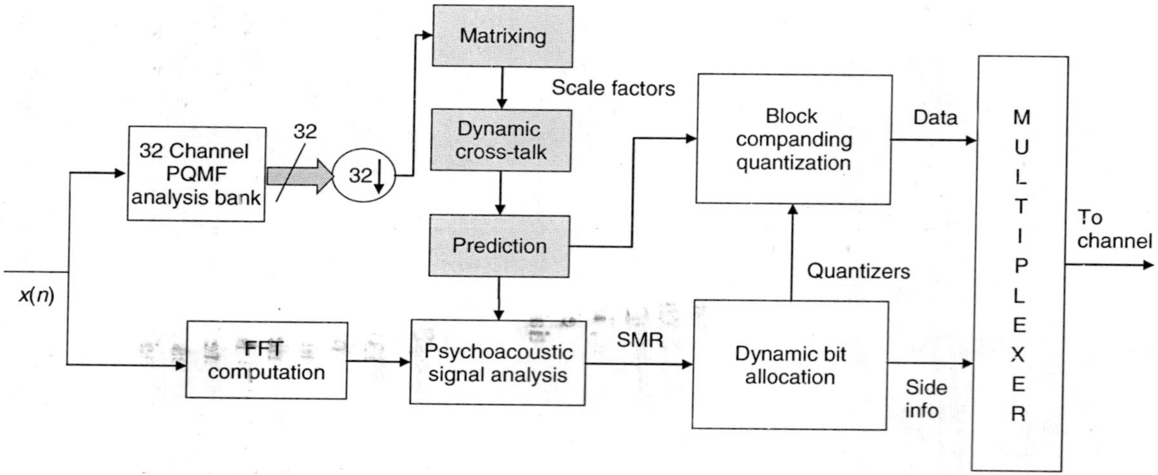

Figure 10.9. ISO/MPEG-2 BC/LSF audio (layer I/II) encoder algorithm.

prediction error is further processed to eliminate interchannel dependencies. JND thresholds are computed in parallel with the subband decomposition for each decimated block. A bit-allocation procedure similar to the one in the MPEG-1 audio standard is used to estimate the number of appropriate bits required for quantization.

10.4.2.3 MPEG-2 BC/LSF Decoder Synchronization followed by error detection and correction are performed first at the decoder. Then the coded audio bitstream is de-multiplexed into the individual subbands of each audio channel. Next, the subband signals are converted to subband PCM signals, based on the instructions in the header and the side information transmitted for every subband. De-matrixing is performed to compute L and R bitstreams as follows:

$$L = \frac{L\_total}{x} - (yC + zL_s) \tag{10.6}$$

$$R = \frac{R\_total}{x} - (yC + zR_s) \tag{10.7}$$

where x, y, and z are constants and are known at the decoder. The inverse-quantized, de-matrixed subband PCM signals are then inverse-filtered to reconstruct the full-band time-domain PCM signals for each channel.

The second MPEG-2 standard, i.e., the MPEG-2 NBC/AAC, sacrificed backward MPEG-1 compatibility to eliminate quantization noise unmasking artifacts [tenK94], which are potentially introduced by the forced backward compatibility.

10.4.3 MPEG-2 NBC/AAC (ISO/IEC-13818-7)

The 11172-3 MPEG-1 and IS13818-3 MPEG-2 BC/LSF are standardized algorithms for high-quality coding of monaural and stereophonic program material. By the early 1990s, however, the demand for high-quality coding of multichannel audio at reduced bit rates had increased significantly. The backwards compatibility constraints imposed on the MPEG-2 BC/LSF algorithm made it impractical to code 5-channel program material at rates below 640 kb/s. As a result, MPEG began standardization activities for a nonbackward compatible advanced coding system targeting "indistinguishable" quality [ITUR91] [ISOI96a] at a rate of 384 kb/s for five full-bandwidth channels. In less than three years, this effort led to the adoption of the IS13818-7 MPEG-2 Non-backward Compatible/Advanced Audio Coding (NBC/AAC) algorithm [ISOI97b], a system that exceeded design goals and produced the desired quality at 320 kb/s for five full-bandwidth channels. While similar in many respects to its predecessors, the AAC algorithm [Bosi96] [Bosi97] [Bran97] [John99] achieves performance improvements by incorporating coding tools previously not found in the standards such as filter-bank window shape adaptation, spectral coefficient prediction, temporal noise shaping (TNS), and bandwidth- and bit-rate-scaleable operation. Bit rate and quality improvements are also realized through the use of a sophisticated noiseless coding scheme integrated with a two-stage bit allocation procedure. Moreover, the AAC algorithm contains scalability and complexity management tools not previously included with the MPEG algorithms. As far as applications are concerned, the AAC algorithm is embedded in the atob™ and LiquidAudio™ players for streaming of high-fidelity stereophonic audio. It is also a candidate for standardization in the United States Digital Audio Radio (US DAR) project. The remainder of this section describes some of the features unique to MPEG-2 AAC.

The MPEG-2 AAC algorithm (Figure 10.10) is organized as a set of coding tools. Depending upon available CPU or channel resources and desired quality, one can select from among three complexity "profiles," namely main, low, and scalable sample rate profiles. Each profile recommends a specific combination of tools. Our focus here is on the complete set of tools available for main profile coding, which works as follows.

10.4.3.1 Filter Bank First, a high-resolution MDCT filter bank obtains a spectral representation of the input. Like previous MPEG coders, the AAC filter-bank resolution is signal adaptive. Quasi-stationary segments are analyzed with a 2048-point window, while transients are analyzed with a block of eight 256-point windows to maintain time synchronization for channels using different filter-bank resolutions during multichannel operations. The frequency resolution is therefore 23 Hz for a 48-kHz sample rate, and the time resolution is 2.6 ms. Unlike previous MPEG coders, however, AAC eliminates the hybrid filter bank and relies on the MDCT exclusively. The AAC filter bank is also unique in its ability to switch between two distinct MDCT analysis window shapes, i.e., a sine window (Eq. (10.8)) and a Kaiser-Bessel designed (KBD) window (Eq. (10.9)). Given specific input signal characteristics, the idea behind window shape adaptation is to optimize filter-bank frequency selectivity in order to localize the supra-masking threshold signal energy in the fewest spectral coefficients. This strategy seeks essentially to maximize the perceptual coding gain of the filter bank. While both windows satisfy the perfect reconstruction and aliasing cancellation constraints of the MDCT, they offer different spectral analysis properties. The sine window is given by

$$w(n) = \sin\left[\left(n + \frac{1}{2}\right)\frac{\pi}{2M}\right] \tag{10.8}$$

for $0 \leqslant n \leqslant M - 1$, where M is the number of subbands. This particular window is perhaps the most popular in audio coding. In fact, this window has become standard in MDCT audio applications, and its properties are typically referenced as performance benchmarks when new windows are proposed. The so-called KBD window was obtained in a procedure devised at Dolby Laboratories, by applying a transformation of the form

$$w_a(n) = w_s(n)\sqrt{\frac{\sum_{j=0}^{n} v(j)}{\sum_{j=0}^{M} v(j)}},\ 0 \leqslant n < M, \tag{10.9}$$

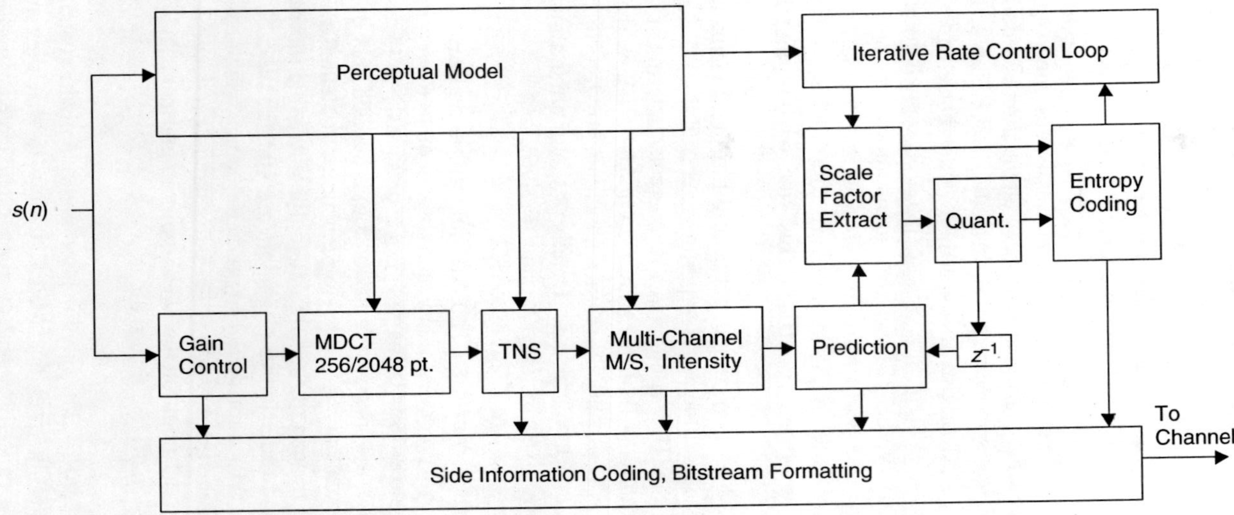

Figure 10.10. ISO/IEC IS13818-7 (MPEG-2 NBC/AAC) encoder ([Bosi96]).

where the sequence $v(n)$ represents the symmetric kernel. The resulting identical analysis and synthesis windows, $w_a(n)$ and $w_s(n)$, respectively, are of length $M+1$ and symmetric, i.e., $w(2M-n-1) = w(n)$. More detailed explanation on the MDCT windows is given in Chapter 6, Section 6.7.

A filter-bank simulation exemplifying the performance of the two windows, sine and KBD, for the MPEG-2 AAC algorithm follows. A sine window is selected when narrow pass-band selectivity is more beneficial than strong stop-band attenuation. For example, sounds characterized by a dense harmonic structure (less than 140 Hz spacing) such as harpsichord or pitch pipe benefit from a sine window. On the other hand, a Kaiser-Bessel designed (KBD) window is selected in cases for which stronger stop-band attenuation is required, or for situations in which strong components are separated by more than 220 Hz. The KBD window in AAC has its origins in the MDCT filter bank window designed at Dolby Labs for the AC-3 algorithm using explicit perceptual criteria. By sacrificing pass-band selectivity, the KBD window gains improved stop-band attenuation relative to the sine window. In fact, the stop-band magnitude response is below a conservative composite minimum masking threshold for a tonal masker at the center of the pass-band. A KBD versus sine window simulation example (Figure 10.11) for a signal containing 300 Hz plus 3 harmonics shows the KBD potential for reduced bit allocation. A masking threshold estimate generated by MPEG-1 psychoacoustic model 2 is superimposed (red line). It can be seen that, for the given input, the KBD window is advantageous in terms of supra-threshold component minimization. All of the MDCT components below the superimposed masking threshold will potentially require allocations of zero bits. This tradeoff can ultimately lead to a lower bit rate. Details of the minimum masking template design procedure are given in [Davi94] and [Fiel96].

10.4.3.2 Spectral Prediction The AAC algorithm realizes improved coding efficiency relative to its predecessors by applying prediction over time to the transform coefficients below 16 kHz, as was done previously in [Mahi89] [Fuch93] [Fuch95]. In this case, the spectral prediction tool is applied only during long analysis windows and then only if a bit-rate reduction is obtained when coding the prediction residuals instead of the original coefficients. Side information is minimal, since the second-order lattice predictors are updated on each frame using a backward adaptive LMS algorithm. The predictor banks, which can be selectively activated for individual quantization scale-factor bands, produced an improvement for a fixed bit rate of +1 point on the ITU 5-point impairment scale for the critical pitch pipe and harpsichord test material.

10.4.3.3 Bit Allocation The bit allocation and quantization strategies in AAC bear some similarities to previous MPEG coders in that they make use of a nested-loop iterative procedure, and in that psychoacoustic masking thresholds are obtained from an analysis model similar to MPEG-1, model recommendation number two. Both lossy and lossless coding blocks are integrated into the rate-control loop structure so that redundancy removal and irrelevancy reduction are

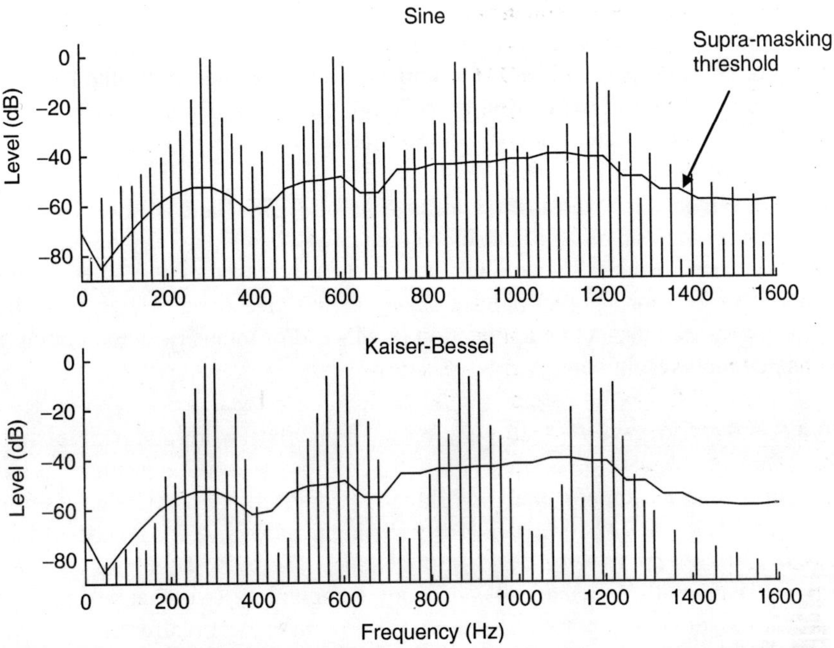

Figure 10.11. Comparison of the MPEG-2 AAC MDCT analysis filter-bank outputs for the sine window vs. the KBD window.

simultaneously affected in a single analysis-by-synthesis process. The scheme works as follows. As in the case of MPEG-1, layer III, the AAC coefficients are grouped into 49 scale-factor bands that mimic the auditory system's frequency resolution.

In the nested-loop allocation procedure, the inner loop adjusts scale-factor quantizer step sizes in increments of 1.5 dB (approximates intensity difference limen (DL)) and obtains Huffman codewords for both quantized scale factors and quantized coefficients until the desired bit rate is achieved. Then, in the outer loop, the quantization noise introduced by the inner loop is compared to the masking threshold in order to assess noise audibility. Undercoded scale factor bands are amplified to force increased coding precision, and then the inner loop is called again for compliance with the desired bit rate. A best result is stored after each iteration since the two-loop process is not guaranteed to converge. As with other algorithms such as the MPEG-1 layer III and the Lucent Technologies PAC [John96c], a bit reservoir is maintained to compensate for time-varying perceptual bit-rate requirements.

10.4.3.4 Noiseless Coding The noiseless coding block [Quac97] embedded in the rate-control loop has several innovative features as well. Twelve Huffman code books are available for 2- and 4-tuple blocks of quantized coefficients. Sectioning and merging techniques are applied to maximize redundancy reduction.

Individual code books are applied to time-varying "sections" of scale-factor bands, and the sections are defined on each frame through a greedy merge algorithm that minimizes the bit rate. Grouping across time and intraframe frequency interleaving of coefficients prior to code-book application are also applied to maximize zero coefficient runs and further reduce bit rates.

10.4.3.5 Other Enhancements Relative to MPEG-1 and MPEG-2 BC/LSF, other enhancements have also been embedded in AAC. For example, the AAC algorithm has an embedded TNS module [Herr96] for pre-echo control (Section 6.9), a special profile for sample-rate scalability (SSR), and time-varying as well as frequency subband selective application of MS and/or intensity stereo coding for 5-channel inputs [John96b].

10.4.3.6 Performance Incorporation of the nonbackward compatible coding enhancements proved to be a judicious strategy for the AAC algorithm. In independent listening tests conducted worldwide [ISOI96d], the AAC algorithm met the strict ITU-R BS.1116 criteria for indistinguishable quality [ITUR94b] at a rate of 320 kb/s for five full-bandwidth channels [Kirb97]. This level of quality was achieved with a manageable decoder complexity. Two-channel, real-time AAC decoders were reported to run on 133-MHz Pentium platforms using 40% and 25% of available CPU resources for the main and low-complexity profiles, respectively [Quac98a]. MPEG-2 AAC maintained its presence as the core "time-frequency" coder reference model for the MPEG-4 standard.

10.4.3.7 Reference Model Validation (RM) Before proceeding with a discussion of MPEG-4, we first consider a significant system-level aspect of MPEG-2 AAC that also propagated into MPEG-4. Both algorithms are structured in terms of so-called *reference models* (RMs). In the RM approach, generic coder blocks or tools (e.g., perceptual model, filter bank, rate-control loop, etc.) adhere to a set of defined interfaces. The RM therefore facilitates the testing of incremental single block improvements without disturbing the existing macroscopic RM structure. For instance, one could devise a new psychoacoustic analysis model that satisfies the AAC RM interface and then simply replace the existing RM perceptual model in the reference software with the proposed model. It is then a straightforward matter to construct performance comparisons between the RM method and the proposed method in terms of quality, complexity, bit rate, delay, or robustness. The RM definitions are intended to expedite the process of evolutionary coder improvements.

In fact, several practical AAC improvements have already been analyzed within the RM framework. For example, a backward predictor was proposed [Yin97] as a replacement for the existing backward adaptive LMS predictors. This method that relies upon a block LPC estimation procedure rather than a running LMS estimation, was reported to achieve comparable quality with a 38% (instruction) complexity reduction [Yin97]. This contribution was significant in light of the fact that the spectral prediction tool in the AAC main profile decoder constitutes 40%

of the computational complexity [Yin97]. Decoder complexity is further reduced since the block predictors only require updates when the prediction module has been enabled rather than requiring sample-by-sample updating regardless of activation status. Forward adaptive predictors have also been investigated [Ojan99]. In another example of RM efficacy, improvements to the AAC noiseless coding module were reported in [Taka97]. A modification to the greedy merge sectioning algorithm was proposed in which high-magnitude spectral peaks that tended to degrade Huffman coding efficiency were coded separately. The improvement yielded consistent bit-rate reductions up to 11%. In informal listening tests it was found that the bit savings resulted in higher quality at the same bit rate. In yet another example of RM innovation aimed at improving quality for a given bit rate, product code VQ techniques [Gers92] were applied to increase AAC scale-factor coding efficiency [Sree98a]. In the proposed scheme, scale-factors are decorrelated using a DCT and then grouped into subvectors for quantization by a product code VQ. The method is intended primarily for low-rate coding, since the side information bit burden rises from roughly 6% at 64 kb/s to in some cases 25% at 16 kb/s. As expected, subjective tests reflected an insignificant quality improvement at 64 kb/s. On the other hand, the reduction in bits allocated to side information at low rates (e.g., 16 kb/s), allowed more bits for spectral coefficient coding, and therefore produced mean improvements of +0.52 and +0.36 on subjective differential improvement tests at bit rates of 16 and 40 kb/s, respectively [Sree98b]. Additionally, noise-to-mask ratios (NMRs) were reduced by as much as −2.43 for the "harpsichord" critical test item at 16 kb/s. Several architectures for MPEG-2 AAC real-time implementations were proposed. Some of these include [Chen99] [Hilp98] [Geye99] [Saka00] [Hong01] [Rett01] [Taka01] [Duen02] [Tsai02].

10.4.3.8 Enhanced AAC in MPEG-4 The next section is concerned with the multimodal MPEG-4 audio standard, for which the MPEG-2 AAC RM core was selected as the "time-frequency" audio coding RM with some improvements. For example, perceptual noise substitution (PNS) was included [Herr98a] as part of the MPEG-4 AAC RM. Moreover, the long-term prediction (LTP) [Ojan99] and transform-domain weighted interleave VQ (TwinVQ) [Iwak96] modules became part of the MPEG-4 audio. LTP after the MPEG-2 AAC prediction block provides a higher coding precision for tonal signals, while the TwinVQ provided scalability and ultra-low bit-rate audio coding.

10.4.4 MPEG-4 Audio (ISO/IEC 14496-3)

The MPEG-4 ISO/IEC-14496 Part 3 audio was adopted in December 1998 after many proposed algorithms were tested [Cont96] [Edle96a] [ISOI96b] [ISOI96c] for compliance with the program objectives [ISOI94b] established by the MPEG committee. MPEG-4 audio (Figure 10.12) encompasses a great deal more functionality than just perceptual coding [Koen96] [Koen98] [Koen99]. It comprises an integrated family of algorithms with wide-ranging provisions for scalable,

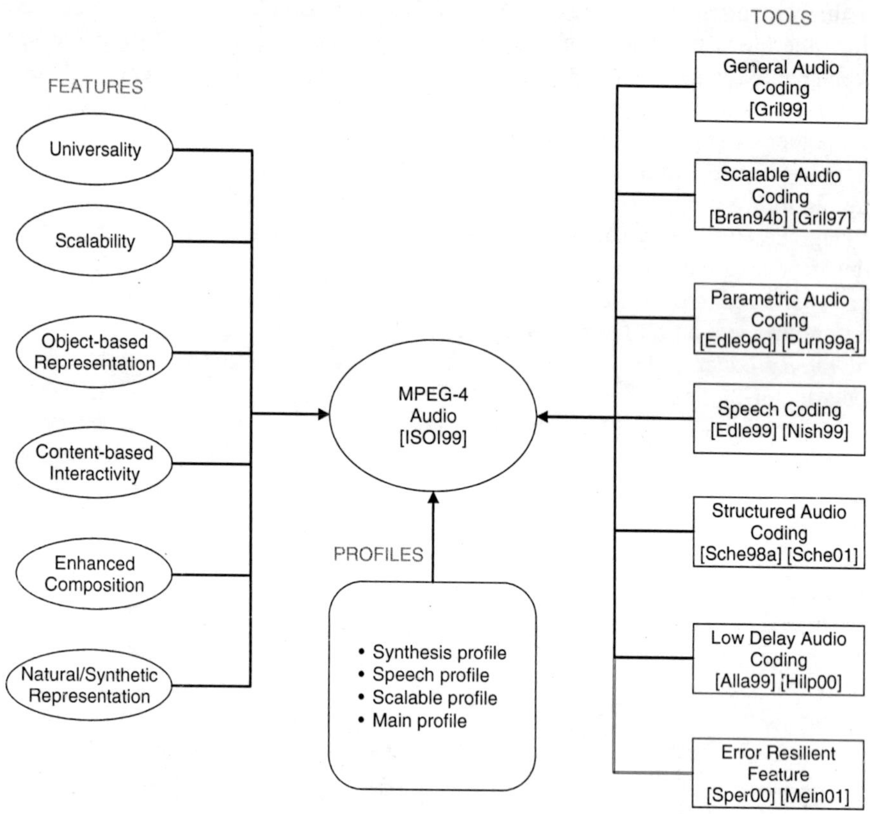

Figure 10.12. An overview of the MPEG-4 audio coder.

object-based speech and audio coding at bit rates from as low as 200 b/s up to 64 kb/s per channel. The distinguishing features of MPEG-4 relative to its predecessors are extensive scalability, object-based representations, user interactivity/object manipulation, and a comprehensive set of coding tools available to accommodate almost any desired tradeoff between bit rate, complexity, and quality. Efficient and flexible coding of different content (objects) such as natural audio/speech and synthetic audio/speech became indispensable for some of the innovative multimedia applications. To facilitate this, MPEG-4 audio provides coding and composition of natural and synthetic audio/speech content at various bit rates. Very low rates are achieved through the use of structured representations for synthetic speech and music, such as text-to-speech and MIDI. For higher bit rates and "natural audio" speech and music, the standard provides integrated coding tools that make use of different signal models, the choice of which is made depending upon desired bit rate, bandwidth, complexity, and quality. Coding tools are also specified in terms of MPEG-4 "profiles" that essentially recommend tool sets for a given level of functionality and complexity. Beyond its provisions specific to coding of speech and audio, MPEG-4 also specifies numerous sophisticated system-level functions for media-independent transport, efficient buffer management, syntactic bitstream descriptions, and time-stamping for synchronization of audiovisual information units.

10.4.4.1 MPEG-4 Audio Versions The MPEG-4 audio standard was released in several steps due to timing constraints. This resulted in two different versions of MPEG-4. Version 1 [ISOI99] was standardized in February 1999, followed by version 2 [ISOI00] (also referred as Amendment 1 to version 1) in February 2000. New amendments for bandwidth extension, parametric audio extension, MP3 on MP4, audio lossless coding, and scalable to lossless coding have also been considered in the MPEG-4 audio standard.

MPEG-4 Audio Version 1. The MPEG-4 audio version 1 comprises the majority of the MPEG-4 audio tools. These are general audio coding, scalable coding, speech coding techniques, structured audio coding, and text-to-speech synthetic coding. These techniques can be grouped into two main categories, i.e., *natural* [Quac98b] and *synthetic* audio coding [Vaan00]. The MPEG-4 *natural audio coding* part describes traditional type speech coding and high-quality audio coding algorithms at bit rates ranging from 2 kb/s to 64 kb/s and above. Three types of coders enable hierarchical (scalable) coding in MPEG-4 Audio version-1 at different bit rates. Firstly, at lower bit rates ranging from 2 kb/s to 6 kb/s, parametric speech coding is employed. Secondly, a code excited linear predictive (CELP) coding is used for medium bit rates between 6 kb/s and 24 kb/s. Finally, for the higher bit rates typically ranging from 24 kb/s, transform-based (time-frequency) general audio coding techniques are applied. The MPEG-4 *synthetic audio coding* part describes the text-to-speech (TTS) and structured audio synthesis tools. Typically, the structured tools are used to provide effects like echo, reverberation, and chorus effects; the TTS synthetic tools generate synthetic speech from text parameters.

MPEG-4 Audio Version 2. While remaining backwards compatible with MPEG-4 version 1, version 2 adds new profiles that incorporate a number of significant system-level enhancements. These include error robustness, low-delay audio coding, small-step scalability, and enhanced composition [Purn99b]. At the system level, version 2 includes a media independent bit stream format that supports streaming, editing, local playback, and interchange of contents. Furthermore in version 2, an MPEG-J programmatic system specifies an application programming interface (API) for interoperation of MPEG players with JAVA. Version 2 offers improved audio realism in sound rendering. New tools allow parameterization of the acoustical properties of an audio scene, enabling features such as immersive audiovisual rendering, room acoustical modeling, and enhanced 3-D sound presentation. New error resilience techniques in version 2 allow both equal and unequal error protection for the audio bit streams. Low-delay audio coding is employed at low bit rates where the coding delay is significantly high. Moreover, to facilitate the bit rate scalability in small steps, version 2 provides a highly desirable tool called small-step scalability or fine-grain scalability. Text-to-speech (TTS) interfaces from version 1 are enhanced in version 2 with a mark-up TTS intended for applications such as speech-enhanced web browsing, verbal email, and story-teller on demand. Markup TTS has the ability to process HTML, SABLE, and facial animation parameter (FAP) bookmarks.

10.4.4.2 MPEG-4 Audio Profiles Although many coding and processing tools are available in MPEG-4 audio, cost and complexity constraints often dictate that it is not practical to implement all of them in a particular system. Version 1 therefore defines four complexity-ranked audio profiles intended to help system designers in the task of appropriate tool subset selection. In order of bit rate, they are as follows. The *low rate synthesis audio profile* provides only wavetable-based synthesis and a text-to-speech (TTS) interface. For natural audio processing capabilities, the *speech audio profile* provides a very-low-rate speech coder and a CELP speech coder. The *scalable audio profile* offers a superset of the first two profiles. With bit rates ranging from 6 to 24 kb/s and bandwidths from 3.5 to 9 kHz, this profile is suitable for scalable coding of speech, music, and synthetic music in applications such as Internet streaming or narrow-band audio digital broadcasting (NADIB). Finally, the *main audio profile* is a superset of all other profiles, and it contains tools for both natural and synthetic audio.

10.4.4.3 MPEG-4 Audio Tools Unlike MPEG-1 and MPEG-2, the MPEG-4 audio describes not only a set of compression schemes but also a complete functionality for a broad range of applications from low-bit-rate speech coding to high-quality audio coding or music synthesis. This feature is called the *universality*. MPEG-4 enables scalable audio coding, i.e., variable rate encoding is provided to adapt dynamically to the varying transmission channel capacity. This property is called *scalability*. One of the main features of the MPEG-4 audio is its ability to represent the audiovisual content as a set of objects. This enables the *content-based interactivity*.

Natural Audio Coding Tools. MPEG-4 audio [Koen99] integrates a set of tools (Figure 10.13) for coding of natural sounds [Quac98b] at bit rates ranging from as low as 200 b/s up to 64 kb/s per channel. For speech and audio, three distinct algorithms are integrated into the framework. These include parametric coding, CELP coding, and transform coding. The parametric coding is employed for bit rates of 2–4 kb/s and 8 kHz sampling rate as well as 4–16 kb/s and 8 or 16 kHz sampling rates (Section 9.4). For higher quality, narrow-band (8 kHz sampling rate) and wideband (16 kHz) speech is handled by a CELP speech codec operating between 6 and 24 kb/s. For generic audio at bit rates above 16 kb/s, a time/frequency perceptual coder is employed, and, in particular, the MPEG-2 AAC algorithm with extensions for fine-grain bit-rate scalability [Park97] is specified in MPEG-4 version 1 RM as the time-frequency coder. The multimodal framework of MPEG-4 audio allows the user to tailor the coder characteristics to the program material.

Synthetic Audio Coding Tools. While the earlier MPEG standards treated only natural audio program material, the MPEG-4 audio achieves very-low-rate coding by supplementing its natural audio coding techniques with tools for synthetic audio processing [Sche98a] [Sche01] and interfaces for structured, high-level

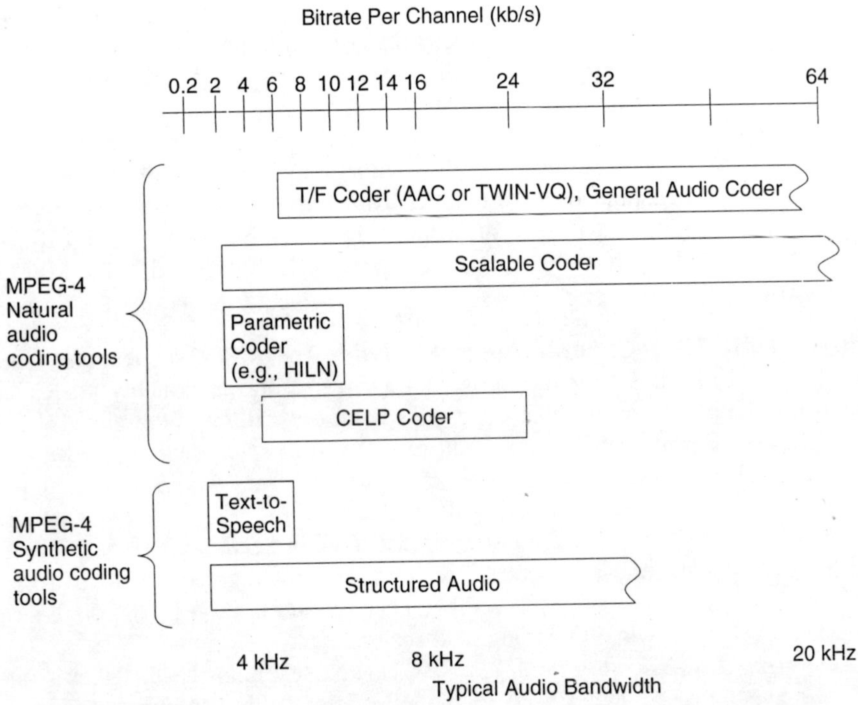

Figure 10.13. ISO/IEC MPEG-4 integrated tools for audio coding ([Koen99]).

audio representations. Chief among these are the text-to-speech interface (TTSI) and methods for score-driven synthesis. The TTSI provides the capability for 200–1200 b/s transmission of synthetic speech that can be represented in terms of either text only or text plus prosodic parameters such as a pitch contour or a set of phoneme durations. Also, one can specify the age, gender, and speech rate of the speaker. Additionally, there are facilities for lip synchronization control, international language, and dialect support, as well as controls for pause, resume, and jump forward/backward. The TTSI specifies only an interface rather than a normative speech synthesis methodology in order to maximize implementation flexibility.

Beyond speech, general music synthesis capabilities in MPEG-4 are provided by a set of structured audio tools [Sche98a] [Sche98d] [Sche98e]. Synthetic sounds are represented using the structured audio orchestra language (SAOL). SAOL [Sche98d] treats music as a collection of instruments. Instruments are then treated as small networks of signal-processing primitives, all of which can be downloaded to a decoder. Some of the available synthesis methods include wavetable, FM, additive, physical modeling, granular synthesis, or nonparametric hybrids of any of these methods [Sche98c]. An excellent tutorial on these and other structured audio methods and applications appeared in [Verc98]. The SAOL instruments are controlled at the decoder by "scores" or scripts in the structured audio score language (SASL). A score is a time-sequenced set of commands that invokes various instruments at specific times to contribute their outputs to an overall performance. SASL provides significant flexibility in that not only can instruments be controlled, but the existing sounds can be modified. For those situations in which fine control is not required, structured audio in MPEG-4 also provides backward compatibility with the MIDI protocol. Moreover, a standardized "wavetable bank format" is available for low-functionality terminals [Koen99]. In the next seven subsections, i.e., 10.4.4.4 through 10.4.4.10, we describe in detail the features and tools (Figure 10.12) integrated in the MPEG-4 audio.

10.4.4.4 MPEG-4 General Audio Coding The MPEG-4 General Audio Coder (GAC) [Gril99] has the most vital and versatile functionality associated with the MPEG-4 tool-set that covers the arbitrary natural audio signals. The MPEG-4 GAC is often called as the "all-round" coding system among the MPEG-4 audio schemes and operates at bit rates ranging from 6 to 300 kb/s and at sampling rates between 7.35 kHz and 96 kHz. The MPEG-4 GA coder is built around the MPEG-2 AAC (Figure 10.10 discussed in Section 10.4.3) along with some extended features and coder configurations highlighted in Figure 10.14. These features are given by the perceptual noise substitution (PNS), long-term prediction (LTP), Twin VQ coding, and scalability.

Perceptual Noise Substitution (PNS). The PNS exploits the fact that a random noise process can be used to model efficiently transform-coefficients in noise-like frequency subbands, provided the noise vector has an appropriate temporal fine structure [Schu96]. Bit-rate reduction is realized since only a compact, parametric representation is required for each PNS subband (i.e., noise energy) rather than full quantization and coding of subband transform coefficients. The PNS technique was integrated into the existing AAC bitstream definition in a backward-compatible manner. Moreover, PNS actually led to reduced decoder complexity since pseudo-random sequences are less expensive to compute than Huffman decoding operations. Therefore, in order to improve the coding efficiency, the following principle of PNS is employed.

The PNS acronym is composed from the following: *perceptual* coding + *substitute* parametric form of *noise*-like signals, i.e., PNS allows frequency-selective parametric encoding of noise-like components. These noise-like components are detected based on a scale-factor band and are grouped into separate categories. The spectral coefficients corresponding to these categories are not quantized and are excluded from the coding process. Furthermore, only a noise substitution flag along with the total power of these spectral coefficients are transmitted for each band. At the decoder, the spectral coefficients are replaced by the pseudo-random vectors with the desired target noise power. At a bit rate of 32 kb/s, a mean improvement due to PNS of +0.61 on the comparison mean opinion score (CMOS) test (for critical test items such as speech, castanets, and complex sound mixtures) was reported in [Herr98a]. The multichannel PNS modes include some provisions for binaural masking level difference (BMLD) compensation.

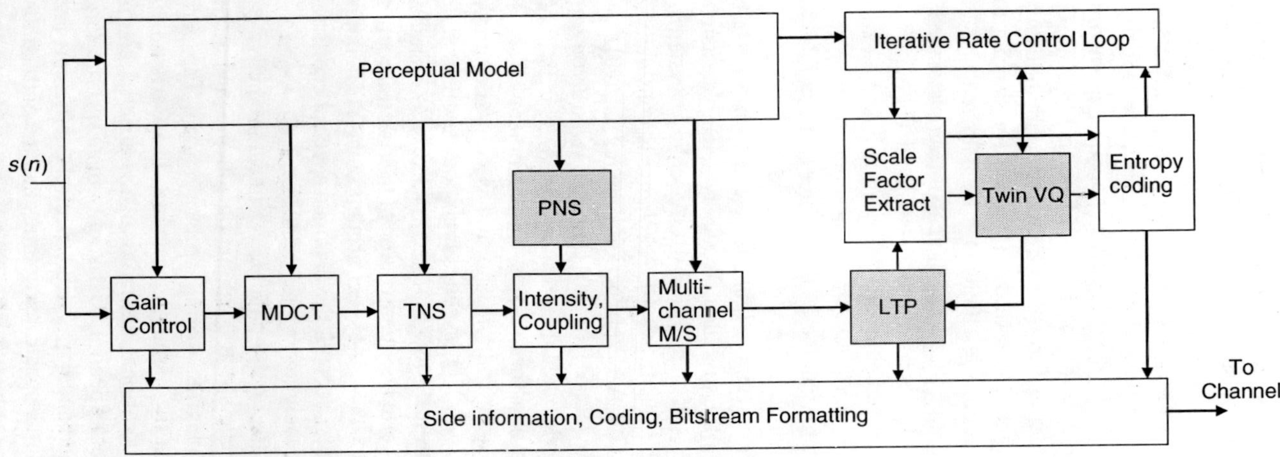

Figure 10.14. MPEG-4 GA coder.

Long-Term Prediction (LTP). Unlike the noise-like signals, the tonal signals require higher coding precision. In order to achieve a required coding precision (20 dB for tone-like and 6 dB for noise-like signals), the long-term prediction (LTP) technique [Ojan99] is employed. In particular, since the tonal signal components are predictable, the speech coding pitch prediction techniques [Span94] can be used to improve the coding precision. The only significant difference between the prediction techniques performed in a common speech coder and in the MPEG-4 GA coder is that in the latter case, the LTP is performed in the frequency domain, while in speech codecs the LTP is carried out in the time domain. A brief description of the LTP scheme in MPEG-4 GA coder follows. First, the input audio is transformed to frequency domain using an analysis filter bank and later a TNS analysis filter is employed for shaping the noise artifacts. Next, the processed spectral coefficients are quantized and encoded. For prediction purposes, these quantized coefficients are transformed back to the time domain by a synthesis filter bank and the associated TNS operation. The optimum pitch lag and the gain parameters are determined based on the residual and the input signal. In the next step, both the input signal and the residual are mapped to a spectral representation via the analysis filter bank and the forward TNS filter bank. Depending on which alternative is more favorable, coding of either the difference signal or the original signal is selected on a scale-factor basis. This is achieved by means of a so-called frequency-selective switch (FSS), which is also used in the context of the MPEG-4 GA scalable systems. The complexity associated with the LTP in MPEG-4 GA scheme is considerably (50%) reduced compared to the MPEG-2 AAC prediction scheme [Gril99].

Twin VQ. Twin VQ [Iwak96] [Hwan01] [Iwak01] is an acronym of the T*ransform-domain* W*eighted* In*terleave* V*ector* Q*uantization*. The Twin VQ performs vector quantization of the transformed spectral coefficients based on a perceptually weighted model. The quantization distortion is controlled through a perceptual model [Iwak96]. The Twin VQ provides high coding efficiencies even for music and tonal signals at extremely low bit rates (6–8 kb/s), which CELP coders fail to achieve. The Twin VQ performs quantization of the spectral coefficients in two steps as shown in Figure 10.15. First, the spectral coefficients are flattened and normalized across the frequency axis. Second, the flattened spectral coefficients are quantized based on a perceptually weighted vector quantizer.

From Figure 10.15, the first step includes a linear predictive coding, periodicity computation, a Bark scale spectral estimation scheme, and a power computation block. The LPC provides the overall spectral shape. The periodic component includes information on the harmonic structure. The Bark-scale envelope coding provides the required additional flattening of the spectral coefficients. The normalization restricts these spectral coefficients to a specific target range. In the second step, the flattened and normalized spectral coefficients are interleaved into subvectors. Based on some spectral properties and a weighted distortion measure, perceptual weights are computed for each subvector. These weights are applied to the vector quantizer (VQ). A conjugate-structure VQ that uses a pair of code books is employed. More detailed information on the conjugate structure VQ can

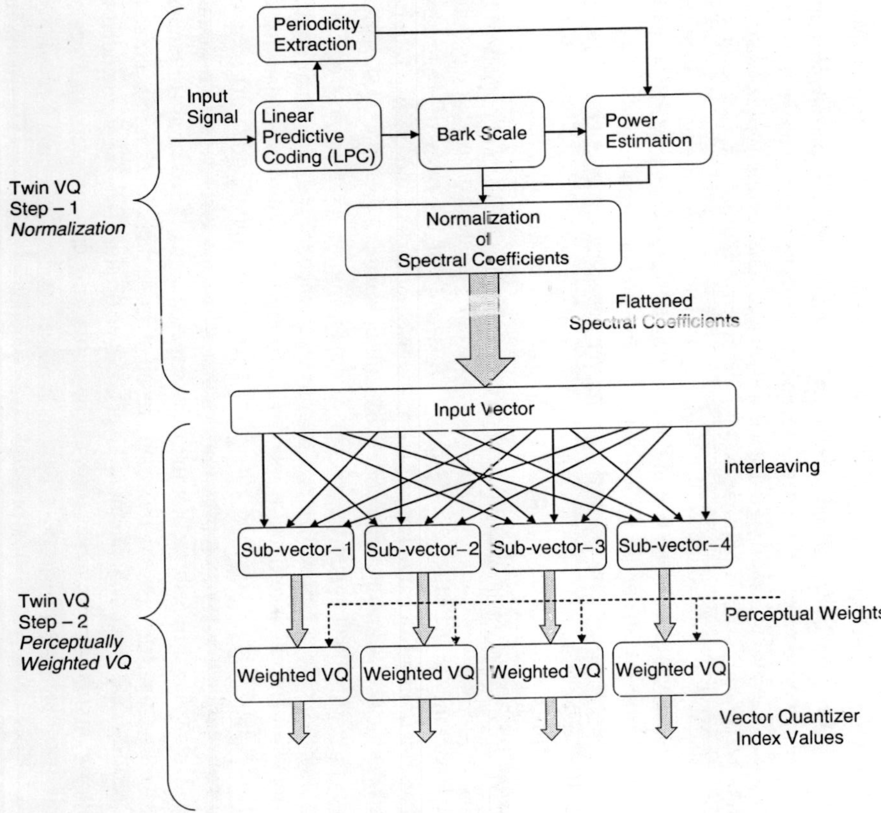

Figure 10.15. Twin VQ scheme in MPEG-4 GA coder.

be obtained from [Kata93] [Kata96]. The MPEG-4 Audio Twin VQ scheme provides audio coding at ultra-low bit rates (6–8 kb/s) and supports the perceptual control of the quantization distortion. Comparative tests of MPEG AAC with and without Twin VQ tool were performed and are given in [ISOI98]. Furthermore, the Twin VQ tool has provisions for scalable audio coding, which will be discussed next.

10.4.4.5 MPEG-4 Scalable Audio Coding MPEG-4 scalable audio coding implies a variable rate encoding/decoding of bitstreams at bit rates that can be adapted dynamically to the varying transmission channel capacity [Gril97] [Park97] [Herr98b] [Creu02]. Scalable coding schemes [Bran94b] generate partial bitstreams that can be decoded separately. Therefore, encoding/decoding of a subset of the total bitstream will result in a valid signal at a lower bit rate. The various types of scalability [Gril97] are given by, signal-to-noise ratio (SNR) scalability, noise-to-mask ratio (NMR) scalability, audio bandwidth scalability, and bit-rate scalability. The bit-rate scalability is considered to be one of the core functionalities of the MPEG-4 audio standard. Therefore, in our discussion on the MPEG-4 scalable audio coding, we will consider only the bit-rate scalability and the various scalable coder configurations described in the standard.

The MPEG-4 bit-rate scalability scheme (Figure 10.16) allows an encoder to transmit bitstreams at a high bit rate, while decoding successfully a low-rate bitstream contained within the high-rate code. For instance, if an encoder transmits bitstreams at 64 kb/s, the decoder can decode at bit rates of 16, 32, or 64 kb/s according to channel capacity, receiver complexity, and quality requirements. Typically, scalable audio coders constitute several layers, i.e., a core layer and a series of enhancement layers. For example, Figure 10.16 depicts one core layer and two enhancement layers. The core layer encodes the core (main) audio stream, while the enhancement layers provide further resolution and scalability. In particular, in the first stage, the core layer encodes the input audio, $s(n)$, based on a conventional lossy compression scheme. Next, an error signal (residual), $E_1(n)$ is calculated by subtracting the reconstructed signal, $\hat{s}(n)$ (that is obtained by decoding the compressed bitstream locally) from the input signal, $s(n)$. In the second stage (first enhancement layer), the error signal $E_1(n)$ is encoded to obtain the compressed residual, $e_1(n)$. The above sequence of steps is repeated for all the enhancement layers.

To further demonstrate this principle we consider an example (Figure 10.16) where the core layer uses 32 kb/s, and the two enhancement layers employ bit rates of 16 kb/s and 8 kb/s, and the final sink layer supports 8 kb/s coding. Therefore, if no side information is encoded, then the coding rate associated with the codec is 64 kb/s. At the decoder, one can decode this multiplexed audio bitstream at various rates, i.e., 64, 32, or 40 kb/s, etc., depending up on the bit-rate requirements, receiver complexity, and channel capacity. In particular, the core bitstream guarantees reconstruction of the original input audio with minimum artifacts. On top of the core layer, additional enhancement layers are added to increase the quality of the decoded signal.

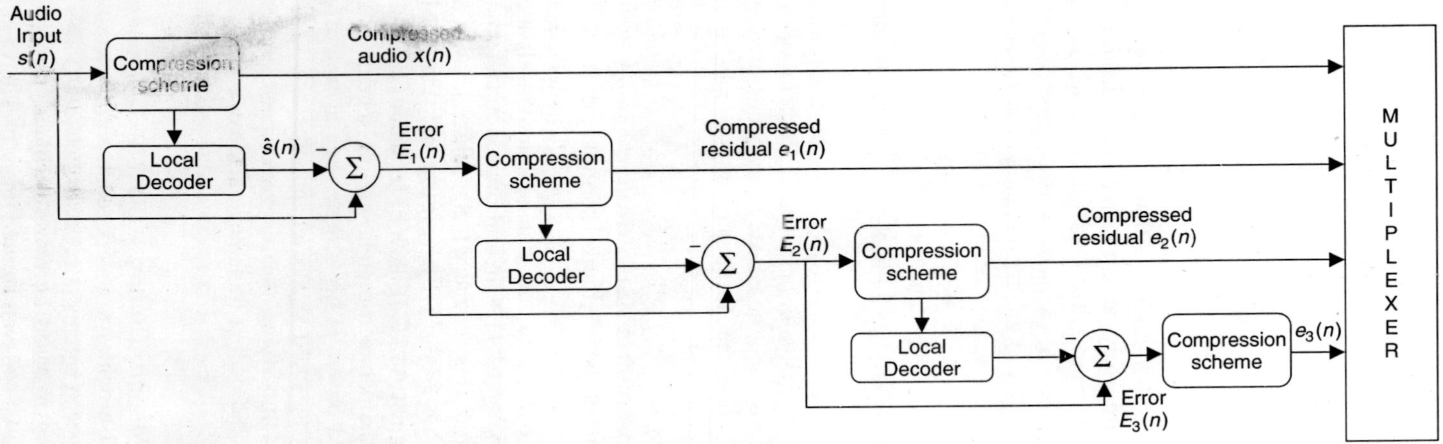

Figure 10.16. MPEG-4 scalable audio coding.

Scalable audio coding finds potential applications in the fields of digital audio broadcasting, mobile multimedia communication, and streaming audio. It supports real-time streaming with a low buffer delay. One of the significant extensions of the MPEG-4 scalable audio coding is the *fine-grain scalability* [Kim01], where a *bit-sliced arithmetic coding* (BSAC) [Kim02a] is used. In each frame, bit planes are coded in the order of significance, beginning with the most significant bits (MSBs) and progressing to the LSBs. This results in a fully embedded coder containing all lower-rate codecs. The BSAC and fine-grain scalability concepts are explained below in detail.

Fine-Grain Scalability. It is important that bit-rate scalability is achieved without significant coding efficiency penalty compared to fixed-bit-rate systems, and with low computational complexity. This can be achieved using the fine-grain scalability technique [Purn99b] [Kim01]. In this approach, *bit-sliced arithmetic coding* is employed along with the combination of advanced audio coding tools (Section 10.4.2). In particular, the noiseless coding of spectral coefficients and the scale-factor selection scheme is replaced by the BSAC technique that provides scalability in steps of 1 kb/s/channel. The BSAC scheme works as follows. First, the quantized spectral values are grouped into frequency bands, each of these groups contain the quantized spectral values in the binary form. Then the bits of each group are processed in slices and in the order of significance, beginning with the MSBs. These bit-slices are then encoded using an arithmetic coding technique (Chapter 3). Usually, the BSAC technique is used in conjunction with the MPEG-4 GA tool, where the Huffman coding is replaced by this special type of arithmetic coding.

10.4.4.6 MPEG-4 Parametric Audio Coding In research proposed as part of an MPEG-4 "core experiment" [Purn97], Purnhagen at the University of Hannover developed in conjunction with Deutsche Telekom Berkom an object-based algorithm. In this approach, harmonic sinusoid, individual sinusoid, and colored noise objects were combined in a hybrid source model to create a parametric signal representation. The enhanced algorithm, known as the "Harmonic and Individual Lines Plus Noise" (HILN) [Purn00a] [Purn00b] is architecturally very similar to the original ASAC [Edle96b] [Edle96c] [Purn98] [Purn99a], with some modifications. The parametric audio coding scheme is a part of MPEG-4 version 2, and is based on the HILN scheme (see also Section 9.4). This technique involves coding of audio signals at bit rates of 4 kb/s and above based on the possibilities of modifying the playback speed or pitch during decoding. The parametric audio coding tools have also been extended to high-quality audio [Oom03].

10.4.4.7 MPEG-4 Speech Coding The MPEG-4 natural speech coding tool [Edle99] [Nish99] provides a generic coding framework for a wide range of applications with speech signals at bit rates between 2 kb/s and 24 kb/s. The MPEG-4 speech coding is based on two algorithms, namely, harmonic vector excitation coding (HVXC) and code excited linear predictive coding (CELP). The

HVXC algorithm, essentially based on the parametric representation of speech, handles very low bit rates of 1.4–4 kb/s at a sampling rate of 8 kHz. On the other hand, the CELP algorithm employs multipulse excitation (MPE) and regular-pulse excitation (RPE) coding techniques (Chapter 4); and supports higher bit rates of 4–24 kb/s operating at sampling rates of 8 kHz and 16 kHz. The specifications of MPEG-4 Natural Speech Coding Tool Set are summarized in Table 10.4.

In all the aforementioned algorithms, i.e., HVXC, CELP-MPE, and CELP-RPE, the idea is that an LP analysis filter models the human vocal tract while an excitation signal models the vocal chord and the glottal activity. All the three configurations share the same LP analysis method, while they generally differ only in the excitation computation. In the LP analysis, first, the autocorrelation coefficients of the input speech are computed once every 10 ms and are converted to LP coefficients using the Levinson-Durbin algorithm. The LP coefficients are transformed to line spectrum pairs using Chebyshev polynomials [Kaba86]. These are later quantized using a two-stage, split-vector quantizer. The excitation signal is chosen in such a way that the error between the original and reconstructed signal is minimized according to a perceptually weighted distortion measure.

Multiple Bit Rates/Sampling Rates, Scalability. The speech coder family in MPEG-4 audio is different from the standard speech coding algorithms (e.g., ITU-T G.723.1, G.729, etc.). Some of the salient features and functionalities (Figure 10.17) of the MPEG-4 speech coder include multiple sampling rates and bit rates, bit-rate scalability [Gril97], and bandwidth scalability [Nomu98].

The *multiple bit rates/sampling rates* functionality provides flexible bit rate selection among multiple available bit rates (1.4–24 kb/s) based on the channel conditions and the bandwidth availability (8 kHz and 16 kHz). At lower bit rates, an algorithmic delay of the order of 30–40 ms is expected, while at higher bit

Table 10.4. MPEG-4 speech coding sampling rates and bandwidth specifications [Edle99].

Specification	HVXC	CELP-MPE	CELP-RPE
Sampling frequency (kHz)	8	8, 16	16
Bit rate (kb/s)	1.4–4	3.85–23.8 58 Bit rates	10.9–23.8 30 Bit rates
Frame size (ms)	10–40	10–40	10–20
Delay (ms)	33.5–56	∼15–45	∼20–25
Features	Multi-bit-rate coding, bit-rate scalability	Multi-bit-rate coding, bit-rate scalability, bandwidth scalability	Multi-bit-rate coding, bit-rate scalability

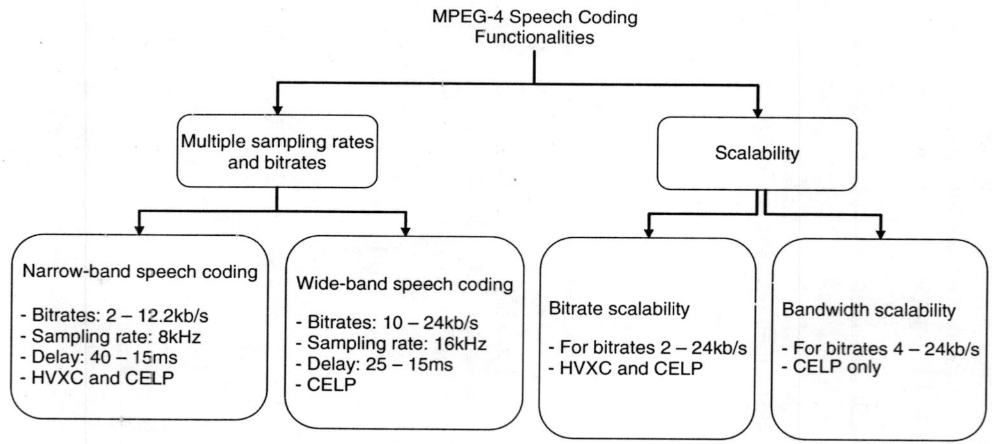

Figure 10.17. MPEG-4 speech coder functionalities.

rates, 15–25 ms delay is common. The *bit-rate scalability* feature allows a wide range of bit rates (2–24 kb/s) in step sizes of as low as 100 b/s. Both HVXC and CELP tools can be used to realize bit-rate scalability by employing a core layer and a series of enhancement layers at the encoder. When the HVXC encoding is used, one enhancement layer is preferred, while three bit-rate scalable enhancement layers may be used for the CELP codec [Gril97]. *Bandwidth scalability* improves the audio quality by adding an additional coding layer that extends the transmitted audio bandwidth. Only CELP-RPE and CELP-MPE schemes allow bandwidth scalability in the MPEG-4 audio. Furthermore, only one bandwidth scalable enhancement layer is possible [Nomu98] [Herr00a].

10.4.4.8 MPEG-4 Structured Audio Coding Structured audio (SA), introduced by Vercoe *et al.*, [Verc98] presents a new dimension to MPEG-4 audio, primarily due to its ability to represent and encode efficiently the synthetic audio and multimedia content. The MPEG-4 SA tool [Sche98a] [Sche98c] [Sche99a] [Sche99b] was developed based on a synthesizer-description language called the *Csound* [Verc95], developed by Vercoe at the MIT Media Labs. Moreover, the MPEG-4 SA tool inherits its features from "Netsound" [Case96], a structured audio experiment carried out by Casey *et al.* based on the Csound synthesis language. Instead of specifying a synthesis method, the MPEG-4 SA describes a *special language* that defines synthesis methods. In particular, the MPEG-4 SA tool defines a set of syntax and semantic rules corresponding to the synthesis-description language called the Structured Audio Orchestra Language (SAOL) [Sche98d]. A control (score) language called the Structured Audio Score Language (SASL) was also defined to describe the details of the SAOL code compaction. Another component, namely, the Structured Audio Sample Bank Format (SASBF) is used for the transmission of data samples in blocks. These blocks contain sample data as well as details of the parameters used for selecting optimum wave-table synthesizers and facilitate algorithmic modifications. A theoretical basis for the SA coding was established in [Sche01] based on the Kolmogorov complexity theory. Also, in [Sche01], Scheirer proposed a new paradigm called the *generalized audio coding* in which SA encompasses all other audio coding techniques. Furthermore, treatment of structured audio in view of both lossless coding and perceptual coding is also given in [Sche01].

The SA bitstream available at the MPEG-4 SA decoder (Figure 10.18) consists of a header, sample data, and score data. The *SAOL decoder* block acts as an interpreter and reads the header structure. It also provides the information required to reconfigure the synthesis engine. The header carries descriptions of several instruments, synthesizers, control algorithms, and routing instructions. The *Event List and Data* block obtains the actual stream of data samples, and parameters controlling algorithmic modifications. In particular, the bitstream data consists of access units that primarily contain the list of events. Furthermore, each event refers to an instrument described (e.g., in the orchestra chunk) in the header [Sche01]. The *SASL decoder* block compiles the score data from the SA bitstream and provides control sequences and signals to the synthesis engine via a *run-time scheduler*. This control information determines the time at which the events (or commands) are to be dispatched in order to create notes (or instances) of an instrument. Each note produces some sound output. Finally, all these sound outputs (corresponding to each note) are added, in order to create the overall orchestra output. In Figure 10.18, we represented the *run-time scheduler* and *reconfigurable synthesis engine* blocks separately, however, in practice they are usually combined into one block.

As mentioned earlier, the structured audio tool and the text-to-speech (TTS) fall in the synthetic audio coding group. Recall that the structured audio tools convert structured representation into synthetic sound, while the TTS tools translate text to synthetic speech. In both these methods, the particular synthesis method or implementation is not defined by the MPEG-4 audio standard; however, the input-output relation for SA and the TTS interface are standardized. The next question that arises is how the natural and synthetic audio content can be mixed. This is typically carried out based on a special format specified by the MPEG-4 namely, the Audio Binary Format for Scene Description (AudioBIFS) [Sche98e]. AudioBIFS enables sound mixing, grouping, morphing, and effects like echo (delay), reverberation (feedback delay), chorus, etc.

10.4.4.9 MPEG-4 Low-Delay Audio Coding Significantly large algorithmic delays (of the order of 100–200 ms) in the MPEG-4 GA coding tool (discussed in Section 10.4.4.4) hinder its applications in two-way, real-time communication. These algorithmic delays in the GA coder can be attributed primarily to the analysis/synthesis filter bank window, the look-ahead, the bit-reservoir, and the frame length. In order to overcome large algorithmic delays, a simplified version of the GA tool, i.e., the MPEG-4 low-delay (LD) audio coder has been proposed [Herr98c] [Herr99]. One of the main reasons for the wide proliferation of this tool is the low algorithm delay requirements in voice-over Internet protocol (VoIP) applications. In contrast to the ITU-T G.728 speech standard that is based

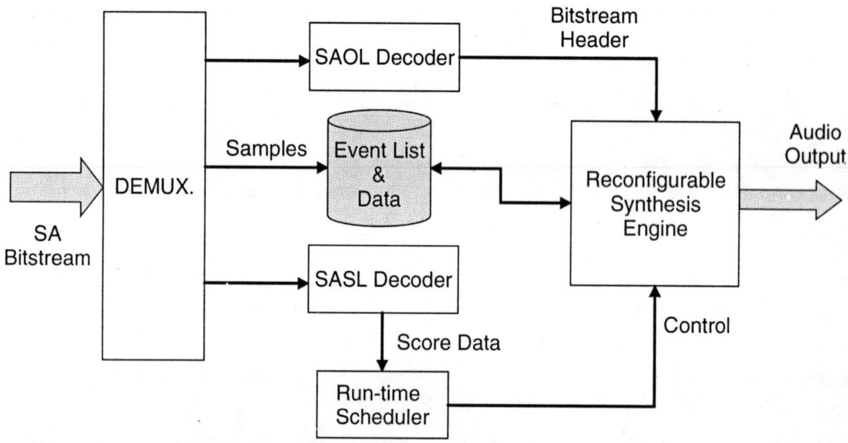

Figure 10.18. MPEG-4 SA decoder (after [Sche98a]).

on the LD-CELP [G728], the MPEG-4 LD audio coder [Alla99] is derived from the GA coder and MPEG-2 AAC. The ITU-T G.728 LD-CELP algorithm operates on speech frames of 2.5 ms (20 samples) at a sampling rate of 8 kHz and results in an algorithmic delay of 0.625 ms (5 samples). On the other hand, the MPEG-4 LD audio coding tool operates on 512 or 480 samples at a sampling rate of up to 48 kHz with an overall algorithmic delay of 20 ms. Recall that the GA tool that is based on the MPEG-2 AAC operates on frames of 1024 or 960 samples.

The delays due to the analysis/synthesis filter-bank window can be reduced by employing shorter windows. The look-ahead delays can be avoided by not employing the block switching. To reduce pre-echo distortions (Sections 6.9 and 6.10), TNS is employed in conjunction with window shape adaptation. In particular, for nontransient parts of the signal, a sine window is used, while a so-called low-overlap window is used in case of transient signals to achieve optimum TNS performance [Purn99b] [ISOI00]. Although most algorithms are fixed rate, the instantaneous bit rates required to satisfy masked thresholds on each frame are in fact time-varying. Thus, the idea behind a bit reservoir is to store surplus bits during periods of low demand, and then to allocate bits from the reservoir during localized periods of peak demand, resulting in a time-varying instantaneous bit rate but at the same time a fixed average bit rate. However, in MPEG-4 LD audio codec, the use of the bit reservoir is minimized in order to reach the desired target delay.

Based on the results published in [Alla99] [Herr99] [Purn99b] [ISOI00], the MPEG-4 LD audio codec performs relatively well compared to the MP3 coder at a bit rate of 64 kb/s/channel. It can also be noted from the MPEG-4 version 2 audio verification test [ISOI00], the quality measures of MPEG-2 AAC at 24 kb/s and MPEG-4 LD audio codec at 32 kb/s can be favorably compared. Moreover, the MPEG-4 LD audio codec [Herr98c] [Alla99] [Herr99] outperformed the ITU-T G.728 LD CELP [G728] for the case of coding both music and speech signals. However, as expected, the coding efficiency in the case of MPEG-4 LD codec is slightly reduced compared to its predecessors, MPEG-2 AAC and MPEG-4 GA. It should be noted that this reduction in the coding efficiency is attributed to the low coding delay achieved.

10.4.4.10 MPEG-4 Audio Error Robustness Tool One of the key issues in achieving reliable transmission over noisy and fast time-varying channels is the bit-rate scalability feature (discussed in Section 10.4.4.5). The bit-rate scalability enables flexible selection of coding features and dynamically adapts to the channel conditions and the varying channel capacity. However, the bit-rate scalability feature alone is not adequate for reliable transmission. The error resilience and error protection tools are also essential to obtain high quality audio. To this end, the MPEG-4 audio version 2 is fitted with codec-specific error robustness techniques [Purn99b] [ISOI00]. In this subsection, we will review the error robustness and equal and unequal error protection (EEP and UEP) tools in the MPEG-4 audio version 1. In particular, we discuss the *error resilience* [Sper00] [Sper02], *error protection* [Purn99b] [Mein01], and *error concealment* [Sper01] functionalities that are primarily designed for mobile applications.

The main idea behind the error resilience and protection tools is to provide better protection to sensitive and priority (important) bits. For instance, the audio frame header requires maximum error robustness; otherwise, transmission errors in the header will seriously impair the entire audio frame. The codewords corresponding to these priority bits are called the priority codewords (PCW). The error resilience tools available in the MPEG-4 audio version 2 are classified into three groups: the Huffman codeword reordering (HCR), the reversible variable length coding (RVLC), and the virtual codebooks (VCB11). In the HCR technique, some of the codewords, e.g., the PCWs, are sorted in advance and placed at known positions. First, a presorting procedure is employed that reorders the codewords based on their priority. The resulting PCWs are placed such that an error in one codeword will not affect the subsequent codewords. This can be achieved by defining segments of known length (L_{SEG}) and placing the PCWs at the beginning of these segments. The non-PCWs are filled into the gaps left by the PCWs, as shown in Figure 10.19.

The various applications of reversible variable length codes (RVLC) [Taki95] [Wen98] [Tsai01] in image coding have inspired researchers to consider them in error-resilient techniques for MPEG-4 audio. RVLC codes are used instead of Huffman codes for packing the scale factors in an AAC bitstream. The RVLC codes are (symmetrically) designed to enable both forward and backward decoding without affecting the coding efficiency. In particular, RVLCs allow instantaneous decoding in both directions that provides error robustness and significantly reduces the effects of bit errors in delay-constrained real-time applications. The next important tool employed for error resilience is the virtual codebook 11 (VCB11). Virtual codebooks are used to detect serious errors within spectral data [Purn99b] [ISOI00]. The error robustness techniques are codec specific (e.g., AAC and BSAC bitstreams). For example, AAC supports the HCR, the RVLC, and the VCB11 error-resilient tools. On the other hand, BSAC supports segmented binary arithmetic coding [ISOI00] to avoid error propagation within spectral data.

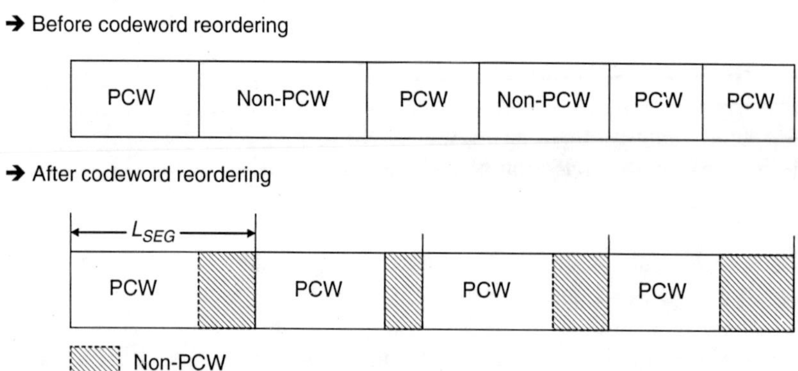

Figure 10.19. Huffman codeword reordering (HCR) algorithm to minimize error propagation in spectral data.

The MPEG-4 audio error protection tools include cyclic redundancy check (CRC), forward error correction (FEC), and interleaving. Note that these tools are inspired by some of the error correcting/detecting features inherent in the convolutional and block codes that essentially provide the controlled redundancy desired for error protection. Unlike the error-resilient tools that are limited only to the AAC and BSAC bitstreams, the error protection tools can be used in conjunction with a variety of MPEG-4 audio tools, namely, General Audio Coder (LTP and TwinVQ), Scalable Audio Coder, parametric audio coder (HILN), CELP, HVXC, and low-delay audio coder. Similar to the error-resilient tools, the first step in the EP tools is to reorder the bits based on their priority and error sensitiveness. The bits are sorted and grouped into different classes (usually 4 or 5) according to their error sensitivities. For example, consider that there are four error-sensitive classes (ESC), namely, ESC-0, ESC-1, ESC-2, and ESC-3. Usually, header bitstream or other very important bits that control the syntax and the global gain are included in the ESC-0. While the scale factors and spectral data (spectral envelope) are grouped in ESC-1 and ESC-2, respectively. The remaining side information and indices of MDCT coefficients are classified in ESC-3. After reordering (grouping) the bits, each error sensitive class receives a different error protection depending on the overhead allowed for each configuration. CRC and systematic rate-compatible punctured convolutional code (SRCPC) enable error detection and forward error correction (FEC). The SRCPC codes also aid in adjusting the redundancy rates in small steps. An interleaver is employed typically to deconcentrate or spread the burst errors. Shortened Reed-Solomon (SRS) codes are used to protect the interleaved data. Details on the design of Reed-Solomon codes for MPEG AAC are given in [Huan02]. For an in-depth treatment on the error correcting and error detecting codes refer to [Lin82] [Wick95].

10.4.4.11 MPEG-4 Speech Coding Tool Versus ITU-T Speech Standards It is noteworthy to compare the MPEG-4 speech coding tool against the ITU-T speech coding standards. While the latter applies source-filter configuration to model the speech parameters, the former employs a variety of techniques in addition to the traditional parametric representation. The MPEG-4 speech coding tool allows bit-rate scalability and real-time processing as well as applications related to storage media. The MPEG-4 speech coding tool incorporates algorithms such as the TwinVQ, the BSAC, and the HVXC/CELP. The MPEG-4 speech coding tool also accommodates multiple sampling rates. Error protection and error resilient techniques are provided in the MPEG-4 speech coding tool to obtain improved performance over error-prone channels. Other important features that distinguish the MPEG-4 tool from the ITU-T speech standards are the content-based interactivity and the ability to represent the audiovisual content as a set of objects.

10.4.4.12 MPEG-4 Audio Applications The MPEG-4 audio standard finds applications in low-bit-rate audio/speech compression, individual coding of natural and synthetic audio objects, low-delay coding, error-resilient transmission, and real-time audio transmission over packet-switching networks such as the Internet [Diet96] [Liu99]. MPEG-4 tools allow parameterization of the acoustical properties of an audio scene, with features such as immersive audiovisual rendering (virtual 3-D environments [Kau98]), room acoustical modeling, and enhanced 3-D sound presentation. MPEG-4 finds interesting applications in remote robot control system design [Kim02b]. Streaming audio codecs have also been proposed as a result of the MPEG-4 standardization efforts.

Applications of MPEG-4 audio in DRM digital narrowband broadcasting (DNB) and digital multimedia broadcasting (DMB) are given in [Diet00] and [Grub01], respectively. The general audio coding tool provides the necessary infrastructure for the design of error-robust scalable coders [Mori00b] and delivers improved speech/audio quality [Moori00a]. The "bit rate scalability" and "error resilience/protection" tools of the MPEG-4 audio standard dynamically adapt to the channel conditions and the varying channel capacity. Other important application-oriented features of MPEG-4 audio include low-delay bi-directional audio transmission, content-based interactivity, and object-based representation. Real-time implementation of the MPEG-4 audio is reported in [Hilp00] [Mesa00] [Pena01].

10.4.4.13 Spectral Band Replication and Parametric Stereo Spectral band replication (SBR) [Diet02] and parametric stereo (PS) [Schu04] are the two new compression techniques recently added to the MPEG 4 audio standard [ISOI03c]. The SBR technique is used in conjunction with a conventional coder such as the MP3 or the MPEG AAC. The audio signal is divided into low- and high-frequency bands. The underlying core coder operates at a reduced sampling rate and encodes the low-frequency band. The SBR technique operates at the original sampling rate to estimate the spectral envelope associated with the input audio. The spectral envelope along with a set of control parameters are encoded and transmitted to the decoder. The control parameters contain information regarding the gain and the spectral envelope level adjustment of the high frequency components. At the decoder, the SBR reconstructs the high frequencies based on the transposition of the lower frequencies.

aacPlus v1 is the combination of AAC and SBR and is standardized as the MPEG 4 high-efficiency (HE)-AAC [ISOI03c] [Wolt03]. Relative to the conventional AAC, the MPEG 4 HE-AAC results in bit rate reductions of about 30% [Wolt03]. The SBR has also been used to enhance the performance of MP3 [Zieg02] and the MPEG layer 2 digital audio broadcasting systems [Gros03].

aacPlus v2 [Purn03] adds the parametric stereo coding to the MPEG 4 HE-AAC standard. In the PS encoding [Schu04], the stereo signal is represented as a monaural signal plus ancillary data that describe the stereo image. The stereo image is described using four different PS parameters, i.e., inter-channel intensity differences (IID), inter-channel phase differences (IPD), inter-channel coherence (IC), and overall phase difference (OPD). These PS parameters can capture the perceptually relevant spatial cues at bit rates as low as 10 kb/s [Bree04].

10.4.5 MPEG-7 Audio (ISO/IEC 15938-4)

MPEG-7 audio standard targets content-based multimedia applications [ISOI01b]. MPEG-7 audio supports a broad range of applications [ISOI01d] that include multimedia indexing/searching, multimedia editing, broadcast media selection, and multimedia digital library sorting. Moreover, it provides ways for efficient audio file retrieval and supports both text-based and context-based queries. It is important to note that MPEG-7 will not replace MPEG-1, MPEG-2 BC/LSF, MPEG-2 AAC, or MPEG-4. It is intended to provide complementary functionality to these MPEG standards. If MPEG-4 is considered as the first object-based multimedia representation standard, then MPEG-7 can be regarded as the first content-based standard that incorporates multimedia interfaces through *descriptions*. These descriptions are the means of linking the audio content features and attributes with the audio itself. Figure 10.20 presents an overview of the MPEG-7 audio standard. This figure depicts the various audio tools, features, and profiles associated with the MPEG-7 audio. Publications on the MPEG-7 Audio Standard include [Lind99] [Nack99a] [Nack99b] [Lind00] [ISOI01b] [ISOI01e] [Lind01] [Quac01] [Manj02].

Motivated by the need to exchange multimedia content through the World Wide Web, in 1996, the ISO/IEC MPEG workgroup worked on a project called "Multimedia Content Description Interface" (MCDI) – MPEG-7. A working draft was formed in December 1999 followed by a final committee draft in February 2001. Seven months later, MPEG-7 ISO/IEC 15938: Part 4 Audio, an international standard (IS) for *content-based multimedia applications* was published along with seven other parts of the MPEG-7 standard (Figure 10.20). Figure 10.20 shows a summary of various features, applications, and profiles specified by the MPEG-7 audio coding standard.

10.4.5.1 MPEG-7 Parts MPEG-7 defines the following eight parts [MPEG] (Figure 10.20): MPEG-7 Systems, MPEG-7 DDL, MPEG-7 Visual, MPEG-7 Audio, MPEG-7 MDS, MPEG-7 Reference Software (RS), MPEG-7 Conformance Testing (CT), and MPEG-7 Extraction and use of Descriptions.

MPEG-7 Systems (Part I) specifies the binary format for encoding MPEG-7 Descriptions; MPEG-7 DDL (Part II) is the language for defining the syntax of the Description Tools. MPEG-7 Visual (Part III) and MPEG-7 Audio (Part IV) deal with the visual and audio descriptions, respectively. MPEG-7 MDS (Part V) defines the structures for multimedia descriptions. MPEG-7 RS (Part VI) is a unique software implementation of certain parts of the MPEG-7 Standard with noninformative status. MPEG-7 CT (Part VII) provides the essential guidelines/procedures for conformance testing of MPEG-7 implementations. Finally, the eighth part, addresses the use and formulation of a variety of description tools that we will discuss later in this section.

In our discussion on MPEG-7 Audio, we refer to MPEG-7 DDL and MPEG-7 MDS parts quite regularly, mostly due to their interconnectivity within the MPEG-7 Audio Framework. Therefore, it is necessary that we introduce these two parts first, before we move on to the MPEG-7 Audio Description Tools.

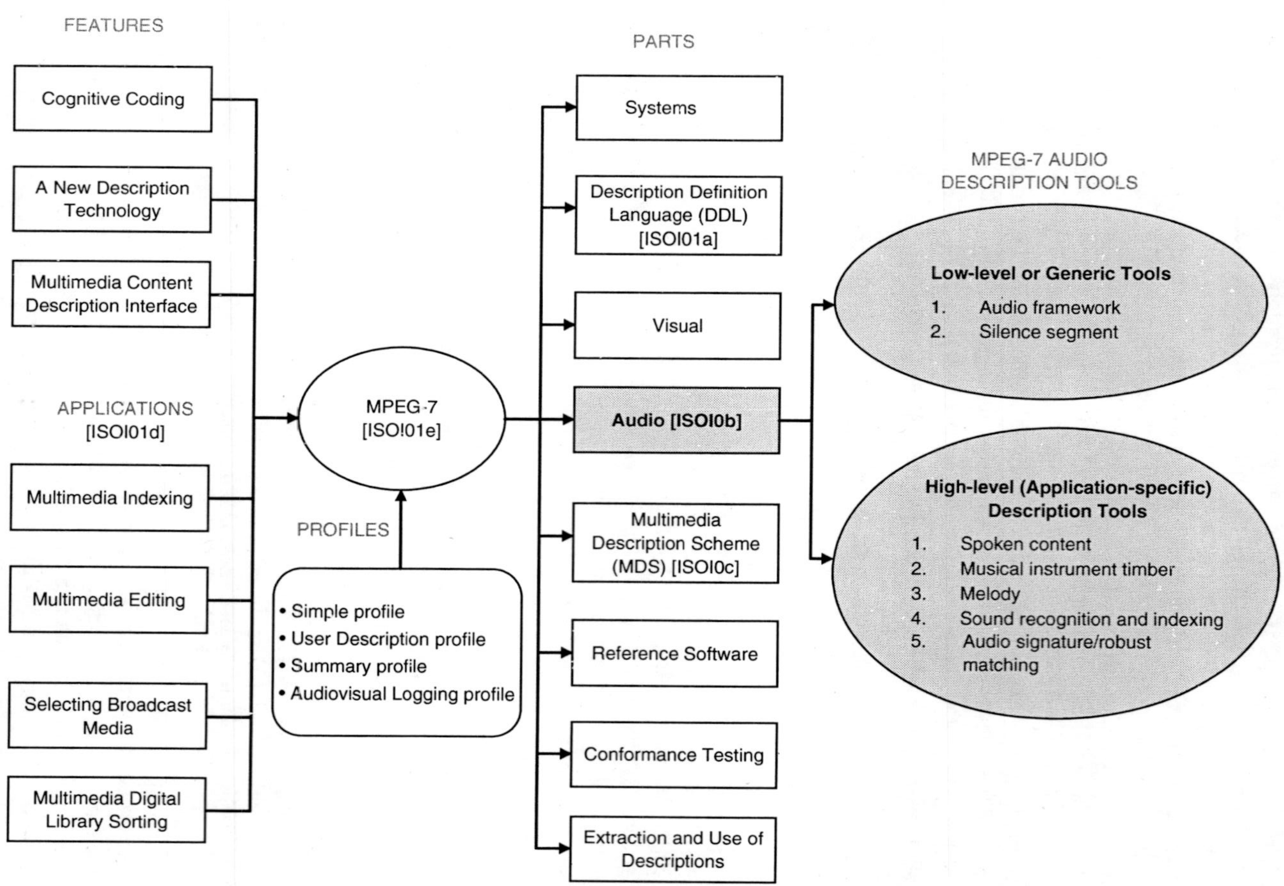

Figure 10.20. An overview of the MPEG-7 standard.

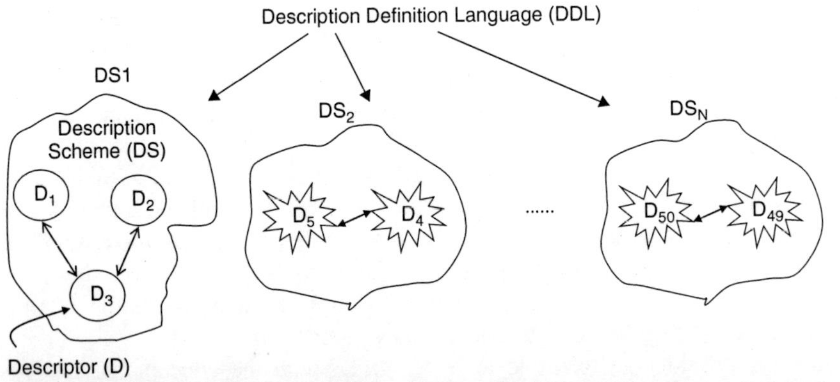

NOTE:

1. Descriptors (Ds) are the features and attributes associated with an audio waveform
2. The structure and the relationships among the descriptors are defined by a Description Scheme (DS)
3. Description Definition Language (DDL) defines the syntax necessary to create, extend, and combine avariety of DSs and Ds

Figure 10.21. Some essential building blocks of MPEG-7 standard: descriptors (Ds), description schemes (DSs), and description definition language (DDL).

MPEG-7 Description Definition Language (DDL) – Part II. We mentioned earlier that MPEG-7 incorporates multimedia interfaces through *descriptors*. These descriptors are the features and attributes associated with the audio. For example, descriptors in the case of MPEG-7 Visual part describe the visual features such as color, resolution, contour, mapping techniques, etc. A group of descriptors related in a manner suitable for a specific application, forms a *description scheme* (DS). The standard [ISOI01e] defines the description scheme as one that specifies a structure for the descriptors and semantics of their relationships.

MPEG-7 in its entirety has been built around these descriptors (Ds) and description schemes (DSs), and most importantly on a language called the description definition language (DDL). The DDL defines the syntax necessary to create, extend, and combine a variety of DSs and Ds. In particular, the DDL forms "the core part" of the MPEG-7 standard and will also be invoked by other parts (i.e., Visual, Audio, and MDS) to create new Ds and DSs. The DDL follows a set of programming rules/structure similar to the ones employed in the eXtensible Markup Language (XML). It is important to note that the DDL is not a modeling language but a schema language that is based on the WWW Consortium's XML schema [XML] [ISOI01e]. Several modifications were needed before adopting the XML schema language as the basis for the DDL. We refer to [XML] [ISOI01a] [ISOI01e] for further details on the XML schema language and its liaison with MPEG-7 DDL [ISOI01a].

MPEG-7 Multimedia Description Schemes (MDS) – Part V. Recall that a description scheme (DS) specifies structures for descriptors; similarly, a multimedia description scheme (MDS) [ISOI01c] provides details on the structures for describing multimedia content (in particular audio, visual, and textual data). MPEG-7 MDS defines two classes of description tools, namely, the basic (or low-level) and multimedia (or high-level) tools [ISOI01c]. Figure 10.22 shows the classification of MDS elements. The *basic tools* specified by the MPEG-7 MDS are the generic entities, usually associated with simple descriptors, such as the basic data types, textual database, etc. On the other hand, the *high-level multimedia tools* deal with the content-specific entities that are complex and involve signal structures, semantics, models, efficient navigation, and access. The high-level (complex) tools are further subdivided into five groups (Figure 10.22), i.e., content description, content management, content organization, navigation and access, and user interaction.

Let us consider an example to better understand the concepts of DDL and MDS framework. Suppose that an audio signal, $s(n)$, is described using three descriptors, namely, spectral features D_1, parametric models D_2, and energy D_3. Similarly, visual $v(i, j)$ and textual content can also be described as shown in Table 10.5. We arbitrarily chose four description schemes (DS_1 through DS_4) that link these multimedia features (audio, visual, and textual) in a structured manner. This linking mechanism is performed through DDL, a schema language designed specifically for MPEG-7. From Table 10.5, the descriptors D_2, D_8, D_9 are related using the description scheme DS_2. The melody descriptor D_8 provides the melodic information (e.g., rhythmic, high-pitch, etc.), and the timbre descriptor D_9 represents some perceptual features (e.g., pitch/loudness details, bass/treble adjustments in audio, etc.). The parametric model descriptor D_2 describes the audio encoding model and related encoder details (e.g., MPEG-1 layer III, sampling rates, delay, bit rates, etc.). While the descriptor D_2 provides details on the encoding procedure, the descriptors D_8 and D_9 describe audio morphing, echo/reverberation, tone control, etc.

MPEG-7 Audio – Part IV. MPEG-7 Audio represents part IV of the MPEG-7 standard and provides structures for describing the audio content. Figure 10.23 shows the organization of MPEG-7 audio framework.

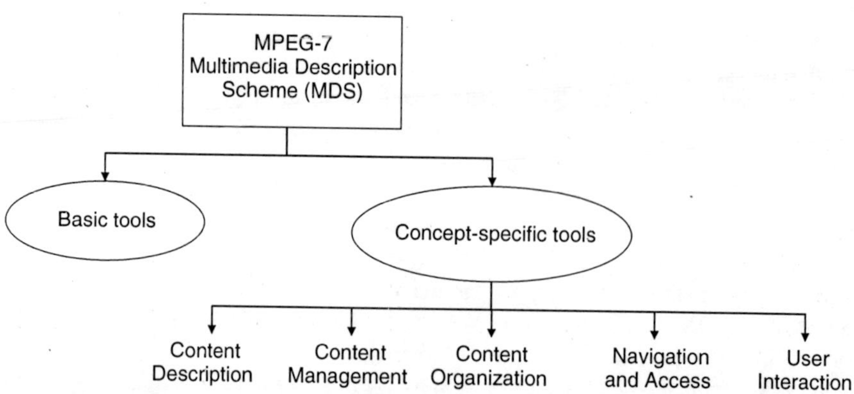

Figure 10.22. Classification of multimedia description scheme (MDS) tools.

Table 10.5. A hypothetical example that gives a broader perspective on multimedia descriptors; i.e., audio, visual, and textual features to describe a multimedia content.

Group	Descriptors	Description schemes
Audio content, $s(n)$	D_1: Spectral features	
	D_2: Parametric models	
	D_3: Energy of the signal	
		DS_1: D_1, D_3
Visual content, $v(i, j)$	D_4: Color	
		DS_2: D_2, D_8, D_9
	D_5: Shape	
		DS_3: DS_2, D_4, D_5
Textual descriptions	D_6: Title of the clip	
		DS_4: DS_1, D_6, D_7
	D_7: Author information	
	D_8: Melody details	
	D_9: Timbre details	

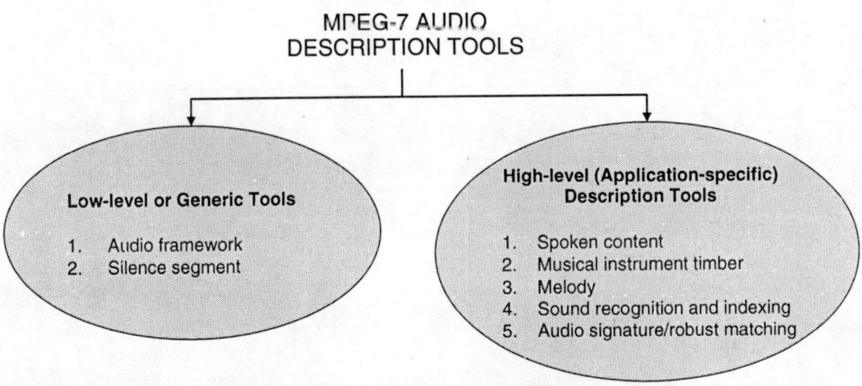

Figure 10.23. MPEG-7 audio description tools.

10.4.5.2 MPEG-7 Audio Versions and Profiles New extensions (Amendment 1) for the existing MPEG-7 Audio are being considered. Some of the extensions are in the areas of application-specific spoken content, tempo description, and specification of precision for low-level data types. This new amendment will be standardized as MPEG-7 Audio Version 2 (Final drafts of International Standard (FDIS) for Version 2 were finalized in March 2003).

Although many description tools are available in MPEG-7 audio, it is not practical to implement all of them in a particular system. MPEG-7 Version 1 therefore defines four complexity-ranked profiles (Figure 10.20) intended to help system designers in the task of tool subset selection. These include simple profile, user description profile, summary profile, and audiovisual logging profile.

10.4.5.3 MPEG-7 Audio Description Tools The MPEG-7 Audio framework comprises two main categories; namely, *generic* tools and a set of *application-specific* tools (see Figure 10.20 and Figure 10.23).

10.4.5.3.1. Generic Tools The generic toolset consists of 17 low-level audio descriptors and a silence segment descriptor (Table 10.6).

MPEG-7 Audio Low-level Descriptors. MPEG-7 audio [ISOI01b] defines two ways of representing the low-level audio features, i.e., segmenting and sampling. In segmentation, usually, common datatypes or scalars are grouped together (e.g., energy, power, bit rate, sampling rate, etc.). On the other hand, sampling enables discretization of audio features in a vector form (e.g., spectral features, excitation samples, etc.). Recently, a unified framework called the *scalable series* [Lind99] [Lind00] [ISOI01b] [Lind01] has been proposed to manipulate these discretized values. This is somewhat similar to MPEG-4 scalable audio coding that we discussed in Section 10.4.4. A list of low-level audio descriptors defined by the MPEG-7 Audio standard [ISOI01b] is summarized in Table 10.6. These

Table 10.6. Low-level audio descriptors (17 in number) and the silence descriptor supported by the MPEG-7 generic toolset [ISOI01b].

	Generic toolset	Descriptors
Low-level audio descriptors group	1. Basic	D_1: Audio waveform
		D_2: Power
	2. Basic spectral	D_3: Spectrum envelope
		D_4: Spectral centroid
		D_5: Spectral spread
		D_6: Spectral flatness
	3. Signal parameters	D_7: Harmonicity
		D_8: Fundamental frequency
	4. Spectral basis	D_9: Spectrum basis
		D_{10}: Spectrum projection
	5. Timbral spectral	D_{11}: Harmonic spectral centroid
		D_{12}: Harmonic spectral deviation
		D_{13}: Harmonic spectral spread
		D_{14}: Harmonic spectral variation
		D_{15}: Spectral centroid
	6. Timbral temporal	D_{16}: Log attack time
		D_{17}: Temporal centroid
Silence	7. Silence segment	D_{18}: Silence descriptor

descriptors can be classified into the following groups: basic, basic spectral, signal parameters, spectral basis, timbral spectral, and timbral temporal.

MPEG-7 Silence Segment. The MPEG-7 silence segment attaches a semantic of silence to an audio segment. The silence descriptor provides ways to specify threshold levels (e.g., the level of silence).

10.4.5.3.2. High-Level or Application-Specific MPEG-7 Audio Tools Besides the aforementioned generic toolset, the MPEG-7 audio standard describes five specialized high-level tools (Table 10.7). These application-specific description tools can be grouped as spoken content, musical instrument, melody, sound recognition/indexing, and robust audio matching.

Spoken Content Description Tool (SC-DT). The SC-DT provides descriptions of spoken words in an audio clip, thereby enabling speech recognition and speech parameter indexing/searching. Spoken content lattice and spoken content header are the two important parts of the SC-DT (see Table 10.7). While the SC header carries the lexical information (i.e., wordlexicon, phonelexicon, ConfusionInfo, and SpeakerInfo descriptors), the SC-lattice DS represents lattice-structures to connect words or phonemes chosen from the corresponding lexicon. The idea of using *lattice structures* in the SC-lattice DS is similar to the one employed in a typical continuous automatic speech recognition scenario [Rabi89] [Rabi93].

Musical Instrument Timbre Description Tool (MIT-DT). The MIT-DT describes the timbre features (i.e., perceptual attributes) of sounds from musical instruments. Timbre can be defined as the collection of perceptual attributes that make two

Table 10.7. Application-specific audio descriptors and description schemes [ISOI01b].

High-level descriptor toolset		Descriptor details
SC-DT	1. SC-header	D_1: Word lexicon D_2: Phone lexicon D_3: Confusion info D_4: Speaker info
	2. SC-lattice DS	Provides structures to connect or link the words/phonemes in the lexicon.
MIT-DT	3. Timbre (perceptual) features of musical instruments	D_1: Harmonic Instrument Timbre D_2: Percussive Instrument Timbre
M-DT	4. Melody features	DS_1: Melody contour DS_2: Melody sequence
SRI-DT	5. Sound recognition and indexing application	D_1: Sound Model State Path D_2: Sound Model State Histogram DS_1: Sound model DS_2: Sound classification model
AS DT	6. Robust audio identification	DS_1: Audio signature DS

audio clips having the same pitch and loudness sound different [ISOI01b]. Musical instrument sounds, in general, can be classified as harmonic-coherent-sustained, percussive-nonsustained, nonharmonic-coherent-sustained, and non-coherent-sustained. The standard defines descriptors for the first two classes of musical sounds (Table 10.7). In particular, MIT-DT defines two descriptors, namely, the harmonic instrument timbre (HIT) descriptor and the percussive instrument timbre (PIT) descriptor. The HIT descriptor was built on the four harmonic low-level descriptors (i.e., D_{11} through D_{14} in Table 10.6) and the Logattacktime descriptor. On the other hand, the PIT descriptor is based on the combination of the timbral temporal low-level descriptors (i.e., Logattacktime and Temporalcentroid) and the spectral centroid descriptor.

Melody Description Tool (M-DT). The M-DT represents the melody details of an audio clip. The melodycontourDS and the melodysequenceDS are the two schemes included in M-DT. While the former scheme enables simple and robust melody contour representation, the latter approach involves detailed and expanded melody/rhythmic information.

Sound Recognition and Indexing Description Tool (SRI-DT). The SRI-DT is on automatic sound identification/recognition and indexing. Recall that the SC-DT employs lexicon descriptors (Table 10.7) for SC recognition in an audio clip. In the case of SRI, classification/indexing of sound tracks are achieved through sound models. These models are constructed based on the spectral basis low-level descriptors, i.e., spectral basis (D_9) and spectral projection (D_{10}), listed in Table 10.6. Two descriptors, namely the sound model state path descriptor and the sound model state histogram descriptor, are defined to keep track of the active paths in a trellis.

Robust Audio Identification and Matching. Robust matching and identification of audio clips is one of the important applications of MPEG-7 audio standard [ISOI01d]. This feature is enabled by the low-level spectral flatness descriptor (Table 10.6). A description scheme, namely, the Audio Signature DS defines the semantics and structures for the spectral flatness descriptor. Hellmuth *et al.* [Hell01] proposed an advanced audio identification procedure based on content descriptions.

10.4.5.4 MPEG-7 Audio Applications Being the first metadata standard, MPEG-7 audio provides new ideas for audio-content indexing and archiving [ISOI01d]. Some of the applications are in the areas of multimedia searching, audio file indexing, sharing and retrieval, and media selection for digital audio broadcasting (DAB). We discussed most of these applications while addressing the high-level audio descriptors and description schemes. A summary of these applications follows. Unlike in an automatic speech recognition scenario where word or phoneme lattices (based on feature vectors) are employed for identifying speech, in MPEG-7 these lattice structures are denoted as Ds and DSs. These description data enable *spoken content retrieval.* MPEG-7 audio version 2 includes new tools and specialized enhancements to spoken content search. *Musical instrument timbre search* is another important application that targets content-based editing. *Melody search* enables query by humming [Quac01]. *Sound recognition/indexing* and *audio identification/fingerprinting* form two other important applications of the MPEG-7. We will next address the concepts of "interoperability" and "universal multimedia access" (UMA) in the context of the new work initiated by the ISO/IEC MPEG workgroup in June 2000, called the *Multimedia Framework – MPEG* 21 [Borm03].

10.4.6 MPEG-21 Framework (ISO/IEC-21000)

Motivated by the need for a standard that enables multimedia content access and distribution, the ISO/IEC MPEG workgroup addressed the 21st Century Multimedia Framework – MPEG-21: ISO/IEC 21000 [Spen01] [ISOI02a] [ISOI03a] [ISOI03b] [Borm03] [Burn03]. This multimedia standard should be interoperable and highly automated [Borm03]. The MPEG-21 multimedia framework envisions creating a platform that encompasses a great deal of functionalities for both content-users and content-creators/providers. Some of these functions include the multimedia resource delivery to a wide range of networks and terminals (e.g., personal computers (PCs), PDAs and other digital assistants, mobile phones, third-generation cellular networks, digital audio/video broadcasting (DAB/DVB), HDTVs, and several other home entertainment systems); protection of intellectual property rights through digital rights management (DRM) systems.

Content creators and service providers face several challenging tasks in order to satisfy simultaneously the conflicting demands of "interoperability" and "intellectual property management and protection" (IPMP). To this end, MPEG-21 defines a multimedia framework that comprises seven important parts [ISOI02a], as shown in Table 10.8. Recall that the MPEG-7 ISO/IEC-15938 standard defines a fundamental unit called "Descriptors" (Ds) to define/declare the features and attributes of multimedia content. In a manner analogous to this, MPEG-21 ISO/IEC-21000: Part 1 defines a basic unit called the "Digital Item" (DI). Besides DI, MPEG-21 specifies another entity called the "User" interaction [ISOI02a] [Burn03] that provides details on how each "User" interacts with other users via objects called the "Digital Items." Furthermore, MPEG-21 Parts 2 and 3 define the declaration and identification of the DIs, respectively (see Table 10.8). MPEG-21 ISO/IEC-21000 Parts 4 through 6 enables interoperable digital content distribution and transactions that take into account the IPMP requirements. In particular, a machine-readable language called the Rights Expression Language (REL) is specified in MPEG-21 ISO/IEC-21000: Part 5 that defines the rights and permissions for the access and distribution of multimedia resources across a variety of heterogeneous terminals and networks. MPEG-21 ISO/IEC-21000: Part 6 defines a dictionary called the Rights Data Dictionary (RDD) that contains information on content protection and rights.

MPEG-7 and MPEG-21 standards provide an open framework on which one can build application-oriented interfaces or tools that satisfy a specific criterion (e.g., a query, an audio file indexing, etc.). In particular, the MPEG-7 standard provides an *interface* for indexing, accessing, and distribution of multimedia content; and the MPEG-21 defines an *interoperable framework* to access the multimedia content.

Table 10.8. MPEG-21 multimedia framework and the associated parts [ISOI02a].

Parts in the MPEG-21: ISO/IEC 21000 Standard [ISOI02a]		Details
Part 1	Vision, technologies, and strategy	Defines the vision, requirements, and applications of the standard; and provides an overview of the multimedia framework. Introduces two new terms, i.e., *digital item* (DI) and *user interaction*.
Part 2	Digital item declaration	Defines the relationship between a variety of multimedia resources and provides information regarding the declaration of Dis.
Part 3	Digital item identification	Provides ways to identify different types of digital items (DIs) and descriptors/description schemes (Ds/DSs) via uniform resource identifiers (URIs).
Part 4	IPMP	Defines a framework for the intellectual property management and protection (IPMP) that enables interoperability.
Part 5	Rights expression language	A syntax language that enables multimedia content distribution in a way that protects the digital content. The rights and the permissions are expressed or declared based on the terms defined in the rights data dictionary.

Table 10.8. (*continued*)

Parts in the MPEG-21: ISO/IEC 21000 Standard [ISOI02a]		Details
Part 6	Rights data dictionary	A database or a dictionary that contains the information regarding the rights and permissions to protect the digital content.
Part 7	Digital item adaptation	Defines the concept of an adapted digital item.

Until now, our focus was primarily on ISO/IEC MPEG Audio Standards. In the next few sections, we will attend to company-oriented perceptual audio coding algorithms, i.e., the Sony Adaptive Transform Acoustic Coding (ATRAC), the Lucent Technologies Perceptual Audio Coder (PAC), the Enhanced PAC (EPAC), the Multichannel PAC (MPAC), Dolby Laboratories AC-2/AC-2A/AC-3, Audio Processing Technology (APT-x100), and the Digital Theater Systems (DTS) Coherent Acoustics (CA) encoder algorithms.

10.4.7 MPEG Surround and Spatial Audio Coding

MPEG spatial audio coding began receiving attention during the early 2000s [Fall01] [Davis03]. Advances in joint stereo coding of multichannel signals [Herr04b], binaural cue coding [Fall01], and the success of the recent low complexity parametric stereo encoding in MPEG 4 HE-AAC/PS standard [Schu04] generated interest in the MPEG surround and spatial audio coding [Herr04a] [Bree05]. Unlike the discrete 5.1-channel encoding as used in Dolby Digital or DTS, the MPEG spatial audio coding, captures the "spatial image" of a multichannel audio signal. The spatial image is represented using a compact set of parameters that describe the perceptually relevant differences among the channels. Typical parameters include the interchannel level difference (ICLD), the interchannel coherence (ICC), and the interchannel time difference (ICTD). The multichannel signal is first downmixed to a stereo signal and then a conventional MP3 coder is used. Spatial image parameters are computed using the binaural cue coding (BCC) technique and are transmitted to the decoder as side information [Herr04a]. At the decoder, a one-to-two (OTT) or two-to-three (TTT) channel mapping is used to synthesize the multichannel surround sound.

10.5 ADAPTIVE TRANSFORM ACOUSTIC CODING (ATRAC)

The ATRAC algorithm, developed by Sony for use in its rewriteable MiniDisc system [Yosh94], combines subband and transform coding to achieve nearly CD-quality coding of 44.1 kHz 16-bit PCM input data at a bit rate of 146 kb/s per

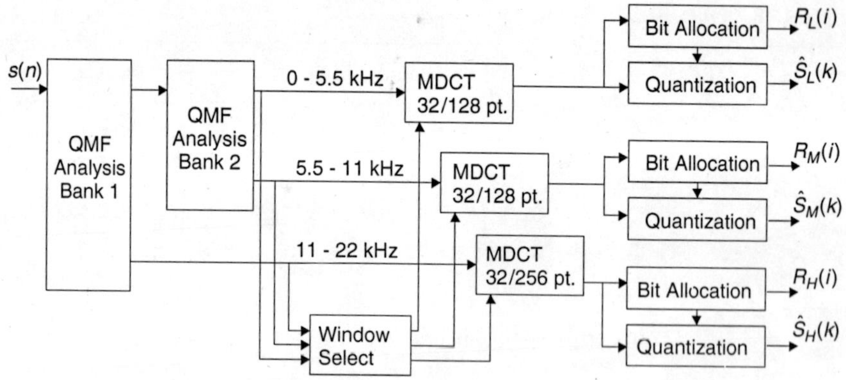

Figure 10.24. Sony ATRAC (embedded in MiniDisc, SDDS).

channel [Tsut98]. Using a tree-structured QMF analysis bank (Section 6.5), the ATRAC encoder (Figure 10.24) first splits the input signal into three subbands of 0–5.5 kHz, 5.5–11 kHz, and 11–22 kHz. Like MPEG-1 layer III, the ATRAC QMF bank is followed by signal-adaptive MDCT analysis in each subband. Next, a window-switching scheme is employed that can be summarized as follows. During steady-state input periods, high-resolution spectral analysis is attained using 512 sample blocks (11.6 ms). During input attack or transient periods, short block sizes of 1.45 ms in the high-frequency band and 2.9 ms in the low- and mid-frequency bands are used for pre-echo cancellation.

After MDCT analysis, spectral components are clustered into 52 nonuniform subbands (block floating units or BFUs) according to critical band spacing. The BFUs are block-companded, quantized, and encoded according to a psychoacoustically derived bit allocation. For each analysis frame, the ATRAC encoder transmits quantized MDCT coefficients, subband window lengths, BFU scalefactors, and BFU word lengths to the decoder. Like the MPEG family, the ATRAC architecture decouples the decoder from psychoacoustic analysis and bit-allocation details. Evolutionary improvements in the encoder bit allocation strategy are therefore possible without modifying the decoder structure. An added benefit of this architecture is asymmetric complexity, which enables inexpensive decoder implementations.

Suggested bit allocation techniques for ATRAC are of lower complexity than those found in other standards since ATRAC is intended for low-cost, battery-powered devices. One proposed method distributes bits between BFUs according to a weighted combination of fixed and adaptive bit allocations [Tsut96]. For the k-th BFU, bits are allocated according to the relation

$$r(k) = \alpha r_a(k) + (1 - \alpha) r_f(k) - \beta, \tag{10.10}$$

where $r_f(k)$ is a fixed allocation, $r_a(k)$ is a signal-adaptive allocation, the parameter β is a constant offset computed to guarantee a fixed bit rate, and the parameter α is a tonality estimate ranging from 0 (noise-like) to 1 (tone-like). The fixed allocations, $r_f(k)$, are the same for all inputs and concentrate more bits at the lower frequencies. The signal adaptive bit allocations, $r_a(k)$, assign bits according to the strength of the MDCT components. The effect of Eq. (10.10) is that more bits are allocated to BFUs containing strong peaks for tonal signals. For noise-like signals, bits are allocated according to a fixed allocation rule, with low bands receiving more bits than high bands.

Sony Dynamic Digital Sound (SDDS). In addition to providing near CD-quality on a MiniDisc medium, the ATRAC algorithm has also been deployed as the core of Sony's digital cinematic sound system, SDDS. SDDS integrates eight independent ATRAC modules to carry the program information for the left (L), left center (LC), center (C), right center (RC), right (R), subwoofer (SW), left surround (LS), and right surround (RS) channels typically present in a modern theater. SDDS data is recorded using optical black and white dot-matrix technology onto two thin strips along the right and left edges of the film, outside of the sprocket holes. Each edge contains four channels. There are 512 ATRAC bits per channel associated with each movie frame, and each optical data frame contains a matrix of 52×192 bits [Yama98]. SDDS data tracks do not interfere with or replace the existing analog sound tracks. Both Reed-Solomon error correction and redundant track information are delayed by 18 frames and employed to make SDDS robust to bit errors introduced by run-length scratches, dust, splice points, and defocusing during playback or film printing. Analog program information is used as a backup in the event of uncorrectable digital errors.

10.6 LUCENT TECHNOLOGIES PAC, EPAC, AND MPAC

The pioneering research contributions on perceptual entropy [John88b], monophonic PXFM [John88a], stereophonic PXFM [John92a], and ASPEC [Bran91] strongly influenced not only the MPEG family architecture but also evolved at AT&T Bell Laboratories into the Perceptual Audio Coder (PAC). The PAC algorithm eventually became property of Lucent. AT&T, meanwhile, became active in the MPEG-2 AAC research and standardization. The low-complexity profile of AAC became the AT&T coding standard.

Like the MPEG coders, the Lucent PAC algorithm is flexible in that it supports monophonic, stereophonic, and multiple channel modes. In fact, the bit stream definition will accommodate up to 16 front side, 7 surround, and 7 auxiliary channel pairs, as well as 3 low-frequency effects (LFE or subwoofer) channels. Depending upon the desired quality, PAC supports several bit rates. For a modest increase in complexity at a particular bit rate, improved output quality can be realized by enabling enhancements to the original system. For example, whereas 96 kb/s output was judged to be adequate with stereophonic PAC, near CD quality was reported at 56–64 kb/s for stereophonic enhanced PAC [Sinh98a].

10.6.1 Perceptual Audio Coder (PAC)

The original PAC system described in [John96c] achieves very-high-quality coding of stereophonic inputs at 96 kb/s. Like the MPEG-1 layer III and the ATRAC, the PAC encoder (Figure 10.25) uses a signal-adaptive MDCT filter bank to analyze the input spectrum with appropriate frequency resolution. A long window of 2048 points (1024 subbands) is used during steady-state segments, or else a series of short 256-point windows (128 subbands) is applied for segments containing transients or sharp attacks. In contrast to MPEG-1 and ATRAC, however, PAC relies on the MDCT alone rather than a hybrid filter-bank structure, thus realizing a complexity reduction. As noted previously [Bran88a] [Mahi90], the MDCT lends itself to compact representation of stationary signals, and a 2048-point block size yields sufficiently high-frequency resolution for most sources. This segment length was also associated with the maximum realizable coding gain as a function of block size [Sinh96]. Filter-bank resolution switching decisions are made on the basis of PE (high complexity) or signal energy (low complexity) criteria.

The PAC perceptual model derives noise masking thresholds from filter-bank output samples in a manner similar to MPEG-1 psychoacoustic model recommendation 2 [ISOI92] and the PE calculation in [John88b]. The PAC model, however, accounts explicitly for both simultaneous and temporal masking effects. Samples are grouped into 1/3 critical band partitions, tonality is estimated in each band, and then time and frequency spreading functions are used to compute a masking threshold that can be related to the filter-bank outputs. One can observe that PAC realizes some complexity reduction relative to MPEG by avoiding parallel frequency analysis structures for quantization and perceptual modeling. The masking thresholds are used to select one of 128 exponentially distributed quantization step sizes in each of 49 or 14 coder bands (analogous to ATRAC BFUs) in high-resolution and low-resolution modes, respectively. The coder bands are quantized using an iterative rate control loop in which thresholds are adjusted to satisfy simultaneously bit-rate constraints and an equal loudness criterion that attempts to shape quantization noise such that its absolute loudness is constant relative to the masking threshold. The rate control loop allows time-varying instantaneous bit rates so that additional bits are available in times of peak demand, much like the bit reservoir of MPEG-1 layer III. Remaining statistical redundancies are removed from the stream of quantized spectral samples prior to bit stream formatting using eight structured, multidimensional Huffman codebooks. These codebooks are applied to DPCM-encoded quantizer outputs. By clustering coder bands into sections and selecting only one codebook per section, the system minimizes the overhead.

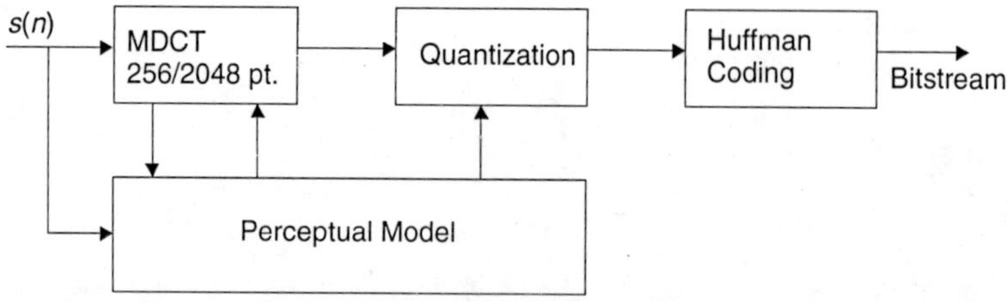

Figure 10.25. Lucent Technologies Perceptual Audio Coder (PAC).

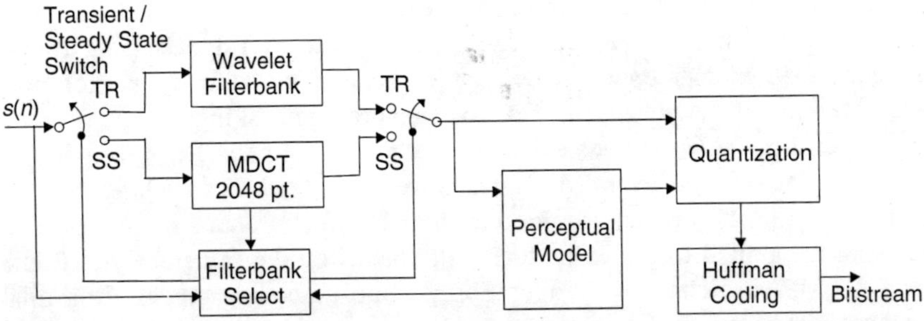

Figure 10.26. Lucent Technologies Enhanced Perceptual Audio Coder (EPAC).

10.6.2 Enhanced PAC (EPAC)

In an effort to enhance PAC output quality at low bit rates, Sinha and Johnston [Sinh96] introduced a novel signal-adaptive MDCT/WP[1] switched filter bank scheme that resulted in nearly transparent coding for CD-quality source material at 64 kb/s per stereo pair. EPAC (Figure 10.26) is unique in that it switches between two distinct filter banks rather than relying upon hybrid [Tsut98] [ISOI92] or nonuniform cascade [Prin95] structures.

A 2048-point MDCT decomposition is applied normally, during "stationary" periods. EPAC switches to a tree-structured wavelet packet (WP) decomposition matched to the auditory filter bank during sharp transients. Switching decisions occur every 25 ms, as in PAC, using either PE or energy criteria. The WP analysis offers the dual advantages of more compact signal representation during transient periods than MDCT, as well as improved time resolution at high frequencies for accurate estimation of the time/frequency distribution of masking power contained in sharp attacks. In contrast to the uniform time-frequency tiling associated with MDCT window-length switching schemes (e.g., [ISOI92] [Bran94a]), the EPAC WP transform (tree-structured QMF bank) achieves a nonuniform time-frequency tiling. For a suitably designed analysis wavelet and tree-structure, an improvement in time resolution is restricted to the high-frequency regions of interest, while good-frequency resolution is maintained in the low-frequency subbands. The EPAC WP filter bank was specifically designed for time-localized impulse responses at high frequencies to minimize the temporal spread of quantization noise (pre-echo). Novel start and stop windows are inserted between analysis frames during switching intervals to mitigate boundary effects associated with the MDCT-to-WP and WP-to-MDCT transitions. Other than the enhanced filter bank, EPAC is identical to PAC. In subjective tests involving 12 expert and nonexpert listeners with difficult castanets and triangle test signals, EPAC outperformed PAC for a 64 kb/s per stereo pair by an average of 0.4–0.6 on a five-point quality scale.

10.6.3 Multichannel PAC (MPAC)

Like the MPEG, AC-3, and SDDS systems, the PAC algorithm also extends its monophonic processing capabilities into stereophonic and multiple-channel

[1] See Chapter 8, Sections 8.2 and 8.3, for descriptions on wavelet filter bank and WP transforms.

modes. Stereophonic PAC computes individual masking thresholds for the left, right, mono, and stereo (L, R, M = L + R, and S = L − R) signals using a version of the monophonic perceptual model that has been modified to account for binary-level masking differences (BLMD), or binaural unmasking effects [Moor77]. Then, monaural PAC methods encode either the signal pairs (L, R) or (M, S). In order to minimize the overall bit rate, however, a LR/MS switching procedure is embedded in the rate-control loop such that different encoding modes (LR or MS) can be applied to the individual coder bands on the same analysis frame.

In the MPAC 5-channel configuration, composite coding modes are available for the front side left, center, right, and left and right surround (L, C, R, Ls, and Rs) channels. On each frame, the composite algorithm works as follows: First, appropriate window-switched filter-bank frequency resolution is determined separately for the front, side, and surround channels. Next, the four signal pairs LR, MS, LsRs, and MsSs (Ms = Ls + Rs, Ss = Ls − Rs) are generated. The MPAC perceptual model then computes individual BLMD-compensated masking thresholds for the eight LR and MS signals, as well as the center channel, C. Once thresholds have been obtained, a two-step coding process is applied. In step 1, a minimum PE criterion is first used to select either MS or LR coding for the front, side, and surround channel groups in each coder band. Then, step 2 applies interchannel prediction to the quantized spectral samples. The prediction residuals are quantized such that the final quantization noise satisfies the original masking thresholds for each channel (LR or MS). The interchannel prediction schemes are summarized in [Sinh98a]. In pursuit of a minimum bit rate, the composite coding algorithm may elect to utilize either step 1 or step 2, both step 1 and step 2, or neither step 1 nor step 2. Finally, the composite perceptual model computes a global masking threshold as the maximum of the five individual thresholds, minus a safety margin. This threshold is phased in gradually for joint coding when the bit reservoir drops below 20% [Sinh98a]. The safety margin magnitude depends upon the bit reservoir state. Composite modes are applied separately for each coder band on each analysis frame. In terms of performance, the MPAC system was found to produce the best quality at 320 kb/s for 5 channels during a recent ISO test of multichannel algorithms [ISOII94].

Applications. Both 128 and 160 kb/s stereophonic versions of PAC were considered for standardization in the U.S. Digital Audio Radio (DAR) project. In an effort to provide graceful degradation and extend broadcast range in the presence of heavy fading associated with fringe reception areas, perceptually motivated unequal error protection (UEP channel coding) schemes were examined in [Sinh98b]. The proposed scheme ranks bit stream elements into two classes of perceptual importance. Bit stream parameters associated with center channel information and certain mid-frequency subbands are given greater channel protection (class 1) than other parameters (class 2). Subjective tests revealed a strong preference for UEP over equal error protection (EEP), particularly when bit error rates (BER) exceeded 2×10^{-4}. For network applications, acceptable PAC output quality at bit rates as low as 12–16 kb/s per channel in conjunction with the availability of JAVA PAC decoder implementations are reportedly increasing PAC deployment among suppliers of Internet audio program material [Sinh98a]. MPAC has also been considered for cinematic and advanced television applications. Real-time PAC and EPAC decoder implementations have been demonstrated on 486-class PC platforms.

10.7 DOLBY AUDIO CODING STANDARDS

Since the late 1980s, Dolby Laboratories has been active in perceptual audio coding research and standardization, and Dolby researchers have made numerous scientific contributions within the collaborative framework of MPEG audio. On the commercial front, Dolby has developed the AC-2 and the AC-3 algorithms [Fiel91] [Fiel96].

10.7.1 Dolby AC-2, AC-2A

The AC-2 [Davi90] [Fiel91] is a family of single-channel algorithms operating at bit rates between 128 and 192 kb/s for 20 kHz bandwidth input sampled at 44.1 or 48 kHz. There are four available AC-2 variants, all of which share an architecture in which the input is mapped to the frequency domain by an evenly stacked TDAC filter bank [Prin86] with a novel parametric Kaiser-Bessel analysis window (Section 6.7) optimized for improved stop-band attenuation relative to the sine window. The evenly stacked TDAC differs from the oddly stacked MDCT in that the evenly stacked low-band filter is half-band, and its magnitude response wraps around the fold-over frequency (see Chapter 6). A unique mantissa-exponent coding scheme is applied to the TDAC transform coefficients. First, sets of frequency-adjacent coefficients are grouped into blocks (subbands) of roughly critical bandwidth. For each block, the maximum is identified and then quantized as an exponent in terms of the number of left shifts required until overflow occurs. The collection of exponents forms a stair-step spectral envelope having 6 dB (left shift = multiply by 2 = 6.02 dB) resolution, and normalizing the transform coefficients by the envelope generates a set of mantissas. The envelope approximates the short-time spectrum, and therefore a perceptual model uses the exponents to compute both a fixed and a signal-adaptive bit allocation for the mantissas on each frame.

As far as details on the four AC-2 variants are concerned, two versions are designed for low-complexity, low-delay applications, and the other two for higher quality at the expense of increased delay or complexity. In version 1, a 128-sample (64-channel) filter bank is used, and the coder operates at 192 kb/s per channel, resulting in high-quality output with only 7-ms delay at 48 kHz. Version 2 is also for low-delay applications with improved quality at the same bit rate, and it uses the same filter bank but exploits time redundancy across block pairs, thus increasing delay to 12 ms. Version 3 achieves similar quality with the reduced rate of 128 kb/s per channel at the expense of longer delay (45 ms) by using a 512-sample (256 channel) filter bank to improve steady-state coding gain. Finally, version 4 (the AC-2A [Davi92] algorithm) employs a switched 128/512-point TDAC filter bank to improve quality for transient signals while maintaining high coding gain for stationary signals. A 320-sample bridge window preserves PR filter bank properties during mode switching, and a transient detector consisting of an 8-kHz Chebyshev highpass filter is responsible for switching decisions. Order of magnitude peak level increases between 64-sample sub-blocks at the filter output are interpreted as transient events. The Kaiser window parameters used for the KBD windows in each of the AC-2 algorithms appeared in [Fiel96]. For all four algorithms, the AC-2 encoder multiplexes spectral envelope and mantissa parameters into an output bitstream, along with some auxiliary information. Byte-wide Reed-Solomon ECC allows for correction of single byte errors in the exponents at the expense of 1% overhead, resulting in good performance up to a BER of 0.001.

One AC-2 feature that is unique among the standards is that the perceptual model is backward adaptive, meaning that the bit allocation is not transmitted explicitly. Instead, the AC-2 decoder extracts the bit allocation from the quantized spectral envelope using the same perceptual model as the AC-2 encoder. This structure leads to a significant reduction of side information and induces a symmetric encoder/decoder complexity, which was well suited to the original AC-2 target application of single point-to-point audio transport. An example single point-to-point system now using low-delay AC-2 is the DolbyFAX™, a full-duplex codec that carries simultaneously two channels in both directions over four ISDN "B" links for film and TV studio distance collaboration. Low-delay AC-2 codecs have also been installed on 950-MHz wireless digital studio transmitter links (DSTL). The AC-2 moderate delay and AC-2A algorithms have been used for both network and wireless broadcast applications such as cable and direct broadcast satellite (DBS) television. The AC-2A is the predecessor to the now popular multichannel AC-3 algorithm. As the next section will show, the AC-3 coder has inherited and enhanced several facets of the AC-2/AC-2A architecture. In fact, the AC-2 encoder is nearly identical to (one channel of) the simplified AC-3 encoder shown in Figure 10.27, except that AC-2 does not transmit explicitly any perceptual model parameters.

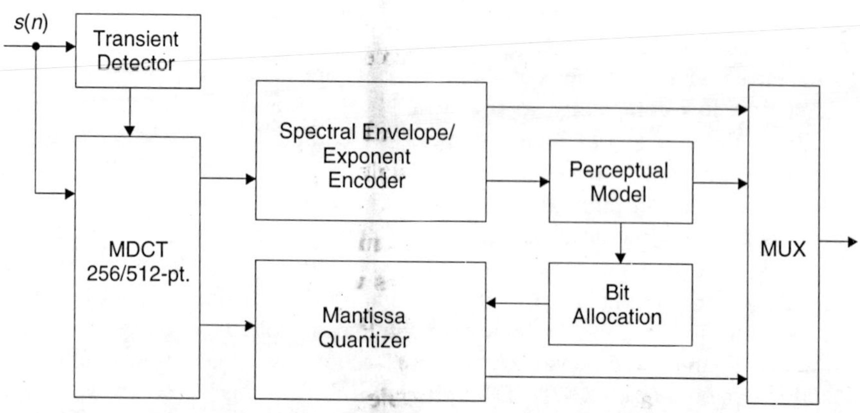

Figure 10.27. Dolby Laboratories AC-3 encoder.

10.7.2 Dolby AC-3/Dolby Digital/Dolby SR · D

The 5.1-channel "surround" format that had become the *de facto* standard in most movie houses during the 1980s was becoming ubiquitous in home theaters of the 1990s that were equipped with matrixed multichannel sound (e.g., Dolby ProLogic™). As a result of this trend, it was clear that emerging applications for perceptual coding would eventually minimally require stereophonic or even multichannel surround-sound capabilities to gain consumer acceptance. Although single-channel algorithms such as the AC-2 can run on parallel independent channels, significantly better performance can be realized by treating multiple channels together in order to exploit interchannel redundancies and irrelevancies. The Dolby Laboratories AC-3 algorithm [Davis93] [Todd94] [Davi98], also known as "Dolby Digital" or "SR · D," was developed specifically for multichannel coding by refining all of the fundamental AC-2 blocks, including the filter bank, the spectral envelope encoding, the perceptual model, and the bit allocation. The coder carries 5.1 channels of audio (left, center, right, left surround, right surround, and a subwoofer), but at the same time it incorporates a flexible downmix strategy at the decoder to maintain compatibility with conventional monaural and stereophonic sound reproduction systems. The ".1" channel is usually reserved for low-frequency effects, and is lowpass bandlimited below 120 Hz. The main features of the AC-3 algorithm are as follows:

- Sample rates: 32, 44.1, and 48 kHz
- Bit rates: 32–640 kb/s, variable
- High-quality output at 64 kb/s per channel
- Delay roughly 100 ms
- MDCT filter bank (oddly stacked TDAC [Prin87]), KBD prototype window
- MDCT coefficients quantized and encoded in terms of exponents, mantissas
- Spectral envelope represented by exponents
- Signal-adaptive exponent strategy with time-varying time-frequency resolution
- Hybrid forward-backward adaptive perceptual model
- Parametric bit allocation
- Uniform quantization of mantissas according to signal-adaptive bit allocation
- Perceptual model improvements possible at the encoder without changing decoder
- Multiple channels processed as an ensemble
- Frequency-selective intensity coding, as well as LR, MS
- Robust decoder downmix functionality from 5.1 to fewer channels
- Integral dynamic range control system
- Board-level real-time encoders available
- Chip-level real-time decoders available.

The AC-3 works in the following way. A signal-adaptive MDCT filter bank with a customized KBD window (Section 6.7) maps the input to the frequency domain. Long windows are applied during steady-state segments, and a pair of short windows is used for transient segments. The MDCT coefficients are quantized and encoded by an exponent/mantissa scheme similar to AC-2. Bit allocation for the mantissas is performed according to a perceptual model that estimates the masked threshold from the quantized spectral envelope. Like AC-2, an identical perceptual model resides at both the encoder and decoder to allow for backward adaptive bit allocation on the basis of the spectral envelope, thus reducing the burden of side information on the bitstream. Unlike AC-2, however, the perceptual model is also forward adaptive in the sense that it is parametric. Model parameters can be changed at the encoder and the new parameters transmitted to the decoder in order to affect modified masked threshold calculations. Particularly at lower bit rates, the perceptual bit allocation may yield insufficient bits to satisfy both the masked threshold and the rate constraint. When this happens, mid-side and intensity coding ("channel coupling" above 2 kHz) reduce the demand for bits by exploiting, respectively, interchannel redundancies and irrelevancies. Ultimately, exponents, mantissas, coupling data, and exponent strategy data are combined and transmitted to the receiver.

The remainder of this section provides details on the major functional blocks of the AC-3 algorithm, including the filter bank, exponent strategy, perceptual model, bit allocation, mantissa quantization, intensity coding, system-level functions, complexity, and applications and standardization activities.

10.7.2.1 Filter Bank Although the high-level AC-3 structure (Figure 10.27) resembles that of AC-2, there are significant differences between the two algorithms. Like AC-2, the AC-3 algorithm first maps input samples to the frequency domain using a PR cosine-modulated filter bank with a novel KBD window (Section 6.7 parameters given in [Fiel96]). Unlike AC-2, however, AC-3 is based on the oddly stacked MDCT. The AC-3 also handles window switching differently than AC-2A. Long, 512-sample (93.75 Hz resolution @ 48 kHz) windows are used to achieve reasonable coding gain during stationary segments. During transients, however, a pair of 256-sample windows replaces the long window to minimize pre-echoes. Also in contrast to the MPEG and AC-2 algorithms, the AC-3 MDCT filter bank retains PR properties during window switching without resorting to bridge windows by introducing a suitable phase shift into the MDCT basis vectors (Chapter 6, Eq. (6.38a) and (6.38b); see also [Shli97]) for one of the two short transforms. Whenever a scheme similar to the one used in AC-2A detects transients, short filter-bank windows may activate independently on any one or more of the 5.1 channels.

10.7.2.2 Exponent Strategy The AC-3 algorithm uses a refined version of the AC-2 exponent/mantissa MDCT coefficient representation, resulting in a significantly improved coding gain. In AC-3, the MDCT coefficients corresponding to 1536 input samples (six transform blocks) are combined into a single frame.

Then, a frame-processing routine optimizes the exponent representation to exploit temporal redundancy, while at the same time representing the stair-step spectral envelope with adequate frequency resolution. In particular, spectral envelopes are formed from partitions of either one, two, or four consecutive MDCT coefficients on each of the six MDCT blocks in the frame. To exploit time-redundancy, the six envelopes can be represented individually, or any or all of the six can be combined into temporal partitions. As in AC-2, the exponents correspond to the peak values of each time-frequency partition, and each exponent is represented with 6 dB of resolution by determining the number of left shifts until overflow. The overall exponent strategy is selected by evaluating spectral stability. Many strategies are possible. For example, all transform coefficients could be transmitted for stable spectra, but time updates might be restricted to 32-ms intervals, i.e., an envelope of single-coefficient partitions might be repeated five times to exploit temporal redundancy. On the other hand, partitions of two or four components might be encoded for transient signals, but the time-partition might be smaller, e.g., updates could occur for every 5.3-ms MDCT block. Regardless of the particular strategy in use for a given frame, exponents are differentially encoded across frequency. Differential coding of exponents exploits knowledge of the filter-bank transition band characteristics, thus avoiding slope overload with only a five-level quantization strategy. The AC-3 exponent strategy exploits in a signal-dependent fashion the time- and frequency-domain redundancies that exist on a frame of MDCT coefficients.

10.7.2.3 Perceptual Model A novel parametric forward-backward adaptive perceptual model estimates the masked threshold on each frame. The forward-adaptive component exists only at the encoder. Given a rate constraint, this block interacts with an iterative rate control loop to determine the best set of perceptual model parameters. These parameters are passed to the backward-adaptive component, which estimates the masked threshold by applying the parameters from the forward-adaptive component to a calculation involving the quantized spectral envelope. Identical backward-adaptive model components are embedded in both the encoder and decoder. Thus, model parameters are fixed at the encoder after several threshold calculations in an iterative rate control process, and then transmitted to the decoder. The decoder only needs to perform one threshold calculation given the parameter values established at the encoder.

The backward-adaptive model component works as follows. First, the quantized spectral envelope exponents are clustered into 50, 0.5-Bark-width subbands. Then, a spreading function is applied (Figure 10.28a) that accounts only for the upward spread of masking. To compensate for filter-bank leakage at low frequencies, spreading is disabled below 200 Hz. Also, spreading is not enabled between 200 and 700 Hz for frequencies below the occurrence of the first significant masker. The absolute threshold of hearing is accounted for after the spreading function has been applied. Unlike other algorithms, AC-3 neglects the downward spread of masking, assumes that masking power is nonadditive, and makes no explicit assumptions about the relationship between tonality and the skirt slopes on the spreading function. Instead, these characteristics are captured in a set of parameters that comprise the forward-adaptive model component. Masking threshold calculations at the decoder are controlled by a set of parameters transmitted by the encoder, creating flexibility for model improvements at the encoder such that improved estimates of the masked threshold can be realized without modifying the embedded model at the decoder.

For example, a parametric (upwards only) spreading function is defined (Figure 10.28a) in terms of two slopes, S_i, and two level offsets, L_i, for $i \in [1, 2]$. While the parameters S_1 and L_1 can be uniquely specified for each channel, the parameters S_2 and L_2 are applied to all channels. The parametric spreading function is advantageous in that it allows the perceptual model at the encoder

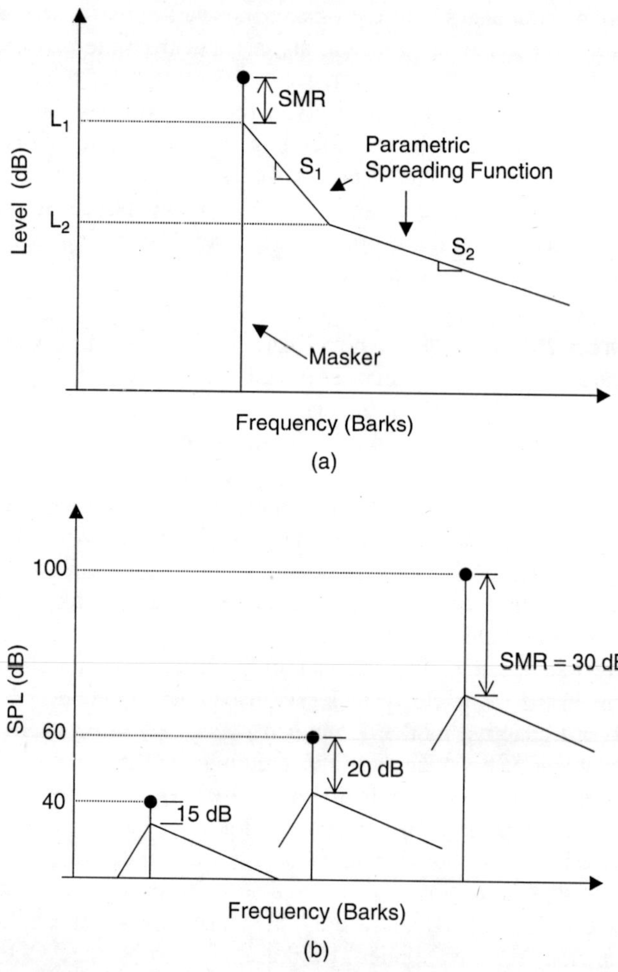

Figure 10.28. Dolby AC-3 parametric perceptual model: (a) prototype spreading function, (b) masker SMR level-dependence.

to account for tonality or dynamic masking patterns without the need to alter the decoder model. A range of values is available for each parameter. With units of dB per 1/2 Bark, the slopes are defined to be within the ranges $-2.95 \leqslant S_1 \leqslant -5.77$, and $-0.7 \leqslant S_2 \leqslant -0.98$. With units of dB SPL, the levels are defined to be within the ranges $-6 \leqslant L_1 \leqslant -48$ and $-49 \leqslant L_2 \leqslant -63$. Ultimately, there are 512 unique spreading function shapes to choose from. The acoustic-level dependence of masking thresholds is also modeled in AC-3. It is in general true that the signal-to-mask ratio (SMR) increases with increasing stimulus level (Figure 10.28b), i.e., the threshold moves closer to the stimulus as the stimulus intensity decreases. In the AC-3 parametric perceptual model, this phenomenon is captured by adding a positive bias to the masked thresholds when the spectral envelope is below a threshold level. Acoustic level threshold biasing is applied on a band-by-band basis. The decision threshold for the biasing is one of the forward adaptive parameters transmitted by the encoder. This function can also be disabled altogether. The parametric perceptual model also provides a convenient upgrade path in the form of a bit allocation delta parameter.

It was envisioned that future, more sophisticated AC-3 encoders might run in parallel two perceptual models, with one being the original reference model, and the other being an enhanced model with more accurate estimates of masked threshold. The delta parameter allows the encoder to transmit a stair-step function for which each tread specifies a masking level adjustment for an integral number of 1/2-Bark bands. Thus, the masking model can be incrementally improved without alterations to the existing decoders. Other details on the hybrid backward-forwards AC-3 perceptual model can be found in [Davi94].

10.7.2.4 Bit Allocation and Mantissa Quantization A bit allocation is determined at the encoder for each frame of mantissas by an iterative procedure that adjusts the mantissa quantizers, the multichannel coding strategies (below), and the forward-adaptive model parameters to satisfy simultaneously the specified rate constraint and the masked threshold. Within the rate-control loop, threshold partitions are formed on the basis of a bit allocation frequency resolution parameter, with coefficient partitions ranging in width between 94 and 375 Hz. In a manner similar to MPEG-1, quantizers are selected for the set of mantissas in each partition based on an SMR calculation. Sufficient bits are allocated to ensure that the SNR for the quantized mantissas is greater than or equal to the SMR. The quantization noise is thus rendered inaudible, below masked threshold. Uniform quantizers are selected from a set of 15 having 0, 3, 5, 7, 11, and 15 levels symmetric about 0, and conventional 2's-complement quantizers having 32, 64, 128, 256, 512, 1024, 2048, 4096, 16384, or 65536 levels. Certain quantizer codewords are group-encoded to make more efficient usage of available bits. Dithering can be enabled optionally on individual channels for 0-bit mantissas. If the bit supply is insufficient to satisfy the masked threshold, then SNRs can be reduced in selected threshold partitions until the rate is satisfied, or intensity coding and MS transformations are used in a frequency-selective fashion to reduce the bit demand. Two variable-rate methods are also available to satisfy peak-rate demands. Within a frame of six MDCT coefficient blocks, bits can be distributed unevenly, such that the instantaneous bit rate is variable but the average rate is constant. In addition, bit rates are adjustable, and a unique rate can be specified for each frame of six MDCT blocks. Unlike some of the other standardized algorithms, the AC-3 does not include an explicit lossless coding stage for final redundancy reduction after quantization and encoding.

10.7.2.5 Multichannel Coding When bit demand imposed by multiple independent channels exceeds the bit budget, the AC-3 ensemble processing of 5.1 channels exploits interchannel redundancies and irrelevancies, respectively, by making frequency-selective use of mid-side (MS) and intensity coding techniques. Although the MS and intensity functions can be simultaneously active on a given channel, they are restricted to nonoverlapping subbands. The MS scheme is carefully controlled [Davi98] to maintain compatibility between AC-3 and matrixed surround systems such as Dolby ProLogic. Intensity coding, also known as "channel coupling," is a multichannel irrelevancy reduction coding technique that exploits properties of spatial hearing. There is considerable experimental evidence [Blau74] suggesting that the interaural time difference of a signal's fine structure has negligible influence on sound localization above a certain frequency. Instead, the ear evaluates primarily energy envelopes. Thus, the idea behind intensity coding is to transmit only one envelope in place of two or more sufficiently correlated spectra from independent channels, together with some side information. The side information consists of a set of coefficients that is used to recover individual spectra from the intensity channel.

A simplified version of the AC-3 intensity coding scheme is illustrated in Figure 10.29. At the encoder (Figure 10.29a), two or more input spectra are added together to form a single intensity channel. Prior to the addition, an optional adjustment is applied to prevent phase cancellation. Then, groups of adjacent coefficients are partitioned into between 1 and 18 separate intensity subbands on both the individual and the intensity channels. A set of coupling coefficients is computed, c_{ij}, that expresses the fraction of energy contributed by the i-th individual channel to the j-th band of the intensity envelope, i.e., $c_{ij} = \beta_{ij}/\alpha_j$, where β_{ij} is the power contained in the j-th band of the i-th channel, and α_j is the power contained in the j-th band of the intensity channel. Finally, the intensity spectrum is quantized, encoded, and transmitted to the decoder. The coupling coefficients, c_{ij}, are transmitted as side information. Once the intensity channel has been recovered at the decoder (Figure 10.29b), the intensity subbands are scaled by the coupling coefficients, c_{ij}, in order to recover an appropriate fraction of intensity energy in the j-th band of the i-th channel. The intensity-coded coefficients are then combined with any remaining uncoupled transform coefficients and passed through the synthesis filter bank to reconstruct the individual channel. The AC-3 coupling coefficients have a dynamic range that spans -132 to $+18$ dB, with quantization step sizes between 0.28 and 0.53 dB. Intensity coding is applied in a frequency-selective manner, parameterized by a start frequency of 3.42 kHz or higher, and a bandwidth expressed in multiples of 1.2 kHz for

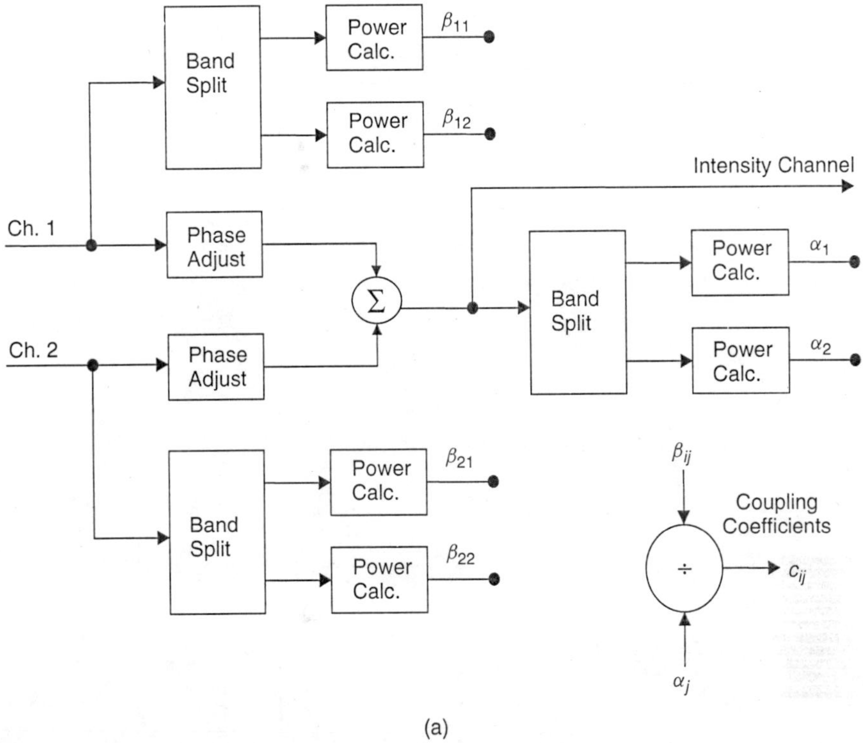

Figure 10.29. Dolby AC-3 intensity coding ("channel coupling"): (a) encoder, (b) decoder.

a 48-kHz system [Davi98]. Note that unlike the simplified system shown in the figure, the actual AC-3 intensity coding scheme may couple the spectra from as many as five channels.

10.7.2.6 System-Level Functions At the system level, AC-3 provides mechanisms for channel down-mixing and dynamic range control. Down-mix capability is essential for the 5.1-channel system since the majority of potential playback systems are still monaural or, at best, stereophonic. Down-mixing is performed at the decoder in the frequency domain rather than the time-domain to reduce complexity. This is possible because of the filter-bank linearity. The bit stream carries some down-mix information since different listening situations call for different down-mix weighting. Dialog-level normalization is also available at the decoder. Finally, the bit stream has available facilities to handle other control and ancillary user information such as copyright, language, production, and time-code data [Davis94].

10.7.2.7 Complexity Assuming the standard HDTV configuration of 384 kb/s with a 48 kHz sample rate and implementation using the Zoran ZR38001 general-purpose DSP instruction set, the AC-3 decoder memory requirements and complexity

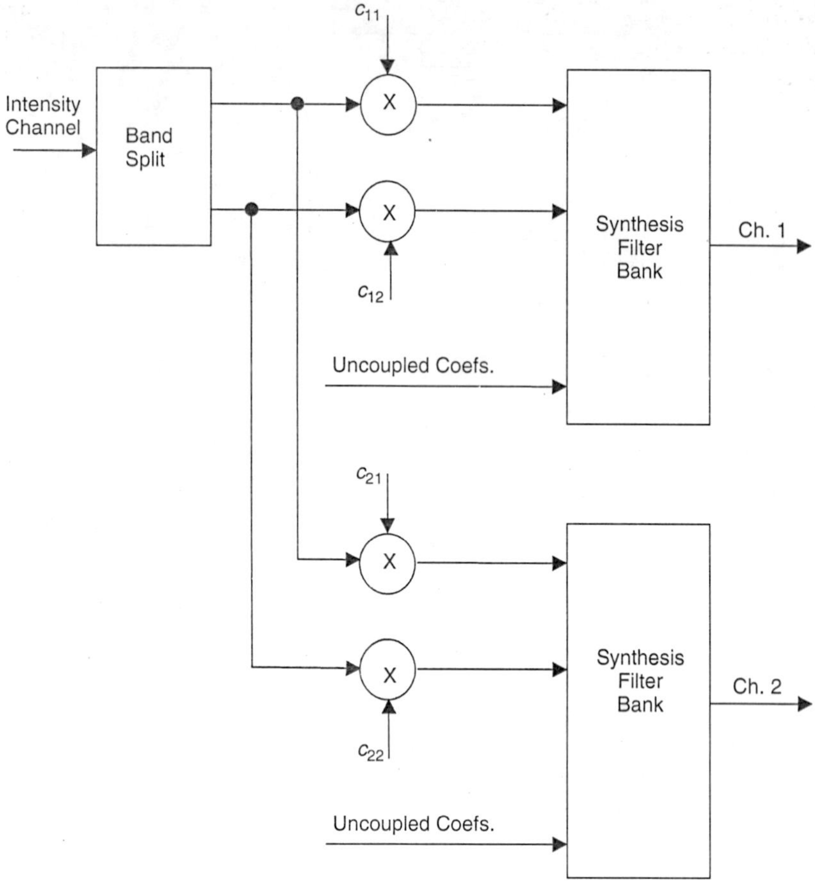

(b)

Figure 10.29. (*continued*).

are as follows: 6.6 kbytes RAM, 5.4 kbytes ROM, 27.3 MIPS for 5.1 channels, and 3.1 kbytes RAM, 5.4 kbytes ROM, and 26.5 MIPS for 2 channels [Vern95]. Note that complexity estimates are processor-dependent. For example, on a Motorola DSP56002, 45 MIPS are required for a 5.1-channel decoder. Encoder complexity varies between two and five times decoder complexity depending on the encoder sophistication [Vern95]. Numerous real-time encoder and decoder implementations have been reported. Early on, for example, a single-chip decoder was implemented on a Zoran DSP [Vern93]. More recently, a DP561 AC-3 encoder (5.1 channels, 44.1- or 48-kHz sample rate) for DVD mastering was implemented in real-time on a PC host with a plug-in DSP subsystem. The computational requirements were handled by an Ariel PC-Hydra DSP array of eight Texas Instruments TMS 320C44 floating-point DSP devices clocked at 50 MHz [Terr96]. Current information on real-time AC-3 implementations is also available online from Dolby Laboratories.

10.7.2.8 Applications and Standardization The first popular AC-3 application was in the cinema. The “Dolby Digital” or “SR D” AC-3 information is interleaved between sprocket holes on one side of the 35-mm film. The AC-3 was first deployed in only three theaters for the film *Star Trek VI* in 1991, after which the official rollout of Dolby SR D occurred in 1992 with *Batman Returns*. By 1997 April, over 900 film soundtracks had been AC-3 encoded. Nowadays, the AC-3 algorithm is finding use in digital versatile disc (DVD), cable television (CATV), and direct broadcast satellite (DBS). Many hi-fidelity amplifiers and receiver units now contain embedded AC-3 decoders and accept an AC-3 digital rather than an analog feed from external sources such as DVD.

In addition, the DP504/524 version of the DolbyFAX system (Section 10.7.1) has added AC-3 stereo and MPEG-1 layer II to the original AC-2-based system. Film, television, and music studios use DolbyFAX over ISDN links for automatic dialog replacement, music collaboration, sound effects delivery, and remote videotape audio playback. As far as standardization is concerned, the United States Advanced Television Systems Committee (ATSC) has adopted the AC-3 algorithm as the A/52 audio compression standard [USAT95b] and as the audio component of the A/52 Digital Television (DTV) Standard [USAT95a]. The United States Federal Communications Commission (US FCC) in 1996 December adopted the ATSC standard for DTV, including the AC-3 audio component. On the international standardization front, the Digital Audio-Visual Council (DAVIC) selected AC-3 and MPEG-1 layer II for the audio component of the DAVIC 1.2 specification [DAVC96].

10.7.2.9 Recent Developments – The Dolby Digital Plus A Dolby digital plus system or the enhanced AC-3 (E-AC-3) [Fiel04] was recently introduced to extend the capabilities of the Dolby AC-3 algorithm. While remaining backward compatible with the Dolby AC-3 standard, the Dolby digital plus provides several enhancements. Some of the extensions include flexibility to encode up to 13.1 channels, extended data rates up to 6.144 Mb/s. The AC-3 filterbank is supplemented with a second stage DCT to exploit the stationary characteristics in the audio. Other coding tools include spectral extension, enhanced channel coupling, and transient pre-noise processing. The E-AC-3 is used in cable and satellite television set-top boxes and broadcast distribution transcoding devices. For a detailed description on the Dolby digital plus refer to [Fiel04].

10.8 AUDIO PROCESSING TECHNOLOGY APT-x100

Without exception, all of the commercial and international audio coding standards described thus far couple explicit models of auditory perception with classical quantization techniques in an attempt to distribute quantization noise over the time-frequency plane such that it is imperceptible to the human listener. In addition to irrelevancy reduction, most of these algorithms simultaneously seek to reduce statistical redundancies. For the sake of comparison and perhaps to better assess the impact of perceptual models on realizable coding gain, it is instructive to next consider a commercially available audio coding algorithm that relies only upon redundancy removal without any explicit regard for auditory perception.

We turn to the Audio Processing Technology APT-x100 algorithm, which has been reported to achieve nearly transparent coding of CD-quality 44.1 kHz 16-bit PCM input at a compression ratio of 4:1, or 176.4 kb/s per monaural channel [Wyli96b]. Like the ITU-T G.722 wideband speech codec [G722], the APT-x100 encoder (Figure 10.30) relies upon subband signal decomposition followed by independent ADPCM quantization of the decimated subband output sequences. Codewords from four uniform bandwidth subbands are multiplexed onto the channel and sent to the decoder where the ADPCM and filter-bank operations are inverted to generate an output. As shown in the figure, a tree-structured QMF filter bank splits the input signal into four subbands. The first and second filter stages have 64 and 32 taps, respectively. Backward adaptive prediction is applied to the four subband output sequences. The resulting prediction residual is quantized with a backward-adaptive Laplacian quantizer. Backward adaptation in the prediction and quantization steps eliminates side information but increases sensitivity to fast transients. On the other hand, both prediction and adaptive quantization were found to significantly improve coding gain for a wide range of test signals [Wyli96b]. Adaptive quantization attempts to track signal dynamics and tends to produce constant SNR in each subband during stationary segments.

Unlike the other algorithms reviewed in this document, APT-x100 contains no perceptual model or rate control loop. The ADPCM output codewords are of fixed resolution (1 bit per sample), and therefore with four subbands the output bit rate is reduced 4:1. A comparison between APT-x100 quantization noise and

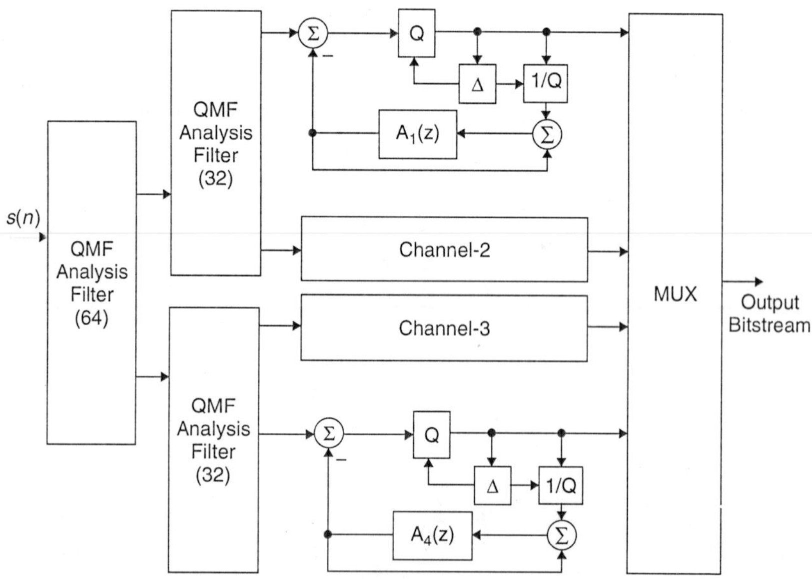

Figure 10.30. Audio Processing Technology APT-x100 encoder.

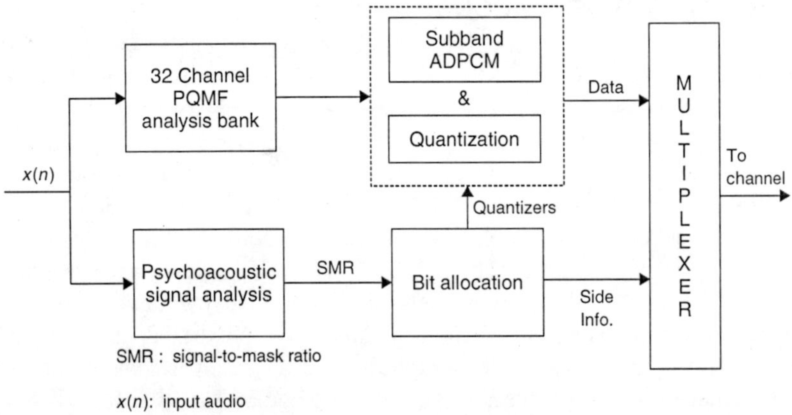

Figure 10.31. DTS-coherent acoustics (DTS-CA) encoding scheme.

noise masking thresholds computed as in [John88a] for a variety of test signals from the SQAM test CD [SQAM88] revealed two trends in the APT-x100 noise floor. First, as expected, it is flat rather than shaped. Second, the noise is below the masking threshold in most critical bands for most stationary test signals, but tends to exceed the threshold in some critical bands for transient signals. In [Wyli96b], however, the fast step-size adaptation in APT-x100 is reported to exploit temporal masking effects and mitigate the audibility of unmasked quantization noise. While the lack of a perceptual model results in an inefficient flat noise floor, it also affords some advantages including reduced complexity, reduced frequency resolution requirements, and low delay of only 122 samples or 2.77 ms at 44.1 kHz.

Several other relevant facts on APT-x100 quality and robustness were also reported in [Wyli96b]. Objective output quality was evaluated in terms of average subband SNRs, which were 30, 15, 10, and 7 dB, respectively, for the lowest to highest subbands, and the authors stated that the algorithm outperformed NICAM [NICAM] in an informal subjective comparison [Wyli96b]. APT-x100 was robust to both random bit errors and tandem encoding. Errors were inaudible for a bit error rate (BER) of 10^{-4}, and speech remained intelligible for a BER of 10^{-1}. In one test, 10 stages of synchronous tandeming reduced output SNR from 45 dB to 37 dB. An auxiliary channel that accommodates up to 1/4 kb/s of the sample rate in buried data (e.g., 24 kb/s for 48-kHz stereo samples) by bit stealing from one of the subbands had a negligible effect on output quality. Finally, real-time APT-x100 encoder and decoder modules were implemented on a single AT&T DSP16A masked ROM DSP. As far as applications are concerned, APT-x100 has been deployed in digital studio-transmitter links, audio storage products, and cinematic surround sound applications. A cursory performance comparison of the nonperceptual algorithms versus the perceptually based algorithms (e.g., NICAM or APT-x100 vs. MPEG or PAC, etc.) confirms that some awareness of peripheral auditory processing is necessary to achieve high-quality coding of CD-quality audio for compression ratios in excess of 4:1.

10.9 DTS – COHERENT ACOUSTICS

The performance comparison of the nonperceptual algorithms versus the perceptually based algorithms (e.g., APT-x100 vs. MPEG or PAC, etc.) given in the earlier section, highlights that some awareness of *peripheral auditory processing* is necessary to achieve high-quality encoding of digital audio for compression ratios in excess of 4:1. To this end, DTS employs an audio compression algorithm based on the principles of "coherent acoustics encoding" [Smyt96] [Smyt99] [DTS]. In coherent acoustics, both ADPCM-subband filtering and psychoacoustic analysis are employed to compress the audio data. The main emphasis in DTS is to improve the precision (and, hence, the quality) of the digital audio. The DTS encoding algorithm provides a resolution of up to 24 bits per sample and at the same time can deliver compression rates in the range of 3 to 40. Moreover, DTS can deliver up to eight discrete channels of multiplexed audio at sampling frequencies of 8–192 kHz and at bit rates of 8–512 kb/s per channel. Table 10.9 summarizes the various bit rates, sampling frequencies, and the bit resolutions employed in the four configurations supported by the DTS-coherent acoustics.

10.9.1 Framing and Subband Analysis

The DTS-CA encoding algorithm (Figure 10.31) operates on 24-bit linear PCM signals. The audio signals are typically analyzed in blocks (frames) of 1024 samples, although frame sizes of 256, 512, 2048, and 4096 samples are also supported depending on the bit rates and sampling frequencies used (Table 10.10). For example, if operating at bit rates of 1024–2048 kb/s and sampling frequencies of 32 or 44.1 or 48 kHz; then the maximum number of samples allowed per frame is 1024. Next, the segmented audio frames are decomposed into 32 critically subsampled subbands using a polyphase realization of a pseudo QMF (PQMF) bank (Chapter 6). Two different PQMF filter-bank structures, namely, perfect reconstructing (PR) and nonperfect reconstructing (NPR) are provided in DTS-coherent acoustics. In the example that we considered above, a frame size of 1024 samples results in 32 PCM samples per subband (i.e., 1024/32). The channels are equally spaced such that a 32 kHz input signal is split into 500 Hz subbands (i.e., 16 kHz/32), with the subbands being decimated at the ratio 32:1.

Table 10.9. A list of encoding parameters used in DTS-coherent acoustics after [Smyt99].

Bit rates (kb/s/channel)	Sampling rates (kHz)	Bit resolution per sample
8–32	$\leqslant$24	16
32–96	$\leqslant$48	20
96–256	$\leqslant$96	24
256–512	$\leqslant$192	24

Table 10.10. Maximum frame sizes allowed in DTS-CA (after [Smyt99]).

	Sampling frequency-set, $f_s = [8/11.05/12]$ (kHz)				
Bit rates (kb/s)	f_s	$2f_s$	$4f_s$	$8f_s$	$16f_s$
0–512	Max. 1024	Max. 2048	Max. 4096	N/A	N/A
512–1024	N/A	Max. 1024	Max. 2048	N/A	N/A
1024–2048	N/A	N/A	Max. 1024	Max. 2048	N/A
2048–4096	N/A	N/A	N/A	Max. 1024	Max. 2048

10.9.2 Psychoacoustic Analysis

While the subband filtering stage minimizes the statistical dependencies associated with the input PCM signal, the psychoacoustic analysis stage eliminates the perceptually irrelevant information. Since we have already established the necessary background on psychoacoustic analysis in Chapters 5, we will not elaborate on these steps. However, we describe next the advantages of combining the differential subband coding techniques (e.g., ADPCM) with the psychoacoustic analysis.

10.9.3 ADPCM – Differential Subband Coding

A block diagram depicting the steps involved in the differential subband coding in DTS-CA is shown in Figure 10.32. A fourth-order forward linear prediction is performed on each subband containing 32 PCM samples. From the above example, we have 32 subbands and 32 PCM samples per subband. Recall that in LP we predict the current time-domain audio sample based on a linearly weighted combination of previous samples. From the LPC analysis corresponding to the i-th subband, we obtain predictor coefficients, $a_{i,k}$ for $k = 0, 1, \ldots, 4$ and the residual error, $e_i(n)$ for $n = 0, 1, 2, \ldots, 31$ samples. The predictor coefficients are usually vector quantized in the line spectral frequency (LSF) domain.

Two stages of ADPCM modules are provided in the DTS-CA algorithm, i.e., the *ADPCM estimation stage* and the *real ADPCM stage*. ADPCM utilizes the redundancy in the subband PCM audio by exploiting the correlation between adjacent samples. First, the "estimation ADPCM" module is used to determine the degree of prediction achieved by the fourth-order linear prediction filter (Figure 10.32). Depending upon the statistical features of audio, a decision to enable or disable the second "real ADPCM" stage is made.

A predictor mode flag, "PMODE" = 1 or 0, is set to indicate if the "real ADPCM" module is active or not, respectively.

$$s_{i,pred}(n) = \sum_{k=0}^{4} a_{i,k} s_i(n-k), \text{ for } n = 0, 1, \ldots, 31 \qquad (10.11)$$

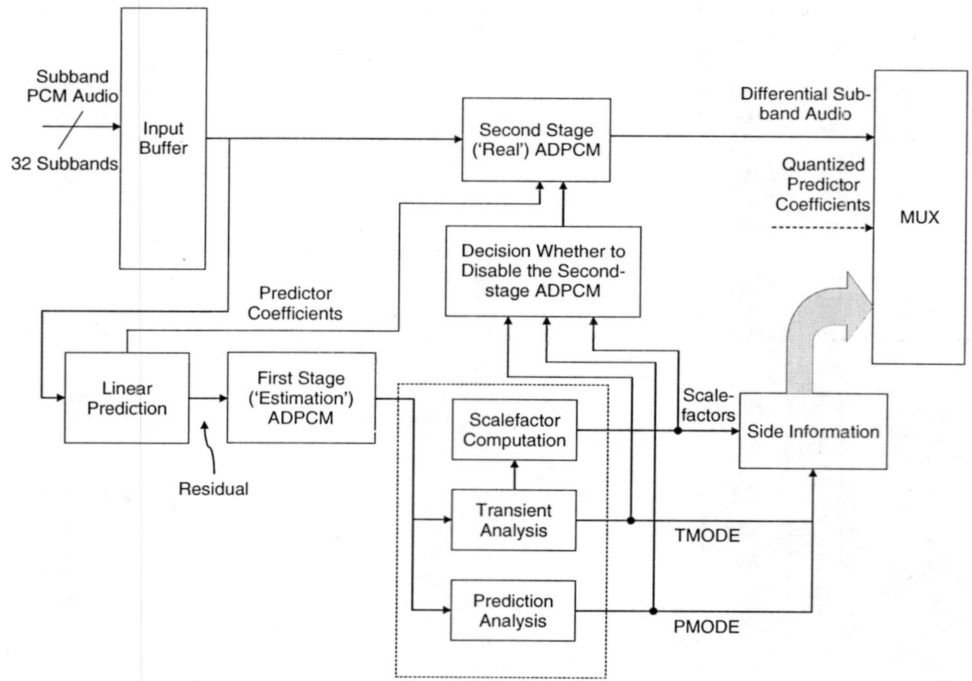

Figure 10.32. Differential subband (ADPCM) coding in the DTS-CA encoder.

$$
\begin{aligned}
e_i(n) &= s_i(n) - s_{i,pred}(n) \\
&= s_i(n) - \sum_{k=0}^{4} a_{i,k} s_i(n-k)
\end{aligned}
\tag{10.12}
$$

While the "prediction analysis" block computes the PMODE flag based on the prediction gain, the "transient analysis" module monitors the transient behavior of the error residual. In particular, when a signal with a sharp attack (i.e., rapid transitions) begins near the end of a transform block and immediately following a region of low energy, pre-echoes occur. Several pre-echo control strategies have been developed (Chapter 6, Section 6.10). These include window switching, gain modification, switched filter banks, including the bit reservoir, and temporal noise shaping. In DTS-CA, the pre-echo artifacts are controlled by dividing the subband analysis buffer into four sub-buffers. A transient mode, "TMODE" = 0, 1, 2, or 3, is set to denote the beginning of a transient signal in sub-buffers 1, 2, 3, or 4, respectively. In addition, two scale factors are computed for each subband (i.e., before and after the transition) based on the peak magnitude of the residual error, $e_i(n)$. A 64-level nonuniform quantizer is usually employed to encode the scale factors in DTS-CA.

Note that the PMODE flag is a "Boolean" and the TMODE flag has four values. Therefore, a total of 15 bits (i.e., 12 bits for two scale factors, 1 bit for the "PMODE" flag, and 2 bits for the "TMODE" flag) are sufficient to encode the entire *side information* in the DTS-CA algorithm. Next, based on the predictor mode flag (1 or 0), the second-stage ADPCM is used to encode the differential subband PCM audio as shown in Figure 10.32. The optimum number of bits (in the sense of minimizing the quantization noise) required to encode the differential audio in each subband is estimated using a bit allocation procedure.

10.9.4 Bit Allocation, Quantization, and Multiplexing

Bit allocation is determined at the encoder for each frame (32 subbands) by an iterative procedure that adjusts the scale-factor quantizers, the fourth-order linear predictive model parameters, and the quantization levels of the differential subband audio. This is done in order to satisfy simultaneously the specified rate constraint and the masked threshold. In a manner similar to MPEG-1, quantizers are selected in each subband based on an SMR calculation. A sufficient number of bits is allocated to ensure that the SNR for the quantized error is greater than or equal to the SMR. The quantization noise is thus rendered inaudible, i.e., below the masked threshold. Recall that the main emphasis in DTS-CA is to improve precision and hence the quality of the digital audio, while giving relatively less importance to minimizing the data rate. Therefore, the DTS-CA bit reservoir will almost always meet the bit demand imposed by the psychoacoustic model. Similar to some of the other standardized algorithms (e.g., MPEG codecs, lossless audio coders), the DTS-CA includes an explicit lossless coding stage for final redundancy reduction after quantization and encoding. A data multiplexer merely packs the differential subband data, the side information, the synchronization details, and the header syntax into a serial bitstream. Details on the structure of the "output data frame" employed in the DTS-CA algorithm are given in [Smyt99] [DTS].

As an extension to the current coherent acoustics algorithm, Fejzo *et al.* proposed a new enhancement that delivers 96 kHz, 24-bit resolution audio quality [Fejz00]. The proposed enhancement makes use of both "core" and "extension" data to reproduce 96-kHz audio bitstreams. Details on the real-time implementation of the 5.1-channel decoder on a 32-bit floating-point processor are also presented in [Fejz00]. Although much work has been done in the area of encoder/decoder architectures for the DTS-CA codecs, relatively little has been published [Mesa99].

10.9.5 DTS-CA Versus Dolby Digital

The DTS-Coherent Acoustics and the Dolby AC-3 algorithms were the two competing standards during the mid-1990s. While the former employs an adaptive differential linear prediction (ADPCM–subband coding) in conjunction with a perceptual model, the latter employs a unique exponent/mantissa MDCT coefficient encoding technique in conjunction with a parametric forward-backward adaptive perceptual model.

PROBLEMS

10.1. List some of the primary differences between the DTS, the Dolby digital, and the Sony ATRAC encoding schemes.

10.2. Using a block diagram, describe how the ISO/IEC MPEG-1 layer I codec is different from the ISO/IEC MPEG-1 layer III algorithm.

10.3. What are the enhancements integrated into MPEG-2 AAC relative to the MP3 algorithm. State key differences in the algorithms.

10.4. List some of the distinguishing features of the MP4 audio format over the MP3 format. Give bitrates and cite references.

10.5. What is the main idea behind the scalable audio coding? Explain using a block diagram. Give examples.

10.6. What is structured audio coding and parametric audio coding?

10.7. How is ISO/IEC MPEG-7 standard different from the other MPEG standards.

COMPUTER EXERCISE

10.8. The audio files *Ch10aud2L.wav*, *Ch10aud2R.wav*, *Ch10aud2C.wav*, *Ch10aud2Ls.wav*, and *Ch10aud2Rs.wav* correspond to left, right, center, left-surround, and right-surround, respectively, of a 3/2-channel configuration. Using the matrixing technique obtain a stereo output.

CHAPTER 11

LOSSLESS AUDIO CODING AND DIGITAL WATERMARKING

11.1 INTRODUCTION

The emergence of high-end storage formats such as the DVD-audio and the super-audio CD (SACD) created opportunities for lossless audio coding schemes. Lossless audio coding techniques yield high-quality audio without any artifacts. Lossy audio coding (LAC) results, typically, in compression ratios of 10:1–25:1, while the lossless audio coding (L^2AC) algorithms achieve compression ratios of 2:1–4:1. Therefore, L^2AC techniques are not typically used in real-time storage/multimedia processing and Internet streaming but can be of use in storage-rich formats. Several lossless audio coding algorithms, including the SHORTEN [Robi94], the DVD [Crav96] [Oome96] [Crav97], the MUSICompress [Wege97], the AudioPaK [Hans98a] [Hans98b] [Hans01], the lossless transform coding of audio (LTAC) [Pura97], and the IntMDCT [Geig01] [Geig02] have been proposed. The meridian lossless packing (MLP) [Gerz99] and the direct-stream digital (DSD) techniques [Brue97] form a group of high-end lossless compression algorithms that have already become standards in the DVD-Audio [DVD01] and the SACD [SACD02] storage formats, respectively. In Figure 11.1, we taxonomize various lossless audio coding algorithms.

This chapter is organized as follows. In Section 11.2, we describe the principles of lossless audio coding. A survey of various lossless audio coding algorithms is given in Section 11.2.2. This is followed by more detailed descriptions of the DVD-audio (Section 11.3) and the SACD (Section 11.4) formats. Section 11.5 addresses

Audio Signal Processing and Coding, by Andreas Spanias, Ted Painter, and Venkatraman Atti

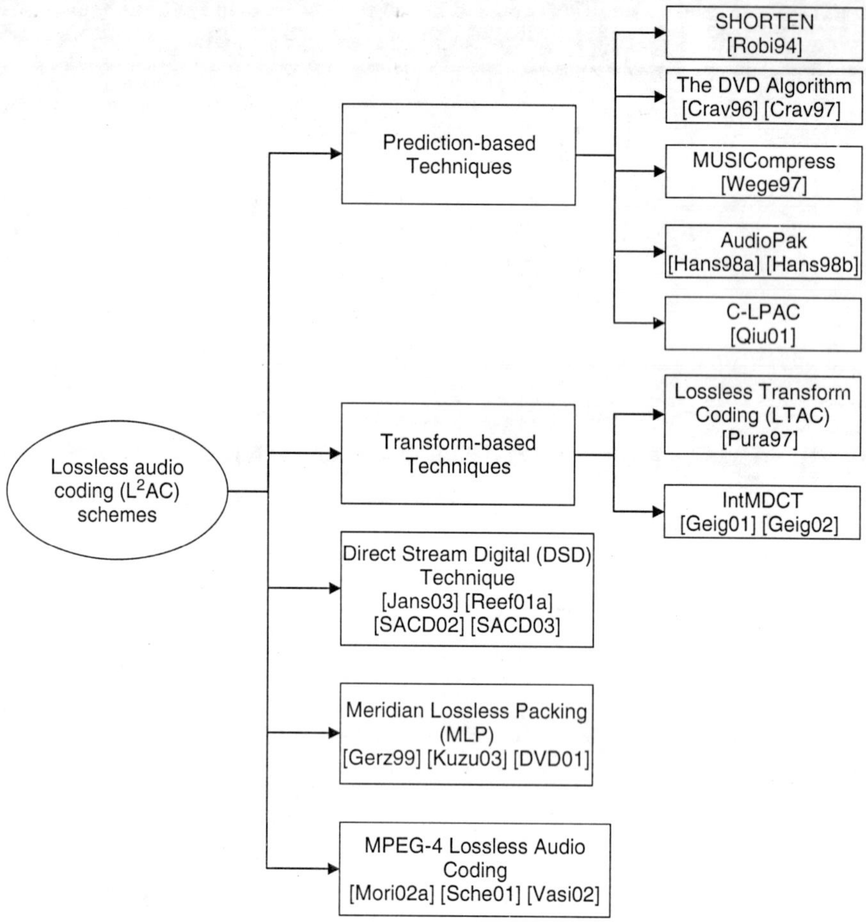

Figure 11.1. A list of lossless audio coding algorithms.

digital audio watermarking. Finally, in Section 11.6, we present commercial applications and recent developments in audio coding standardization.

11.2 LOSSLESS AUDIO CODING (L^2AC)

LAC schemes employ psychoacoustic principles in conjunction with time-frequency mapping techniques to eliminate perceptually irrelevant information. Hence, the encoded signal is not be an exact replica of the input audio and might contain some artifacts. Lossless coding schemes (L^2AC) obtain bit-exact compression by eliminating the statistical dependencies associated with the signal via prediction techniques. Several prediction modeling methods for L^2AC have been proposed, namely, FIR/IIR prediction [Robi94] [Crav96] [Crav97], polynomial approximation/curve fitting methods [Robi94] [Hans98b], transform coding [Pura97] [Geig02], high-order context modeling [Qiu01], backward adaptive prediction method [Angu97], subband linear prediction [Giur98], and set partitioning [Raad02]. L^2AC algorithms are broadly classified into two categories, namely, prediction-based and transform-based.

Prediction-based L^2AC algorithms employ linear predictive (LP) modeling or some form of polynomial approximation to remove redundancies in the waveform. The SHORTEN, the DVD, the MUSICompress, and the C-LPAC use LP analysis, while the AudioPaK is based on curve-fitting methods. Transform-based coding schemes employ specific transforms to decorrelate samples within a frame. The LTAC scheme proposed by Purat *et al.* and the IntMDCT scheme introduced by Geiger *et al.* are two L^2AC schemes that use transform coding.

In all the aforementioned L^2AC methods, the idea is to obtain a close approximation to the input audio and compute a low variance prediction residual that can be coded efficiently. Typically, the residual error is entropy coded using one of the following methods: Huffman coding, Lempel-Ziv coding, run-length coding, arithmetic coding, or Rice coding. Table 11.1 lists the various L^2AC coders and their associated prediction and entropy coding methods.

Before getting into the details of lossless coding of digital audio, let us take a look at some of the lossless *data* compression algorithms, i.e., PkZip, WinZip, WinRAR, gzip, and TAR. Although these compression algorithms achieve a 40–60% bit rate reductions with data and text files, they only achieve a mere 10% in the case of audio. [Hans01] lists an interesting example that illustrates the following. A PostScript file of 4.56 MB can be compressed with PkZip to 696 KB, achieving a bit rate reduction of about 85%, whereas an audio file of 33.72 MB can be compressed to 31.57 MB, resulting in a bit-rate reduction of only about 6.5%. In designing and evaluating L^2AC algorithms, one must address two key issues, i.e., the trade-offs of compression ratio and average residual energy per frame and the selection of the entropy coding method.

11.2.1 L^2AC Principles

A general block diagram of a L^2AC algorithm is depicted in Figure 11.2. First, a conventional lossy audio coding scheme (e.g., MPEG-1 layer III) is used to produce a compressed bitstream. Then, an error signal is calculated by subtracting

Table 11.1. Prediction and entropy coding schemes in some of the L^2AC schemes.

L^2AC scheme	Prediction model	Entropy coding used
SHORTEN	FIR prediction filter	Rice coding
DVD	IIR prediction	Huffman coding
MUSICompress	Adaptively varied approximations	Huffman coding
AudioPak	Curve fitting/polynomial approximations	Golomb coding
LTAC	Orthonormal transform coding	Rice coding
IntMDCT	Integer transform coding	Huffman coding
C-LPAC	FIR prediction	High-order context modeling

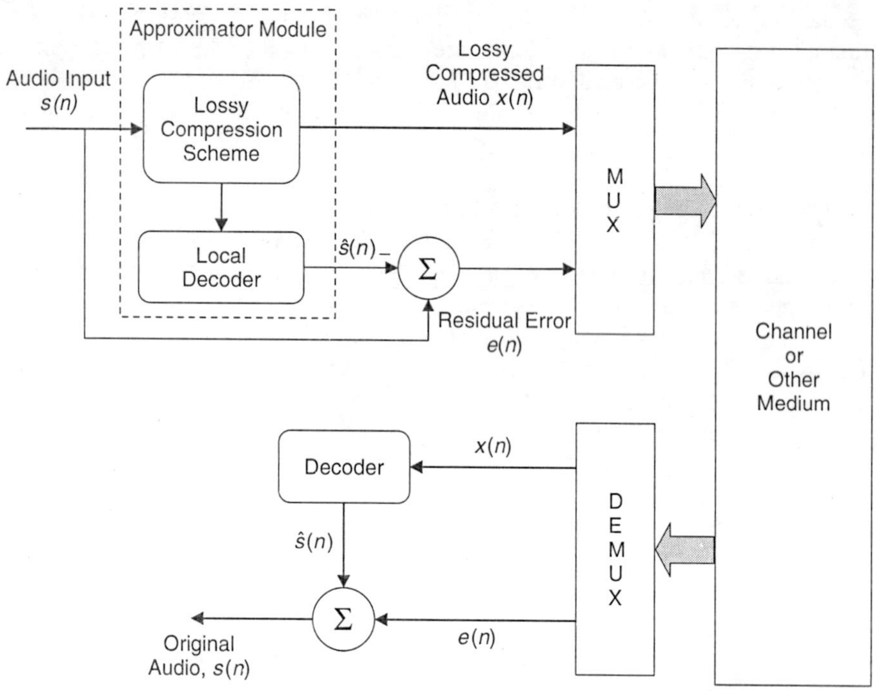

Figure 11.2. Lossless audio coding scheme based on the LAC method.

the reconstructed signal from the input signal. Reconstruction involves decoding the compressed bitstream locally at the encoder from the input signal. The residual is encoded using entropy coding methods and transmitted along with the encoded (lossy) audio bitstream. The transmission of the coding error enables the decoder to obtain an exact replica of the input audio resulting in a lossless scheme. An alternative to this scheme was also proposed, where, instead of using a lossy audio compression scheme, a predictor is used to approximate the input audio signal.

11.2.2 L²AC Algorithms

11.2.2.1 The SHORTEN Algorithm Framing, prediction, and residual coding are the three primary stages associated with the *SHORTEN* algorithm. The algorithm operates on 16-bit linear PCM signals sampled at 16 kHz. The audio signal is divided into blocks of 16 ms corresponding to 256 samples. An L-th order LP analysis is performed using the Levinson-Durbin algorithm. The LP analysis window (usually rectangular) is typically L samples longer than the frame size, i.e., $N = (256 + L)$. Typically, a 10th-order ($L = 10$) LP filter is employed.

A prediction technique based on curve fitting and polynomial approximations has also been proposed in [Robi94] (Figure 11.3). This is a simplified form of

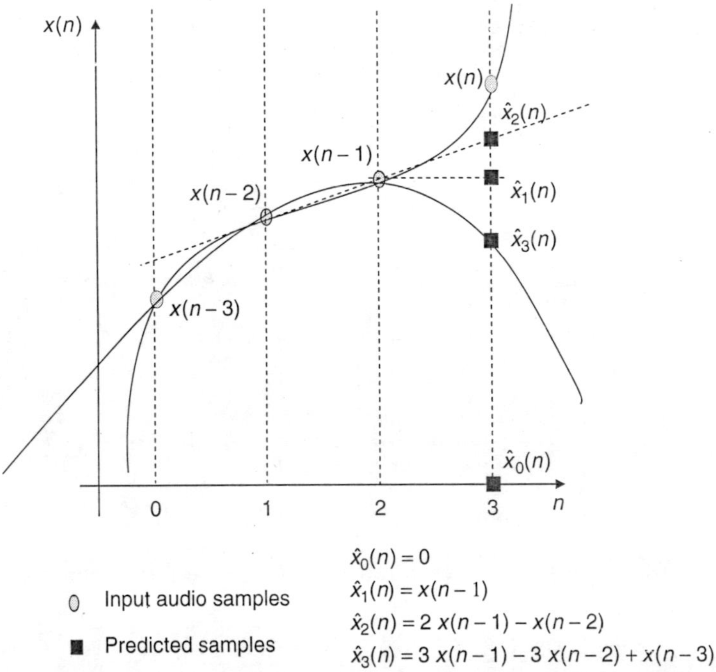

Figure 11.3. Polynomial approximation of $x(n)$ used in the *SHORTEN* [Robi94] and in the *AudioPaK* [Hans98a] algorithms.

linear predictor that uses a selection criterion by fitting an L-th order polynomial to the last L data points. For instance four polynomials, i.e., $i = 0, 1, 2, 3$ are formed as follows:

$$
\begin{aligned}
\hat{x}_0(n) &= 0 \\
\hat{x}_1(n) &= x(n-1) \\
\hat{x}_2(n) &= 2x(n-1) - x(n-2) \\
\hat{x}_3(n) &= 3x(n-1) - 3x(n-2) + x(n-3)
\end{aligned}
\tag{11.1}
$$

Note that the FIR predictor coefficients in the above equation are integers. The residual, $e_i(n)$, is computed by subtracting the corresponding polynomial estimate from the actual signal, i.e.,

$$e_i(n) = x(n) - \hat{x}_i(n). \tag{11.2}$$

From (11.1) and (11.2), we obtain

$$e_0(n) = x(n). \tag{11.3}$$

Similarly, $e_1(n)$ is given by

$$e_1(n) = x(n) - \hat{x}_1(n). \tag{11.4}$$

Finally, from (11.1) and (11.3), we get

$$e_1(n) = e_0(n) - e_0(n-1). \tag{11.5}$$

The above equations lead to an efficient recursive algorithm to compute the polynomial prediction residuals, i.e.,

$$\begin{aligned} e_0(n) &= x(n) \\ e_1(n) &= e_0(n) - e_0(n-1) \\ e_2(n) &= e_1(n) - e_1(n-1) \\ e_3(n) &= e_2(n) - e_2(n-1) \end{aligned} \tag{11.6}$$

Next, the residual energy, E_i corresponding to each of the four residuals is computed.

$$E_i = \sum_n e_i{}^2(n) \qquad \text{for } i = 0, 1, 2, 3 \tag{11.7}$$

The polynomial that results in the smallest residual energy is considered as the best approximation for that particular frame. It should be noted that the prediction, based on polynomial approximations, does not always result in maximum data rate reduction often due to the poor estimation of signal statistics. On the other hand, FIR predictors with real-valued coefficients are more versatile than integer predictors. Based on experiments conducted by Robinson [Robi94], the residual, $e(n)$, exhibits a Laplacian distribution. Usually, a Huffman coder that best fits the Laplacian distribution is employed for residual packing. Note that these codes are also called Rice codes in the entropy coding literature. An error statistic of 0.004 bits/sample has been reported in [Robi94].

11.2.2.2 The AudioPaK Coder The AudioPaK algorithm developed by Hans *et al.* [Hans98a] [Hans98b] uses linear prediction with integer coefficients and Golomb-Rice coding for packing the error. In particular, the AudioPaK coder employs an adaptive polynomial approximation method in order to eliminate the intrachannel dependencies. In addition to the intrachannel decorrelation, the AudioPaK algorithm also allows interchannel decorrelation in the case of the stereo and dual modes[1]. Inter-channel decorrelation in these modes can be achieved as follows. First, the residuals associated with the left and right channels

[1] Recall that in the stereo mode two audio channels that form a stereo pair (left and right) are encoded with one bitstream. On the other hand, in the dual mode, two audio channels with independent program contents, e.g., bilingual are encoded within one bitstream. However, the encoding process is the same for both modes [ISOI92].

are computed followed by the differences between the left and the right channel residuals. An average residual energy, E_i, corresponding to each of the four right-channel residuals and, ΔE_i, corresponding to the four difference residuals are computed (similar to Eq. (11.7)). The residual that results in the smallest residual energy among E_0, E_1, E_2, E_3, ΔE_0, ΔE_1, ΔE_2, and ΔE_3 is considered as the best approximation for that particular frame. Hans and Schafer reported small improvements in the compression rates when the interchannel decorrelation is included [Hans01].

Golomb Coding. The entropy coding techniques employed in AudioPaK are inspired by the Golomb codes used in the lossless image compression scheme [Wein96] [ISOJ97]. First, the frames are classified as silent (constant) or time varying. In the case of silent frames, the residuals $e_0(n)$ and $e_1(n)$ are used as the predictor polynomials. On the other hand, for the varying frames, Golomb coding is employed. In particular, residual $e_0(n)$ is used to encode silent frames made of zeros, and residual $e_1(n)$ is used to encode silent frames made of a constant value. For frames that are classified as time varying, Golomb coding is employed as follows. Golomb codes are optimal for exponentially decaying probability distributions of positive integers. Therefore, a mapping process is required to reorder the negative integers to positive values. This is done as follows,

$$e_i(n) = \begin{cases} 2e_i(n) & \text{if } e_i(n) \geqslant 0 \\ 2|e_i(n)| - 1 & \text{otherwise,} \end{cases} \tag{11.8}$$

where $e_i(n)$ is the residual corresponding to the i-th polynomial. Note that the Golomb codes are prefix codes that can be characterized by a unique parameter "m." An integer "I" can be encoded using a Golomb code as follows. If $m = 2^k$, the codeword for "I" consists of "k" LSBs of "I," followed by the number formed by the remaining MSBs of "I" in unary representation and a stop bit. Therefore, the length of the code is $k + \left\lceil \frac{I}{2^k} \right\rceil + 1$. For example, if $n = 69$, $m = 16$, i.e., $k = 4$; Part 1: 4 LSBs,of [1000101] = 0101; Part 2: $unary(100) = unary(4) = 1110$, $Stopbit = 0$, $GolombCode = [010111100]$; Length of the code = 9. Usually, the value of "k" is constant over an entire frame and is computed based on the expectation $E(|e_i(n)|)$ as follows.

$$k = [\log_2(|e_i(n)|)]. \tag{11.9}$$

The lower and upper bounds of k are given by 0 and 15, respectively, for a 16-bit audio input.

11.2.2.3 The DVD Algorithm Craven *et al.* proposed a L^2AC scheme based on IIR prediction filters [Crav97]. This was later adopted in the DVD standard AUDIO SPECIFICATIONS: Part 4 [DVD99]. An important insight to FIR/IIR prediction techniques towards the lossless compression is given in [Crav96]. The FIR prediction fails to provide satisfactory compression when dealing with signals exhibiting *wide dynamic ranges* (>60 dB). Usually, these wider dynamic ranges in the spectrum of input audio are more likely when the sampling frequencies are in excess of 96 kHz [ARA95]. To this end, Craven *et al.* pointed out that IIR filter prediction techniques [Crav96] [Crav97] perform well for these cases (Figure 11.4).

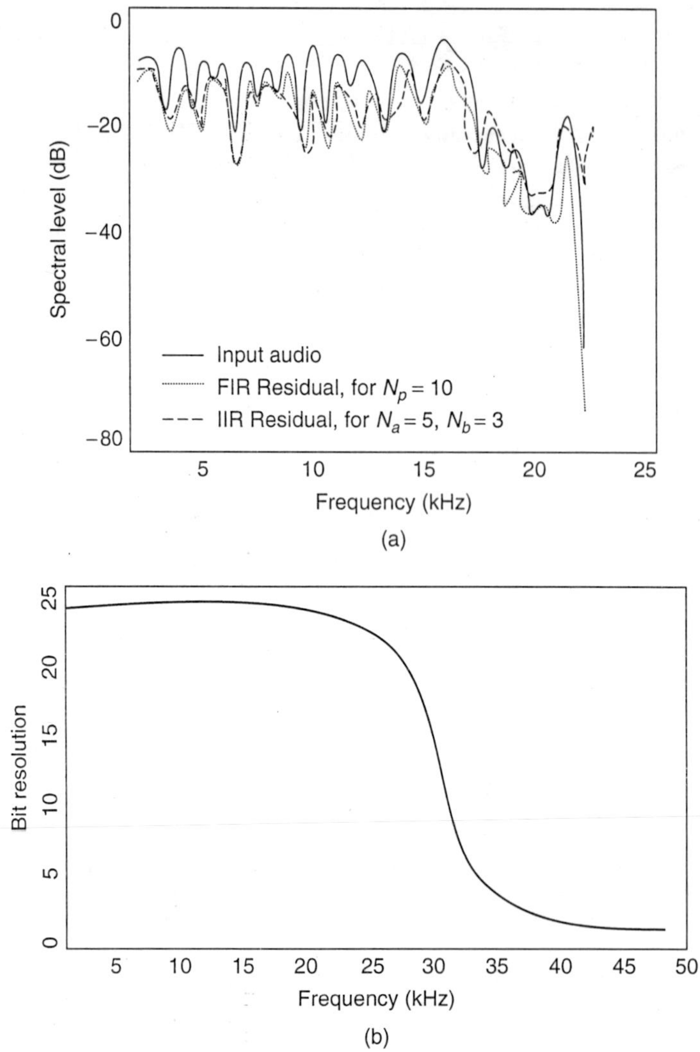

Figure 11.4. (a) A simulation example depicting the advantages of employing IIR predictors over FIR predictors in L^2AC schemes. Note that the FIR predictor fails to flatten the drop in the input spectrum above 20 kHz, while a third-order IIR predictor is able to do this efficiently. (b) Bit resolution as a function of frequency. The shaded area represents the total information content of 96 kHz sampled signal (after [Crav96]).

The DVD algorithm [Crav96] works as follows. First, the input audio is divided into frames of lengths that are integer multiples of 384. A third-order ($N_b = N_a = 3$) IIR predictor with fine coefficient quantization was proposed. Figure 11.4(a) illustrates the advantages of employing an IIR predictor. Note that a 10th-order FIR predictor fails to flatten the drop (~50dB above 20 kHz) in the input spectrum. On the other hand, a third-order IIR predictor is able to do this efficiently. IIR prediction schemes provide superior performance over FIR prediction techniques for the cases where control of both average and peak data rates is equally important. Moreover, IIR predictors are particularly effective when large portions of the spectrum are left unoccupied, i.e., if the filter roll-off is much less than the sampling rate, $f_c \ll F_s$ as shown in Figure 11.4(b). Note that the area under the curve in Figure 11.4(b) represents the total information content, and $f_c \approx 30$ kHz, $F_s = 96$ kHz. Results of 1-bit reduction per sample per channel in the data rate were reported in [Crav96] when the order in the denominator was incremented ($N_b = 3$, $N_a = 4$). In other words, a performance improvement of 6 dB can be realized by simply using an extra order in the denominator of the predictor transfer function. A special quantization structure is proposed in order to avoid precision errors while dealing with fixed-point DSP computations [DVD99]. A Huffman coder that best fits the Laplacian distribution is employed for the residual packing. The Huffman codes are chosen according to an adaptive scheme [Crav97] (based on the input quantization step size) in order to accommodate a wide range of input word lengths.

It is interesting to note that the L^2AC IIR prediction techniques [Crav97] described in this section resulted in new initiatives towards a more efficient lossless packing of digital audio [Gerz99]. In particular, these ideas were integrated in the Meridian lossless packing algorithm (MLP) proposed by Gerzon *et al.* that became an integral part of the DVD-audio specifications [DVD99].

11.2.2.4 The MUSICompress Wegner at Soundspace Audio developed a low complexity, low-MIPS, fast L^2AC scheme called the MUSICompress (MC) algorithm [Wege97]. This algorithm employs an adaptive approximator to estimate

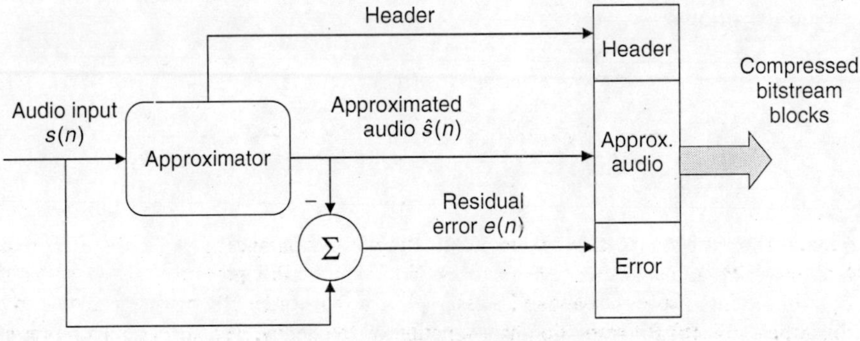

Figure 11.5. The MUSICompress (MC) algorithm L^2AC scheme.

the residual error as shown in Figure 11.5. Each encoded bitstream block contains a header, the approximated audio $\hat{s}(n)$, and the residual $e(n)$. The header block specifies the type of the approximator used for compression. Depending upon the bit-budget and computational complexity, the MC algorithm can also support lossy compression by discarding LSBs from the encoded bitstream. A metric for comparing L^{2}AC algorithms is also proposed in [Wege97]. This is given by

$$\text{L}^2\text{AC metric} = \frac{\text{Compression ratio}}{\text{Number of MIPS}}. \tag{11.10}$$

Although, this metric is not universal, it provides a rough estimate by considering both the compression ratio and the complexity associated with a L^{2}AC algorithm. At the decoder, these compressed bitstream chunks are first demultiplexed. An interpreter reads the header-structure that provides information regarding the approximator used in the encoder. A synthesis operation is performed at the decoder to reconstruct the lossless audio signal. Real-time DSP and hardware implementation aspects of the MC algorithm along with the compression results are given in [Wege97].

11.2.2.5 The Context Modeling – Linear Predictive Audio Coding (C-LPAC) Inspired by applications of context-modeling in image coding [Wu97] [Wu98], Qiu proposed a L^{2}AC scheme, called C-LPAC [Qiu01]. The main idea behind the C-LPAC scheme is that the residual error, $e(n)$, is encoded based on higher-order context modeling that uses local (neighborhood) information to estimate the probability of the occurrence of an event. An FIR linear prediction is employed to compute the prediction coefficients and the residual error. An error feedback is also employed in order to model the random process, $r_p(n)$, i.e.,

$$s_{pred}(n) = \sum_{k=1}^{L} a_k s(n-k) + r_p(n), \tag{11.11}$$

where $s_{pred}(n)$ is the predicted value based on the previous L samples of the input signal $s(n)$, and a_k are the prediction coefficients. The LP coefficients are computed by minimizing the error, $e(n)$, between the predicted and the original signal as follows:

$$e(n) = s(n) - s_{pred}(n) \tag{11.12}$$

$$e(n) = s(n) - \sum_{k=1}^{L} a_k s(n-k) - r(n), \qquad n = 1, 2, \ldots, N. \tag{11.13}$$

The mean square error (MSE) is given by,

$$\varepsilon = \frac{1}{N}\sum_{n=1}^{N} e^2(n), \tag{11.14}$$

where N is the number of samples. Note that the minimization of the MSE, ε, (11.14) fails to model the random process, $r(n)$, since the audio frames assumed to be stationary with constant mean and variance. Therefore, in order to estimate correctly the random process, $r(n)$, the residual error, $e(n)$ is modified as follows:

$$e_f(n) = \begin{cases} e(n) - \Delta(n), \text{ for encoder} \\ e(n) + \Delta(n), \text{ for decoder,} \end{cases} \tag{11.15}$$

where

$$\Delta(n) = \mathcal{F}(e(n-1), e(n-2), \ldots e(n-n_f)), \tag{11.16}$$

where n_f denotes the number of error samples fed back to compute the correction factor and $\Delta(n)$ is based on an arbitrary function $\mathcal{F}$. Although, details on the selection of n_f and the function $\mathcal{F}$ are not included in [Qiu01], we note that the basic idea is to employ an error feedback in order to account for the estimation of the random process, $r_{\text{p}}(n)$. The modified residual error, $e_f(n)$, is entropy coded using the high-order context modeling. In particular, bit-plane encoding is employed [Kunt80] [Cai95]. For example, let $e(i-1)$, $e(i)$, and $e(i+1)$ be the residual samples. Each bit plane (*BP0* through *BP7*) contains a representation of each significant bit corresponding to a sample. The bit-planes $\{BP0, \ldots, BP7\}$ contain the bits corresponding to {MSB, ..., LSB}, respectively, and therefore, the most significant information is included in the top few bit-planes:

$$\begin{array}{rl} & \begin{matrix} BP0 & BP7 \\ \downarrow & \downarrow \end{matrix} \\ e(i-1) = 63, & (00111111) \\ e(i) = 64, & (01000000) \\ e(i+1) = 65, & (01000001) \end{array} \tag{11.17}$$

In the above example, bit-plane 0 contains $(\ldots 0, 0, 0 \ldots)$, bit-plane-1 contains $(\ldots 0, 1, 1 \ldots)$, and so on. First a local energy, $\xi_{e(i)} = \sum_{m_1=1}^{p} |e(i-m_1)| + \sum_{m_2=1}^{q} |e(i+m_2)|$ in the neighborhood of "p" samples before "e_i" and "q" samples after "e_i" is computed. Depending on the energy level and the bit-plane position (0 through 7 in our example), the significance of the coding symbol, e_i, is computed. Expressions for context modeling for "*significant*" coding, "*sign*" coding, and "*refinement*" coding are given in [Qiu01]. The research involving the higher-order context modeling applied to L²AC uses concepts from image coding [Wu97] [Wu98].

11.2.2.6 The Lossless Transform Audio Coding (LTAC) The LTAC algorithm [Pura97] employs a unitary transform ($\text{A}^{\text{H}} = \text{A}^{-1}$). In particular, an orthonormal DCT is used to decorrelate the input signal. Figure 11.6 shows the LTAC block diagram where the input audio, $s(n)$, is first buffered in N_{buff} sample blocks, transformed to the frequency domain using the DCT, and then quantized to obtain the spectral coefficients, $X(k)$. Both fixed and adaptive block length modes are available for DCT block transformation. In the fixed block length mode, the DCT block size is equal to the number of input samples ($N_{buff} = 128$, 256, 1024, and 2048). On the other hand, in the adaptive block length mode, the number of input samples per block, $N_{buff} = 2048$, and the transformation that results in the optimum performance among the 512, 1024, or 2048-point DCT is selected. Higher block lengths often provide good decorrelation of the input signal. However, this is true only when the input signal is not rapidly time varying. The LTAC algorithm works as follows. An estimate of the input audio, $\hat{s}(n)$, is computed at the local decoder by performing the inverse transformation (i.e., IDCT) as shown in Figure 11.6. An error signal (residual) is calculated by subtracting the estimated signal from the input signal

$$e(n) = s(n) - \hat{s}(n) \tag{11.18}$$

where $\hat{s}(n) = Q\langle \text{IDCT}[Q\langle \text{DCT}[s(n)]\rangle]\rangle$ and 'Q' denotes quantization.

The residual, $e(n)$, is entropy coded using Rice codes and transmitted along with the DCT coefficients, $X(k)$, as shown in Figure 11.6. Compression rates in the range of 5.9–7.2, and performance improvements over prediction-based L^2AC schemes have been reported in [Pura97].

11.2.2.7 The IntMDCT Algorithm Inspired by the use of the DCT and other integer transforms in lossless audio coding, Geiger *et al.* proposed the integer-MDCT algorithm [Geig01] [Geig02]. One of the attractive properties that has contributed to the widespread use of the MDCT, in audio standards, is the

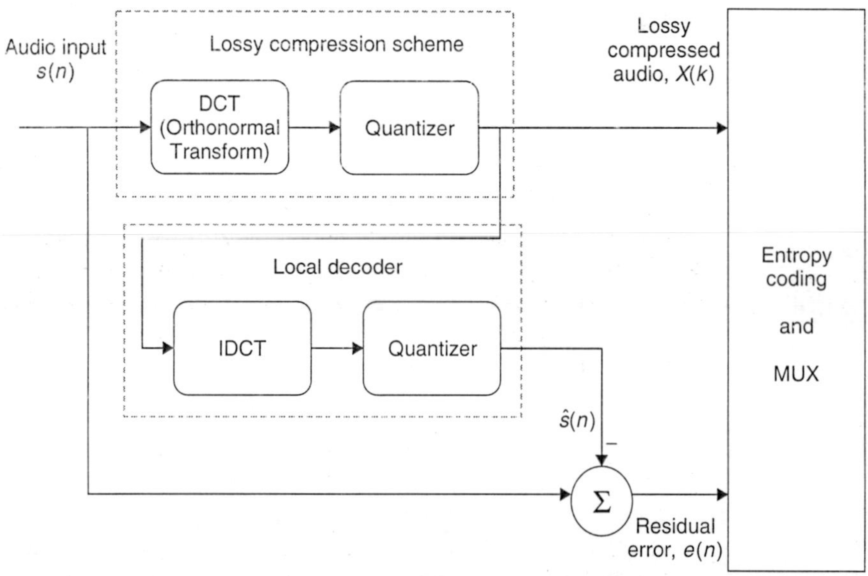

Figure 11.6. Lossless transform audio coding (LTAC).

availability of FFT-based algorithms [Duha91] [Sevi94] that make it viable for real-time applications. MDCT features that are important in audio coding, including its forward and inverse transform interpretations, prototype filter (window) design criteria, and window design examples are described in Chapter 6. Recall that, in the orthogonal case, the generalized perfect reconstruction conditions given in Chapter 6, Eqs. (6.19) through (6.23) can be reduced to linear phase and Nyquist constraints on the window, $w(n)$, i.e.,

$$\begin{aligned} -w(2M-1-n)+w(n)&=0 \text{ and} \\ w^2(n)+w^2(n+M)&=1 \text{ for } 0\leqslant n\leqslant M-1, \end{aligned} \tag{11.19}$$

where M denotes the number of transform coefficients.

Givens Rotations and Lifting Scheme. The IntMDCT algorithm is based on the integer approximation of the MDCT algorithm. This is accomplished using the "lifting scheme" [Daub96] proposed by Daubechies *et al.* First, the MDCT is decomposed [Geig01] into Givens rotations,[2] by choosing c and s in Eq. (11.20) so that a vector $\boldsymbol{X}$ is orthogonally transformed (rotated) into a vector $\boldsymbol{Y}$ as follows:

$$Y=\begin{bmatrix} y_1 \\ y_2 \end{bmatrix}=\begin{bmatrix} c & -s \\ s & c \end{bmatrix}\begin{bmatrix} x_1 \\ x_2 \end{bmatrix}\overset{\Downarrow\, Givens rotation}{=}\begin{bmatrix} y_1 \\ 0 \end{bmatrix}. \tag{11.20}$$

Note from the above equation,

$$\begin{aligned} y_2=-sx_1+cx_2&=0 \text{ and} \\ c^2+s^2&=1. \end{aligned} \tag{11.21}$$

Details on the decomposition of the MDCT into windowing, time-domain aliasing, and DCT-IV and eventually into Givens rotations are given in [Vaid93]. Note the similarities between Eqs. (11.19) and (11.21). The lifting scheme is applied next in order to approximate Givens rotations. The lifting scheme splits the Givens rotations into three lifting steps, Eq. (11.22), that can be easily approximated and mapped onto the integer domain [Daub96]. This integer mapping is reversible and therefore contributes to the lossless coding. Equations (11.23) and (11.24) show that the lifting scheme is indeed a reversible integer transform, since each of the three lifting steps can be inverted uniquely by simply subtracting the value that has been added:

$$\text{Lifting scheme}: \begin{bmatrix} c & -s \\ s & c \end{bmatrix}=\begin{bmatrix} 1 & \dfrac{c-1}{s} \\ 0 & 1 \end{bmatrix}\begin{bmatrix} 1 & 0 \\ s & 1 \end{bmatrix}\begin{bmatrix} 1 & \dfrac{c-1}{s} \\ 0 & 1 \end{bmatrix}. \tag{11.22}$$

[2] Note that c and s represent $\cos(\alpha)$ and $\sin(\alpha)$, where "α" corresponds to the angle of rotation of some arbitrary plane. In particular, the Givens rotation results in an orthogonal transformation without changing any of the norms of the vectors, $\boldsymbol{X}$ and $\boldsymbol{Y}$.

The lifting scheme and its inverting property are described as follows:

$$Y = \begin{bmatrix} c & -s \\ s & c \end{bmatrix} X = \begin{bmatrix} 1 & \dfrac{c-1}{s} \\ 0 & 1 \end{bmatrix} \begin{bmatrix} 1 & 0 \\ s & 1 \end{bmatrix} \begin{bmatrix} 1 & \dfrac{c-1}{s} \\ 0 & 1 \end{bmatrix} X$$

$$X = \begin{bmatrix} c & -s \\ s & c \end{bmatrix}^{-1} Y = \begin{bmatrix} c & s \\ -s & c \end{bmatrix} Y \qquad (11.23)$$

$$= \begin{bmatrix} 1 & -\left(\dfrac{c-1}{s}\right) \\ 0 & 1 \end{bmatrix} \begin{bmatrix} 1 & 0 \\ -s & 1 \end{bmatrix} \begin{bmatrix} 1 & -\left(\dfrac{c-1}{s}\right) \\ 0 & 1 \end{bmatrix} Y.$$

The lifting scheme is a reversible integer transform, i.e.,

Consider the first lifting step, $\begin{pmatrix} 1 & (c-1)/s \\ 0 & 1 \end{pmatrix}$

$$Y^1 = \begin{bmatrix} {y_1}^1 \\ {y_2}^1 \end{bmatrix} = \begin{bmatrix} 1 & (c-1)/s \\ 0 & 1 \end{bmatrix} \begin{bmatrix} {x_1}^1 \\ {x_2}^1 \end{bmatrix} = \begin{bmatrix} {x_1}^1 + {x_2}^1(c-1)/s \\ {x_2}^1 \end{bmatrix} \qquad (11.24)$$

Decoding :

$$X^1 = \begin{bmatrix} {x_1}^1 \\ {x_2}^1 \end{bmatrix} = \begin{bmatrix} 1 & -(c-1)/s \\ 0 & 1 \end{bmatrix} \begin{bmatrix} {y_1}^1 \\ {y_2}^1 \end{bmatrix} = \begin{bmatrix} {y_1}^1 - {y_2}^1(c-1)/s \\ {y_2}^1 \end{bmatrix}$$

The IntMDCT is typically used in conjunction with the MDCT-based perceptual coding scheme as shown in Figure 11.7. First, the input audio, $s(n)$, is compressed using a perceptual coding scheme (e.g., ISO/IEC MPEG-audio algorithms) resulting in a lossy compressed audio, $X(k)$. Later, these coefficients are inverse quantized and integer rounded, and subtracted from the IntMDCT spectrum coefficients. This results in an error residual, $e(n)$. The error essentially carries the enhancement bitstream that contributes to lossless coding. The error spectrum and the MDCT coefficients can be entropy coded using Huffman codes. Bit-sliced arithmetic coding (BSAC) that inherits the properties of fine grain scalability can also be used to encode the lossy audio and the enhancement bitstream.

In BSAC, arithmetic coding replaces Huffman coding. In each frame, bit planes are coded in order of significance, beginning with the most significant bits (MSBs). This results in a fully embedded coder containing all lower-rate codecs. With proper tuning of quantization steps, the IntMDCT algorithm shown in Figure 11.7 can also be used in perceptual (lossy) audio coding applications [Geig02] [Geig03].

11.3 DVD-AUDIO

The *DVD Forum Committee* recommended standardizing the DVD-A specifications [DVD01] and the related coding techniques for DVD-A, i.e., the Meridian

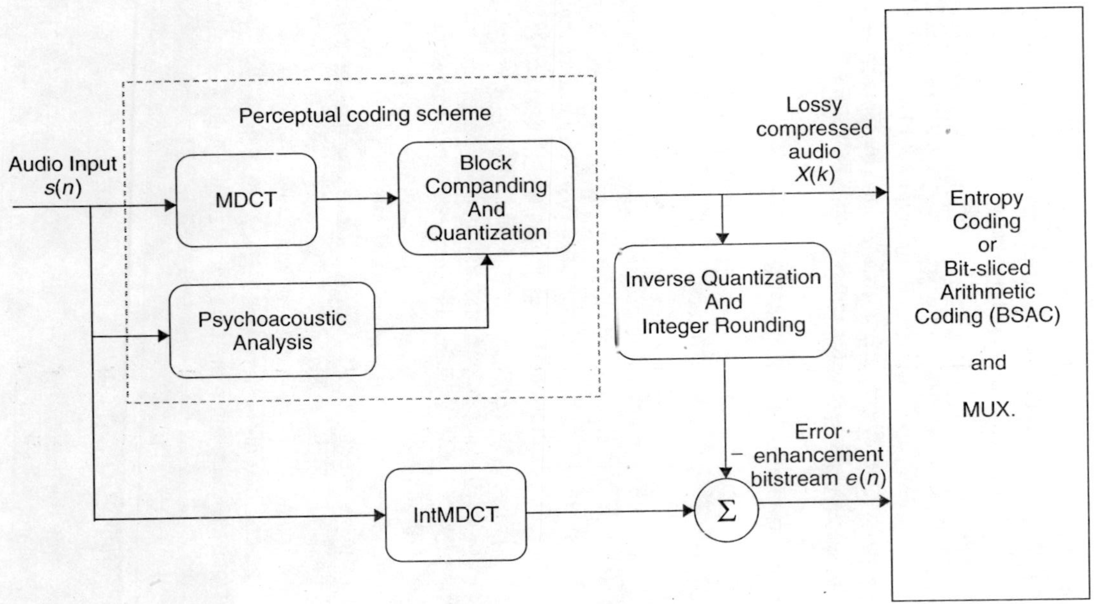

Figure 11.7. Lossless audio coding using IntMDCT algorithm.

lossless packing [DVD99]. DVD-A offers two important features, i.e., multiple sampling rates (44.1, 48, 88.2, 96, 176.4, and 192 kHz) and multiple bit-resolutions (16-, 20-, and 24-bit). The DVD-A format can store up to 6 channels of digital media at 24-bit resolution sampled at 96 kHz. Table 11.2 lists some of the important parameters of the DVD-A format. DVD-A provides a dynamic range of the order of 144 dB over a 96-kHz frequency range; and the maximum input data rate allowed is 9.6 Mb/s. Along with these features DVD-A addresses the digital content protection and copyright issues with a variety of encryption and audio watermarking technologies. A 74-min, 96-kHz, multichannel 20-bit resolution audio will require a minimum of 8.93 GB storage space (see Table 11.2). However, the maximum storage capacity offered by DVD-A is 4.7 GB. Therefore, a lossless compression is required. The Meridian lossless packing (MLP) scheme [Gerz99] [Kuzu03] developed by Gerzon *et al.* has been chosen as the compression standard for the DVD-Audio format.

11.3.1 Meridian Lossless Packing (MLP)

The MLP encoding scheme [Gerz99] was built around the work done by Craven and Gerzon [Crav96] [Crav97] and Oomen, Bruekers, and Vleuten [Oome96]. Furthermore, it inherits some features from the DVD algorithm discussed in section 11.2.2. Figure 11.8 shows the five important stages associated with the MLP encoding scheme. These include channel remapping and shifting, interchannel decorrelation, intrachannel decorrelation (or prediction), entropy coding (lossless packing), and interleaving/MLP bitstream formatting. In the first stage, N channels of PCM audio are remapped and shifted in order to facilitate efficient buffering of MLP substreams and to improve bit-precision. MLP can encode a maximum of 63 audio channels. In the second stage, the interchannel dependencies within these N channels are exploited. Typically, matrixing techniques (see Chapter 10) are employed to eliminate interchannel redundancies/irrelevancies. In particular, MLP employs *lossless matrixing scheme* that is described in Figure 11.9. From this figure, it can be noted that only one PCM audio channel, for instance, x_1 is modified, while the channels x_2 through x_N are unchanged. The coefficients $[a_2, a_3, a_4, \ldots, a_N]$ are computed for each audio frame that results in minimum interchannel redundancy. Stage 3 is concerned with the intrachannel redundancy removal based on the FIR/IIR prediction techniques.

11.4 SUPER-AUDIO CD (SACD)

Almost 20 years after the launch of the audio CD format [PhSo82] [IECA87], Philips and Sony established a new storage format called the "super-audio CD" (SACD) [SACD02] [SACD03]. Unlike the conventional CD format that employs PCM encoding with a 16-bit resolution and a sampling rate of 44.1 kHz, the SACD system uses a superior 1-bit recording technology called direct stream digital (DSD) [Nish96a] [Nish96b] [Reef01a]. SACD operates at a sampling frequency 64 times the rate used in conventional CDs, i.e., 64×44.1 kHz = 2.8224 MHz.

Table 11.2. Conventional CDs versus the next generation storage formats.

Parameter	Conventional CD	SACD	DVD-Audio
Standardized year	1982; [IECA87]	1999; [SACD02]	1999; [DVD01]
Encoding scheme	Pulse code modulation (PCM)	Direct-stream digital (DSD)	Meridian lossless packing (MLP)
Bit resolution	16-bit linear PCM	1-bit DSD	16, 20, or 24-bit linear PCM
Sampling rate	44.1 kHz	2822.4 kHz	44.1, 48, 88.2, 96, 176.4, or 192 kHz
Data rate	1.4 Mb/s	5.6 to 8.4 Mb/s	<9.6 Mb/s
Frequency range	DC to 20 kHz	DC to 100 kHz	DC to 96 kHz
Dynamic range	96 dB	~120 dB	~144 dB
Number of channels	2	2 to 6	1 to 6
Playback time	74 min	~ 74(±5) min	~70 to 800 min (depending upon the bit rate)
Maximum storage capacity available	~750 to 800 MB	~4.38 to 4.7 GB (single-layer)	~4.7 GB
Storage required	$74 \times 60 \times 44100 \times (2) \times \left(\frac{16}{8}\right) \cong 740$ MB	$74 \times 60 \times 2.8224 \times 10^6 \times (2+6) \times \left(\frac{1}{8}\right) \cong 11.67$ GB	$74 \times 60 \times 96 \times 10^3 \times (1+2+6) \times \left(\frac{20}{8}\right) \cong 8.93$ GB
Compression	Not required	Required	Required

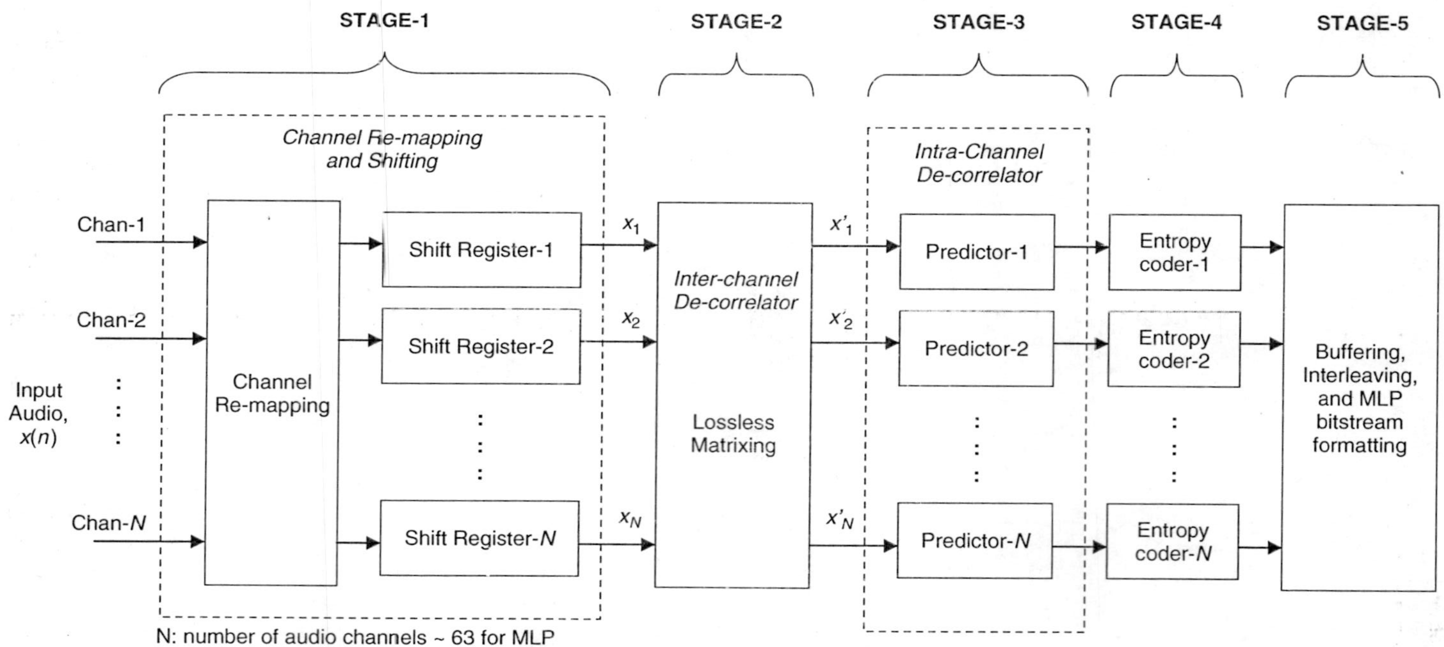

Figure 11.8. MLP encoding scheme [Gerz99].

Lossless Matrixing :

$$\begin{bmatrix} x'_1 \\ x'_2 \\ x'_3 \\ \vdots \\ \vdots \\ x'_N \end{bmatrix} = \begin{bmatrix} 1 & -a_2 & -a_3 & -a_4 & \cdots & -a_N \\ 0 & 1 & 0 & 0 & \cdots & 0 \\ 0 & 0 & 1 & 0 & \cdots & 0 \\ \vdots & \vdots & \vdots & \vdots & \vdots & \vdots \\ \vdots & \vdots & \vdots & \vdots & \vdots & \vdots \\ 0 & 0 & 0 & 0 & \cdots & 1 \end{bmatrix} \begin{bmatrix} x_1 \\ x_2 \\ x_3 \\ \vdots \\ \vdots \\ x_N \end{bmatrix} \rightarrow \text{(I)}$$

$$x'_1 = x_1 - \widehat{a_2 x_2} - \widehat{a_3 x_3} - \widehat{a_4 x_4} - \ldots - \widehat{a_N x_N}; \quad \rightarrow \text{(II)}$$

$$\left.\begin{aligned} x'_2 &= x_2; \\ x'_3 &= x_3; \\ &\vdots \\ x'_N &= x_N; \end{aligned}\right\} \rightarrow \text{(III)}$$

Note: $\widehat{a_2 x_2}$ corresponds to Quantization of $a_2 x_2$

Lossless De – matrixing :

$$\begin{bmatrix} x_1 \\ x_2 \\ x_3 \\ \vdots \\ \vdots \\ x_N \end{bmatrix} = \begin{bmatrix} 1 & a_2 & a_3 & a_4 & \cdots & a_N \\ 0 & 1 & 0 & 0 & \cdots & 0 \\ 0 & 0 & 1 & 0 & \cdots & 0 \\ \vdots & \vdots & \vdots & \vdots & \vdots & \vdots \\ \vdots & \vdots & \vdots & \vdots & \vdots & \vdots \\ 0 & 0 & 0 & 0 & \cdots & 1 \end{bmatrix} \begin{bmatrix} x'_1 \\ x'_2 \\ x'_3 \\ \vdots \\ \vdots \\ x'_N \end{bmatrix} \rightarrow \text{(IV)}$$

(1). Decoding x_2 through x_N is straightforward from (III)

(2). Lets see if decoding x_1 is lossless:

From (IV) => $x_1 = x'_1 + \widehat{a_2 x'_2} + \widehat{a_3 x'_3} + \widehat{a_4 x'_4} + \ldots + \widehat{a_N x'_N}$

$x_1 = x'_1 + \widehat{a_2 x_2} + \widehat{a_3 x_3} + \widehat{a_4 x_4} + \ldots + \widehat{a_N x_N}$ $\quad (\because x'_i = x_i$ from (III))

=> The above equation when re-arranged is same as (II)

=> ∴ Lossless matrix encoding and decoding

Figure 11.9. Lossless matrixing in MLP.

Moreover, SACD enables frequency response from DC to 100 kHz with a dynamic range of 120 dB. SACD systems enable both high-resolution surround sound audio recordings (5.1 or 3/2-channel format) as well as high-quality stereo encoding. Some important characteristics of the red book audio CD format [IECA87] and the SACD and DVD-audio storage formats are summarized in Table 11.2.

The main idea behind the DSD technology is to avoid the decimation and interpolation filters that invariably impart quantization noise in the compression process. To this end, DSD directly records each sample as a 1-bit pulse [Hori96] [Nish96a] [Nish96b] [Reef01a], thereby eliminating the need for down-sampling and up-sampling filters (see Figure 11.10(b)). Moreover, the DSD encoding enables improved lossless compression of digital audio [Reef01a] [Jans03], i.e., direct stream transfer (DST) (compression ratios range from 2 to 4). Figures 11.10, (a) and (b), show the steps involved in the conventional PCM encoding and the DSD technology, respectively. Before describing the DSD encoding, we provide a brief introduction to the SACD storage format [tenK97] and the sigma-delta ($\Sigma\Delta$) modulators [Cand92] [Sore96] [Nors97] in the next two subsections.

11.4.1 SACD Storage Format

SACDs come in three formats based on the number of layers employed [tenK97] [SACD02] [SACD03] [Jans03]. These include the single layer, the dual-layer, and the hybrid formats. In the *single-layer* SACD, a single high-density layer for the high-resolution DSD recording (4.7 GB) is allowed. On the other hand, the *dual-layer* SACD contains one or two layers of DSD content (2×4.7 GB $= 8.4$ GB). In order to allow backwards compatibility with conventional CD players, a *hybrid* disc format is standardized. In particular, the *hybrid* disc contains one inner layer of DSD content and one outer layer of conventional CD content (i.e., 750 MB $+$ 4.7 GB $=$ 5.45 GB). Similar to the conventional CDs, the SACDs have a 12 cm diameter and 1.2 mm thickness. The lens numerical aperture (NA) is 0.6; and the laser wavelengths to read the SACD and the CD content are 650 nm and 780 nm, respectively.

11.4.2 Sigma-Delta Modulators (SDM)

Sigma-Delta ($\Sigma\Delta$) modulators [Cand92] [Sore96] [Nors97] convert analog signals to one-bit digital outputs. Figure 11.11 shows a schematic block diagram of the $\Sigma\Delta$ modulator that consists of a single-bit quantizer and a loop filter (i.e., low-pass filter) in a negative feedback loop. $\Sigma\Delta$ modulators output 1-bit samples that indicate whether the current sample is larger or smaller than the value accumulated from the previous samples. If the amplitude of the current sample is greater than the value accumulated during the negative-feedback-loop, the SDM outputs "1"; or otherwise "0." SDM involves high sampling frequencies and a two-level ("0" or "1") quantization process. Over-sampling relaxes the requirements on the analog antialiasing filter (see Figure 11.10) and, hence, it simplifies analog hardware. Recall from Chapter 3 that multi-bit PCM quantizers (i.e., more than two discrete amplitude levels) suffer from differential nonlinearities, particularly, with analog signals. Several solutions have been proposed [Nors92] [Nors93] [Dunn96] [Angu97] to overcome the nonlinear distortion in the PCM quantization process including, *oversampling* and *noise shaping*.

Nonlinear distortion are eliminated by switching from multi-bit PCM to single-bit quantizers. However, the single-bit quantizer employed in the $\Sigma\Delta$ modulator results in high quantization noise; therefore, a noise shaping process is essential. Noise shaping removes or "moves" the quantization noise from the audible

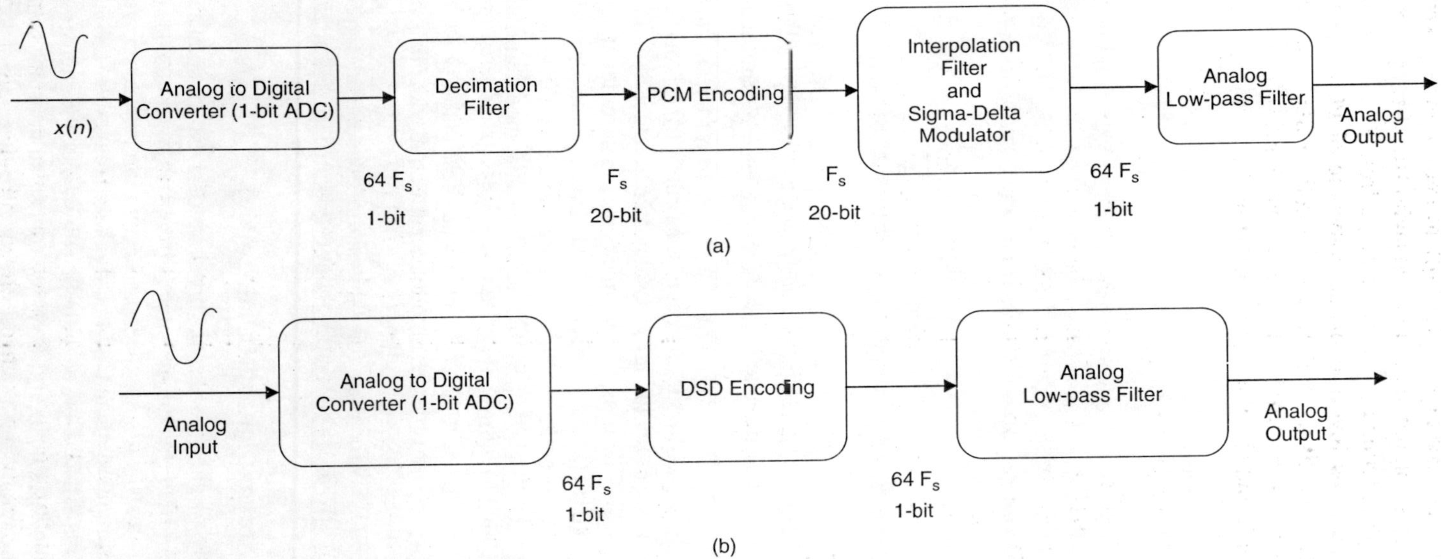

Figure 11.10. (a) Conventional PCM encoding and (b) direct-stream digital (DSD) technology.

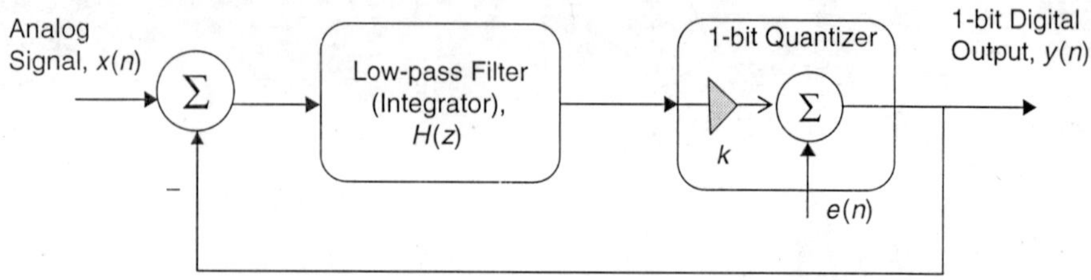

Figure 11.11. Sigma-delta ($\Sigma\Delta$) modulator.

band (<20 kHz) to the high-frequency band (> 50 kHz). From Figure 11.11, the transfer function of the SDM system is given by

$$(X(z) - Y(z))H(z)k + E(z) = Y(z) \tag{11.25}$$

$$Y(z) = \underbrace{\left(\frac{kH(z)}{1 + kH(z)}\right)}_{Signal transfer function} X(z) + \underbrace{\left(\frac{1}{1 + kH(z)}\right)}_{Noise shaping function} E(z), \tag{11.26}$$

where $H(z)$ is the loop-filter, k is the error in gain, and $E(z)$ is the DC offset corresponding to 1-bit quantization. In order to remove the quantization noise from the audible band, the noise shaping function, $NS(z) = \dfrac{1}{1 + kH(z)}$, must be a high-pass filter; and hence the loop-filter, $H(z)$, must be a low-pass filter that acts as an integrator. The next important step is to decide the order of the loop-filter, $H(z)$. A higher-order loop-filter results in higher resolution and larger dynamic ranges. We discussed earlier that SACDs provide dynamic ranges in the order of 120 dB. Experimental results [Reef01a] [Jans03] indicate that at least a *fifth-order* loop-filter must be employed in order to obtain a dynamic range greater than 120 dB. However, due to the negative feedback loop employed in the $\Sigma\Delta$ modulator, depending on the analog input signal amplitude the SDM may become unstable if the order of the loop-filter exceeds *two*. In order to overcome these stability issues, Reefman and Janssen proposed a fifth-order $\Sigma\Delta$ modulator structure with *integrator-clipping* designed specifically for DSD encoding [Reef01a] [Reef02]. Several publications that describe the SDM structures for 1-bit audio encoding [Angu97] [East97] [Angu98] [Angu01] [Reef01a] [Reef02] [Harp03] [Jans03] have appeared.

11.4.3 Direct Stream Digital (DSD) Encoding

Designed primarily to enhance the performance of 1-bit lossless audio coding, the DSD encoding offers several advantages [Reef01b] over standard PCM technology – it eliminates the requirement of decimation and interpolation filters; enables lossless compression and higher dynamic ranges (~120 dB); reduces storage requirements (since 1-bit encoding); allows much higher time resolution (due to higher sampling rates); and involves controlled transient behavior (since it employs simple antialiasing low-pass filters). Moreover, the DSD bitstreams can be easily down-converted to lower frequencies (32 kHz, 44.1 kHz, 96 kHz, etc.). During mid-1990s, Nishio *et al.* provided the first ideas of direct stream digital audio system [Nish96a] [Nish96b] [Nogu97]. At the same time, several other researchers proposed methods for 1-bit (lossless) audio coding [Hori96] [Brue97] [East97] [Ogur97]. In 1999, Philips and Sony released the licensing for the SACD systems and considered the 1-bit DSD encoding as its recording/coding technology. The following year, Reefman and Janssen, at Philips Research Labs, presented a white paper on the signal processing aspects of DSD for SACD recording [Reef01a]. Since then, several advancements towards efficient DSD encoding have been made [Reef01c] [Hawk02] [Kato02] [Reef02] [Harp03] [SACD03].

Figure 11.12 shows the DSD encoding procedure that employs a $\Sigma\Delta$ modulator and the direct stream transfer (DST) technique to perform lossless compression.

From Table 11.2, note that an SACD allows 4.7 GB of storage space; however, the storage required for a 74-min, 1-bit DSD data is approximately 11.67 GB, and hence the need for lossless compression of 1-bit audio. Lossless coding in case of 1-bit digital signals is altogether a different scenario from what we discussed earlier in Section 11.2. Bruekers *et al.*, in 1997, proposed a lossless coding technique for 1-bit audio signals [Brue97]. This was later adopted as the *direct stream transfer* (DST) L^2AC technique for 1-bt DSD encoding by Philips and Sony.

11.4.3.1 Analog to 1-Bit Digital Conversion ($\Sigma\Delta$ Modulator) First, the analog input signal, $x(n)$, is converted to 1-bit digital output bitstream, $y(n)$ (Figure 11.12). The steps involved in the analog to 1-bit digital bitstream conversion have been presented earlier in Section 11.4.2. In this section, we present some simulation results (Figure 11.13) to analyze the dynamic ranges offered in case of 16-bit linear PCM coding (~96 dB) and 1-bit DSD encoding (~120 dB). Figure 11.13(a) shows the time-domain input waveform, $x(n)$, which in this case is a 4 kHz sine wave. Figure 11.13(b) represents the $\Sigma\Delta$ modulator output, i.e., 1-bit digital output bitstream, $y(n)$. Note that, as the amplitude of the input signal increases, the density of "1s" increases (plotted as a bar-chart); and if the amplitude decreases, the density of "0s" increases. Figure 11.13(c) shows the frequency-domain output of the 16-bit linear PCM quantizer for the 1-kHz sine wave input, $x(n)$. Recall from Chapter 3 that the SNR for linear PCM will improve approximately by 6 dB per bit. This is given by

$$SNR_{PCM} = 6.02R_b + K_1, \tag{11.27}$$

where R_b denotes the number of bits and the factor K_1 is a constant that accounts for the step size and loading factors Therefore, for a 16-bit linear PCM, the output SNR is given by $SNR_{PCM} \approx 6.02(16) \approx 96$ dB. This can be associated with the dynamic ranges (~96 dB) obtained in case of conventional CDs that employ 16-bit linear PCM encoding (see Figure 11.13(c)). Figure 11.13(d) shows the spectrum of the 1-bit digital output bitstream. Although there are no mathematical

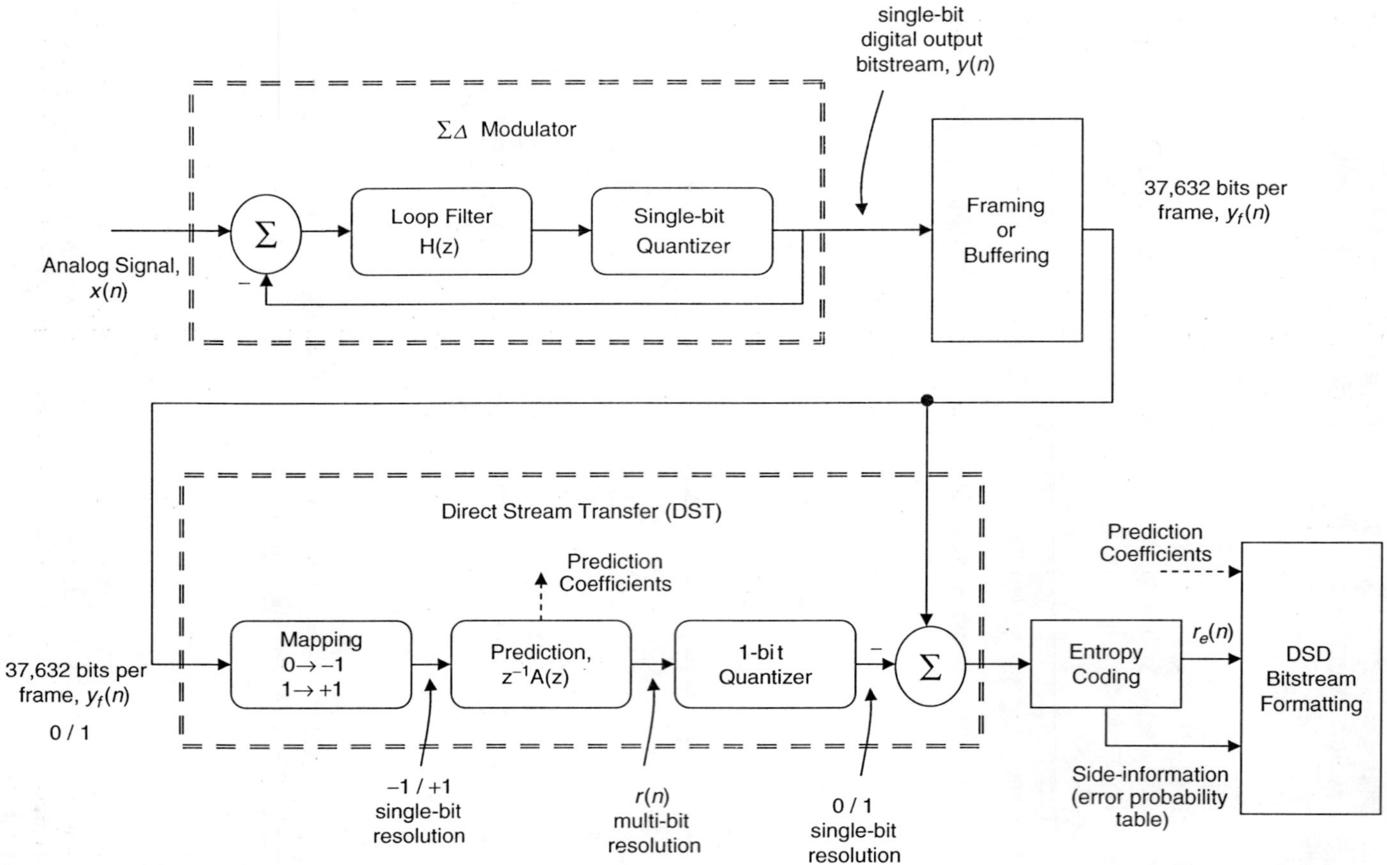

Figure 11.12. DSD encoding using a $\Sigma\Delta$ modulator and the direct stream transfer (DST) technique for 1-bit lossless audio coding in SACDs.

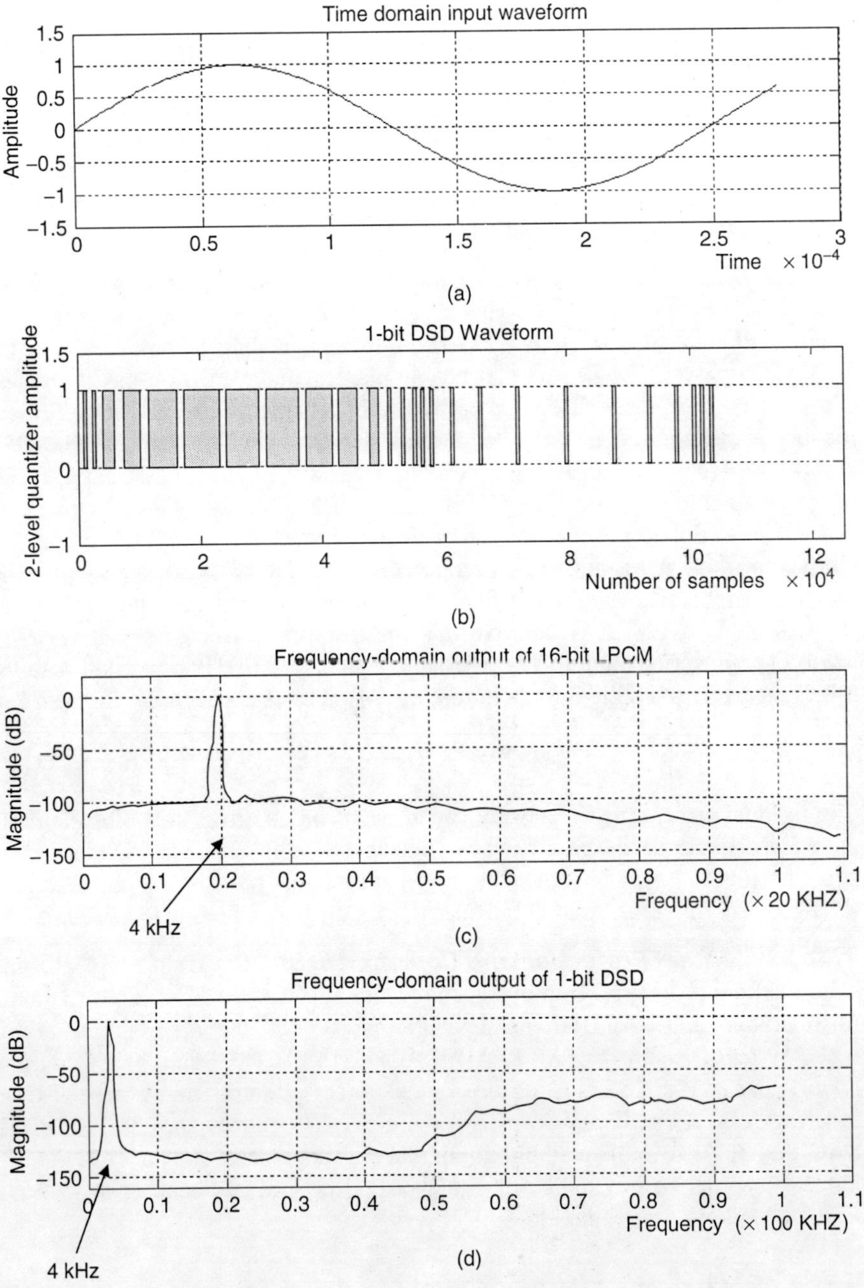

Figure 11.13. Analog to 1-bit digital conversion: (a) time-domain input sinusoid (4 kHz), $x(n)$; (b) $\Sigma\Delta$ modulator output; (c) frequency-domain output of 16-bit linear PCM encoding (DR ~96 dB); and (d) frequency-domain representation of 1-bit DSD (DR ~120 dB).

models to compute the SNR values directly for 1-bit DSD encoding, several experimental and simulation results are available [Reef01a] [Jans03]. The output SNR values in case of 1-bit encoding (and therefore the dynamic ranges) are loop-filter-order dependent. In particular, a fourth-order loop filter provides a dynamic range (DR) of 105 dB; a fifth-order loop filter results in a DR of 120 dB; and a seventh-order loop filter offers a DR of 135 dB. In Figure 11.13(d), slight increase in the noise-floor around the 50-kHz range is due to the noise-shaping employed in case of the $\Sigma\Delta$ modulator.

11.4.3.2 Direct Stream Transfer (DST) The 1-bit digital output bitstream ($\sim$11 GB) available at the output of $\Sigma\Delta$ modulator is to be compressed in order to fit into an SACD ($\sim$4.7 GB). A lossless coding technique called the DST is employed. DST is somewhat similar to the prediction-based lossless compression techniques discussed earlier in this chapter. However, several modifications are required in order to process the 1-bit digital samples. The DST algorithm [Brue97] [Reef01a] works as follows (see Figure 11.12). First, the 1-bit pulse stream $y(n)$ is buffered into frames of 37,632 bits, $y_f(n)$. A mapping procedure is applied in order to reorder the amplitudes of $y_f(n)$. DST employs a prediction filter, $A(z)$, in order to reduce the redundancies associated with samples in a frame. The prediction filter can be adaptive and the filter orders can range from 1 to 128. It is intuitive that the resulting predicted signal, $y_{f,m}(n)$, exhibits multi-bit resolution. The predicted signal, $y_{f,m}(n)$, is converted (using a 1-bit quantizer) to single-bit resolution, $y_{f,s}(n)$. The residual, $r(n)$, is given by,

$$r(n) = y_f(n) - y_{f,s}(n). \tag{11.28}$$

The residual error, $r(n)$, is entropy coded based on an error probability look-up table. For each frame, the prediction coefficients a_i, $i \in [1, 2, \ldots, 128]$, the entropy coded residual $r_e(n)$, and some side information (error probability table, etc.) are arranged according to the DSD bitstream formatting specified by [SACD02].

11.4.3.3 Content Protection and Copyright in SACD Unlike typical audio watermarking techniques employed in conventional CDs, SACDs employ a unique (distinct for each SACD) invisible watermark called the *Pit Signal Processing–Physical Disc Mark* (PSP-PDM) [SACD02] [SACD03] and three important *controls*, i.e., Disc-Access Control, Content-Access Control, and Playback Control. Figure 11.14 provides details on these SACD controls and the PSP-PDM. SACD employs several scrambling (content rearrangement) and encryption techniques. In order to descramble and play back the DSD content, knowledge of PSP-PDM "key" is essential.

11.5 DIGITAL AUDIO WATERMARKING

Audio content protection from unauthorized users and the issue of copyright management are very important these days, considering the vast amount of multimedia resources available. Moreover, with the popularity of efficient search engines and

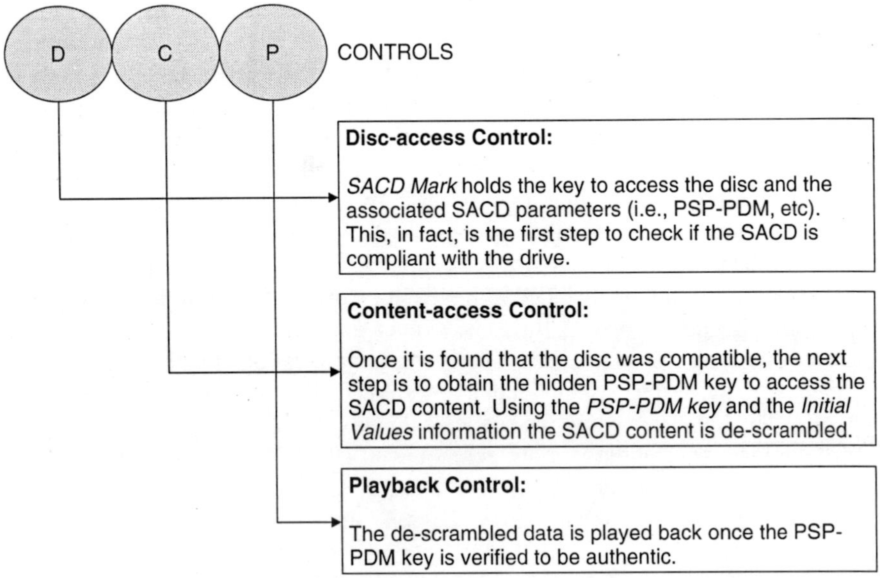

Figure 11.14. D-C-P controls in SACD.

fast Internet connections, it became feasible to search/download an audio clip over the World Wide Web in the MP3 format that can offer CD-quality high-fidelity audio. For example, an audio clip of 6-minute duration in the MP3 format, in general, would require 3–5 MB of storage space; and note that at 20 kB/s, it would take no more than 5 minutes to download the entire audio clip. Because of this, security issues in MP3 audio streaming have become important [Horv02] [Thor02]. Furthermore, with the advancements in the computer industry, any unprotected CD can be reproduced (i.e., copied) with minimal effort. The need for "intellectual property management and protection" (IPMP) [Spec98] [Spec99] [Rose01] [Beck04], motivated content creators and service providers to seek efficient *encryption* and *data hiding* techniques. The main idea behind these data hiding techniques is to insert (or hide) an *auxiliary data string* into the digital media, in order to resolve rightful ownership. Some of the important tasks of these encryption/data-hiding techniques include the ability: to ensure that the encryption is imperceptible and robust to a variety of signal processing manipulations (e.g., down-sampling, up-sampling, filtering, etc,) [Bend96] [Bone96]; to provide reliable information regarding the ownership of a media file [Crav97] [Crav98]; to deactivate the playback option if the media clip was found to be a pirated one; and, to comply with the digital rights management (DRM) principles [Rose01] [Arno02] [Beck04].

This section addresses various data-hiding techniques employed for *audio content* protection. A data hiding technique [Bend95] [Peti99] [Cox01a] [Wu03] involves schemes that insert *signals* of certain characteristics (imperceptible, robust, statistically undetectable, etc,) in the audio. Such *signals* are called *watermarks*. A *digital audio watermark* (DAW) can be defined as a stream of bits embedded (or hidden) in an audio file that offers features such as content protection, intellectual property management, and proof of ownership. Before getting into the details of audio watermarking techniques, we present some of the basic concepts of "content protection" and "copyright management." A simple idea is to include in the audio clip a *password* or a *key* that is relatively difficult to be "hacked" in a given amount of time. Details of this *key* must be essential in order for someone to playback the audio file. This ensures content protection. For copyright management, a data string or some form of a signature that indicates the proof of ownership is included in the audio. However, in both the cases, if the *key* or the *signature string* is lost, one cannot guarantee for either content protection or copyright management. Therefore, in applications related to copyright management, the watermarking techniques are usually combined with encryption schemes. The latter is beyond the scope of this text, and the reader can find more information on several encryption techniques in [Wick95] [Trap02] [Mao03].

Some of the important aspects that pose major challenges in digital audio watermarking are described below. Digital audio and video watermarking techniques rely on the perceptual properties of the human auditory (HAS) and human visual systems (HVS), respectively. In certain ways, the HVS is *less sensitive* compared to the human auditory system (HAS). This indicates that it is more challenging to embed imperceptible audio watermarks. Moreover, the HAS is characterized by larger dynamic ranges in terms of power/amplitudes ($\sim 10^8{:}1$) and frequencies ($\sim 10^4{:}1$). Unlike in video compression where data rates of the order of 1 Mb/s are employed, in audio coding the data rates typically range from 16 to 320 kb/s. This means that the stream of bits embedded in an audio file is much shorter compared to the auxiliary information inserted in a video file for watermarking. Researchers published a variety of sophisticated techniques ([Bend95] [Cox95] [Bend96] [Bone96] [Cox97] [Bass98] [Lacy98] [Swan98a] [Kiro03a] [Zhen03]) in order to meet several conflicting demands ([Crav98] [Swan99] [Cox99] [Barn03a] [Barn03b]) associated with the digital audio watermarking ([Cox01a] [Arno02] [Wu03]).

11.5.1 Background

The advent of ISO/IEC MPEG-4 standardization (1996–2000) [ISOI99] [ISOI00a] established new research goals for high-quality coding of general audio signals even at low bit rates. Moreover, *bit-rate scalability* and *error resilience/protection* tools of the MPEG-4 audio standard enabled flexible selection of coding features and dynamically adapt to the channel conditions and the varying channel capacity. These and other significant strides in the area of digital audio coding have motivated considerable research during the last 10 years towards formulation of several audio watermarking algorithms. Table 11.3 lists some of the important innovations in digital audio watermarking research.

In 1995, Bender *et al.* at the MIT Media Laboratory, proposed a variety of data-hiding techniques for audio [Bend95] [Bend96]. These include *low-bit coding, phase coding, spread-spectrum*, and *echo data hiding*. The pioneering work

Table 11.3. Some of the milestones in the digital audio watermarking research – a historical perspective.

Year	Digital audio watermarking (DAW) scheme	Related references
1995	Phase coding, low-bit coding, echo-hiding, and spread-spectrum techniques	[Bend95] [Bend96]
1995–96	Echo-hiding watermarking scheme	[Gruh96]
1995–96	Spread-spectrum watermarking for multimedia	[Cox95] [Cox96] [Cox97]
1996	Watermarking based on perceptual masking	[Bone96] [Swan98a]
1997	DAW based on vector quantization	[Mori97]
1998	Time-domain DAW technique	[Bass98] [Bass01]
1999	DAW based on bandlimited random sequences	[Xu99a]
1999	DAW based on a psychoacoustic model and spread-spectrum techniques	[Garc99]
1999	Permutation and data scrambling	[Wang99]
2000	DAW in the cepstrum-domain	[Lee00a] [Lee00b]
2000	Watermarking techniques for MPEG-2 AAC bitstreams	[Neub00a] [Neub00b]
2000	Audio watermarking based on the "patchwork" algorithm	[Arno00]
2001	Combined audio compression/watermarking methods	[Herr01a] [Herr01b] [Herr01c]
2001	Modified echo hiding technique	[Oh01] [Oh02]
2001	Adaptive and content-based DAW scheme (improve the accuracy of watermark detection in echo hiding)	[Foo01]
2001	Modified patchwork algorithm	[Yeo01] [Yeo03]
2001	Time-scale modification techniques for DAW	[Mans01a] [Mans01b]
2001	Replica modification	[Petr01]
2001	Security aspects of DAW schemes	[Kalk01a] [Kalk01b] [Mans02]
2001	Synchronization methods of audio watermarks	[Leon01] [Gang02]
2002	Pitch scaling techniques for robust DAW	[Shin02]
2002	DAW using artificial neural networks	[Huij02]
2002	Time-spreading echo hiding using PN sequences	[Ko02]
2002	Improved spread-spectrum audio watermarking	[Kiro01a] [Kiro01b] [Kiro02] [Kiro03a]
2002	Hybrid spread spectrum DAW	[Munt02]
2002	DAW using subband division/QMF banks	[Sait02a] [Sait02b]
2002	Blind cepstrum-domain DAW	[Hsie02]
2002	Blind DAW with self-synchronization	[Huan02]
2003	Time-frequency techniques	[Esma03]
2003	Chirp-based techniques for robust DAW	[Erku03]

(*continued overleaf*)

Table 11.3. (*continued*)

Year	Digital audio watermarking (DAW) scheme	Related references
2003	DAW using turbo codes	[Cvej03]
2003	DAW using sinusoidal patterns based on PN sequences	[Zhen03]
Year	**Tutorial reviews and survey papers**	**Reference**
1996	Techniques for data hiding	[Bend96]
1997	Secure spread spectrum watermarking for multimedia	[Cox97]
1997	Resolving rightful ownerships with invisible watermarks	[Crav97] [Crav98]
1998	Multimedia data-embedding and watermarking technologies	[Swan98]
1998	Special issue on copyright and privacy protection	[Spec98]
1999	Special issue on identification and protection of multimedia information	[Spec99]
1999	Watermarking as communications with side information	[Cox99]
1999	Multimedia watermarking techniques	[Hart99]
1999	Information hiding – a survey	[Peti99]
1999	Current state of the art, challenges and future directions for audio watermarking	[Swan99]
2001	Digital watermarking: algorithms and applications	[Podi01]
2001	Electronic watermarking: the first 50 years	[Cox01]
2003	What is the future of watermarking? Part I & II	[Barn03a] [Barn03b]

of Cox *et al.* resulted in a novel scheme called spread spectrum watermarking for multimedia [Cox95] [Cox96] [Cox97]. In 1996, the first ever international workshop on information hiding was held in Cambridge, U.K., that attracted many researchers working on audio watermarking [Cox96] [Gruh96]. Boney *et al.*, in 1996, presented a watermarking scheme for audio signals based on the perceptual masking properties of the human ear [Bone96] [Swan98a]. This was perhaps the first audio watermarking scheme that employed the psychoacoustic principles and the frequency/temporal masking characteristics of the HAS. A watermarking scheme based on vector quantization was also proposed [Mori97]. All the aforementioned data hiding schemes, except the "echo hiding" and "phase coding" techniques, embed watermarks in the frequency domain. In 1998, Bassia *et al.* presented a time-domain audio watermarking technique [Bass98] [Bass01]. Audio watermarking using bandlimited random sequences [Iked99], multi-bit

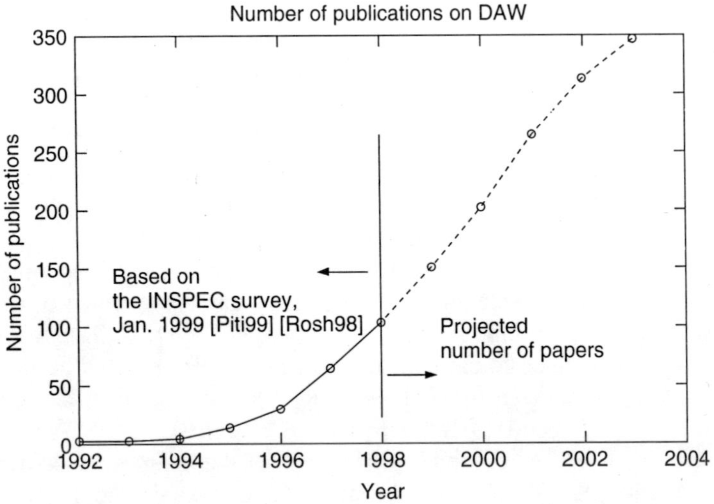

Figure 11.15. Number of publications on digital watermarking (according to INSPEC, Jan. 1999; [Roch98] [Peti99]).

hopping and HAS properties [Xu99a], psychoacoustic models and spread spectrum theory [Garc99], and permutation-based data scrambling [Wang99] have also been proposed. Data hiding research during 1995–1999 [Swan99] [Arno00] laid the foundation for the next generation (2000–2004) watermarking techniques. Figure 11.15 presents a review of the number of publications on digital watermarking appeared over the years, according to INSPEC, Jan. 1999 [Roch98] [Peti99]. Special publication issues on copyright/privacy protection and multimedia information protection have appeared in *Proceedings of the IEEE* and other journals [Spec98] [Spec99] during 1998–1999. Several tutorials and survey papers highlighting watermarking techniques for multimedia have been presented [Bend96] [Cox97] [Swan98a] [Swan98b] [Cox99] [Hart99] [Peti99].

In general, the success of a data hiding scheme is characterized by the watermark *embedding* and *detection* procedures. Researchers began to develop schemes that enable improved watermark detection [Cilo00] [Furo00] [Kim00] [Lean00] [Seok00]. Lee *et al.* presented a digital audio watermarking scheme that works in the cepstrum domain [Lee00a] [Lee00b]. Neubauer and Herre proposed watermarking techniques for MPEG-2 AAC bitstreams [Neub00a] [Neub00b]. Enhanced spread spectrum watermarking for MPEG-2 AAC was proposed later by Cheng *et al.* [Chen02b]. With the advent of sophisticated audio compression schemes, the combined audio compression/watermarking techniques also gained interest [Herr01a] [Herr01b] [Herr01c].

Motivated by the success of the "patchwork" algorithm in image watermarking [Bend96], Arnold proposed an audio watermarking scheme based on the statistical patchwork method that operates in the frequency domain [Arno00]. Later, Yeo *et al.* presented a *modified patchwork algorithm* (MPA) [Yeo01] [Yeo03] that is relatively more robust compared to other patchwork algorithms for DAW. Robust audio watermarking schemes based on time-scale modification [Mans01a] [Mans01b], replica modulation [Petr01], pitch scaling [Shin02], and artificial neural networks [Huij02] have also been proposed.

The echo hiding technique proposed by Gruhl *et al.* [Gruh96] was modified and improved by Oh *et al.* [Oh01] [Oh02] for robust and imperceptible audio watermarking. Ko *et al.* [Ko02] further modified the echo hiding technique by time-spreading the echo using pseudorandom noise (PN) sequences. In order to avoid the audible echoes and further improve the accuracy of watermark detection in the echo hiding technique (i.e., [Gruh96]), an adaptive and content-based audio watermarking technique has been proposed [Foo01]. Costa's pioneering work on the theoretical bounds on the amount of data that can be hidden in a signal, [Cost83] motivated researchers to apply these bounds to the data hiding problem in audio [Chou01] [Kalk02] [Neub02] [Peel03].

Watermark detection becomes particularly challenging when a part of the embedded watermark is lost due to some common signal processing manipulations (e.g., cropping, sampling, shifting, etc). Hence, security and synchronization aspects of DAW became an important issue during the early 2000. Several synchronization methods for audio watermarks have been proposed [Lean01] [Gang02]. Efficient methods to improve the security of watermarks have also been presented [Kalk01a] [Kalk01b] [Gang02] [Mans02]. Kirovski and Malvar presented a robust framework for direct-sequence spread-spectrum (DSSS)-based audio watermarking by incorporating several sophisticated techniques, such as robustness to desynchronization, cepstrum filtering, and chess watermarks [Kiro01a] [Kiro01b] [Kiro02] [Kiro03a] [Kiro03b]. DAW based on hybrid spread-spectrum methods have also been investigated [Munt02]. Saito *et al.* employed a subband division technique based on QMF banks for digital audio watermarking [Sait02a] [Sait02b]. DAW based on time-frequency characteristics and chirp-based techniques have also been proposed [Erku03] [Esma03]. Motivated by the applications of *turbo codes* in encryption techniques, Cvejic *et al.* [Cvej03] developed a robust DAW scheme using turbo codes. DAW using sinusoidal patterns based on PN sequences has been described in [Zhen03].

Blind and asymmetric audio watermarking gained popularity due to their inherent property that the original signal is not required for watermark detection. Several blind audio watermarking algorithms have been reported [Hsie02] [Huan02] [Peti02], however, the research associated with the blind DAW is still in its early stages. *Asymmetric watermarking schemes* involve watermark detection by knowing only a part of the secret information used to encode and embed the watermark.

11.5.2 A Generic Architecture for DAW

During the last decade, researches have proposed several efficient methods for audio watermarking. Most of these algorithms are based on the generic architecture shown in Figure 11.16. The three primary modules associated with a digital

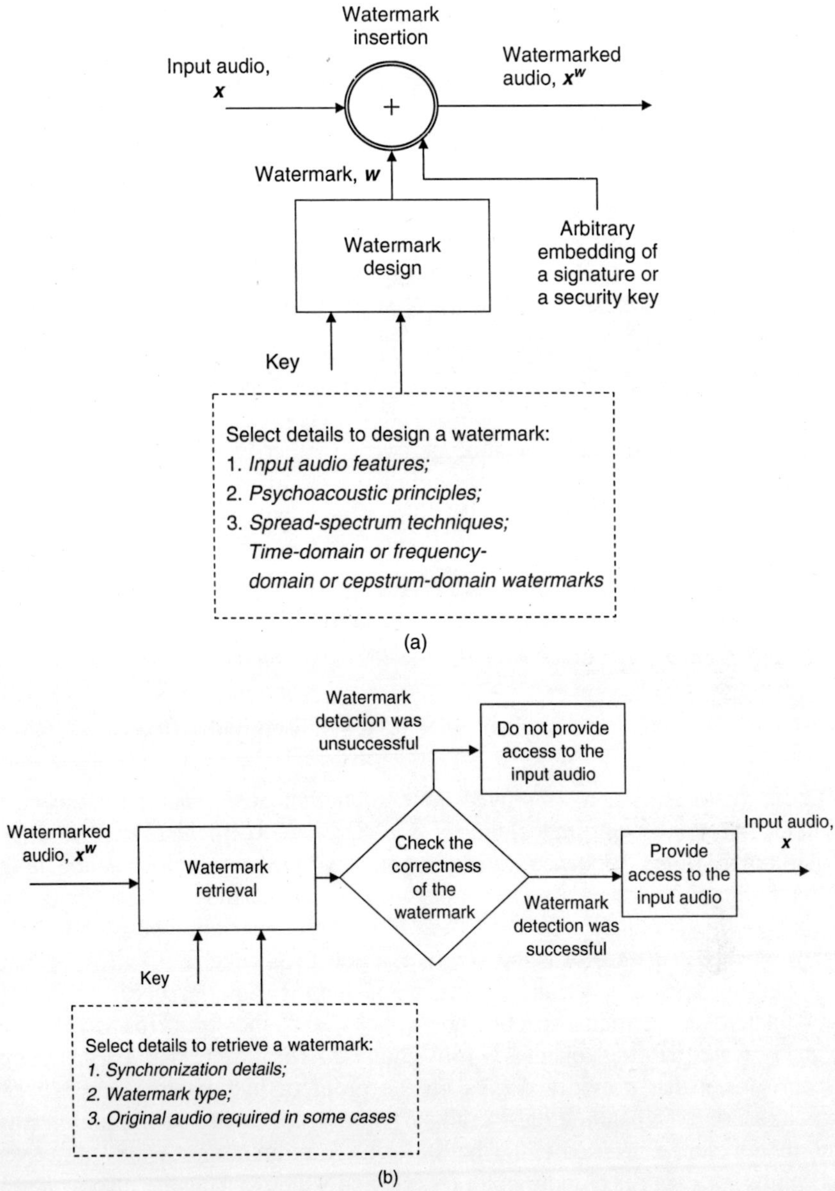

Figure 11.16. A general audio watermarking framework: (a) watermark encoding procedure (watermark design and embedding) and (b) watermark decoder.

audio watermarking scheme include the watermark design, the watermark embedding algorithm, and the watermark retrieval. The first two modules belong to the *watermark encoding* procedure (Figure 11.16(a)), and the watermark retrieval can be viewed as a *watermark decoder* (Figure 11.16(b)).

First, the watermark design section generates an *auxiliary data string* based on a security key and a set of design parameters (e.g., bit rates, robustness, psychoacoustic principles, etc,). Furthermore, the watermark generation technique often depends on the analysis features of the input audio, the human auditory system (HAS), and some data scrambling/spread spectrum technique. Nevertheless, the ultimate objective is to design a watermark that is perceptually inaudible and robust. The watermark embedding algorithm relies in hiding the designed watermark sequence, $\boldsymbol{w}$, in the audio bitstream, $\boldsymbol{x}$. This is given by

$$\boldsymbol{x}^{w} = \wp(\boldsymbol{x}, \boldsymbol{w}), \tag{11.29}$$

where $\boldsymbol{x}^{w}$ is the watermarked audio and $\wp$ represents the watermark embedding function.

The watermark embedding can be performed either in the frequency domain. The idea is to exploit the frequency- and the temporal-masking properties of the human ear. Three key features of the human auditory systems that are used in the audio watermarking are:

i) sounds that are louder typically tend to mask out the quieter sounds;
ii) the human ear perceives only the "relative phase" associated with the sounds; and
iii) the HAS fails to recognize minor reverberations and echoes.

These features, when employed in conjunction with encryption schemes [Mao03] or spread spectrum techniques [Pick82], lead to efficient DAW schemes. In particular, direct sequence spread spectrum (DSSS) technique – due to its inherent suitability towards secure transmission – is the most popular spreading scheme employed in data hiding. The DSSS technique employs pseudo-noise (PN) sequences to transform the original signal to a wideband and noise-like. The resulting spread spectrum signals are resilient to interferences, difficult to demodulate/decode, robust to collusion attacks, and, therefore, forms the basis for many watermarking schemes. From Figure 11.16(a), note that a signature or a security key that provides details on the proof of authorship/ownership can also be inserted. Depending on overall system objectives and design philosophy, watermarks can be inserted either before or after audio compression. Moreover, algorithms for combined audio compression and watermarking are also available [Herr01a] [Herr01b]. The watermark security should lie in the hidden random keys and not in the watermark-embedding algorithm. In other words, details of the watermark-embedding scheme must be available at the decoder for efficient watermark detection.

Figure 11.16(b) illustrates the watermark decoding procedure. The secret key and the design parameters employed in watermark embedding are available during the watermark retrieval process. Two common scenarios of watermark retrieval include (11.30) – when the input audio is employed for decoding and when the input audio is not available for watermark detection. The latter case results in blind watermark detection:

$$\begin{aligned} &\text{Case I: Input audio required for decoding} \\ &\quad \boldsymbol{w}' = \varsigma(\boldsymbol{x}^w, \boldsymbol{x}) \\ &\text{Case II: Input audio NOT required} \\ &\quad \boldsymbol{w}' = \varsigma(\boldsymbol{x}^w), \end{aligned} \tag{11.30}$$

where $\boldsymbol{w}'$ is the decoded watermark and ς represents the watermark retrieval function. Note that, in case of perfect watermark retrieval, $\boldsymbol{w}' = \boldsymbol{w}$. Watermark retrieval must be unambiguous. In other words, the watermark detection must provide reliable details about the content ownership. In particular, an audio watermark must be imperceptible, statistically undetectable, remain in the audio after repeated encoding/copying, and robust to a variety of signal processing manipulations and deliberate attacks. For a DAW scheme to be successful in copyright management and audio content protection applications, it should have some important attributes that are described next.

11.5.3 DAW Schemes – Attributes

11.5.3.1 *Imperceptibility or Perceptual Transparency* Watermarks embedded in the digital audio must be perceptually transparent and should not result in any audible artifacts. Watermarks can be inserted in either perceptually *significant* or *insignificant* regions depending upon the application. If inserted in a perceptually significant region, special care must be taken for watermarks to be inaudible. Inaudibility can be guaranteed by placing watermark signals in audio segments whose low-frequency components have higher energy. Note that this technique is based on the frequency-masking characteristics of the HAS that we discussed in Chapter 5. On the other hand, if embedded in perceptually insignificant regions, watermarks become vulnerable to compression algorithms. Therefore, watermark designers must consider the trade-offs between the *robustness* and the *imperceptibility* features while designing a watermarking scheme. Another important factor is the optimum energy of the watermarked signal required for efficient watermark detection. Watermarks with higher energy possess some inherent features, such as the reliable watermark detection and the robustness to intentional and unintentional signal processing operations. However, it should be noted that increase in the audio watermark energy may result in audible artifacts.

11.5.3.2 *Statistical Transparency* Consider a set of audio signals, $x, y, \ldots, z$ and a watermark signal, w. Let the watermarked signals be denoted as $x^w, y^w, \ldots, z^w$. Watermarked signals must be statistically undetectable. Watermarks are designed such that users are not able to detect the embedded watermark by performing some statistical signal processing operation (e.g., correlation, autoregressive estimation, etc.). This scenario can be treated as a special case of intentional or collusion attacks that employ statistical methods.

11.5.3.3 Robustness to Signal Processing Manipulations Watermarks are embedded directly in the audio stream. In fact, watermark bits are distributed (typically 2–128 b/s) uniformly across the audio bitstream. An audio signal undergoes several changes on its way from a transmitter/encoder to a receiver/decoder. Therefore, watermark robustness to a variety of signal processing operations is very important. Common signal processing operations typically considered for the robustness of an audio watermark include: linear and nonlinear filtering, resampling, requantization, A-to-D and D-to-A conversions, cropping, MP3 compression, low-pass/high-pass filtering, dithering, bass and treble control changes, etc. Watermark robustness can be improved by also increasing the number of watermark bits. Nonetheless, this results in increased encoding bit rates. Spread-spectrum techniques offer good robustness, however, with increased computational complexity and synchronization problems.

11.6 SUMMARY OF COMMERCIAL APPLICATIONS

Some of the popular applications (Table 11.4) for embedded audio coding include digital broadcast audio (DBA) [Stol93b] [Jurg96], direct broadcast satellite (DBS) [Prit90], digital versatile disc (DVD) [Crav96], high-definition television (HDTV) [USAT95b], cinema [Todd94], electronic distribution of music [Bran02] and audio-on-demand over wide area networks such as the Internet [Diet96]. Audio coding has also enabled miniaturization of digital audio storage media such as compact MiniDisc [Yosh94]. With the advent of the MP3 audio format, that denotes audio files that have been compressed using the MPEG-1 layer III algorithm, perceptual audio coding has become of central importance to overnetwork exchange of multimedia information. Moreover, the MP3 algorithm has been integrated into several popular portable consumer audio playback devices that are specifically designed for web compatibility. Streaming audio codecs have been developed by several commercial entities. Example codecs include RealAudio™, and atob™ audio. In addition, DolbyNET™ [DOLBY], a version of the Dolby AC-3 algorithm modified to accommodate Internet streaming [Vern99] has been successfully integrated into streaming audio processors for delivery of audio-on-demand to the desktop web browser. Moreover, some of the MPEG-4 audio tools enable real-time audio transmission over packet-switched networks such as the Internet [Diet96] [Ben99] [Liu99] [Zhan00], transparent compression at lower bit rates with reduced complexity, low delay coding, and enhanced bit-error-robustness. The MPEG-7 audio supports multimedia indexing and searching, multimedia editing, broadcast media selection, and multimedia digital library sorting.

New audio coding algorithms must satisfy several requirements. First, consumers expect high quality at low to medium bit rates. Secondly, the popularity of streaming audio requires that algorithms intended for packet-switched networks

Table 11.4. Audio coding standards and applications.

Algorithm	Applications	References
Sony ATRAC	MiniDisc	[Yosh94]
Lucent PAC	DBA: 128/160 kb/s	[John95]
Philips PASC	Digital audio tape (DAT)	[Lokh91] [Lokh92]
Dolby AC-2	Digital broadcast audio (DBA)	[Todd84]
Dolby AC-3	Cinema, HDTV, DVD	[Todd94]
Audio Processing Technology, APT-X100	Cinema	[Wyli96a] [Wyli96b]
DTS Coherent acoustics	Cinema	[Smyt96] [Bosi00]
MPEG-1, LI-III	"MP3": LIII DBA: LII@256 kb/s DBS: LII@224 kb/s DCC: LI@384 kb/s	[ISOI92]
MPEG-2/BC-LSF	Cinema	[ISOI94a]
MPEG-2/AAC	Internet/www, e.g., LiquidAudio™, atob™ audio	[ISOI96a] [ISOI97b]
MPEG-4	Low bit rate, low delay, and scalable audio encoding; streaming audio applications	[ISOI97a] [ISOI99] [ISOI00]
MPEG-7	Broadcast media selection, multimedia indexing/searching/editing;	[ISOI01b] [ISOI01d]
Direct-stream digital (DSD) encoding	Super audio CD encoding	[Reef01a] [Jans03]
Meridian lossless packing (MLP)	DVD-Audio	[Gerz99] [Kuzu03]

are able to deal with a highly time-varying channel. Therefore, new algorithms must include facilities for *scalable* operation, i.e., the ability to decode only a subset of the overall bitstream with minimal loss of quality. Finally, for any algorithm to be viable, it should be possible to implement the decoder in real time on a general-purpose processor architecture. Some of the important aspects that have been influential, over the years, in the design of an audio codec are listed below:

- Streaming audio applications
- Low delay audio coding
- Error robustness
- Hybrid speech/audio coding structures
- Digital audio watermarking
- Perceptual audio quality measurements
- Bit rate scalability
- Complexity issues
- Psychoacoustic signal analysis techniques
- Parametric models in audio coding
- Lossless audio compression
- Interoperable multimedia framework
- Intellectual property management and protection

Development of sophisticated audio coding algorithms that deliver *rate-scalable* and *quality-scalable* compression in error-prone channels received attention during the late 1990s. Several researchers developed structured sound representations oriented specifically towards low-bandwidth and native-signal-processing sound synthesis applications. For example, NetSound [Case96] is a sound and music description system in which sound streams are described by decomposition into a sound-specification description that represents arbitrarily complex signal processing algorithms, as well as an event list comprising scores or MIDI files. The advantage of this approach over streaming codecs is that only a very small amount of data is required to transmit complex instrument descriptions and appropriately parameterized scores. The trade-off with respect to streaming codecs, however, is that significant client-side computing resources are required during real-time synthesis. Some examples of synthesis techniques used by NetSound include wavetable, FM, phase-vocoder, or additive synthesis. Scheirer proposed a paradigm called the *generalized audio coding* [Sche01] in which Structured Audio (SA) encompasses all other audio coding techniques. Moreover, the MPEG-4 SA tool offers a flexible framework in which both lossy and lossless audio compression can be realized. A variety of lossless audio coding features have been incorporated in the MPEG-4 audio standard [Sche01] [Mori02a] [Vasi02].

Although streaming audio has achieved widespread acceptance on 28.8 kb/s dial-up telephone connections, consistent high-quality output is still difficult.

Network-specific design considerations motivated research into joint source-channel coding [Ben99] for audio over the Internet. Another emerging trend is one of convergence between low-rate audio coding algorithms and speech coders that are increasingly embedding mechanisms to exploit perceptual irrelevancies [Malv98] [Mori98] [Ramp98] [Ramp99] [Trin99a] [Trin99b] [Maki00]. Several wideband speech and audio coders based on the CS-ACELP ITU-T G.729 standard have been proposed [Kois00] [Kuma00]. Low delay [Jbir98] [Alla99] [Dorw01] [Schu02] [Schu05], enhanced bit error robustness [ISOI00] [Sper00] [Sper02], and scalability [Bran94b] [Gril95] [Span97] [Gril98] [Hans98] [Herr98] [Jin99] [Zhou01] [Dong02] [Kim02] [Mori02b] have also become very important. In particular, the MPEG-4 bit rate scalability tool provides encoder with options to transmit bitstreams at variable bit rates. Note that scalable audio coders have several layers, i.e., a core layer and a series of enhancement layers. The core layer encodes the main audio stream, while the enhancement layers provide further resolution and scalability. Scalable algorithms will ultimately be used to accommodate the unique challenges associated with audio transmission over time-varying channels such as the packet-switched networks. Several ideas that provide a link between the perceptual and the lossless audio coding techniques [Geig02] [Mori00] [Schu01] [Mori02a] [Mori02b] [Geig03] [Mori03] have also been proposed.

Hybrid coding algorithms that make use of specific characteristics of a signal while operating over a range of bit rates became popular [Bola98] [Deri99] [Rong99] [Maki00] [Ning02]. Some experimental work performed in the context of MPEG-4 standardization has demonstrated that a cascaded, hybrid sinusoidal/time-frequency coder can not only meet but in some cases even exceed the output quality achieved by the time-frequency coder alone at the same bit rate for certain test signals [Edle96b]. Sinusoidal signal models, due to their inherent suitability towards scalable audio coding, gained significant research interests during the late 1990s [Taor99] [Pain00] [Ferr01a] [Ferr01b] [Pain01] [Raad01] [Herm02] [Heus02] [Pain02] [Purn02] [Shao02] [Pain05].

Potential improvements for the various perceptual coder building blocks, such as novel filter banks for low-delay coding and reduced pre-echo [Schu98] [Schu00], efficient bit-allocation strategies [Yang03], and new psychoacoustic signal analysis techniques [Baum98] [Huan99] [Vande02] have been considered. Researchers also investigated new algorithms for tasks of peripheral interest to perceptual audio coding such as transform-domain signal modifications [Lanc99] and digital watermarking [Neub98] [Tewf99].

In order to offer audiophiles with listening experiences that promise to be more realistic than ever, audio codec designers pursued several sophisticated multichannel audio coding techniques [DOLBY] [Bosi93] [Holm99] [Bosi00]. In the mid-1990s, discrete encoding, i.e., 5.1 separate channels of audio, was introduced by the Dolby Laboratories and the Digital Theater Systems (DTS). The human auditory system allows hearing in substantially more directions, than what current multichannel audio systems offer, therefore, future research will focus on overcoming the limitations of existing multichannel systems. In particular, research involving spatial audio, real-time acoustic source localization, binaural cue coding, and application of head-related transfer functions (HRTF) towards rendering immersive audio [Sand01] [Fall02a] [Fall02b] [Kimu02] [Yang02] have been considered.

PROBLEMS

11.1. What are the differences between the SACD and DVD-A formats? Explain using diagrams. Provide bitrates.

11.2. Use diagrams and describe the MLP algorithm employed in the DVD-A.

11.3. How is direct stream digital (DSD) technology employed in the SACD different from other encoding schemes? Describe using a block diagram.

11.4. What is the main idea behind the Integer MDCT and its application in lossless audio coding? Explain in a concise manner and by using mathematical arguments the difference between Integer MDCT (Int MDCT) and the MDCT. Show the utility of the FFT in both.

11.5. Explain using an example the lossless matrixing employed in the MLP. How is this scheme different from the Givens rotations employed in the IntMDCT?

11.6. List some of the key differences between the lossless transform audio coder (LTAC) and the IntMDCT algorithm?

11.7. Survey the literature for studies on fast algorithms for IntMDCT and describe at least one such algorithm. Show how computational savings are achieved.

11.8. Survey the literature for studies on the properties of IntMDCT. Discuss in particular properties dealing with preservation of energy/power (whichever is appropriate) and transform pair uniqueness.

11.9. Describe the concept of noise shaping. Explain how it is applied in the DSD algorithm.

COMPUTER EXERCISE

11.10. Write MATLAB code to implement an IntMDCT. Profile the algorithm and compare the computational complexity against a standard MDCT.

CHAPTER 12

QUALITY MEASURES FOR PERCEPTUAL AUDIO CODING

12.1 INTRODUCTION

In many situations, and particularly in the context of standardization activities, performance measures are needed to evaluate whether one of the established or emerging techniques is in some sense superior to other alternative methods. Perceptual audio codecs are most often evaluated in terms of bit rate, complexity, delay, robustness, and output quality. Reliable and repeatable output quality assessment (which is related to robustness) presents a significant challenge. It is well known that perceptual coders can achieve transparent quality over a very broad, highly signal-dependent range of segmental SNRs ranging from as low as 13 dB to as high as 90 dB. Classical objective measures of signal fidelity such as signal-to-noise ratio (SNR) or total harmonic distortion (THD) are therefore inadequate [Ryde96]. As a result, time-consuming and expensive subjective listening tests are required to measure the small impairments that often characterize perceptual coding algorithms. Despite some confounding factors, subjective listening tests are nevertheless the most reliable tool available for codec quality evaluation, and standardized listening test procedures have been developed to maximize reliability. Research into improved subjective testing methodologies is also quite active. At the same time, considerable research is being devoted to the development of automatic perceptual measurement systems that can predict accurately the outcomes of subjective listening tests. Several algorithms have been proposed in the last two decades, and the ITU-R has recently adopted an international standard after combining several of the best techniques proposed in the literature.

This chapter offers a perspective on quality measures for perceptual audio coding. The first half of the chapter is concerned with subjective evaluations of perceptual codec quality. Sections 12.2 through 12.5 describe subjective quality measurement techniques, identify confounding factors that complicate subjective tests, and give sample subjective test results that facilitate comparisons between several 2- and 5.1-channel standards. The second half of the chapter reviews fundamental techniques and significant developments in automatic perceptual measurement systems, with particular attention given to several of the candidates from the ITU-R standardization process. In particular, Sections 12.6 and 12.7, respectively, describe perceptual measurement systems and the classes of algorithms that have been proposed for standardization. Section 12.8 describes the recently completed ITU-R TG 10/4, which led to the adoption of the ITU-R Rec. BS.1387 algorithm for perceptual measurement. Section 9.10 also provides some information on the ITU-T P.861. Section 12.9 offers some final remarks on anticipated future developments in quality measures for perceptual codecs.

12.2 SUBJECTIVE QUALITY MEASURES

Although listening tests are often conducted informally, the ITU-R Recommendation BS.1116 [ITUR94b] formally specifies a listening environment and test procedure appropriate for subjective evaluations of the small impairments associated with high-quality audio codecs. The standard procedure calls for grading by expert listeners [Bech92] using the CCIR continuous impairment scale (Figure 12.1-a) [ITUR90] in a double blind, A-B-C triple stimulus hidden reference comparison paradigm. While stimulus A always contains the reference (uncoded) signal, the B and C stimuli contain in random order a repetition of the reference and then the impaired (coded) signal, i.e., either B or C is a hidden reference. After listening to all three, the subject must identify either B or C as the hidden reference, and then grade the impaired stimulus (coded signal) relative to the reference stimulus using the five-category, 41-point "continuous" absolute category rating (ACR) impairment scale shown in the left-hand column of Figure 12.1 (a). From best to worst, the five ACR ranges rate the coding distortion as "imperceptible (5)," "perceptible but not annoying (4.0–4.9)," "slightly annoying (3.0–3.9)", "annoying (2.0–2.9)," or "very annoying (1.0–1.9)." A default grade of 5.0 is assigned to the stimulus identified by the subject as the hidden reference. A subjective difference grade (SDG) is computed by subtracting the score assigned to the actual hidden reference from the score assigned to the actual impaired signal. Thus, negative difference grades are obtained when the subject identifies correctly the hidden reference, and positive difference grades are obtained if the subject misidentifies the hidden reference. Over many subjects and many trials, mean impairment scores are calculated and used to evaluate codec performance relative to the ideal of transparency. It is important to notice the difference between the small impairment subjective measurements in [ITUR94b] and the five-point mean opinion score (MOS) most often associated with speech coding algorithms [ITUT94]. Unlike the small impairment scale, the scale of the speech coding MOS is discrete, and scores are absolute rather than relative to a hidden reference. To emphasize this difference, it has been proposed [Spor96] that "mean subjective score" (MSS) denote the small impairment subjective score for perceptual audio coders. During data reduction, MSS and MSS standard deviations are computed for both the coded signals and the hidden references. An MSS for the hidden reference different from 5.0 means that subjects have difficulty distinguishing the hidden reference from the coded signal.

In fact, nearly transparent quality for the coded signal is implied if the hidden reference MSS lies within the 95% confidence interval of the coded signal and the coded signal MSS lies within the 95% confidence interval of the hidden reference. Unless otherwise specified, the subjective listening test scores cited for the various algorithms described throughout this text are from either the absolute or the differential small impairment scales in Figure 12.1 (a).

It is important to realize that the most reliable subjective evaluation strategy for a given perceptual codec depends on the nature of the coding distortion. Although the small-scale impairments associated with nearly transparent coding are well characterized by measurements relative to a reference standard using a fine-grade scale, some experts have argued that the more-audible distortions associated with nontransparent coding are best measured using a different scale that can better cope with large impairments.

For example, in listening tests [Keyh98] on 16-kb/s codecs for the WorldSpace satellite communications system, it was determined that an ITU-T P.800/P.830

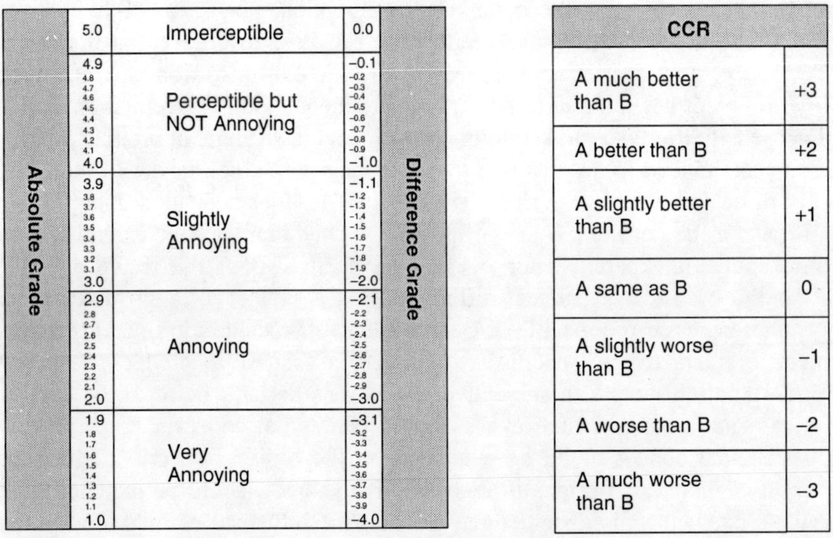

Figure 12.1. Subjective quality scales: (a) ITU-R Rec. BS.1116 [ITUR94b] small impairment scale for absolute and differential subjective quality grades, (b) ITU-T Rec. P.800/P.830 [ITUR96] large impairment comparison category rating.

seven-point comparison category rating (CCR) method [ITUR96] was better suited to the evaluation task (Figure 12.1b) than the scale of BS.1116 because of the nontransparent quality associated with the test signal. Investigators preferred the CCR over both the small impairment scale as well as the five-point absolute category rating (ACR) commonly used in tests of speech codecs. A listening test standard for large scale impairments analogous to BS.1116 does not yet exist for audio codec evaluation. This is likely to change in light of the growing popularity for network and wireless applications of low-rate, nontransparent audio codecs characterized at times by distinctly audible (not small-scale) artifacts.

12.3 CONFOUNDING FACTORS IN SUBJECTIVE EVALUATIONS

Regardless of the particular grading scale in use, subjective test outcomes generated using even rigorous methodologies such as the ITU-R BS.1116 are still influenced by factors such as context, site selection, and individual listener acuity (physical) or preference (cognitive). Before comparing subjective test results on particular codecs, therefore, one should be prepared to interpret the subjective scores with some care. For example, consider the variability of "expert" listeners. A study of decision strategies [Prec97] using multidimensional scaling techniques [Schi81] found that subjects disagree on the relative importance with which to weigh perceptual criteria during impairment detection tasks. In another study [Shli96], Shlien and Soulodre presented experimental evidence that can be interpreted as a repudiation of the "golden ear." Expert listeners were tasked with discriminating between clean audio and audio corrupted by low-level artifacts typically induced by audio codecs (five types were analyzed in [Miln92]), including pre-echo distortion, unmasked granular (quantization) noise, and high-frequency boost or attenuation. Different experts were sensitive to different artifact types. For instance, a subject strongly sensitive to pre-echo distortion was not necessarily proficient when asked to detect high-frequency gain or attenuation. In fact, artifact sensitivities were linked to psychophysical measurement profiles. Because test subjects tended to be sensitive to one impairment type but not to others, the ability of an individual to perform as an expert listener depended upon the type of artifacts to be detected. Sporer [Spor96] reached similar conclusions after yet a third study of expert listeners. A statistical analysis of the results indicated that significant site and subject differences exist, that even well-trained listeners cannot necessarily repeat their own results (self-recalibration), and that some subjects are sensitive to specific artifacts but insensitive to others. The numerous confounding factors in subjective evaluations perhaps point to the need for a fundamental change in the evaluation of perceptual codecs. Sporer [Spor96] suggested that codecs could be evaluated on the basis of specific artifact perceptibility. Then, appropriate artifact-sensitive subjects could be matched to the test material. It has also been suggested that the influence of individual nonrepeatability could be minimized with larger subject pools. Nonhuman factors also influence subjective listening test outcomes. For example, playback level (SPL) and background noise, respectively, can influence excitation pattern shapes and introduce undesired masking effects. While both SPL and background noise can be replicated in different listening test environments, other environmental factors are not as easy to replicate. The presentation method can strongly influence perceived quality, because loudspeakers introduce distortions on their own and in conjunction with a listening room. These effects can introduce site dependencies. In short, although they have proven effective, existing subjective test procedures for audio codecs are clearly suboptimal. Recent research into more reliable tools for subjective codec evaluations has shown promise and is continuing. Confounding factors must be considered carefully when subjective test results are compared between codecs and between different test sites. These considerations have motivated a number of subjective tests in which multiple algorithms were tested on the same site under identical listening conditions. In Sections 12.4 and 12.5, we will consider two example tests, one for 2-channel codecs, and the other for 5.1-channel algorithms.

12.4 SUBJECTIVE EVALUATIONS OF TWO-CHANNEL STANDARDIZED CODECS

The influence of site and subject dependencies on subjective listening tests can potentially invalidate direct comparisons between independent test results for different algorithms. Ideally, fair intercodec comparisons require that scores are obtained from a single site with the same test subjects. Soulodre *et al.* conducted a formal ITU-R BS.1116-compliant [ITUR94b] listening test that compared several standardized two-channel stereo codecs [Soul98], including the MPEG-1 layer II [ISOI92], the MPEG-1 layer III [ISOI92], the MPEG-2 AAC [ISOI96a], the Lucent Technologies PAC [John95], and the Dolby AC-3 [Fiel96] codecs over a variety of bit rates between 64 and 192 kb/s per stereo pair. The AAC algorithm was tested in the main complexity profile, but with the dynamic window shape switching and intensity coding tools disabled. The MPEG-1 layer II codec was tested in both software simulation and in a real-time implementation ("ITIS"). In all, 17 algorithm/bit rate combinations were examined in the tests. Listening material was selected from a library of 80 items deemed critical by experts, and ultimately the two most critical items were chosen for each codec tested. Mean difference grades were computed over the set of results from 21 expert listeners after three of the original subjects were eliminated from consideration due to their statistically unreliable performance (t-test disqualification).

The test results, reproduced in Table 12.1, clearly show eight performance classes. The AAC and AC-3 codecs at 128 and 192 kb/s, respectively, exhibited best performance with mean difference grades better than -1.0. The MPEG-2 AAC algorithm at 128 kb/s, however, was the only codec that satisfied the quality requirements defined by ITU-R Rec. BS.1115 [ITUR97] for perceptual audio coding systems in broadcast applications, namely that there may not be any audio materials rated below -1.00. Overall, the ranking of the families from best to worst with respect to quality was AAC, PAC, MPEG-1 layer III, AC-3, MPEG-1 layer II, and ITIS (MPEG-1, LII, hardware implementation). The trend is exemplified in Table 12.2, where the codec families are ranked in terms of SDGs for a fixed bit rate of 128 kb/s. The class three results can be interpreted to mean that bit rate increases of 32, 64, and 96 kb/s per stereo pair are required for the PAC, AC-3, and layer II codec families, respectively, to match the output quality of the MPEG-2 AAC at 96 kb/s per stereo pair.

12.5 SUBJECTIVE EVALUATIONS OF 5.1-CHANNEL STANDARDIZED CODECS

In addition to two-channel stereo systems, multichannel perceptual audio coders are increasingly in demand for multimedia, cinema, and home theater

Table 12.1. Comparison of standardized two-channel algorithms (after [Soul98]).

Group	Algorithm	Rate (kb/s)	Mean diff. grade	Transparent items	Items below −1.00
1	AAC	128	−0.47	1	0
	AC-3	192	−0.52	1	1
2	PAC	160	−0.82	1	3
3	PAC	128	−1.03	1	4
	AC-3	160	−1.04	0	4
	AAC	96	−1.15	0	5
	MP-1 L2	192	−1.18	0	5
4	IT IS	192	−1.38	0	6
5	MP-1 L3	128	−1.73	0	6
	MP-1 L2	160	−1.75	0	7
	PAC	96	−1.83	0	6
	IT IS	160	−1.84	0	6
6	AC-3	128	−2.11	0	8
	MP-1 L2	128	−2.14	0	8
	IT IS	128	−2.21	0	7
7	PAC	64	−3.09	0	8
8	IT IS	96	−3.32	0	8

Table 12.2. Comparison of standardized two-channel algorithms at 128 kb/s (after [Soul98]).

Rank	Algorithm (128 kb/s)	Mean diff. grade
1	AAC	−0.47
2	PAC	−1.03
3	MP-1 L3	−1.73
4	AC-3	−2.11
5	MP-1 L2	−2.14
6	IT IS	−2.21

applications. As a result, the European Broadcasting Union (EBU) sponsored Deutsche Telekom Berkom in a formal subjective evaluation [Wust98] that compared the output quality for real-time implementations of the 5.1-channel Dolby AC-3 and the matrixed 5.1-channel MPEG-2/BC layer II algorithms at bit rates between 384 and 640 kb/s (Table 12.3). The tests adhered to the methodologies outlined in ITU BS.1116, and the five-channel listening environment was configured according to ITU-R Rec. BS.775 [ITUR93a]. Mean subjective difference grades were computed over the scores from 32 experts after a t-test analysis was used to discard inconsistent results from seven subjects.

The resulting difference grades given in Table 12.3 represent averages of the mean grades reported for a collection of eight critical test items. Even though none of the tested codec configurations satisfied "transparency," the MPEG-2/BC layer II had only one nontransparent test item at the 640 kb/s bit rate. On the other hand, the Dolby AC-3 outperformed MPEG-2/BC layer II at 384 kb/s by more than half a difference grade. In terms of specific weaknesses, the "applause around" item was most critical for AC-3 (e.g., −2.41 @ 384 kb/s), whereas the "pitch pipe" item was most critical for MPEG-2/BC layer II (e.g., −3.56 @ 384 kb/s).

12.6 SUBJECTIVE EVALUATIONS USING PERCEPTUAL MEASUREMENT SYSTEMS

Even well-designed and carefully controlled experiments [ITUR94b] are susceptible to problems with reliability, repeatability, site-dependence, and subject-dependence [Grew91]. Moreover, subjective tests are difficult, time consuming, tedious, and expensive. These circumstances have inspired researchers to seek automatic perceptual measurement techniques that can evaluate coding distortion on a subjective impairment scale, with the ultimate objective being the replacement of human test subjects. Ideally, a perceptual measurement system should answer the following questions:

- Is the codec transparent for all sources?
- If so, what is the "coding margin," or the degree of distortion inaudibility?
- If not, how disturbing are the artifacts on a meaningful subjective scale?

Table 12.3. Comparison of standardized 5.1-channel algorithms.

Group	Algorithm	Rate (kb/s)	Mean diff. grade
1	MP-2 BC	640	−0.51
2	AC-3	448	−0.93
	MP-2 BC	512	−0.99
A3	AC-3	384	−1.17
	MP-2 BC	384	−1.73

In other words, the system should quantify overall subjective quality, provide a distance measure for the coding margin ("perceptual headroom"), and quantify the audibility of unmasked distortion. Many researchers have proposed measurement systems that attempt to answer these questions. The resulting system architectures can be broadly categorized into two classes (Figure 12.2), namely, systems that compare internal auditory representations (CIR), and systems that perform noise signal evaluation (NSE).

12.6.1 CIR Perceptual Measurement Schemes

The CIR systems (Figure 12.2a) generate independent auditory images of the original and coded signals and then compare them. The degree of audible difference between the auditory images is then evaluated using either a deterministic difference thresholding scheme or a probabilistic detector. The auditory images in the figure could represent excitation patterns, loudness patterns, or some other perceptually relevant physiological quantity, such as neural firing rates, for example. CIR systems avoid explicit masking threshold calculations by capturing the essence of cochlear signal processing in a filter bank combined with a series of post-filter-bank operations on each cochlear channel. Moreover, CIR systems do not require any explicit knowledge of an error signal.

12.6.2 NSE Perceptual Measurement Schemes

The NSE systems (Figure 12.2b), on the other hand, require an explicit representation of the coding distortion. The idea behind NSE systems is to analyze the coding distortion with respect to properties of the auditory system. Typical NSE systems estimate explicitly the masked threshold for each short time signal

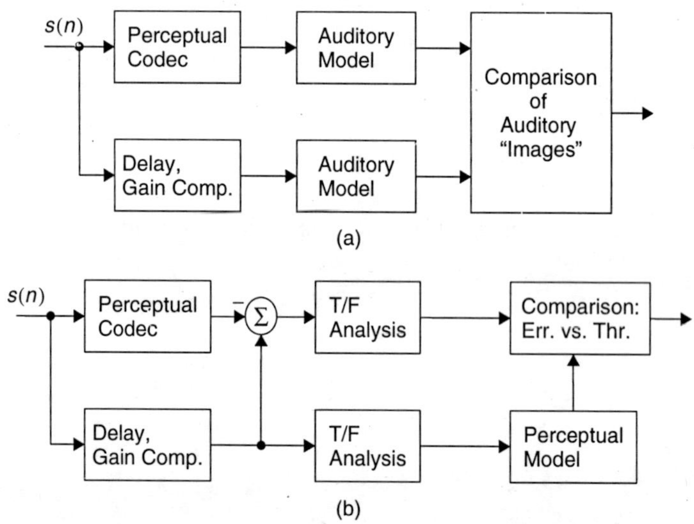

Figure 12.2. Perceptual measurement system architectures: (a) comparison of internal representation (CIR), (b) noise signal evaluation (NSE).

segment using the properties of simultaneous and temporal masking, as is usually done in perceptual audio coders. Then, the distribution of noise energy over time and frequency is compared against the corresponding masked threshold. In both the CIR and NSE scenarios, the automatic measurement system seeks to emulate the signal processing behavior of the human ear and/or the cognitive processing of auditory stimuli in the human listener. Emulation of the human ear for the purpose of making measurements requires a filter bank with time and frequency resolution similar to the auditory filter bank [Spor95a]. Although it is possible to model the characteristics of individual listeners [Treu96], the usual approach is to match the acuity of average subjects. Frequency resolution should be sufficient to model excitation pattern or hair cell selectivity for NSE and CIR algorithms, respectively. For modeling temporal effects, time resolution should be at least 3 ms below 300 Hz and at least 1.5 ms at higher frequencies. Fortunately, a great deal of information about the auditory filter bank is available to assist in the modeling task. The primary challenge is to design a system of manageable complexity while adequately emulating the ear's time-frequency analysis properties. Cognitive effects, however, are less well understood and more difficult to model. Cognitive weighting of perceptual criteria can be quite different across listeners [Prec97]. While some subjects might interpret single, strong distortions such as clicks to be less annoying than small omnipresent distortions, the converse can also be true. Moreover, impairments are asymmetrically weighted. Subjects tend to forget distortions at the beginning of a sample but remember those occurring at the end. Other subjective preferences can also affect outcomes. Some subjects prefer bandwidth limiting ("dullness") over low-level artifacts at high frequencies, but others would prefer greater "brilliance" at the expense of low-level artifacts. While these issues present significant modeling challenges, they can also undermine the utility of subjective listening tests, whereas an automatic measurement system guarantees a repeatable output. The next few sections are concerned with examples of particular CIR and NSE perceptual measurement schemes that have been proposed, including several that have been integrated into international standards.

12.7 ALGORITHMS FOR PERCEPTUAL MEASUREMENT

Numerous examples of both CIR- and NSE-based perceptual measurement schemes for speech and audio codec output quality evaluation have appeared in the literature since the late 1970s, with particular emphasis placed on those systems that can predict accurately subjective listening test results. In 1979, Schroeder, Atal, and Hall at Bell Labs proposed an NSE technique called the "noise loudness" (NL) metric for assessment of vocoder output quality. The NL system [Schr79] estimates output quality by computing a ratio of the coding distortion loudness to the loudness of the original signal. For both signal and noise, loudness patterns are derived every 20 ms from frequency-domain excitation patterns that are in turn computed from short-time critical band densities. The critical band densities are obtained from FFT-based spectral estimates. The approach is hampered by insufficient analysis resolution at low frequencies. Karjaleinen in 1985 proposed a CIR technique [Karj85] known as the "auditory spectrum distance" (ASD). Instead of using an FFT for spectral analysis, the ASD front-end models the cochlear filter-bank using a 48-channel bank of overlapping FIR filters with roughly 0.5 Bark spacing. The filter bank channel magnitude responses are designed to model the smearing of spectral energy that produces the usual psychophysical excitation patterns. Then, a square-law post-processor and two low-pass filters are used to model hair cell rectification and temporal masking, respectively, for each channel. The ASD measure is derived from the maximum deviation between the post-processed filter-bank outputs for the reference and test signals. The ASD profile is analyzed over time and frequency to extract some estimate of perceptual quality.

As the field of perceptual audio coding matured rapidly and created greater demand for listening tests, there was a corresponding growth of interest in perceptual measurement schemes. Several NSE and CIR algorithms appeared in the space of the next few years, namely, the noise-to-mask Ratio [Bran87a] (NMR, 1987), the perceptual audio quality measure (PAQM, 1991) [Beer91], the peripheral internal representation [Stua93] (PIR, 1991), the Bark transform (BT, 1992) [Kapu92], the perceptual evaluation (PERCEVAL, 1992) [Pail92], the perceptual objective measure (POM, 1995) [Colo95], the distortion index (DIX, 1996) [Thie96], and the objective audio signal evaluation (OASE, 1997) [Spor97]. After a proposal phase, the proponents of several of these measurement systems ultimately collaborated on the development of the international standard on perceptual measurement, the ITU-T BS.1387 [ITUR98] (1998). Additional schemes were reported in a special session on perceptual measurement systems for audio codecs during a 1992 Audio Engineering Society conference [Cabo92] [Stau92]. An NSE measure based on quantifying the dissonances associated with coding distortion was proposed in [Bank96].

The remainder of this chapter is intended to provide some functional insights on perceptual measurement schemes. Selected details of the perceptual audio quality measure (PAQM), the noise-to-mask ratio (NMR), and the objective audio signal evaluation (OASE) algorithms are given as representative examples of both the CIR and NSE methodologies.

12.7.1 Example: Perceptual Audio Quality Measure (PAQM)

Beerends and Stemerdink in 1991 proposed a perceptual measurement system [Beer91] known as the "perceptual audio quality measure" (PAQM) for evaluating audio device output quality relative to a reference signal. It was later shown that the parameters of PAQM can be optimized for predicting the results of subjective listening tests on speech [Beer94a] and perceptual audio [Beer92a][Beer92b] codecs. The PAQM (Figure 12.3) is a CIR-type algorithm that maps independently the original, $s(n)$, and coded signal, $\hat{s}(n)$, from the time domain to a corresponding pair of internal, psychophysical representations. In particular, original and coded signals are processed by a two-stage auditory signal processing model. This model captures mostly the peripheral

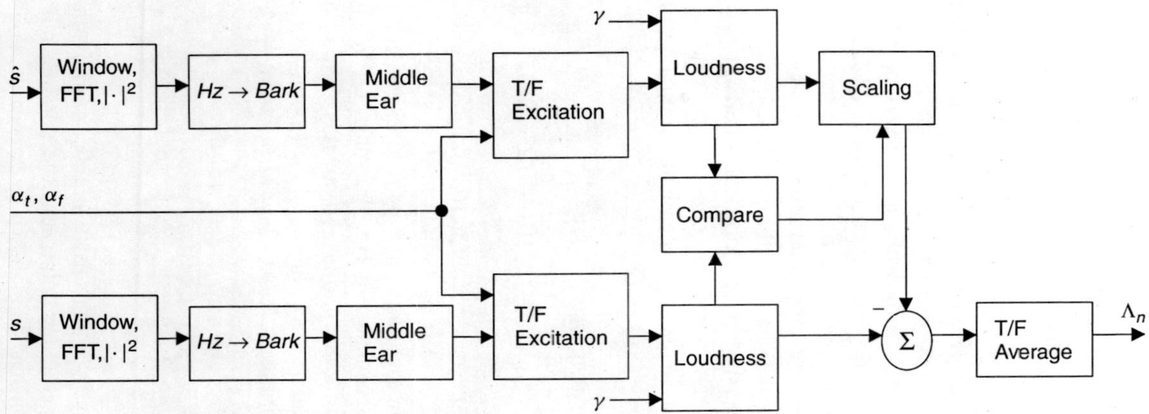

Figure 12.3. Perceptual audio quality measure (PAQM).

process of nonlinear time-frequency energy smearing [Vier86] followed by the predominantly central process of nonlinear intensity compression [Pick88]. A modified mapping derived from Zwicker's loudness model [Zwic67] [Zwic77] is used to transform the two internal representations to a compressed loudness scale. Then a difference is computed to quantify the noise disturbance associated with the coding distortion. Ultimately, a second transformation can be applied that maps the noise disturbance compressed loudness, Λ_n, to a predicted listening test subjective difference grade. Precise computational details of the PAQM algorithm appeared in the appendices of [Beer92b]. The important steps, however, can be summarized as follows.

Note that the sequence of operations is the same for both the reference and impaired signals. First, short-time spectral estimates are obtained using a 2048-point Hann-windowed FFT with 50% block overlap, yielding time resolution close to 20 ms and frequency resolution of about 22 Hz. A middle-ear transfer function is applied at the output of the FFT, and the frequency axis is warped from Hertz to Barks. To simulate the excitation patterns generated along the basilar membrane, the PAQM next smears the middle-ear signal energy across time and frequency in accordance with empirically derived psychophysical models. For frequency-domain smearing within one analysis window, Terhardt's [Terh79] narrow-band level- and frequency-dependent excitation pattern approximation, with slopes defined by

$$S_1 = 31 \quad \text{dB/Bark}, \ f_m > f \tag{12.1a}$$

$$S_2 = 22 + \min(230/f_m, 10) - 0.2L\text{dB/Bark}, \ \ f_m < f \tag{12.1b}$$

is used to spread the spectral energy across critical bands. Thus, the slope S_1 controls the downward spread of excitation, the slope S_2 controls the upward spread of excitation, and the parameters, f_m, L, and f correspond, respectively, to the masker center frequency in Hz, the masker level in units of dB SPL, and the frequencies of the excitation components. Within an analysis window, individual excitation patterns, M_i, are combined point-wise to form a composite pattern, M_c, using a parametric power law of the form

$$M_c = \left(\sum_{i=1}^{N} M_i^{\alpha}\right)^{1/\alpha}, \alpha < 2, \tag{12.2}$$

where the parameter N corresponds to the number of individual excitation patterns being combined at a particular point, and the parameter α controls the power law. Power law addition has been shown to govern the combination of multiple simultaneous [Luft83] [Luft85] and temporal masking [Penn80] patterns, but the parameter α may be different along the frequency and time axes. We denote by α_f the value of the parameter α used in PAQM for combining frequency-domain excitation patterns. The PAQM also models the temporal spread of excitation. Energy from the k-th block is scaled by an exponentially decaying gain, $g(t)$, with a frequency-dependent time constant, $\tau(f)$, i.e.,

$$g(t) = e^{-t/\tau(f)}, \tag{12.3}$$

where the time parameter t corresponds to the time distance between blocks, e.g., 20 ms. The scaled version of the k-th block energy is then combined with the energy of the $(k+1)$-th (overlapping) block using a power law of the form given in Eq. (12.2), and the parameter α is set for the time domain, i.e., $\alpha = \alpha_t$. It was found that the level dependence of temporal excitation spread could be disregarded for the purposes of PAQM without affecting significantly the measurement quality. Thus, the parameters S_1, S_2, and α_f control the spread of frequency-domain signal energy, while the parameters $\tau(f)$ and α_t control the signal energy dispersion in time. Once excitation patterns have been obtained, internal original and coded signal representations are generated by transforming at each frequency the excitation energy, E, into a measure of loudness, Λ, using Zwicker's compressive excitation-to-loudness mapping [Zwic67], i.e.,

$$\Lambda = k\left(\frac{E_0}{s}\right)^{\gamma}\left[\left(1 - s + s\frac{E}{E_0}\right)^{\gamma} - 1\right], \tag{12.4}$$

where the parameter γ controls the compression characteristic, the parameter k is an arbitrary scaling constant, the constant E_0 represents the amount of excitation in a tone of the same frequency at the absolute hearing threshold, and the constants is a so-called "schwell" factor as defined in [Zwic67]. Although the mapping in Eq. (12.4) was proposed in the context of loudness estimation, the particular value of the parameter γ used in PAQM is different from the value empirically determined in Zwicker's experiments, and therefore the internal representations are not, strictly speaking, loudness measurements but are, rather, compressed loudness estimates. Indexed over Bark frequencies, z, and time, t, the compressed loudness of the original signal, $\Lambda_s(z,t)$, and the coded signal, $\Lambda_{\hat{s}}(z,t)$, are compared in three Bark regions. Gains are extracted for each Bark subband in order to scale $\Lambda_{\hat{s}}(z,t)$ such that the matching between $\Lambda_s(z,t)$ and the scaled version of $\Lambda_{\hat{s}}(z,t)$ is maximized. The instantaneous, frequency-localized noise disturbance, $\Lambda_n(z,t) = \Lambda_{\hat{s}}(z,t) - \Lambda_s(z,t)$, i.e., the difference between internal representations is integrated over the Bark scale and then across a suitable time interval to obtain a composite noise disturbance estimate, Λ_n. For the purposes of predicting the results of subjective listening tests, the PAQM parameters of, α_f, α_t, and γ can be adapted to minimize the standard deviation of a third-order regression line that fits the absolute performance grades from actual subjective listening tests to the PAQM output, $\log(\Lambda_n)$.

For example, in a fitting experiment using listening test results from the ISO/IEC MPEG 1990 database [ISOI90] for loudspeaker presentation of the test material, the values of 0.8, 0.6, and 0.04, respectively, were obtained for the parameters, α_f, α_t, and γ. This yielded a correlation coefficient of 0.968 and standard deviation of 0.351 performance grades. With a correlation coefficient of 0.91 and standard deviation of 0.48 MSS, prediction was not quite as good when the same parameters were used to validate PAQM on subjective listening test results from the ISO/IEC MPEG 1991 database [ISOI91]. Some of the reduced prediction accuracy, however, was attributed to inexperienced listeners, particularly for some very high quality items for which the grading task is more difficult. In fact, better prediction accuracy was realized during a later comparison with more experienced listeners on the same database [Beer92b]. In addition, the model was also validated using the results of the ITU-R Task Group 10/2 1993 audio codec test [ITUR93b] on the contribution distribution emission (CDE) database, resulting in a correlation coefficient of 0.88 and standard deviation of 0.29 MSS. As reported in [ITUR93c], these results were later verified by the Swedish Broadcasting Corporation. It is also interesting to note that the PAQM was optimized for listening test predictions on narrowband speech codecs. Several modifications were required to achieve a correlation coefficient of 0.99 and standard deviation of 0.14 points relative to a 5-point discrete MOS. In particular, time and frequency energy dispersion had to be eliminated, the loudness compression parameter was changed to $\gamma = 0.001$, and silence intervals required a zero weighting [Beer94a]. In an effort to improve PAQM performance, cognitive effect modeling was investigated in [Beer94b] and [Beer95]. It was concluded that a unified PAQM configuration for both narrowband and wideband codecs must account for subjective phenomena such as the asymmetry in perceived distortion between audible artifacts and the perceived distortion when signal components are eliminated rather than distorted. The signal-dependent importance of modeling informational masking effects [Leek84] was also demonstrated [Beer96], although a general formulation was not proposed. More details on cognitive modeling improvements for PAQM appeared recently in [Beer98]. Future PAQM enhancements will involve better cognitive models as well as binaural processing capabilities.

12.7.2 Example: Noise-to-Mask Ratio (NMR)

In the late 1980s, Brandenburg proposed an NSE perceptual measurement scheme known as the "noise-to-mask ratio" (NMR) [Bran87a]. The system was originally intended to assess output quality and quantify noise margin during the development of the early perceptual transform coders such as OCF [Bran87b] and ASPEC [Bran91], but it has also been widely used in the development of more recent and sophisticated algorithms (e.g., MPEG-2 AAC [ISOI96a]). The NMR output measurement, η, represents a time- and frequency-averaged dB distance between the coding distortion, $s(n) - \hat{s}(n)$, and the masked threshold associated with the original signal, $s(n)$. The NMR system also sets a binary masking flag [Bran92b] whenever the noise density in any critical band exceeds the masked threshold. Tracking the masking flag over time helps designers to identify critical items for a particular codec. In a procedure markedly similar to the perceptual entropy front-end calculation (Chapter 5, Section 5.5), the NMR (Figure 12.4) is measured as follows. First, the original signal, $s(n)$, is time-delayed and in some cases amplitude scaled to maximize the temporal matching with the coded signal, $\hat{s}(n)$, so that the coding distortion can be estimated through a simple difference operation, i.e., $s(n) - \hat{s}(n)$. Then, FFT-based spectral analysis is performed on Hann-windowed 1024-sample blocks (23.3 ms at 44.1 kHz) of the signal and the noise sequences. Although not shown in the figure, the difference is sometimes computed in the spectral domain (after the FFT) to reduce the impact of phase errors on estimated audible noise. New spectral estimates are computed every 512 samples (50% block overlap) to improve the effective temporal resolution of the measurement. Next, critical band signal and noise densities are estimated on a set of Bark-like subbands by summing and then normalizing the energy contained in 27 blocks of adjacent FFT bins that approximate 1-Bark bandwidth [Bran92b]. The masked threshold is derived from the Bark-scale signal density by means of simplified simultaneous and temporal masking models. For simultaneous masking, a conservatively estimated, level-independent prototype spreading function (specified in [Bran92b]) is convolved with the Bark density. Unlike other perceptual models, NMR assumes that the in-band masked threshold is fixed at 3 dB below the masker peak, independent of tonality. While this assumption apparently overestimates the overall excitation level, the NMR model compensates by ignoring masking additivity. The absolute threshold of hearing in quiet is also accounted for in the Bark-domain masking calculation. As far as temporal masking is concerned, only postmasking is accounted for explicitly, with the simultaneous masked threshold decreased for each critical band independently by 6 dB per analysis window (11.65 ms at 44.1 kHz). Finally, the local NMR, η_{loc}, is defined as

$$\eta_{loc} = 10\log_{10}\left(\frac{1}{27}\sum_{i=1}^{27}\frac{\sigma_n(i)}{\sigma_m(i)}\right), \tag{12.5}$$

i.e., the mean of the ratio between the critical band noise density, $\sigma_n(i)$, and the critical band mask density, $\sigma_m(i)$. For a signal containing N blocks, the global NMR, η, is defined as the mean local NMR, i.e.,

$$\eta = 10\log_{10}\left(\frac{1}{N}\sum_{i=1}^{N}10^{(0.1\eta_{loc}(i))}\right). \tag{12.6}$$

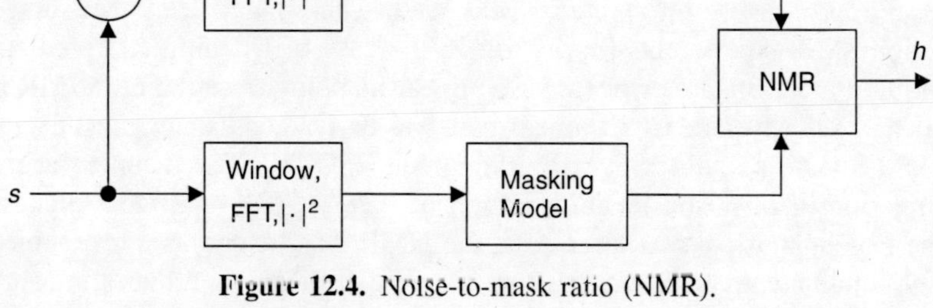

Figure 12.4. Noise-to-mask ratio (NMR).

Finally, a masking flag (MF) is set once every 512 samples if the noise in at least one critical band exceeds the masked threshold. One appealing feature of the NMR algorithm is its modest complexity, making it amenable to real-time implementation. In fact, a real-time NMR system for 44.1 kHz signals has been implemented on a set of three AT&T DSP32C 50 MHz devices [Herr92a] [Herr92b]. Processor-specific tasks were partitioned as follows: 1) correlator for delay and gain determination, 2) FFT and windowing, and 3) perceptual model, NMR calculations. We will consider next how the raw NMR measurements are applied in codec performance evaluation tasks.

Given that NMR updates occur once every 512 samples (11.7 ms at 44.1 kHz), the raw NMR measurements of η_{loc} and the masking flag can create a considerable volume of data. To simplify the evaluation task, several postprocessed, data reducing metrics have been proposed for codec evaluation purposes [Beat96]. These secondary measures, including the global NMR (12.6), worst-case local NMR ($\max(\eta_{loc})$), the percentage of nonzero masking flag (MF), and the "NMR fingerprint," have been demonstrated to predict quite accurately the presence of coding artifacts. For example, a global NMR, η, of less than or equal to -10 dB typically implies an artifact-free signal. High quality is also implied by a MF that occurs less than 3% of the time. An NMR fingerprint shows the time-averaged distribution of NMR over frequency (critical bands). A well-balanced fingerprint has been shown to correspond with better output quality than an unbalanced, "peaky" fingerprint for which high NMRs are localized to a just few critical bands. NMR equivalent bit rate has also been proposed as a quality metric, particularly in the case of impaired channel performance characterization. In this measurement scheme, an impaired codec is quantified in terms of an equivalent bit rate for which an unimpaired reference codec would achieve the same NMR. For example, a 64-kb/s codec in a three-stage tandem configuration might achieve the same NMR as a 32-kb/s version of the same codec in the absence of any tandeming. These and other NMR-based figures of merit have been used successfully to evaluate the performance of perceptual audio codecs in consumer recording equipment [Keyh93], tandem coding configurations [Keyh94], and network applications [Keyh95].

The various NMR-based measures have proven to be useful evaluation tools for many phases of codec research and development. The ultimate objective of TG 10/4, however, is to standardize automatic evaluation tools that can predict subjective quality on an impairment scale. In fact, NMR measurements can be correlated directly with the results of subjective listening tests [Herr92b]. Moreover, an appropriately interpreted NMR output can emulate a human subject in psychophysical experiments [Spor95b]. In the case of listening test predictions, a minimum mean square error (MMSE) linear mapping from the raw NMR measurements to the 5-grade CCIR impairment was derived on listening test data from the 1990 Swedish Radio ISO/MPEG database [ISOI90]. The mapping achieved a correlation of 0.94 and mean deviation of 0.28 CCIR impairment grades. As for the psychophysical measurements, the NMR was treated as a test subject in several simultaneous and temporal masking experiments, and then the resulting psychometric functions were compared against average results for actual human subjects. In the experiments, the NMR "subject" was judged to detect a given probe if the masking flag was set for more than half of the blocks, and then the threshold was interpreted as the lowest level probe for which more than 50% of the masking flags were set. In experiments on noise-masking-tone and tone-masking-noise, the NMR "subject" exhibited masking patterns similar to average human subjects, with some artifacts at low frequencies evident. In tone-masking-tone experiments, however, the NMR "subject" diverged from human performance because the NMR perceptual model fails to account for the masking release caused by beating and difference tones.

The level of performance achieved by the NMR system on listening test prediction and psychometric measurement emulation tasks resulted in its inclusion in the ITU TG 10/4 evaluation process. Although the NMR perceptual model had been "frozen" in 1994 [Spor95b] to facilitate comparisons between NMR results generated at different sites for different algorithms, it was clear that perceptual model enhancements relative to [Bran92b] were necessary in order to improve masked threshold predictions and, ultimately, overall performance. Thus, while maintaining a complexity level compatible with real-time constraints, the TG 10/4 NMR submission was improved relative to the original NMR in several areas. For instance, the excitation model is now dependent on tonality and presentation level. Furthermore, the additivity of masking is now modeled explicitly. Finally, level dependencies are now matched to the actual presentation SPL. The system might eventually realize further performance gains by using higher resolution spectral analysis at low frequencies and by estimating the BLMD for stereo signals.

12.7.3 Example: Objective Audio Signal Evaluation (OASE)

None of the NMR improvements cited above change the fact the uniform bandwidth channels of the FFT filter bank are fundamentally different from the auditory filter bank. In spite of other NMR modeling improvements, the FFT handicaps measurement system because of inadequate frequency resolution at low frequencies and inadequate temporal resolution at high frequencies. To overcome such limitations, an alternative CIR-type algorithm for perceptual measurement known as the objective audio signal evaluation (OASE) [Spor97] was also submitted to the ITU TG 10/4 as one of the three variant proposals from the NMR proponents. Unlike the other NMR variants, the OASE system (Figure 12.5) is a CIR system that generates independent auditory feature sets for the coded and reference signals, denoted by $\hat{s}(n)$ and $s(n)$, respectively. Then, a probabilistic detection stage determines the perceptual distance between the two auditory feature sets.

The OASE was designed to rectify perceptual measurement shortcomings inherent in FFT-based systems by achieving analysis resolution in both time and frequency that is greater than or equal to auditory resolution. In particular, the OASE filter bank approximates a Mel frequency scale by using 241 highly overlapping, 128-tap FIR bandpass filters centered at integral multiples of 0.1 Bark.

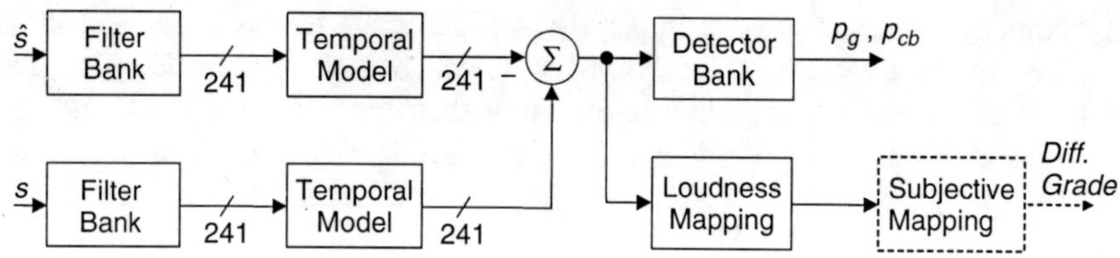

Figure 12.5. Objective audio signal evaluation (OASE).

Meanwhile, the time resolution is $f_s/32$ or 0.67 ms at 48 kHz. Moderate complexity is achieved through an efficient, structured multirate realization that exploits fast convolutions. Individual filter shapes are level- and frequency-dependent, modeled after experimental results on excitation patterns. The bandpass filters also simulate outer and middle ear transfer characteristics. A square-law rectifier post-processes the output from each channel of the filter bank, and then the auditory "internal noise" associated with the absolute threshold of hearing is simulated by adding to each output a predefined constant. Finally, the nonlinear filters proposed in [Karj85] are applied to simulate temporal integration and smoothing. The output of the temporal processing block constitutes a 241-element auditory feature vector for each signal. A Decibel difference is computed on each channel between the features associated with the reference and coded signals. This time-frequency difference vector forms the input to a bank of detectors. At each time index, a frequency-localized difference detection probability, $p_i(t)$, is assigned to the $i = 1, 2, \ldots, 241$ features as a function of the difference magnitude using a signal-dependent detection profile. For example, artificial signals use a detection profile different than speech and music. The time index, t, is discrete and takes on values that are integer multiples of 0.67 ms at 48 kHz ($f_s/32$). Finally, a global probability of difference detection, $p_g(t)$, is computed as the complement of the combined probability that no single channel differences are detected, i.e.,

$$p_g(t) = 1 - \prod_{i=l}^{241}(1 - p_i(t)). \tag{12.7}$$

Alternatively, a critical band detection probability, $p_{cb}(t)$, for a particular 1-Bark subband can be computed by combining the frequency-localized detection probabilities as follows

$$p_{cb}(t) = 1 - \prod_{i=k-4}^{k+5}(1 - p_i(t)), \tag{12.8}$$

where the critical band of interest is centered on any of the eligible Mel-scale features, i.e., $5 \leqslant k \leqslant 236$. In cases where it is desirable for OASE to predict the results of a subjective listening test, a noise loudness mapping can be applied to the feature difference vectors available at the output of the temporal processing stages. Such a configuration estimates the loudness of the difference between the original and coded signals. Then, as is possible with many other perceptual measurement schemes, the noise loudness can be mapped to a predicted subjective difference grade. The ITU TG 10/4 has in fact used OASE model outputs to predict subjective scores. The results were reported in [Spor97] to have been reasonable, and similar to other perceptual measurement systems, but no correlation coefficients or standard deviations were given.

12.8 ITU-R BS.1387 AND ITU-T P.861: STANDARDS FOR PERCEPTUAL QUALITY MEASUREMENT

In 1995, the Radio Communication Sector of the International Telecommunications Union (ITU-R) convened Task Group (TG) 10/4 to select a perceptual measurement scheme capable of predicting the results of subjective listening tests in the form of an objective difference grade (ODG). Six proposals were submitted to TG 10/4 in 1995, including the distortion index (DIX) from TU-Berlin [Thie96], the noise-to-mask ratio (NMR) from FhG-IIS Erlangen [Bran87a], the perceptual audio quality measure (PAQM) from KPN [Beer92a], the perceptual evaluation (PERCEVAL) from CRC Canada [Pail92], the perceptual objective measure (POM) from CCETT, France [Colo95], and the Toolbox (TB) from IRT Munich. During initial TG 10/4 tests in 1996, three variants of each proposal were required to predict the results of an unknown listening test. A lengthy comparison and verification process ensued (e.g., [ITUR94a] compares NMR, PAQM, and PERCEVAL). Ultimately, the six TG 10/4 proponents in conjunction with Opticom and Deutsche Telekom Berkom collaborated to design a single system combining the most successful features from each individual methodology. This effort resulted in the recently adopted ITU-R Recommendation BS.1387 [ITUR98], which specifies two versions of the measurement scheme known as perceptual evaluation of audio quality (PEAQ). The basic, low-complexity version is based on an FFT filter bank, much like the PAQM, PERCEVAL, POM, and NMR algorithms. The advanced version incorporates a more sophisticated filter bank that better approximates the auditory system as in, for example, the DIX and OASE algorithms. Nearly all features extracted from the perceptual models to estimate quality are similar to the feature sets previously used in the individual proposals. Unlike the regression polynomials used in earlier tests, however, in BS.1387 a neural network performs the final mapping from the objective output to an ODG. Standardization activities are ongoing, and the ITU has formed a successor group to the now dissolved TG 10/4 under the designation JWP10-11Q [Spor99a]. Research into perceptual measurement techniques has also resulted in a standardized algorithm for speech quality measurements. In particular, Study Group 12 within ITU Telecommunications Sector (ITU-T) in 1996 adopted the Recommendation P.861 [ITUT96] for perceptual quality measurements of telephone band signals (300–3400 Hz). The technique is essentially a speech-specific version of the PAQM algorithm in which the time-frequency smearing blocks are usually disabled, the outer ear transfer function is eliminated, and simplifications are made in both the intensity-to-loudness mapping and the modeling of cognitive asymmetry effects.

12.9 RESEARCH DIRECTIONS FOR PERCEPTUAL CODEC QUALITY MEASURES

This chapter has provided a perspective on current performance evaluation techniques for perceptual audio codecs, including both subjective and objective quality measures. At present, the subjective listening test remains the most reliable technique for measurement of the small impairments associated with high-quality codecs, and a formal test methodology optimized for maximum reliability has been adopted as an international standard, the ITU-R BS.1116 [ITUR94b]. Nevertheless, even the results of BS.1116-compliant subjective tests continue to exhibit site and subject dependencies. Future research will continue to seek new methods for improving the reliability of subjective test outcomes. For example, Lucent Technologies sponsored an investigation at Moulton Laboratories [Moul98a][Moul98b] into the effectiveness of multifacet Rasch models [Lina94] for improved reliability of subjective listening tests on high-quality audio codecs. Developed initially in the context of intelligence testing, the Rasch model [Rasc80] is a statistical analysis technique designed to remove the effects of local disturbances on test outcomes. This is accomplished by computing an expected data value for each data point that can be compared with the actual collected value to determine whether or not the data point is consistent with the demands of reproducible measurement. The Rasch methodology involves an iterative data pruning process of suspending misfitting data from the analysis, recomputing the measures and expected values, and then repeating the pruning process until all significant misfits are removed from the data set. The Rasch analysis proponents have indicated [Moul98a] [Moul98b] that future work will attempt to demonstrate the elimination of site dependencies and other local data disturbances. Research along these lines will continue as long as subjective listening tests exhibit site and subject dependence.

Meanwhile, the considerable research that has been devoted to development of automatic perceptual measurement schemes has resulted in the adoption of an international standard, the ITU-R BS.1387 [ITUR98]. Experts do not consider the standardized algorithm to be a human subject replacement, however, research into improved perceptual measurement schemes will continue (e.g., ITU-R JWP10-11Q). While the signal processing models of the auditory periphery have reached a high level of sophistication with manageable complexity, it is likely that future perceptual measurement schemes will realize the greatest performance gains as a result of improved models for binaural listening effects and general cognitive effects. As described in [Beer98], several central, cognitive effects are important in subjective audio quality assessment, including:

Informational Masking. The masked threshold of a complex target masked by a complex masker may decrease by more than 40 dB [Leek84] after subject training. Modeling this effect has been shown to cause both increases and decreases in correlation between ODGs and SDGs, depending on the listening test database [Beer98]. Future work will concentrate on the development of satisfactory models for informational masking, i.e., models that do not lead to reduced correlation.

Separation of Linear from Nonlinear Distortions. Linear distortions tend to be less objectionable than nonlinear distortions. Experimental attempts to quantify this effect in perceptual measurement schemes, however, have not yet succeeded.

Auditory Scene Analysis. This describes the cognitive mechanism that groups different auditory events into distinct objects. During a successful application of this principle to perceptual measurement systems [Beer94b], it was found that there is a significant asymmetry between the perceived distortion associated with the introduction of a new time-frequency component versus the perceived distortion associated with the removal of an existing time-frequency component. Moreover, the effect was shown to improve SDG/ODG correlation for some listening test results.

Naturally, there is some debate over the feasibility of creating an automatic system that can predict the responses of human test subjects with sufficient precision to allow for differentiation between competing high-quality algorithms. Even recent systems tend to have difficulty predicting some observable psychoacoustic phenomena. For example, most systems fail to model the beat and combination tone detection cues that generate a masking release (reduced threshold) in tone masking tone experiments. For these and other reasons, the large correlation coefficients that are obtained when mapping particular objective measures to certain subjective test sets have been shown to be data dependent (e.g., [Beer92b]). Keyhl *et al.* calibrated the perceptual audio quality measure (PAQM) to match a set of subjective test results for an MPEG codec in the WorldSpace satellite system [Keyh98]. The idea was to later use the automatic measure in place of expensive subjective tests for quality assurance over time. In spite of any apparent shortcomings, considerable interest in perceptual measurement schemes will persist as long as subjective tests remain time-consuming, expensive, and marginally reliable. For additional information, the interested reader can consult several survey papers on perceptual measurement schemes that appeared in an AES collection [Beat96] [Koml96] [Ryde96], as well as the textbook on perceptual measurement techniques [Beer98]. Two popular standards published include the perceptual evaluation of speech quality (PESQ) [Rix01] and the perceptual evaluation of audio quality (PEAQ) standard [PEAQ].

REFERENCES

[Adou87] J.P. Adoul et al., "Fast CELP coding based on algebraic codes," in *Proc. IEEE ICASSP-87,* vol. 12, pp. 1957–1960, Dallas, TX, Apr. 1987.

[Adou95] J.-P. Adoul and R. Lefebvre, "Wideband speech coding," in *Speech Coding and Synthesis,* Eds., W.B. Kleijn and K.K. Paliwal, Kluwer Academic Publishers, Norwell, MA, 1995.

[AES00] AES Technical Council Document AESTD1001.1.01–10, *Multichannel Surround Sound Systems and Operations*, Audio Engineering Society, New York, NY.

[Ager94] F. Agerkvist, *Time-Frequency Analysis and Auditory Models*, Ph.D. Thesis, The Acoustics Laboratory, Technical University of Denmark, 1994.

[Ager96] F. Agerkvist, "A time-frequency auditory model using wavelet packets," *J. Aud. Eng. Soc.*, vol. 44, no. 1/2, pp. 37–50, Feb. 1996.

[Akag94] K. Akagiri, "Technical Description of Sony Preprocessing," ISO/IEC JTC1/SC29/WG11 MPEG Input Document, 1994.

[Akai95] K. Akaigiri, "Detailed Technical Description for MPEG-2 Audio NBC," Technical Report, ISO/IEC JTC1/SC29/WG11, 1995.

[Akan92] A. Akansu and R. Haddad, *Multiresolution Signal Decomposition: Transforms, Subbands, Wavelets.*, Academic Press, San Diego, CA, 1992.

[Akan96] A. Akansu and M.J.T. Smith, eds., *Subband and Wavelet Transforms, Design and Applications.*, Kluwer Academic Publishers, Norwell, MA, 1996.

[Ali96] M. Ali, *Adaptive Signal Representation with Application in Audio Coding,* Ph.D. Thesis, University of Minnesota, Mar. 1996.

Audio Signal Processing and Coding, by Andreas Spanias, Ted Painter, and Venkatraman Atti

[Alki95] O. Alkin and H. Calgar, "Design of efficient M-band coders with linear phase and perfect-reconstruction properties," *IEEE Trans. Sig. Proc.*, vol. 43, no. 7, pp. 1579–1589, July 1995.

[Alla99] E. Allamanche, R. Geiger, J. Herre, and T. Sporer, "MPEG-4 low-delay audio coding based on the AAC codec," in *Proc. 106th Conv. Aud. Eng. Soc.,* preprint #4929, May 1999.

[Alle77] J.B. Allen and L.R. Rabiner, "A unified theory of short-time spectrum analysis and synthesis," *Proc. IEEE*, vol. 65, no. 11, pp. 1558–1564, Nov. 1977.

[Alme83] L. Almeida, "Nonstationary spectral modeling of voiced speech," *IEEE Trans. Acous. Speech and Sig. Proc.*, vol. 31, pp. 374–390, June 1983.

[Angu01] J.A. Angus, "Achieving effective dither in delta-sigma modulation systems," in *Proc. 110th Conv. Aud. Eng. Soc.*, preprint #5393, May 2001.

[Angu97a] J.A. Angus and N.M. Casey, "Filtering $\Sigma\Delta$ signals directly," in *Proc. 102nd Conv. Aud. Eng. Soc.*, preprint #4445, Mar. 22–25, 1997.

[Angu97b] J.A.S. Angus, "Backward adaptive lossless coding of audio signals," in *Proc. 103rd Conv. Aud. Eng. Soc.*, preprint #4631, New York, Sept. 1997.

[Angu98] J.A. Angus and S. Draper, "An improved method for directly filtering $\Sigma\Delta$ audio signals," in *Proc. 104th Conv. Aud. Eng. Soc.*, preprint #4737, May 16–19, 1998.

[Arno00] M. Arnold, "Audio watermarking: features, applications and algorithms," in *Int. Conf. on Multimedia and Expo (ICME-2000)*, vol. 2, pp. 1013–1016, 30 July–2 Aug. 2000.

[Arno02] M. Arnold et al., *Techniques and Applications of Digital Watermarking and Content Protection*, Artech House, Norwood, MA, Dec. 2002.

[Arno02a] M. Arnold and O. Lobisch, "Real-time concepts for block-based watermarking schemes," in *Proc. of 2nd Int. Conf. on WEDELMUCSIC-02*, pp. 156–160, Dec. 2002.

[Arno02b] M. Arnold, "Subjective and objective quality evaluation of watermarked audio tracks," in *Proc. of 2nd Int. Conf. on WEDELMUCSIC-02*, pp. 161–167, Dec. 2002.

[Arno03] M. Arnold, "Attacks on digital audio watermarks and countermeasures," in *Proc. of 3rd Int. Conf. on WEDELMUCSIC-03*, pp. 55–62, Sept. 2003.

[Atal82a] B.S. Atal, "Predictive coding of speech at low bit rates," *IEEE Trans. on Comm.,* vol. 30, no. 4, p. 600, Apr. 1982.

[Atal82b] B.S. Atal and J. Remde, "A new model for LPC excitation for producing natural sounding speech at low bit rates," in *Proc. IEEE ICASSP-82*, pp. 614–617, Paris, France, April 1982.

[Atal84] B. Atal and M.R. Schroeder, "Stochastic coding of speech signals at very low bit rates," in *Proc. Int. Conf. Comm.*, pp. 1610–1613, May 1984.

[Atal90] B. Atal, V. Cuperman, and A. Gersho, *Advances in Speech Coding*, Kluwer, Norwell, MA, 1990.

[Atti05] V. Atti and A. Spanias, "Speech analysis by estimating perceptually – relevant pole location," in *Proc. IEEE ICASSP-05* vol. 1, pp. 217–220, Philadelphia, PA, March 2005.

[Bald96] A. Baldini, "DSP and MIDI applications in loudspeaker design," in *Proc. 100th Conv. Aud. Eng. Soc.*, preprint #4189, May 1996.

[Bamb94] R. Bamberger et al, "Generalized symmetric extension for size-limited multirate filter banks," *IEEE Trans. Image Proc.*, vol. 3, no. 1, pp. 82–87, Jan. 1994.

[Bank96] M. Bank et al., "An objective method for sound quality estimation of compression systems," in *Proc. 101st Conv. Aud. Eng. Soc.*, preprint #4373, 1996.

[Barn03a] M. Barni, "What is the future of watermarking? Part-I," *IEEE Signal Processing Magazine,* vol. 20, no. 5, pp. 55–60, Sept. 2003

[Barn03b] M. Barni, "What is the future of watermarking? Part-II," *IEEE Signal Processing Magazine,* vol. 20, no. 6, pp. 53–59, Nov. 2003

[Bass98] P. Bassia and I. Pitas, "Robust audio watermarking in the time-domain," in *Proc. EUSIPCO*, vol. 1, pp. 25–28, Greece, Sept. 1998.

[Bass01] P. Bassia, I. Pitas, and N. Nikolaidis, "Robust audio watermarking in the time domain," *IEEE Trans. on Multimedia*, vol. 3, no. 2, pp. 232–241, June 2001.

[Baum95] F. Baumgarte, C. Ferekidis, and H. Fuchs, "A nonlinear psychoacoustic model applied to the ISO MPEG layer 3 coder," in *Proc. 99th Conv. Audio Eng. Soc.*, preprint #4087, New York, Oct. 1995.

[Bayl97] J. Baylis, *Error Correcting Codes: A Mathematical Introduction*, CRC Press, Boca Raton, FL, Dec. 1997.

[Beat89] R.J. Beaton, "High quality audio encoding within 128 kbit/s," in *IEEE Pacific Rim Conf. on Comm., Comp. and Sig. Proc.*, pp. 388–391, June 1989.

[Beat96] R. Beaton et al., "Objective perceptual measurement of audio quality," in *Collected Papers on Digital Audio Bit-Rate Reduction*, N. Gilchrist and C. Grewin, Eds., Aud. Eng. Soc., pp. 126–152, 1996.

[Bech92] S. Bech, "Selection and training of subjects for listening tests on sound-reproducing equipment," *J. Aud. Eng. Soc.*, pp. 590–610, Jul/Aug. 1992.

[Beck04] E. Becker et al., *Digital Rights Management: Technological, Economic, and Legal and Political Aspects*, Springer Verlag, Berlin, Jan. 2004.

[Beer91] J. Beerends and J. Stemerdink, "Measuring the quality of audio devices," in *Proc. 90th Conv. Aud. Eng. Soc.*, preprint #3070, Feb. 1991.

[Beer92a] J. Beerends and J. Stemerdink, "A perceptual audio quality measure," in *Proc. 92nd Conv. Aud. Eng. Soc.*, preprint #3311, Mar. 1992.

[Beer92b] J. Beerends and J. Stemerdink, "A perceptual audio quality measure based on a psychoacoustic sound representation," *J. Aud. Eng. Soc.*, vol. 40, no. 12, pp. 963–978, Dec. 1992.

[Beer94a] J. Beerends and J. Stemerdink, "A perceptual speech-quality measure based on a psychoacoustic sound representation," *J. Aud. Eng. Soc.*, vol. 42, no. 3, pp. 115–123, Mar. 1994.

[Beer94b] J. Beerends and J. Stemerdink, "Modeling a cognitive aspect in the measurement of the quality of music codecs," in *Proc. 96th Conv. Aud. Eng. Soc.*, preprint #3800, 1994.

[Beer95] J. Beerends and J. Stemerdink, "Measuring the quality of speech and music codecs: an integrated psychoacoustic approach," in *Proc. 98th Conv. Aud. Eng. Soc.*, preprint #3945, 1995.

[Beer96] J. Beerends et al., "The role of informational masking and perceptual streaming in the measurement of music codec quality," in *Proc. 100th Conv. Aud. Eng. Soc.*, preprint #4176, May 1996.

[Beer98] J. Beerends, "Audio quality determination based on perceptual measurement techniques," in *Applications of Digital Signal Processing to Audio and Acoustics*, M. Kahrs and K. Brandenburg, Eds., Kluwer Academic Publishers, Boston, pp. 1–83, 1998.

[Beke60] G. von Bekesy, *Experiments in Hearing*. McGraw Hill, New-York, 1960.

[Bell76] M. Bellanger, et al., "Digital filtering by polyphase network: application to sample rate alteration and filter banks," *IEEE Trans. Acous., Speech, and Sig. Proc.*, vol. 24, no. 4, pp. 109–114, Apr. 1976.

[Ben99] L. Ben and M. Sandler, "Joint source and channel coding for Internet audio transmission," in *Proc. 106th Conv. Aud. Eng. Soc.*, preprint #4932, May 1999.

[Bend95] W. Bender, D. Gruhl, and N. Morimoto, "Techniques for data hiding," in *Proc. of the SPIE*, 1995.

[Bend96] W. Bender, D. Gruhl, N. Morimoto, and A. Lu, "Techniques for data hiding," *IBM Syst. J.*, vol. 35, nos. 3 and 4, pp. 313–336, 1996.

[Bene86] N. Benevuto et al., "The 32 kb/s coding standard," *AT&T Technical Journal*, vol. 65(5), pp. 12–22, Sept.–Oct. 1986

[Berm02] G. Berman and I. Jouny, "Temporal watermarking invertible only through source separation," in *Proc. IEEE ICASSP-2002,* vol. 4, pp. 3732–3735, Orlando, FL, May 2002.

[Bess02] B. Bessette, et al., "The adaptive multi-rate wideband speech codec (AMR-WB)," *IEEE Trans. on Speech and Audio Proc.,* vol. 10, no. 8, pp. 620–636, Nov. 2002.

[Blac95] M. Black and M. Zeytinoglu, "Computationally efficient wavelet packet coding of wideband stereo audio signals," in *Proc. IEEE ICASSP-95*, pp. 3075–3078, Detroit, MI, May 1995.

[Blau74] J. Blauert, *Spatial Hearing*. The MIT Press, Cambridge, MA, 1974.

[Bola95] S. Boland and M. Deriche, "High quality audio coding using multipulse LPC and wavelet decomposition," in *Proc. IEEE ICASSP-95*, pp. 3067–3069, Detroit, MI, May 1995.

[Bola96] S. Boland and M. Deriche, "Audio coding using the wavelet packet transform and a combined scalar-vector quantization," in *Proc. IEEE ICASSP-96*, pp. 1041–1044, Atlanta, GA, May 1996.

[Bola97] S. Boland and M. Deriche, "New results in low bitrate audio coding using a combined harmonic-wavelet representation," in *Proc. IEEE ICASSP-97*, pp. 351–354, Munich, Germany, April 1997.

[Bola98] S. Boland and M. Deriche, "Hybrid LPC and discrete wavelet transform audio coding with a novel bit allocation algorithm," in *Proc. IEEE ICASSP-98*, vol. 6, pp. 3657–3660, Seattle, WA, May 1998.

[Bone96] L. Boney, A. Tewfik, and K. Hamdy, "Digital watermarks for audio signals," in *Proc. of Multimedia '96*, pp. 473–480, Piscataway, NJ, IEEE Press, 1996.

[Borm03] J. Bormans, J. Gelissen, and A. Perkis, "MPEG-21: the 21st century multimedia framework," *IEEE Sig. Proc. Mag.*, vol. 20, no. 2, pp. 53-62–, Mar. 2003.

[Bosi00] M. Bosi, "High-quality multichannel audio coding: trends and challenges," *J. Audio Eng. Soc.*, vol. 48, no. 6, pp. 588, 590–595, June 2000

[Bosi93] M. Bosi et al., "Aspects of current standardization activities for high-quality low rate multichannel audio coding," *IEEE Workshop on Applications of Signal Processing to Audio and Acoustics*, New Paltz, NY, Oct. 1993.

[Bosi96] M. Bosi, K. Brandenburg, S. Quackenbush, L. Fielder, K. Akagiri, and H. Fuchs, "ISO/IEC MPEG-2 advanced audio coding," in *Proc. 101st Conv. Aud. Eng. Soc.,* preprint #4382, Los Angeles, Nov. 1996.

[Bosi96a] M. Bosi et al., "ISO/IEC MPEG-2 advanced audio coding," in *Proc. 101st Conv. Audio Eng. Soc.*, preprint #4382, 1996.

[Bosi97] M. Bosi et al., "ISO/IEC MPEG-2 advanced audio coding," *J. Audio Eng. Soc.*, pp. 789–813, Oct. 1997.

[Boyd88] I. Boyd and C. Southcott, "A speech codec for the Skyphone service," *Br. Telecom Technical J.*, vol. 6(2), pp. 51–55, April 1988.

[Bran87a] K. Brandenburg, "Evaluation of quality for audio encoding at low bit rates," in *Proc. 82nd Conv. Aud. Eng. Soc.*, preprint #2433, Mar. 1987.

[Bran87b] K. Brandenburg, "OCF – a new coding algorithm for high quality sound signals," in *Proc. IEEE ICASSP-87*, vol. 12, pp. 141–144, Dallas, TX, May 1987.

[Bran88a] K. Brandenburg, "High quality sound coding at 2.5 bits/sample," in *Proc. 84th Conv. Aud. Eng. Soc.*, preprint #2582, Mar. 1988.

[Bran88b] K. Brandenburg, "OCF: coding high quality audio with data rates of 64 kbit/sec," in *Proc. 85th Conv. Aud. Eng. Soc.*, preprint #2723, Mar. 1988.

[Bran90] K. Brandenburg and J. Johnston, "Second generation perceptual audio coding: the hybrid coder," in *Proc. 88th Conv. Aud. Eng. Soc.*, preprint #2937, Mar. 1990.

[Bran91] K. Brandenburg, et al., "ASPEC: adaptive spectral entropy coding of high quality music signals," in *Proc. 90th Conv. Aud. Eng. Soc.*, preprint #3011, Feb. 1991.

[Bran92a] K. Brandenburg et al., "Comparison of filter banks for high quality audio coding," in *Proc. IEEE ISCAS*, 1992.

[Bran92b] K. Brandenburg and T. Sporer, "'NMR' and 'Masking Flag': Evaluation of quality using perceptual criteria," in *Proc. 11th Int. Conf. Aud. Eng. Soc.*, pp. 169–179, May 1992.

[Bran94a] K. Brandenburg and G. Stoll, "ISO-MPEG-1 audio: a generic standard for coding of high-quality digital audio," *J. Audio Eng. Soc.*, pp. 780–792, Oct. 1994.

[Bran94b] K. Brandenburg and B. Grill, "First ideas on scalable audio coding," in *Proc. 97th Conv. Aud. Eng. Soc.*, San Francisco, preprint #3924, 1994.

[Bran95] K. Brandenburg and M. Bosi, "Overview of MPEG-audio: current and future standards for low bit-rate audio coding," in *Proc. 99th Conv. Aud. Eng. Soc.*, preprint #4130, Oct. 1995.

[Bran97] K. Brandenburg and M. Bosi, "ISO/IEC MPEG-2 advanced audio coding: overview and applications," in *Proc. 103th Conv. Aud. Eng. Soc.*, preprint #4641, Sept. 1997.

[Bran98] K. Brandenburg, "Perceptual coding of high quality digital audio," in *Applications of Digital Signal Processing to Audio and Acoustics*, M. Kahrs and K. Brandenburg, Eds., Kluwer Academic Publishers, Boston: 1998.

[Bran02] K. Brandenburg, "Audio coding and electronic distribution of music," in *Proc. of 2nd International Conference on Web Delivering of Music (WEDELMUSIC-02)*, pp. 3–5, Dec. 2002.

[Bree04] J. Breebaart, et al., "High-quality parametric spatial audio coding at low bitrates," in *116th Conv. Aud. Eng. Soc.*, Berlin, Germany, May 2004.

[Bree05] J. Breebart, et al., "MPEG spatial audio coding/MPEG surround: Overview and current status," in *Proc. 119th AES convention*, paper 6599, New York, New York, Oct 2005.

[Breg90] A. S. Bregman, *Auditory Scene Analysis*, MIT Press, Cambridge, MA, 1990.

[Bros03] P.M. Brossier, M.B. Sandler, and M.D. Plumbley, "Real time object based coding," in *Proc. 101st Conv. Aud. Eng. Soc.*, preprint #5809, Mar. 2003.

[Brue97] F. Bruekers, W. Oomen, R. van der Vleuten, and L. van de Kerkhof, "Improved lossless coding of 1-bit audio signals," in *Proc. 103rd Conv. Aud. Eng. Soc.*, preprint #4563, New York, Sept. 1997.

[Burn03] I. Burnett et al., "MPEG-21: goals and achievements," in *IEEE Multimedia*, vol. 10, no. 4, pp. 60–70, Oct.–Dec. 2003

[Buzo80] A. Buzo et al., "Speech coding based upon vector quantization, " *IEEE Trans. Acoust., Speech, Signal Process.*, vol. 28, no. 5, pp. 562–574, Oct. 1980.

[BWEB] Specification of the Bluetooth System, available online at: www.bluetooth.com.

[Cabo92] R. Cabot, "Measurements on low-rate coders," in *Proc. 11th Int. Conf. Aud. Eng. Soc.*, pp. 216–222, May 1992.

[Cacc01] G. Caccia, R. Lancini, M. Ludovico, F. Mapelli, and S. Tubaro, "Watermarking for musical pieces indexing used in automatic cue sheet generation systems," *IEEE 4th Workshop on Mult. and Sig. Proc.* pp. 487–492, Oct. 2001.

[Cagl91] H. Caglar et al., "Statistically optimized PR-QMF design," in *Proc. SPIE Vis. Comm. and Img. Proc.*, pp. 86–94, Nov. 1991.

[Cai95] H. Cai and G. Mirchandani, "Wavelet transform and bit-plane coding," in *Proc. ICIP*, pp. 578–581, 1995.

[Camp86] J. Campbell and T.E. Tremain, "Voiced/unvoiced classification of speech with applications to the U.S. government LPC-10e algorithm," in *Proc. IEEE ICASSP-86*, vol. 11, pp. 473–476, Tokyo, Japan, April 1986.

[Camp90] J. Campbell, T.E. Tremain, and V. Welch, "The Proposed Federal Standard 1016 4800 bps voice coder: CELP," *Speech Tech. Mag.*, pp. 58–64, Apr. 1990.

[Cand92] J.C. Candy and G.C. Temes, "Oversampling methods for A/D and D/A conversion," in *Oversampling Delta-Sigma Converters*, Eds. J.C. Candy and G.C. Temes, Piscataway, NJ, IEEE Press, 1992.

[Casa98] A. Casal et al., "Testing a flexible time-frequency mapping for high frequencies in TARCO (Tonal Adaptive Resolution COdec)," in *Proc. 104th Conv. Aud. Eng. Soc.*, preprint #4676, Amsterdam, May. 1998.

[Case96] M.A. Casey and P. Smaragdis, "Netsound," in *Proc. Int. Computer Music Conf.*, Hong Kong, 1996.

[CD82] Philips and Sony, "Audio CD System Description," Philips Licensing, Eindhoven, The Netherlands, 1982.

[Chan90] W-Y. Chan and A. Gersho, "High fidelity audio transform coding with vector quantization," in *Proc. IEEE ICASSP-90*, pp. 1109–1112, Albuquerque, NM, May 1990.

[Chan91a] W. Chan and A. Gersho, "constrained-storage quantization of multiple vector sources by codebook sharing," *IEEE Trans. Comm.*, vol. 39, no. 1, pp. 11–13, Jan 1991.

[Chan91b] W. Chan and A. Gersho, "Constrained-storage vector quantization in high fidelity audio transform coding," in *Proc. IEEE ICASSP-91*, pp. 3597–3600, Toronto, Canada, May 1991.

[Chan96] W. Chang and C. Want, "A masking-threshold-adapted weighting filter for excitation search," *IEEE Trans. on Speech and Audio Proc.*, vol. 4, no. 2, pp. 124–132, Mar. 1996.

[Chan97] W. Chang et al., "Audio coding using sinusoidal excitation representation," in *Proc. IEEE ICASSP-97*, vol. 1, pp. 311–314, Munich, Germany, April 1997.

[Char88] A. Charbonnier and J.P. Petit, "Sub-band ADPCM coding for high quality audio signals," in *Proc. IEEE ICASSP-88*, pp. 2540–2543, New York, NY, May 1988.

[Chen92] J. Chen, R. Cox, Y. Lin, N. Jayant, and M. Melchner, "A low-delay CELP coder for the CCITT 16 kb/s speech coding standard," *IEEE Trans. Selected Areas Commun.*, vol. 10, no. 5, pp. 830–849, June 1992.

[Chen95] B. Chen et al., "Optimal signal reconstruction in noisy filterbanks: multirate Kalman synthesis filtering approach," *IEEE Trans. Sig. Proc.*, vol. 43, no. 11, pp. 2496–2504, Nov. 1995.

[Chen99] J. Chen and H-M. Tai, "MPEG-2 AAC coder on a fixed-point DSP," in *Proc. of Int. Conf. on Consumer Electronics,* ICCE-99, pp. 24–25, June 1999.

[Chen02a] S. Cheng, H. Yu, and Z. Xiong, "Error concealment of MPEG-2 AAC audio using modulo watermarks," in *Proc. IEEE ISCAS-2002,* vol. 2, pp. 261–264, May 2002.

[Chen02b] S. Cheng, H. Yu, and X. Zixiang, "Enhanced spread spectrum watermarking of MPEG-2 AAC," in *Proc. IEEE ICASSP-2002,* vol. 4, pp. 3728–3731, Orlando, FL, May 2002.

[Chen04] L-J. Chen, R. Kapoor, K. Lee, M.Y. Sanadidi, M. Gerla, "Audio streaming over Bluetooth: an adaptive ARQ timeout approach," in *6th Int. Workshop on Mult. Network Syst. and App.(MNSA 2004)*, Tokyo, Japan, 2004.

[Cheu95] S. Cheung and J. Lim, "Incorporation of biorthogonality into lapped transforms for audio compression," in *Proc. IEEE ICASSP-95*, pp. 3079–3082, Detroit, MI, May 1995.

[Chia96] H-C Chiang and J-C Liu, "Regressive implementations for the forward and inverse MDCT in MPEG audio coding," *IEEE Sig. Proc. Letters*, vol. 3, no. 4, pp. 116–118, Apr. 1996.

[Chou01a] J. Chou, K. Ramchandran, A. Ortega, "Next generation techniques for robust and imperceptible audio data hiding," in *Proc. IEEE ICASSP-2001,* vol. 3, pp. 1349–1352, Salt Lake City, UT, May 2001.

[Chou01b] J. Chou, K. Ramchandran, and A. Ortega, "High capacity audio data hiding for noisy channels," in *Proc. Int. Conf. on Inf. Tech.: Coding and Computing,* pp. 108–112, Apr. 2001.

[Chow73] J. Chowing, "The synthesis of complex audio spectra by means of frequency modulation," *J. Audio Eng. Soc.*, pp. 526–529, Sep. 1973.

[Chu85] P.L. Chu, "Quadrature mirror filter design for an arbitrary number of equal bandwidth channels," *IEEE Trans. Acous., Speech, and Sig. Proc.*, vol. 33, no. 1, pp. 203–218, Feb. 1985.

[Cilo00] T. Ciloglu and S.U. Karaaslan, "An improved all-pass watermarking scheme for speech and audio," in *Proc. Int. Conf. on Mult. and Expo. (ICME-2000),* vol. 2, pp. 1017–1020, July–Aug. 2000.

[Coif92] R. Coifman and M. Wickerhauser, "Entropy based algorithms for best basis selection," *IEEE Trans. Information Theory*, vol. 38, no. 2, pp. 712–718, Mar. 1992.

[Colo95] C. Colomes, et al., "A perceptual model applied to audio bit-rate reduction," *J. Aud. Eng. Soc.*, vol. 43, no. 4, pp. 233–240, Apr. 1995.

[Cont96] L. Contin et al., "Tests on MPEG-4 audio codec proposals," *J. Sig. Proc. Image Comm.*, Oct. 1996.

[Conw88] J. Conway and N. Sloane, *Sphere Packing, Lattices, and Groups*. Springer-Verlag, New York, 1988.

[Cove91] T. Cover and J. Thomas, *Elements of Information Theory*, Wiley-Interscience, New York, Aug. 1991.

[Cox86] Cox, R., "The design of uniformly and nonuniformly spaced pseudo QMF," *IEEE Trans. Acous., Speech, and Sig. Proc.*, vol. 34, pp. 1090–1096, Oct. 1986.

[Cox95] I.J. Cox, J. Kilian, T. Leighton, and T. Shamoon, "Secure spread spectrum watermarking for multimedia," *Tech. Rep. 95–10*, NEC Research Institute, Princeton, NJ, 1995.

[Cox96] I.J. Cox, J. Kilian, T. Leighton, and T. Shamoon, "A secure, robust watermark for multimedia," in *Proc. of Information Hiding Workshop '96*, Cambridge, U.K., pp. 147–158, May 1996.

[Cox97] I.J. Cox, J. Kilian, T. Leighton, and T. Shamoon, "Secure spread spectrum watermarking for multimedia," *IEEE Trans. Image Processing,* vol. 6, no. 12, pp. 1673–1687, 1997.

[Cox99] I.J. Cox, M.L. Miller, A.L. McKellips, "Watermarking as communications with side information," *Proc. of the IEEE*, vol. 87, no. 7, pp. 1127–1141, July 1999.

[Cox01a] I.J. Cox, M.L. Miller, and J.A. Bloom, *Digital Watermarking*, Morgan Kaufmann Publishers, Sanfransisco, 2001.

[Cox01b] I.J. Cox and M.L. Miller, "Electronic watermarking: the first 50 years," *IEEE 4th Workshop on Mult. Sig. Proc.* pp. 225–230, Oct. 2001.

[Crav96] P.G. Craven and M. Gerzon, "Lossless coding for audio discs," *J. Audio Eng. Soc.*, vol. 44, no. 9, pp. 706–720, Sept. 1996.

[Crav97] S. Craver, N. Memon, B-L. Yeo, and M. Yeung, "Can invisible watermarks resolve rightful ownership?," *IBM Research Report,* RC 20509, Feb. 1997.

[Crav97] P.G. Craven, M. Law, and J. Stuart, "Lossless compression using IIR prediction filters," in *Proc. 102nd Conv. Aud. Eng. Soc.*, preprint #4415, Munich, March 1997.

[Crav98] S. Craver, N. Memon, B-L. Yeo, and M. Yeung, "Resolving rightful ownerships with invisible watermarking techniques: limitations, attacks, and implications," in *IEEE Journal on Selected Areas in Communications,* vol. 16, no. 4, pp. 573–586, 1998.

[Crav01] S.A. Craver, M. Wu, and B. Liu, "What can we reasonably expect from watermarks?," in *2001 IEEE Workshop on the, Applications of Signal Processing to Audio and Acoustics*, 21–24 Oct. 2001 pp. 223–226.

[Creu96] C. Creusere and S. Mitra, "Efficient audio coding using perfect reconstruction noncausal IIR filter banks," *IEEE Trans. Speech and Aud. Proc.*, vol. 4, no. 2, pp. 115–123, Mar. 1996.

[Creu02] C.D. Creusere, "An analysis of perceptual artifacts in MPEG scalable audio coding," *Proc. of Data Compression Conference* (DCC-02), pp. 152–161, Apr. 2002.

[Croc76] R.E. Crochiere, et al., "Digital coding of speech in subbands," *Bell Sys. Tech. J.*, vol. 55, no. 8, pp. 1069–1085, Oct. 1976.

[Croc83] R.E. Crochiere and L. R. Rabiner, *Multirate Digital Signal Processing.*, Prentice-Hall, Englewood Cliffs, NJ, 1983.

[Croi76] A. Croisier et al., "Perfect channel splitting by use of interpolation/decimation/tree decomposition techniques," in *Proc. Int. Conf. on Inf. Sci. and Syst.*, Patras, 1976.

[Cumm73] P. Cummiskey et al., "Adaptive quantization in differential PCM coding of speech," *Bell Syst. Tech. J.*, vol. 52, no. 7, p. 1105, Sept. 1973.

[Cupe85] V. Cuperman and A. Gersho, "Vector predictive coding of speech at 16 kb/s," *IEEE Trans. on Commun.*, vol. 33, no. 7, pp. 685–696, Jul. 1985.

[Cupe89] V. Cuperman, "On adaptive vector transform quantization for speech coding," *IEEE Trans. on Commun.*, vol. 37, no. 3, pp. 261–267, Mar. 1989.

[Cvej01] N. Cvejic, A. Keskinarkaus, T. Seppanen, "Audio watermarking using m-sequences and temporal masking," *IEEE Workshop on App. of Sig. Proc. to Audio and Acoustics,* pp. 227–230, Oct. 2001.

[Cvej03] N. Cvejic, D. Tujkovic, T. Seppanen, "Increasing robustness of an audio watermark using turbo codes," in *Proc. ICME-03*, vol. 1, pp. 217–220, July 2003.

[Daub88] I. Daubechies, "Orthonormal bases of compactly supported wavelets," *Comm. Pure Appl. Math.*, pp. 909–996, Nov. 1988.

[Daub92] I. Daubechies, *Ten Lectures on Wavelets*, Society for Industrial and Applied Mathematics, Philadelphia, PA, 1992.

[Daub96] I. Daubechies and W. Sweldens, "Factoring wavelet transforms into lifting steps," *Tech. Rep.*, Bell Laboratories, Lucent Technologies, 1996.

[DAVC96] Digital Audio-Visual Council (DAVIC), DAVIC Technical Specification 1.2, Part 9, "Information Representation," Dec. 1996. Available online from http://www.davic.org.

[Davi86] G. Davidson and A. Gersho, "Complexity reduction methods for vector excitation coding," in *Proc. IEEE ICASSP-86*, vol. 11, pp. 3055–3058, Tokyo, Japan, April 1986.

[Davi90] G. Davidson et al., "Low-complexity transform coder for satellite link applications," in *Proc. 89th Conv. Aud. Eng. Soc.*, preprint #2966, 1990.

[Davi92] G. Davidson and M. Bosi, "AC-2: high quality audio coding for broadcasting and storage," in *Proc. 46th Annual Broadcast Eng. Conf.*, pp. 98–105, Apr. 1992.

[Davi94] G. Davidson et al., "Parametric bit allocation in a perceptual audio coder," in *Proc. 97th Conv. Aud. Eng. Soc.*, preprint 3921, Nov. 1994.

[Davi98] G. Davidson, "Digital audio coding: Dolby AC-3," in *The Digital Signal Processing Handbook*, V. Madisetti and D. Williams, Eds., CRC Press, pp. 41.1–41.21, 1998.

[Davis93] M. Davis, "The AC-3 multichannel coder," in *Proc. 95th Conv. Aud. Eng. Soc.*, preprint #3774, Oct. 1993.

[Davis94] M. Davis and C. Todd, "AC-3 operation, bitstream syntax, and features," in *Proc. 97th Conv. Aud. Eng. Soc.*, preprint #3910, 1994.

[Davis03] M. Davis, "History of spatial coding," *J. Audio Eng., Soc.,* vol. 51, no. 6, pp. 554-569, June 2003.

[Dehe91] Y.F. Dehery et al., "A MUSICAM source codec for digital audio broadcasting and storage," in *Proc. IEEE ICASSP-91*, pp. 3605–3608, Toronto, Canada, May 1991.

[Delo96] A. Delopoulos and S. Kollias, "Optimal filterbanks for signal reconstruction from noisy subband components," *IEEE Trans. Sig. Proc.*, vol. 44, no. 2, pp. 212–224, Feb. 1996.

[Deri99] M. Deriche, D. Ning, and S. Boland, "A new approach to low bit rate audio coding using a combined harmonic-multiband-wavelet representation," in *Proc. ISSPA-99* vol. 2, pp. 603–606, Aug. 1999.

[Diet96] M. Dietz et al., "Audio compression for network transmission," *J. Aud. Eng. Soc.*, pp. 58–71, Jan., Feb. 1996.

[Diet00] M. Dietz and T. Mlasko, "Using MPEG-4 audio for DRM digital narrowband broadcasting," in *Proc. ISCAS-2000.* vol. 3, pp. 205–208, May 2000.

[Diet02] M Dietz, et al., "Spectral band replication, a novel approach in audio coding," in *112th Conv. Aud. Eng. Soc.,* preprint # 5553, Munich, Germany, May 2002.

[Ditt00] J. Dittmann, A. Mukherjee, and M. Steinebach, "Media-independent watermarking classification and the need for combining digital video and audio watermarking for media authentication," in *Proc. Int. Conf. on Inf. Tech.: Coding and Computing*, pp. 62–67, Mar. 2000.

[Ditt01] J. Dittmann and M. Steinebach, "Joint watermarking of audio-visual data," *IEEE 4th Workshop on Mult. Sig. Proc.*, pp. 601–606, Oct. 2001.

[Dobs97] W. Dobson et al., "High quality low complexity scalable wavelet audio coding," in *Proc. IEEE ICASSP-97*, pp. 327–330, Munich, Germany, April 1997.

[DOLBY] "Surround Sound and Multimedia Dolby Surround and Dolby Digital for video games, CD ROM, DVD ROM, the Internet – and more." Available online at http://www.dolby.com/multi/surmulti.html.

[Dong02] H. Dong, J.D. Gibson, and M.G. Kokes, "SNR and bandwidth scalable speech coding," in *Proc. IEEE ICASSP-02*, vol. 2, pp. 859–862, Orlando, FL, May 2002.

[Dorw01] S. Dorward, D. Huang, S.A. Savari, G. Schuller, and B. Yu, "Low delay perceptually lossless coding of audio signals," in *Proc. of Data Compression Conference (DCC-01)*, pp. 312–320, Mar. 2001.

[DTS] The Digital Theater Systems (DTS). Available online at www.dtsonline.com.

[Duen02] A. Duenas, R. Perez, B. Rivas, E. Alexandre, and A. Pena, "A robust and efficient implementation of MPEG-2/4 AAC natural audio coders," in *Proc. 112th Conv. Aud. Eng. Soc.*, preprint #5556, May 2002.

[Duha91] P. Duhamel et al., "A fast algorithm for the implementation of filter banks based on time domain aliasing cancellation," in *Proc. IEEE ICASSP-91*, pp. 2209–2212, Toronto, Canada, May 1991.

[Dunn96] C. Dunn and M. Sandler, "A comparison of dithered and chaotic sigma-delta modulators," *J. Audio Eng. Soc.*, vol. 44, no. 4, pp. 227–244, Apr. 1996.

[Duro96] X. Durot and J-B. Rault, "A new noise injection model for audio compression algorithms," in *Proc. 101st Conv. Aud. Eng. Soc.*, preprint #4374, Nov. 1996.

[DVD01] DVD Forum, "DVD Specifications for Read-only Disc," Part 4, Audio Specifications, Version 1.2, Tokyo, Japan, Mar. 2001.

[DVD96] DVD Consortium, "DVD-video and DVD-ROM specifications," 1996.

[DVD99] DVD Forum, "DVD Specifications for Read-Only Disc," Part 4, Audio Specifications, Packed PCM: MLP Reference Information Version 1.0, Tokyo, Japan, Mar. 1999.

[East97] P.C. Eastty, C. Sleight, and P.D. Thorpe, "Research on cascadable filtering, equalization, gain control, and mixing of 1-bit signals for professional audio applications," in *Proc. 102th Conv. Aud. Eng. Soc.*, preprint #4444, Mar. 22–25, 1997.

[EBU99] EBU Technical Recommendation R96-1999, *Formats for production and delivery of multichannel audio programs*, European Broadcasting Union, Geneva, Switzerland, 1999.

[Edle89] B. Edler, "Codierung von Audiosignalen mit überlappender Transformation und adaptiven Fensterfunktionen (Coding of Audio Signals with Overlapping Block Transform and Adaptive Window Functions)," *Frequenz*, pp. 252–256, 1989.

[Edle95] B. Edler, "Technical Description of the MPEG-4 Audio Coding Proposal from University of Hannover and Deutsche Bundespost Telekom," ISO/IEC JTC1/SC29/WG11 MPEG95/MO414, Oct. 1995.

[Edle96a] B. Edler and H. Purnhagen, "Technical Description of the MPEG-4 Audio Coding Proposal from University of Hannover and Deutschet Telekom AG," ISO/IEC JTC1/SC29/WG11 MPEG96/MO632, Jan. 1996.

[Edle96a] B. Edler, "Current status of the MPEG-4 audio verification model development," in *Proc. 101st Conv. Aud. Eng. Soc.*, Preprint #4376, Nov. 1996.

[Edle96b] B. Edler and L. Contin, "MPEG-4 Audio Test Results (MOS Test)," ISO/IEC JTC1/SC29/WG11 N1144, Jan. 1996.

[Edle96b] B. Edler et al., "ASAC – Analysis/Synthesis Audio Codec for very low bit rates," in *Proc. 100th Conv. Aud. Eng. Soc.*, preprint #4179, May 1996.

[Edle96c] B. Edler and H. Purnhagen, "Parametric audio coding," in *Proc. International Conference on Signal Processing* (ICSP 2000), Beijing, 2000.

[Edle97] B. Edler, "Very low bit rate audio coding development," in *Proc. 14th Audio Eng. Soc. Int. Conf.*, June 1997.

[Edle98] B. Edler and H. Purnhagen, "Concepts for hybrid audio coding schemes based on parametric techniques," in *Proc. 105th Conv. Audio Eng. Soc.*, preprint #4808, 1998.

[Edle99] B. Edler, "Speech coding in MPEG-4," *Int. J. of Speech Technology*, vol. 2, no. 4, pp. 289–303, May 199.

[Egan50] J. Egan and H. Hake, "On the masking pattern of a simple auditory stimulus," *J. Acoust. Soc. Am.*, vol. 22, pp. 622–630, 1950.

[Ehre04] A. Ehret, et al., "aacPlus, only a low-bitrate codec?," in *117th Conv. Aud. Eng. Soc.,* preprint # 6199, San Francisco, CA, Oct. 2004.

[Ekud99] R. Ekudden, R. Hagen, I. Johansson, and J. Svedburg, "The adaptive multi-rate speech coder," in *Proc. IEEE Workshop on Speech Coding*, pp. 117–119, June 1999.

[Elli96] P.W. Ellis, "Prediction-Driven Computational Auditory Scene Analysis," Ph.D. Thesis, Massachusetts Institute of Technology, June 1996.

[Erku03] S. Erkucuk, S. Krishnan, and M. Zeytinoglu, "Robust audio watermarking using a chirp based technique," in *Proc. ICME-03*, vol. 2, pp. 513–516, July 2003.

[Esma03] S. Esmaili, S. Krishnan, and K. Raahemifar, "A novel spread spectrum audio watermarking scheme based on time-frequency characteristics," in *Proc. of CCECE-03*, vol. 3, pp. 1963–1966, May 2003.

[Este77] D. Esteban and C. Galand, "Application of quadrature mirror filters to split band voice coding schemes," in *Proc. IEEE ICASSP-77*, vol. 2, pp. 191–195, Hartford, CT, May 1977.

[ETSI98] ETSI AMR Qualification Phase Documentation, 1998.

[Fall01] C. Faller and F. Baumgarte, "Efficient representation of spatial audio using perceptual parameterization," *IEEE Workshop on Applications of Signal Processing to Audio and Acoustics,* New Paltz, New York, 2001.

[Fall02a] C. Faller and F. Baumgarte, "Binaural cue coding: a novel and efficient representation of spatial audio," in *Proc. IEEE ICASSP-02*, vol. 2, pp. 1841–1844, Orlando, FL, May 2002.

[Fall02b] C. Faller and F. Baumgarte, "Estimation of auditory spatial cues for binaural cue coding," in *Proc. IEEE ICASSP-02*, vol. 2, pp. 1801–1804, Orlando, FL, May 2002.

[Fall02b] C. Faller, B.-H. Juang, P. Kroon, H.-L. Lou, S.A. Ramprashad, C.-E.W. Sundberg, "Technical advances in digital audio radio broadcasting," *Proc. of the IEEE*, vol. 90, no. 8, pp. 1303–1333, Aug. 2002.

[Feit98] B. Feiten et al., "Dynamically scalable audio internet transmission," in *Proc. 104th Conv. Audio Eng. Soc.*, preprint #4686, May 1998.

[Fejz00] Z. Fejzo, S. Smyth, K. McDowell, and Y.-L. You, "Backward compatible enhancement of DTS multi-channel audio coding that delivers 96-kHz/24-bit audio quality," in *Proc. of 109th Conv. Aud. Eng. Soc.*, preprint #5259, Sept. 22–25, 2000.

[Ferr96a] A. Ferreira, "Convolutional effects in transform coding with TDAC: an optimal window," *IEEE Trans. on Speech and Audio Proc.*, vol. 4, no 2, pp. 104–114, Mar. 1996.

[Ferr96b] A.J.S. Ferreira, "Perceptual coding of harmonic signals," in *Proc. 100th Conv. Audio Eng. Soc.*, preprint #4746, May 1996.

[Ferr01a] A.J.S. Ferreira, "Combined spectral envelope normalization and subtraction of sinusoidal components in the ODFT and MDCT frequency domains," *IEEE Workshop on the Applications of Signal Processing to Audio and Acoustics*, pp. 51–54, Oct. 2001.

[Ferr01b] A.J.S. Ferreira, "Accurate estimation in the ODFT domain of the frequency, phase and magnitude of stationary sinusoids," *IEEE Workshop on the Applications of Signal Processing to Audio and Acoustics*, pp. 47–51, Oct. 2001.

[Fiel91] L. Fielder and G. Davidson, "AC-2: a family of low complexity transform-based music coders," in *Proc. 10th AES Int. Conf.*, Sep. 1991.

[Fiel96] L. Fielder et al., "AC-2 and AC-3: low-complexity transform-based audio coding," in *Collected Papers on Digital Audio Bit-Rate Reduction*, N. Gilchrist and C. Grewin, Eds., Aud. Eng. Soc., pp. 54–72, 1996.

[Fiel04] L. Fielder, et al., "Introduction to Dolby digital plus, an enhancement to the Dolby digital coding system," in *Proc. 117th AES convention,* paper 6196, San Francisco, CA, Oct. 2004.

[Flet37] H. Fletcher and W. Munson, "Relation between loudness and masking," *J. Acoust. Soc. Am.*, vol. 9, pp. 1–10, 1937.

[Flet40] H. Fletcher, "Auditory patterns," *Rev. Mod. Phys.*, pp. 47–65, Jan. 1940.

[Flet91] N. Fletcher and T. Rossing, *The Physics of Musical Instruments*, Springer, New York, 1991.

[Flet98] J. Fletcher, "ISO/MPEG layer 2 – optimum re-encoding of decoded audio using a MOLE signal," in *Proc. 104th Conv. Aud. Eng. Soc.*, preprint 4706, Amsterdam, 1998. See also http://www.bbc.co.uk/atlantic.

[Foo01] S.W. Foo, T.H. Yeo, and D.Y. Huang, "An adaptive audio watermarking system," in *Proc. Int. Conf. on Electrical and Electronic Technology,* vol. 2, pp. 509–513, Aug. 2001.

[Foss95] R. Foss and T. Mosala, "Routing MIDI messages over Ethernet," in *Proc. 99th Conv. Aud. Eng. Soc.,* preprint #4057, Oct. 1995.

[FS1015] Federal Standard 1015, Telecommunications: Analog to Digital Conversion of Radio Voice by 2400 bit/second Linear Predictive Coding, National Communication System – Office Technology and Standards, Nov. 1985.

[FS1016] Federal Standard 1016, Telecommunications: Analog to Digital Conversion of Radio Voice by 4800 bit/second Code Excited Linear Prediction (CELP), National Communication System – Office Technology and Standards, Feb. 1991.

[Fuch93] H. Fuchs, "Improving joint stereo audio coding by adaptive inter-channel prediction," in *Proc. IEEE ASSP Workshop on Apps. of Sig. Proc. to Aud. and Acous.*, 1993.

[Fuch95] H. Fuchs, "Improving MPEG audio coding by backward adaptive linear stereo prediction," in *Proc. 99th Conv. Aud. Eng. Soc.*, preprint #4086, Oct. 1995.

[Furo00] T. Furon, N. Moreau, and P. Duhamel, "Audio public key watermarking technique," in *Proc. IEEE ICASSP*-2000, vol. 4, pp. 1959–1962, Istanbul, Turkey, June 2000.

[G.722] ITU Recommendation G.722, "7 kHz Audio Coding within 64 kb/s," Blue Book, vol. III, Fascicle III, Oct. 1988.

[G.722.2] ITU Recommendation G.722.2, "Wideband coding of speech at around 16 kb/s using Adaptive Multi-rate Wideband (AMR-WB)," Jan. 2001.

[G.723.1] ITU Recommendation G.723.1, "Dual Rate Speech Coder for Multimedia Communications transmitting at 5.3 and 6.3 kb/s," Draft, 1995.

[G.726] ITU Recommendation G.726, "24, 32, 40 kb/s Adaptive Differential Pulse Code Modulation (ADPCM)," Blue Book, vol. III, Fascicle III.3, Oct. 1988.

[G.728] ITU Draft Recommendation G.728, "Coding of Speech at 16 kbit/s Using Low-Delay Code Excited Linear Prediction (LD-CELP)," 1992.

[G.729] ITU Study Group 15 Draft Recommendation G.729, "Coding of Speech at 8 kb/s using Conjugate-Structure Algebraic-Code-Excited Linear-Prediction (CS-ACELP)," 1995.

[G721] CCITT Recommendation G.721, "32 kb/s adaptive differential pulse code modulation (ADPCM)," in Blue Book, vol. III, Fascicle III, Oct. 1988.

[G722] CCITT (ITU-T) Rec. G.722, "7 kHz audio coding within 64 kbit/s," Blue Book, vol. III, fasc. III.4, pp. 269–341, 1988.

[G726] ITU Recommendation G.726 (formerly G.721), "24, 32, 40 kb/s Adaptive differential pulse code modulation (ADPCM)," Blue Book, vol. III, Fascicle III.3, Oct. 1988.

[G728] International Telecommunications Union Rec. G.728, "Coding of Speech at 16 kbps Using Low-Delay Code Excited Linear Prediction," Sept. 1992.

[G729] ITU-T Recommendation G.729, "Coding of speech at 8 kbps using conjugate-structure algebraic code excited linear prediction (CS-ACELP)", 1995.

[Gang02] L. Gang, A.N. Akansu, and M. Ramkumar, "Security and synchronization in watermark sequence," in *Proc. IEEE ICASSP-2002,* vol. 4, pp. 3736–3739, Orlando, FL, May 2002.

[Gao01a] Y. Gao, et al., "The SMV algorithm selected by TIA and 3GPP2 for CDMA applications," in *Proc. IEEE ICASSP-01,* vol. 2, pp. 709–712, Salt Lake City, UT, May 2001.

[Gao01b] Y. Gao, et al., "EX-CELP: A speech coding paradigm," in *Proc. IEEE ICASSP-01,* vol. 2, pp. 689–692, Salt Lake City, UT, May 2001.

[Garc99] R.A. Garcia, "Digital watermarking of audio signals using a psychoacoustic auditory model and spread spectrum theory," in *Proc. 107th Conv. Aud. Eng. Soc.*, preprint #5073, New York, Sept. 1999.

[Gäss54] G. Gässler, "Uber die hörschwelle für schallereignisse mit verschieden breitem frequenzspektrum," *Acustica*, vol. 4, pp. 408–414, 1954.

[Gbur96] U. Gbur, M. Werner, and M. Dietz, "Realtime implementation of an ISO/MPEG layer 3 encoder on Pentium PCs," in *Proc. 101st Conv. Aud. Eng. Soc.,* preprint #4386, Nov. 1996.

[Geig01] R. Geiger, T. Sporer, J. Koller, and K. Brandenburg, "Audio coding based on integer transforms," in *Proc. 111th Conv. Aud. Eng. Soc.*, Preprint #5471, New York, 2001.

[Geig02] R. Geiger, J. Herre, J. Koller, and K. Brandenburg, "IntMDCT – A link between perceptual and lossless audio coding," in *Proc. IEEE ICASSP-2002*, vol. 2, pp. 1813–1816, Orlando, FL, May 2002.

[Geig03] R. Geiger, J. Herre, G. Schuller, and T. Sporer, "Fine grain scalable perceptual and lossless audio coding based on IntMDCT," in *Proc. IEEE ICASSP-2003*, Hong Kong, April 2003.

[Geor87] E.B. George and M.J.T. Smith, "A new speech coding model based on a least-squares sinusoidal representation," in *Proc. IEEE ICASSP-87*, pp. 1641–1644, Dallas, TX, Apr. 1987.

[Geor90] E.B. George and M.J.T. Smith, "Perceptual Considerations in a low bit rate sinusoidal vocoder," in *Proc. IEEE Int. Phoenix Conf. on Comp. in Comm.*, pp. 268–275, Mar. 1990.

[Geor92] E.B. George and M.J.T. Smith, "Analysis-by-synthesis/overlap-add sinusoidal modeling applied to the analysis and synthesis of musical tones," *J. Audio Eng. Soc.*, pp. 497–516, June 1992.

[Geor97] E.B. George and M.J.T. Smith, "Speech analysis/synthesis and modification using and analysis-by-synthesis/overlap-add sinusoidal model," *IEEE Trans. on Speech and Audio Proc.*, pp. 389–406, Sept. 1997.

[Gers83] A. Gersho and V. Cuperman, "Vector quantization: a pattern matching technique for speech coding," *IEEE Commun. Mag.*, vol. 21, no. 9, pp. 15–21, Dec. 1983.

[Gers90a] A. Gersho and S. Wang, "Recent trends and techniques in speech coding," in *Proc. 24th Asilomar Conf.*, Pacific Grove, CA, Nov. 1990.

[Gers90b] I.A. Gerson and M.A. Jasiuk, "Vector sum excited linear prediction (VSELP) speech coding at 8 kb/s," in *Proc. of ICASSP-90*, vol. 1, pp. 461–464, Albuquerque, NM, Apr. 1990.

[Gers91] I. Gerson and M. Jasiuk, "Techniques for improving the performance of CELP-type speech coders," in *Proc. IEEE ICASSP-91*, pp. 205–208, Toronto, Canada, May 1991.

[Gers92] A. Gersho and R. Gray, *Vector Quantization and Signal Compression*, Kluwer Academic Publishers, Norwell, MA, 1992.

[Gerz99] M.A. Gerzon, J.R. Stuart, R.J. Wilson, P.G. Craven, and M.J. Law, "The MLP Lossless Compression System," in *AES 17th Int. Conf. High-Quality Audio Coding*, pp. 61–75 Florence, Italy, Sept. 1999.

[Geye99] S. Geyersberger et al., "MPEG-2 AAC multichannel realtime implementation on floating-point DSPs," in *Proc. 106th Conv. Aud. Eng. Soc.*, preprint #4977, May 1999.

[Gibs74] J. Gibson, "Adaptive prediction in speech differential encoding systems," *Proc. of the IEEE*, vol. 68, pp. 1789–1797, Nov. 1974.

[Gibs78] J. Gibson, "Sequentially adaptive backward prediction in ADPCM speech coders." *IEEE Trans. Commun.*, pp. 145–150, Jan. 1978.

[Gibs90] J. Gibson et al., "Backward adaptive prediction in multi-tree speech coders," in *Advances in Speech Coding*, B. Atal, V. Cuperman, and A. Gersho, Eds., Kluwer, Norwell, MA, 1990, pp. 5–12.

[Gilc98] N. Gilchrist, "ATLANTIC audio: preserving technical quality during low bit rate coding and decoding," in *Proc. 104th Conv. Aud. Eng. Soc.*, preprint #4694, Amsterdam, May 1998.

[Giur98] C.D. Giurcaneanu, I. Tabus, and J. Astola, "Linear prediction from subbands for lossless audio compression," in *Proc. Norsig-98, 3rd IEEE Nordic Signal Processing Symposium*, pp. 225–228, Vigso, Denmark, June 1998.

[Glas90] B. Glasberg and B.C.J. Moore, "Derivation of auditory filter shapes from notched-noise data," *Hearing Res.*, vol. 47, pp. 103–138, 1990.

[Golo66] S.W. Golomb, "Run-length encodings," in *IEEE Trans. on Info. Theory*, vol. 12, pp. 399–401, 1966.

[Good96] M. Goodwin, "Residual modeling in music analysis-synthesis," *Proc. IEEE ICASSP-96*, Atlanta, GA, May 1996.

[Good97] M. Goodwin, "Adaptive Signal Models: Theory, Algorithms, and Audio Applications," Ph.D. Dissertation, University of California at Berkeley, 1997.

[Gray84] R. Gray, "Vector quantization," *IEEE ASSP Mag.*, vol. 1, no. 2, pp. 4–29, Apr. 1984.

[Gree61] D.D. Greenwood, "Critical bandwidth and the frequency coordinates of the basilar membrane," *J. Acous. Soc. Am.*, pp. 1344–1356, Oct. 1961.

[Gree90] D.D. Greenwood, "A cochlear frequency-position function for several species – 29 years later," *J. Acoust. Soc. Am.*, vol. 87, pp. 2592–2605, June 1990.

[Grew91] C. Grewin and T. Ryden, "Subjective assessments on low bit-rate codecs," in *Proc. 10th Int. Conf. Aud. Eng. Soc.*, pp. 91–102, 1991.

[Gril02] B. Grill et al., "Characteristics of audio streaming over IP networks within the ISMA Standard," in *Proc. 112th Conv. Aud. Eng. Soc.*, preprint #5514, May 2002.

[Gril94] B. Grill et al., "Improved MPEG-2 audio multi-channel encoding," in *Proc. 96th Conv. Aud. Eng. Soc.*, preprint #3865, Feb. 1994.

[Gril97] B. Grill, "A bit-rate scalable perceptual coder for MPEG-4 audio," in *Proc. 103rd Conv. Aud. Eng. Soc.*, preprint #4620, Sept. 1997.

[Gril99] B. Grill, "The MPEG-4 general audio coder," in *Proc. AES 17th Int. Conf.*, Sept. 1999.

[Gros03] A. Groschel, et al., "Enhancing audio coding efficiency of MPEG layer-2 with spectral band replication for digitalradio (DAB) in a backwards compatible way," in *114th Conv. Aud. Eng. Soc.,* preprint # 5850, Amsterdam, Netherlands, Mar. 2003.

[Grub01] M. Grube, P. Siepen, C. Mittendorf, J. Friess, and S. Bauer, "Applications of MPEG-4: digital multimedia broadcasting," in *Proc. of Int. Conf. on Consumer Electronics,* ICCE-01, pp. 136–137, June 2001.

[Gruh96] D. Gruhl, A. Lu, and W. Bender, "Echo hiding," in *Proc. of Information Hiding Workshop '96*, Cambridge, UK, pp. 295–316, May 1996.

[GSM89] GSM 06.10, "GSM full-rate transcoding," *Technical Report Version 3.2*, ETSI/GSM, July 1989.

[GSM96a] GSM 06.60, "GSM Digital Cellular Communication Standards: Enhanced Full-Rate Transcoding," ETSI/GSM, 1996.

[GSM96b] GSM 06.20, "GSM Digital Cellular Communication Standards: Half Rate Speech; Half Rate Speech Transcoding," ETSI/GSM, 1996.

[Hadd95] R. Haddad and K. Park, "Modeling, analysis, and optimum design of quantized M-band filterbanks," *IEEE Trans. Sig. Proc.*, vol. 43, no. 11, pp. 2540–2549, Nov. 1995.

[Hall97] J.L. Hall, "Asymmetry of masking revisited: generalization of masker and probe bandwidth," *J. Acoust. Soc. Am.*, vol. 101, pp. 1023–1033, Feb. 1997.

[Hall98] J.L. Hall, "Auditory psychophysics for coding applications," in *The Digital Signal Processing Handbook*, V. Madisetti and D. Williams, Eds., CRC Press, Boca Raton, FL, pp. 39.1–39.25, 1998.

[Hamd96] K.N. Hamdy et al., "Low bit rate high quality audio coding with combined harmonic and wavelet representations," in *Proc. IEEE ICASSP-96*, pp. 1045–1048, Atlanta, GA, May 1996.

[Hans96] M.C. Hans and V. Bhaskaran, "A fast integer-based, CPU scalable, MPEG-1 layer-II audio decoder," in *Proc. 101st Conv. Aud. Eng. Soc.,* preprint #4359, Nov. 1996.

[Hans98a] M. Hans, *Optimization of Digital Audio for Internet Transmission*, Ph.D. dissertation, School Elect. Comp. Eng., Georgia Inst. Technol., Atlanta, 1998.

[Hans98b] M. Hans and R.W. Schafer, "AudioPaK – an integer arithmetic lossless audio codec," in *Proc. Data Compression Conf.*, p. 550, Snowbird, UT, 1998.

[Hans01] M. Hans and R.W. Schafer, "Lossless compression of digital audio," *IEEE Sig. Proc. Mag.*, vol. 18, no. 4, pp. 21–32, July 2001.

[Harm96] A. Harma, et al., "Warped linear prediction (WLP) in audio coding," in *Proc. NORSIG '96*, pp. 367–370, Sep. 1996.

[Harm97a] A. Harma, et al., "An experimental audio codec based on warped linear prediction of complex valued signals," in *Proc. IEEE ICASSP-97*, pp. 323–326, Munich, Germany, April 1997.

[Harm97b] A. Harma, et al., "WLPAC – a perceptual audio codec in a nutshell," *Proc. 102nd Conv. Audio Eng. Soc.*, preprint #4420, Munich, 1997.

[Harm98] A. Harma, et al., "A warped linear predictive stereo codec using temporal noise shaping," in *Proc. NORSIG-98*, pp. 229–232, May 1998.

[Harm00] A. Harma, "Implementation of frequency-warped recursive filters," *signal processing*, vol. 80, pp. 543–548, Feb. 2000.

[Harm01] A. Harma and U.K. Laine, "A comparison of warped and conventional linear prediction," *IEEE Trans. Speech Audio Proc.*, vol. 9, no. 5, pp. 579–588, July 2001.

[Harp03] P. Harpe, D. Reefman, and E. Janssen, "Efficient trellis-type sigma-delta modulator," in *Proc. 114th Conv. Aud. Eng. Soc.*, preprint #5845, Amsterdam, Mar. 22–25, 2003.

[Hart99] F. Hartung and M. Kutter, "Multimedia watermarking techniques," *Proc. of the IEEE*, vol. 87, no. 7, pp. 1079–1107, July 1999.

[Hask97] B.G. Haskell, A. Puri, and A.N. Netravali, *Digital Video: An Introduction to MPEG-2*, Chapman & Hall, New York 1997.

[Hawk02] M.O.J. Hawksford, "Time-quantized frequency modulation with time dispersive codes for the generation of sigma-delta modulation," in *Proc. 112th Conv. Aud. Eng. Soc.*, preprint #5618, Munich, Germany, May 10–13, 2002.

[Hay96] K. Hay et al., "A low delay subband audio coder (20 Hz – 15 kHz) at 64 kbit/s," in *Proc. IEEE Int. Symp. on Time-Freq. and Time-Scale Anal.*, pp. 265–268, June. 1996.

[Hay97] K. Hay et al., "The D_5 lattice quantization for a 64 kbit/s low-delay subband audio coder with a 15 kHz bandwidth," in *Proc. IEEE ICASSP-97*, pp. 319–322, Munich, Germany, Apr. 1997.

[Hede81] P. Hedelin, "A tone-oriented voice-excited vocoder," in *Proc. IEEE ICASSP-81*, pp. 205–208, Atlanta, GA, Mar. 1981.

[Hedg97] R. Hedges, "Hybrid wavelet packet analysis," in *Proc. 31st Asilomar Conf. on Sig., Sys., and Comp.*, Oct. 1997.

[Hedg98a] R. Hedges and D. Cochran, "Hybrid wavelet packet analysis," in *Proc. IEEE SP Int. Sym. on Time-Freq. and Time-Scale Anal.*, Oct. 1998.

[Hedg98b] R. Hedges and D. Cochran, "Hybrid wavelet packet analysis: a top down approach," in *Proc. 32nd Asilomar Conf. on Sig., Sys., and Comp.*, Nov. 1998.

[Hell72] R. Hellman, "Asymmetry of masking between noise and tone," *Percep. and Psychphys.*, vol. 11, pp. 241–246, 1972.

[Hell01] O. Hellmuth et al., "Advanced audio identification using MPEG-7 content description," in *Proc. 111th Conv. Aud. Eng. Soc.*, preprint #5463, Nov–Dec. 2001.

[Herl95] C. Herley, "Boundary filters for finite-length signals and time-varying filter banks," *IEEE Trans. Circ. Sys. II*, vol. 42, pp. 102–114, Feb. 1995.

[Herm90] H. Hermansky, "Perceptual linear predictive (PLP) analysis of speech," *J. Acoust. Soc. Amer.,* vol. 87, no. 4, pp. 1738–1752, Apr. 1990.

[Herm02] K. Hermus, W. Verhelst, and P. Wambacq, "Psychoacoustic modeling of audio with exponentially damped sinusoids," *Proc. IEEE ICASSP-02,* vol. 2, pp. 1821–1824, Orlando, FL, 2002.

[Herr92a] J. Herre et al., "Advanced audio measurement system using psychoacoustic properties," in *Proc. 92nd Conv. Aud. Eng. Soc.*, preprint #3321, Mar. 1992.

[Herr92b] J. Herre et al., "Analysis tool for realtime measurements using perceptual Criteria," in *Proc. 11th Int. Conf. Aud. Eng. Soc.*, pp. 180–190, May 1992.

[Herr95] J. Herre, K. Brandenburg, E. Eberlein, and B. Grill, "Second-generation ISO/MPEG-audio layer III coding," in *Proc. 98th Conv. Aud. Eng. Soc.*, preprint #3939, Feb. 1995.

[Herr96] J. Herre and J.D. Johnston, "Enhancing the performance of perceptual audio coders by using temporal noise shaping (TNS)," in *Proc. 101st Conv. Aud. Eng. Soc.,* preprint #4384, Nov. 1996,

[Herr98] J. Herre and D. Schulz, "Extending the MPEG-4 AAC codec by perceptual noise substitution," in *Proc. 104th Conv. Audio Eng. Soc.*, preprint #4720, Amsterdam, 1998.

[Herr98a] J. Herre and D. Schulz, "Extending the MPEG-4 AAC codec by perceptual noise substitution," in *Proc. 104th Conv. Aud. Eng. Soc.*, preprint #4720, May 1998.

[Herr98b] J. Herre et al., "The integrated filterbank-based scalable MPEG-4 audio coder," in *Proc. 105th Conv. Aud. Eng. Soc.*, preprint #4810, Sept. 1998.

[Herr98c] J. Herre, E. Allamanche, R. Geiger, and T. Sporer, "Proposal for a low delay MPEG-4 audio coder based on AAC," ISO/IEC JTC1/SC29/WG11, M4139, Oct. 1998.

[Herr99] J. Herre, E. Allamanche, R. Geiger, and T. Sporer, "Information and proposed enhancements for MPEG-4 low delay audio coding," ISO/IEC JTC1/SC29/WG11, M4560, Mar. 1998.

[Herr00a] J. Herre and B. Grill, "Overview of MPEG-4 audio and its applications in mobile communications," in *Proc. of 5th Int. Conf. on Sig. Proc.,* (ICSP-00), vol. 1, pp. 11–20, Aug. 2000.

[Herr01a] J. Herre, C. Neubauer, and F. Siebenhaar, "Combined compression/watermarking for audio signals," in *Proc. 110th Conv. Aud. Eng. Soc.*, preprint #5344, Amsterdam, May 2001.

[Herr01b] J. Herre, C. Neubauer, F. Siebenhaar, and R. Kulessa, "New results on combined audio compression/watermarking," in *Proc. 111th Conv. Aud. Eng. Soc.*, preprint 5442, New York, Dec. 2001.

[Herr01c] J. Herre, R. Kulessa, and C. Neubauer, "A compatible family of bitstream watermarking schemes for MPEG-Audio," in *Proc. 110th Conv. Audio Eng. Soc.*, preprint #5346, Amsterdam, May 2001.

[Herr04a] J. Herre, et al., "MP3 surround: Efficient and compatible coding of multichannel audio," in *Proc. 116^{th}AES convention,* paper 6049, Berlin, Germany, May 2004.

[Herr04b] J. Herre, et al., "From joint stereo to spatial audio coding – recent progress and standardization," in *6th International Conference on Digital Audio Effects,* Naples, Italy, Oct. 2004.

[Herr05] J. Herre, et al., "The reference model architecture for MPEG spatial audio coding," in *Proc. 118th AES convention,* paper 6447, Barcelona, Spain, May 2005.

[Herre98] J. Herre, et al., "The integrated filterbank based scalable MPEG-4 audio coder," in *Proc. 105th Conv. Aud. Eng. Soc.*, preprint #4810, Sep. 1998.

[Heus02] R. Heusdens and S. van de Par, "Rate-distortion optimal sinusoidal modeling of audio and speech using psychoacoustical matching pursuits," in *Proc. ICASSP-02*, vol. 2, pp. 1809–1812, 2002.

[Hilp98] J. Hilpert, M. Braun, M. Lutzky, S. Geyersberger, and R. Buchta, "Implementing ISO/MPEG-2 advanced audio coding realtime on a fixed-point DSP," in *Proc. 105th Conv. Aud. Eng. Soc.,* preprint #4822, Sept. 1998.

[Hilp00] J. Hilpert et al., "Real-time implementation of the MPEG-4 low delay advanced audio coding algorithm (AAC-LD) on Motorola DSP56300," in *Proc. 108th Conv. Aud. Eng. Soc.,* preprint #5081, Feb. 2000.

[Holm99] T. Holman, "New factors in sound for cinema and television," *J. Aud. Eng. Soc.*, vol. 39, pp. 529–539, July-Aug. 1999.

[Hong01] J-W. Hong, T-J. Lee, and J.S. Lucas, "Design and implementation of a real-time audio service using MPEG-2 AAC and streaming technology," in *Proc. 110th Conv. Aud. Eng. Soc.,* preprint #5343, May 2001.

[Hoog94] A. Hoogendoorn, "Digital compact cassette," *Proc. of the IEEE*, vol. 82, no. 10, pp. 1479–1489, Oct. 1994.

[Hori96] N. Horikawa and P.C. Eastty, "One-bit audio recording," in *AES UK Audio for New Media Conference*, Apr. 1996, London.

[Horv02] P. Horvatic and N. Schiffner, "Security issues in high quality audio streaming," in *Proc. of 2nd Int. Conf. on Web Delivering of Music (WEDELMUSIC-02)*, p. 224, Dec. 2002.

[Howa94] P. Howard and J. Vitter, "Arithmetic coding for data compression," *Proc. of the IEEE,* vol. 82, no. 6, pp. 857–865, June 1994.

[Hsie02] C.T. Hsieh and P.Y. Sou, "Blind cepstrum domain audio watermarking based on time energy features," in *Proc. Int. Conf. on DSP-2002,* vol. 2, pp. 705–708, July 2002.

[Huan02] D.Y. Huang, J.N. Al Lee, S.W. Foo, and L. Weisi, "Reed-Solomon codes for MPEG advanced audio coding," in *Proc. 113th Conv. Aud. Eng. Soc.*, preprint #5682, Oct. 2002.

[Huan02] J. Huang, Y. Wang, and Y.Q. Shi, "A blind audio watermarking algorithm with self-synchronization," in *Proc. IEEE ISCAS-2002,* vol. 3, pp. 627–630, Orlando, FL, May 2002.

[Hube98] D.M. Huber, *The MIDI Manual*, second edition, Focal Press, MA 1998.

[Huff52] D. Huffman, "A method for the construction of minimum redundancy codes," in *Proc. of the Institute of Radio Engineers,* vol. 40, pp. 1098–1101, Sept. 1952.

[Huij02] Y. Huijuan, J.C. Patra, and C. W. Chan, "An artificial neural network-based scheme for robust watermarking of audio signals," in *Proc. IEEE ICASSP-2002,* vol. 1, pp. 1029–1032, Orlando, FL, May 2002.

[Hwan01] Y-T. Hwang, N-J. Liu, and M-C. Tsai, "An MPEG-4 Twin-VQ based high quality audio codec design," in *IEEE Workshop on Sig. Proc. Systems,* pp. 321–331, Sept. 2001.

[IECA87] International Electrotechnical Commission/American National Standards Institute (IEC/ANSI) CEI-IEC 908, "Compact-Disc Digital Audio System" ("red book"), 1987.

[Iked95] K. Ikeda, et al., "Error protected TwinVQ audio coding at less than 64 kbit/s," in *Proc. IEEE Speech Coding Workshop,* pp. 33–34, 1995.

[Iked99] M. Ikeda, K. Takeda, and F. Itakura, "Audio data hiding by use of band-limited random sequences," in *Proc. IEEE ICASSP-99*, vol. 4, pp. 2315–2318, Phoenix, AZ, Mar. 1999.

[ISO-V96] Multifunctional Ad hoc Group, "Core Experiments Description," ISO/IEC JTC1/SC29/WG11 N1266, March 1996.

[ISOI90] ISO/IEC JTC1/SC2/WG11-MPEG, "MPEG/Audio Test Report, 1990," Doc. MPEG90/N0030, 1990.

[ISOI91] ISO/IEC JTC1/SC2/WG11, "MPEG/Audio Test Report, 1991," Doc. MPEG91/0010, 1991.

[ISOI92] ISO/IEC JTC1/SC29/WG11 MPEG, IS11172-3 "Information Technology – Coding of Moving Pictures and Associated Audio for Digital Storage Media at up to About 1.5 Mbit/s, Part 3: Audio" 1992. ("MPEG-1").

[ISOI94] ISO/IEC JTC1/SC29/WG11 MPEG, IS13818–3 "Information Technology – Generic Coding of Moving Pictures and Associated Audio, Part 3: Audio" 1994. ("MPEG-2 BC-LSF").

[ISOI94a] ISO/IEC JTC1/SC29/WG11 MPEG, IS13818-3 "Information Technology – Generic Coding of Moving Pictures and Associated Audio, Part 3: Audio" 1994. ("MPEG-2 BC-LSF").

[ISOI94b] ISO/IEC JTC1/SC29/WG11 MPEG94/443, "Requirements for Low Bitrate Audio Coding/MPEG-4 Audio", 1994. ("MPEG-4").

[ISOI96] ISO/IEC JTC1/SC29/WG11 MPEG, Committee Draft 13818-7 "Generic Coding of Moving Pictures and Associated Audio: Audio (non backwards compatible coding, NBC)" 1996. ("MPEG-2 NBC/AAC").

[ISOI96a] ISO/IEC JTC1/SC29/WG11 MPEG, Committee Draft 13818-7 "Generic Coding of Moving Pictures and Associated Audio: Audio (non backwards compatible coding, NBC)" 1996. ("MPEG-2 NBC/AAC").

[ISOI96b] ISO/IEC JTC1/SC29/WG11, "MPEG-4 Audio Test Results (MOS Tests)," ISO/IEC JTC1/SC29/WG11/N1144, Munich, Jan. 1996.

[ISOI96c] ISO/IEC JTC1/SC29/WG11, "Report of the Ad Hoc Group on the Evaluation of New Audio Submissions to MPEG-4," ISO/IEC JTC1/SC29/WG11/MPEG96/M0680, Munich, Jan. 1996.

[ISOI96d] ISO/IEC JTC1/SC29/WG11 N1420, "Overview of the Report on the Formal Subjective Listening Tests of MPEG-2 AAC Multichannel Audio Coding," Nov. 1996.

[ISOI97a] ISO/IEC JTC1/SC29/WG11, "MPEG-4 Audio Committee Draft 14496-3," ISO/IEC JTC1/SC29/WG11 N1903, Oct. 1997. (Available on http://www.tnt.uni-hannover.de/project/mpeg/audio/documents.)

[ISOI97b] ISO/IEC 13818-7, "Information Technology – Generic Coding of Moving Pictures and Associated Audio, Part 7: Advanced Audio Coding," 1997.

[ISOI98] ISO/IEC JTC1/SC29/WG11 (MPEG) document N2011, "Results of AAC and TwinVQ Tool Comparative Tests," San Jose, CA, 1998.

[ISOI99] ISO/IEC JTC1/SC29/WG11 (MPEG), International Standard ISO/IEC 14496-3: "Coding of Audio-Visual Objects – Part 3: Audio," 1999. ("MPEG-4").

[ISOI00] ISO/IEC JTC1/SC29/WG11 (MPEG), International Standard ISO/IEC 14496-3 AMD-1: "Coding of Audio-Visual Objects – Part 3: Audio," 2000. ("MPEG-4 version-2").

[ISOI01a] ISO/IEC JTC1/SC29/WG11 MPEG, IS15938-2 "Multimedia Content Description Interface: Description Definition Language (DDL)," International Standard, Sept. 2001. ("MPEG-7: Part-2").

[ISOI01b] ISO/IEC JTC1/SC29/WG11 MPEG, IS15938-4 "Multimedia Content Description Interface: Audio," International Standard, Sept. 2001. ("MPEG-7: Part-4").

[ISOI01c] ISO/IEC JTC1/SC29/WG11 MPEG, IS15938-5 "Multimedia Content Description Interface: Multimedia Description Schemes (MDS)," International Standard, Sept. 2001. ("MPEG-7: Part-5").

[ISOI01d] ISO/IEC JTC1/SC29/WG11 MPEG, "MPEG-7 Applications Document," Doc. N3934, Jan. 2001. Also available online: http://www.cselt.it/mpeg.

[ISOI01e] ISO/IEC JTC1/SC29/WG11 MPEG, "Introduction to MPEG-7," Doc. N4032, Mar. 2001. Also available online: http://www.cselt.it/mpeg.

[ISOI02a] MPEG Requirements Group, MPEG-21 Overview, ISO/MPEG N5231, Oct. 2002.

[ISOI03a] MPEG Requirements Group, MPEG-21 Requirements, ISO/MPEG N5873, July 2003.

[ISOI03b] MPEG Requirements Group, MPEG-21 Architecture, Scenarios and IPMP Requirements, ISO/MPEG N5874, July 2003.

[ISOI03c] ISO/IEC JTC1/SC29/WG11 MPEG, IS14496-3:2001/Amd.1:2003 "Information technology - coding of audio-visual objects, Part 3: Audio, Amendment 1: Bandwidth Extension," 2003. (MPEG 4 SBR).

[ISOII94] ISO-II, "Report on the MPEG/Audio Multichannel Formal Subjective Listening Tests," ISO/MPEG doc. MPEG94/063, ISO/MPEG-II Audio Committee, 1994.

[ISOJ97] ISO/IEC JTC1/SC29/WG1. (1997). JPEG-LS: Emerging lossless/near-lossless compression standard for continuous-tone images JPEG Lossless.

[IS-54] TIA/EIA-PN 2398 (IS-54), "The 8 kbit/s VSELP Algorithm," 1989.

[IS-96] TIA/EIA/IS-96, "QCELP," Speech Service Option 3 for Wideband Spread Spectrum Digital Systems," TIA 1992.

[IS-127] TIA/EIA/IS-127, "Enhanced Variable Rate Codec," Speech Service Option 3 for Wideband Spread Spectrum Digital Systems, TIA 1997.

[IS-641] TIA/EIA/IS-641, "Cellular/PCS Radio Interface – Enhanced Full-Rate Speech Codec," TIA 1996.

[IS-893] TIA/EIA/IS-893, "Selectable Mode Vocoder," *Service Option for Wideband Spread Spectrum Communications Systems*, ver. 2.0, Dec. 2001.

[ITUR90] International Telecommunications Union Radio Communications Sector (ITU-R), "Subjective Assessment of Sound Quality," CCIR Rec. 562-3, vol. X, Part 1, Dusseldorf, 1990.

[ITUR91] ITU-R Document TG10-2/3-E only, "Basic Audio Quality Requirements for Digital Audio Bit Rate Reduction Systems for Broadcast Emission and Primary Distribution," Oct. 1991.

[ITUR92] International Telecommunications Union Rec. G.728, "Coding of Speech at 16 kbps Using Low-Delay Code Excited Linear Prediction," Sept. 1992.

[ITUR93a] International Telecommunications Union, Radio Communications Sector (ITU-R), Recommendation BS.775-1, "Multi-Channel Stereophonic Sound System With and Without Accompanying Picture," Nov. 1993.

[ITUR93b] ITU-R Task Group 10/2, "CCIR Listening Tests, Network Verification Tests without Commentary Codecs, Final Report," Delayed Contribution 10-2/51, 1993.

[ITUR93c] ITU-R Task Group 10/2, "PAQM Measurements," Delayed Contribution 10-2/51, 1993.

[ITUR94b] ITU-R BS.1116, "Methods for Subjective Assessment of Small Impairments in Audio Systems Including Multichannel Sound Systems," 1994.

[ITUR94c] ITU-R BS.775-1, *Multichannel Stereophonic Sound System with and Without Accompanying Picture*, International Telecommunication Union, Geneva, Switzerland, 1994.

[ITUR96] International Telecommunications Union Rec. P 830, "Subjective Performance Assessment of Telephone-Band and Wideband Digital Codecs," 1996.

[ITUR97] International Telecommunications Union, Radio Communications Sector (ITU-R), Recommendation BS.1115, "Low Bit Rate Audio Coding," Geneva, Switzerland, 1997.

[ITUR98] International Telecommunications Union, Radiocommunications Sector (ITU-R), Recommendation BS.1387, "Method For Objective Measurements of Perceived Audio Quality," Dec. 1998.

[ITUT94] International Telecommunications Union, Telecommunications Sector (ITU-T), Recommendation P.80, "Telephone Transmission Quality Subjective Opinion Tests," 1994.

[ITUT96] International Telecommunications Union, Telecommunications Sector (ITU-T), Recommendation P.861, "Objective Quality Measurement of Telephone-Band (300–3400 Hz) Speech Codecs," 1996.

[Iwad92] M. Iwadare, et al., "A 128 kb/s hi-fi audio codec based on adaptive transform coding with adaptive block size MDCT," *IEEE J. Sel. Areas in Comm.*, vol. 10, no. 1, pp. 138–144, Jan. 1992.

[Iwak95] N. Iwakami, et al., "High-quality audio-coding at less than 64 kbit/s by using transform-domain weighted interleave vector quantization (TWINVQ)," in *Proc. IEEE ICASSP-95*, pp. 3095–3098, Detroit, MI, May 1995.

[Iwak96] N. Iwakami and T. Moriya, "Transform domain weighted interleave vector quantization (TwinVQ)," in *Proc. 101st Conv. Aud. Eng. Soc.*, preprint #4377, 1996.

[Iwak01] N. Iwakami et al., "Fast encoding algorithms for MPEG-4 TwinVQ audio tool," in *Proc. IEEE ICASSP-01*, vol. 5, pp. 3253–3256, Salt Lake City, UT, May 2001.

[Jako96] C. Jakob and A. Bradley, "Minimising the effects of subband quantisation of the time domain aliasing cancellation filter bank," in *Proc. IEEE ICASSP-96*, pp. 1033–1036, Atlanta, GA, May 1996.

[Jans03] E. Janssen and D. Reefman, "Super-audio CD: an introduction," in *IEEE Sig. Proc. Mag.*, vol. 20, no. 4, pp. 83–90, July 2003.

[Jawe95] B. Jawerth and W. Sweldens, "Biorthogonal smooth local trigonometric bases," *J. Fourier Anal. Appl.*, vol. 2, no. 2, pp. 109–133, 1995.

[Jaya76] N.S. Jayant, *Waveform Quantization and Coding*. IEEE Press, New York, 1976.

[Jaya84] N. Jayant and P. Noll, *Digital Coding of Waveforms Principles and Applications to Speech and Video*, Prentice-Hall, Englewood Cliffs, NJ, 1984.

[Jaya90] N.S. Jayant, V. Lawrence, and D. Prezas, "Coding of speech and wideband audio," *AT&T Tech. J.*, vol. 69(5), pp. 25–41, Sept.–Oct. 1990.

[Jaya92] N. Jayant, J. Johnston, and Y. Shoham, "Coding of wideband speech," *Speech Comm.,* vol. 11, no. 2–3, 1992.

[Jaya93] N. Jayant et al., "Signal compression based on models of human perception," *Proc. IEEE*, pp. 1385–1422, Oct. 1993.

[Jbir98] A. Jbira, N. Moreau, and P. Dymarski, "Low delay coding of wideband audio (20 Hz–15 kHz) at 64 kbps," in *Proc. IEEE ICASSP-98*, vol. 6, pp. 3645–3648, Seattle, WA, May 1998.

[Jess99] P. Jessop "The business case for audio watermarking," in *Proc. IEEE ICASSP*-99, vol. 4, pp. 2077–2078, Phoenix, AZ, Mar. 1999.

[Jest82] W. Jesteadt, S. Bacon, and J. Lehman, "Forward masking as a function of frequency, masker level, and signal delay," *J. Acoust. Soc. Am.*, vol. 71, pp. 950–962, 1982.

[Jetz79] J. Jetzt, "Critical distance measurement of rooms from the sound energy spectrum response," *J. Acoust. Soc. Am.,* vol. 65 pp. 1204–1211, 1979.

[Ji02] Z. Ji, Q. Zhang, W. Zhu, and J. Zhou, "Power-efficient distortion-minimized rate allocation for audio broadcasting over wireless networks," in *Proc. of ICME-2002,* vol. 1, pp. 257–260, Aug. 2002.

[Joha01] P. Johansson, M. Kazantzidls, R. Kapoor and M. Gerla, "Bluetooth: an enabler for personal area networking", *IEEE Network*, vol. 15, no. 5, pp. 28–37, Sept.–Oct. 2001,

[John80] J. Johnston, "A filter family designed for use in quadrature mirror filter banks," in *Proc. IEEE ICASSP-80*, pp. 291–294, Denver, CO, 1980.

[John88a] J. Johnston, "Transform coding of audio signals using perceptual noise criteria," *IEEE J. Sel. Areas in Comm.*, vol. 6, no. 2, pp. 314–323, Feb. 1988.

[John88b] J. Johnston, "Estimation of perceptual entropy using noise masking criteria," in *Proc. IEEE ICASSP-88*, pp. 2524–2527, New York, NY, May 1988.

[John89] J. Johnston, "Perceptual transform coding of wideband stereo signals," in *Proc. IEEE ICASSP-89*, pp. 1993–1996, Glasgow, Scotland, May 1989.

[John92a] J. Johnston and A. Ferreira, "Sum-difference stereo transform coding," in *Proc. IEEE ICASSP-92*, vol. 2, pp. 569–572, San Francisco, CA, May 1992.

[John95] J.D. Johnston et al., "The AT&T Perceptual Audio Coder (PAC)," Presented at the AES convention, New York, Oct. 1995.

[John96] J. Johnston, et al., "AT&T Perceptual Audio Coding (PAC)," in *Collected Papers on Digital Audio Bit-Rate Reduction*, N. Gilchrist and C. Grewin, Eds., *Aud. Eng. Soc.*, pp. 73–81, 1996.

[John96a] J. Johnston, "Audio coding with filter banks," in *Subband and Wavelet Transforms*, A. Akansu and M.J.T. Smith, Eds., Kluwer Academic Publishers, pp. 287–307, Norwell, MA, 1996.

[John96b] J. Johnston, J. Herre, M. Davis and U. Gbur, "MPEG-2 NBC audio-stereo and multichannel coding methods," in *Proc. 101st Conv. Aud. Eng. Soc.*, preprint #4383, Los Angeles, Nov. 1996.

[John96c] J. Johnston et al., "AT&T Perceptual Audio Coding (PAC)," in *Collected Papers on Digital Audio Bit-Rate Reduction*, N. Gilchrist and C. Grewin, Eds., Aud. Eng. Soc., pp. 73–81, 1996.

[John99] J. Johnston et al., "MPEG Audio Coding," in *Wavelet, Subband, and Block Transforms in Communications and Multimedia.*, A. Akansu and M. Medley, eds., Kluwer Academic Publishers, Boston, 1999.

[Juan82] B. Juang and A. Gray, "Multiple stage vector quantization for speech coding," in *Proc. IEEE ICASSP-82*, pp. 597–600, Paris, France, 1982.

[Jurg96] R.K. Jurgen, "Broadcasting with digital audio," *IEEE Spectrum*, pp. 52–59, Mar. 1996.

[Kaba86] P. Kabal and R.P. Ramachandran, "The computation of line-spectral frequencies using Chebyshev polynomials," *IEEE Trans. Acoust., Speech, Signal Processing,* vol. 34, pp. 1419–1426, 1986.

[Kalk01a] T. Kalker, "Considerations on watermarking security," *IEEE 4th Workshop on Mult. Sig. Proc.*, pp. 201–206, Oct. 2001.

[kalk01b] T. Kalker, W. Oomen, A.N. Lemma, J. Haitsma, F. Bruekers, and M. Veen, "Robust, multi-functional, and high quality audio watermarking technology," in *Proc. 110th Conv. Aud. Eng. Soc.*, preprint #5345, Amsterdam, May 2001.

[Kalk02] T. Kalker and F.M.J. willems, "Capacity bounds and constructions for reversible data-hiding," in *Proc. DSP-2002*, vol. 1, pp. 71–76, July 2002.

[Kapu92] R. Kapust, "A human ear related objective measurement technique yields audible error and error margin," in *Proc. 11th Int. Conf. Aud. Eng. Soc.*, pp. 191–201, May 1992.

[Karj85] M. Karjaleinen, "A new auditory model for the evaluation of sound quality of audio systems," in *Proc. IEEE ICASSP-85*, pp. 608–611, Tampa, FL, May 1985.

[Kata93] A. Kataoka, et al., "Conjugate structure CELP coder for the CCITT 8 kb/s standardization candidate," in *IEEE Workshop on Speech Coding for Telecommunications*, pp. 25–26, 1993.

[Kata96] A. Kataoka, et al., "An 8-kb/s conjugate structure CELP (CS-CELP) speech coder," *IEEE Trans. on Speech and Audio Processing*, vol. 4, no. 6, pp. 401–411, Nov. 1996.

[Kato02] H. Kato, "Trellis noise-shaping converters and 1-bit digital audio," in *Proc. 112th Conv. Aud. Eng. Soc.*, preprint #5615, Munich, Germany, May 10–13, 2002.

[Kau98] P. Kauff, J.-R. Ohm, S. Rauthenberg, and T. Sikora, "The MPEG-4 standard and its applications in virtual 3D environments," in *32nd ASILOMAR Conf. on Sig. Syst. and Computers,* vol. 1, pp. 104–107, Nov. 1998.

[Kay89] S. Kay, "A fast and accurate single frequency estimator," *IEEE Trans. Acous. Speech and Sig. Proc.*, pp. 1987–1990, Dec. 1989.

[Keyh93] M. Keyhl et al., "NMR measurements of consumer recording devices which use low bit-rate audio coding," in *Proc. 94th Conv. Aud. Eng. Soc.*, preprint #3616, 1993.

[Keyh94] M. Keyhl et al., "NMR measurements on multiple generations audio coding," in *Proc. 96th Conv. Aud. Eng. Soc.*, preprint #3803, 1994.

[Keyh95] M. Keyhl et al., "Maintaining sound quality – experiences and constraints of perceptual measurements in today's and future networks," in *Proc. 98th Conv. Aud. Eng. Soc.*, preprint #3946, 1995.

[Keyh98] M. Keyhl et al., "Quality assurance tests of MPEG encoders for a digital broadcasting system (Part II) – minimizing subjective test efforts by perceptual measurements," in *Proc. 104th Conv. Aud. Eng. Soc.*, preprint #4753, May 1998.

[Kim00] H. Kim, "Stochastic model based audio watermark and whitening filter for improved detection," in *Proc. IEEE ICASSP-00,* vol. 4, pp. 1971–1974, Istanbul, Turkey, June 2000.

[Kim01] S-W. Kim, S-H. Park, and Y-B. Kim, "Fine grain scalability in MPEG-4 Audio," in *Proc. 111th Conv. Aud. Eng. Soc.,* preprint #5491, Nov.–Dec. 2001.

[Kim02] K.T. Kim, S.K. Jung, Y.C. Park, and D.H. Youn, "A new bandwidth scalable wideband speech/audio coder," in *Proc. IEEE ICASSP-02*, vol. 1, pp. 13–17, Orlando, FL, May 2002.

[Kim02a] D-H. Kim, J-H. Kim, and S-W. Kim, "Scalable lossless audio coding based on MPEG-4 BSAC," in *Proc. 113th Conv. Aud. Eng. Soc.*, preprint #5679, Oct. 2002.

[Kim02b] J-W. Kim, B-D. Choi, S-H. Park, K-K. Kim, S-J Ko, "Remote control system using real-time MPEG-4 streaming technology for mobile robot," in *Proc. of Int. Conf. on Consumer Electronics,* ICCE-02, pp. 200–201, June 2002.

[Kimu02] T. Kimura, K. Kakehi, K. Takeda, and F. Itakura, "Spatial compression of multi-channel audio signals using inverse filters," in *Proc. ICASSP-02*, vol. 4, p. 4175, May 2002.

[Kirb97] D. Kirby and K. Watanabe, "Formal subjective testing of the MPEG-2 NBC multichannel coding algorithm," in *Proc. 102nd Conv. Aud. Eng. Soc.*, preprint 4418, Munich, 1997.

[Kirk83] S. Kirkpatrick et al., "Optimization by simulated annealing," *Science*, pp. 671–680, May 1983.

[Kiro01a] D. Kirovski and H. Malvar, "Robust spread-spectrum audio watermarking," in *Proc. IEEE ICASSP-01,* vol. 3, pp. 1345–1348, Salt Lake City, UT, May 2001.

[Kiro01b] D. Kirovski and H. Malvar, "Spread-spectrum audio watermarking: requirements, applications, and limitations," *IEEE 4th Workshop on Mult. Sig. Proc.*, pp. 219–224, Oct. 2001.

[Kiro02] D. Kirovski and H. Malvar, "Embedding and detecting spread-spectrum watermarks under estimation attacks," in *Proc. IEEE ICASSP-02,* vol. 2, pp. 1293–1296, Orlando, FL, May 2002.

[Kiro03a] D. Kirovski, and H.S. Malvar, "Spread-spectrum watermarking of audio signals," *Trans. on Sig. Proc.* vol. 51, no. 4, pp. 1020–1033, Apr. 2003.

[Kiro03b] D. Kirovski, and F.A.P. Petitcolas, "Blind pattern matching attack on watermarking systems," *Trans. on Sig. Proc.* vol. 51, no. 4, pp. 1045–1053, Apr. 2003.

[Kiya94] H. Kiya et al., "A development of symmetric extension method for subband image coding," *IEEE Trans. Image Proc.*, vol. 3, no. 1, pp. 78–81, Jan. 1994.

[Klei90a] W.B. Kleijn et al., "Fast methods for the CELP speech coding algorithm," *IEEE Trans. on Acoustics, Speech, and Signal Proc.*, vol. 38, no. 8, p. 1330, Aug. 1990.

[Klei90b] W.B. Kleijn, "Source-dependent channel coding and its application to CELP," *Advances in Speech Coding*, Eds. B. Atal, V. Cuperman, and A. Gersho, pp. 257–266, Kluwer Academic Publishers, Norwell, MA, 1990.

[Klei92] W.B. Kleijn et al., "Generalized analysis-by-synthesis coding and its application to pitch prediction," in *Proc. IEEE ICASSP-92*, vol. 1, pp. 337–340, San Francisco, CA, Mar. 1992.

[Klei95] W.B. Kleijn and K.K. Paliwal, "An introduction to speech coding," in *Speech Coding and Synthesis*, Eds., W.B. Kleijn and K.K. Paliwal, Kluwer Academic Publishers, Norwell, MA, Nov. 1995.

[Knis98] D.N. Knisely, S. Kumar, S. Laha, and S. Navda, "Evolution of wireless data services: IS-95 to CDMA2000," *IEEE Comm. Mag.*, vol. 36, pp. 140–149, Oct. 1998.

[Ko02] B-S. Ko, R. Nishimura, and Y. Suzuki, "Time-spread echo method for digital audio watermarking using PN sequences," in *Proc. IEEE ICASSP-02*, vol. 2, pp. 2001–2004, Orlando, FL, May 2002.

[Koen96] R. Koenen et al., "MPEG-4: context and objectives," *J. Sig. Proc.: Image Comm.*, Oct. 1996.

[Koen98] R. Koenen, "Overview of the MPEG-4 Standard," ISO/IEC JTC1/SC29/WG11 N2323, Jul. 1998. Available online at http://www.cselt.it/mpeg/standards/mpeg-4/mpeg-4.html.

[Koen99] R. Koenen, "Overview of the MPEG-4 Standard," ISO/IEC JTC1/SC29/WG11 N3156, Dec. 1999. Available online at http://www.cselt.it/mpeg/standards/mpeg-4/mpeg-4.htm.

[Koil91] R. Koilpillai and P.P. Vaidyanathan, "New results on cosine-modulated FIR filter banks satisfying perfect reconstruction," in *Proc. IEEE ICASSP-91*, pp. 1793–1796, Toronto, Canada, May 1991.

[Koil92] R. Koilpillai and P.P. Vaidyanathan, "Cosine-modulated FIR filter banks satisfying perfect reconstruction," *IEEE Trans. Sig. Proc.*, vol. 40, pp. 770–783, Apr. 1992.

[Kois00] K. Koishida, V. Cuperman, and A. Gersho, "A 16-kbit/s bandwidth scalable audio coder based on the G.729 standard," *Proc. IEEE ICASSP-00*, vol. 2, pp. 49–52, Istanbul, Turkey, June 2000.

[Koll99] J. Koller, T. Sporer, and K. Brandenburg, "Robust coding of high quality audio signals," in *Proc. 103rd Conv. Aud. Eng. Soc.*, preprint #4621, New York, 1997.

[Koml96] A. Komly, "Assessing the performance of low-bit-rate audio codecs: an account of the work of ITU-R Task Group 10/2," in *Collected Papers on Digital Audio Bit-Rate Reduction*, N. Gilchrist and C. Grewin, Eds., Aud. Eng. Soc., pp. 105–114, 1996.

[Kons94] K. Konstantinides, "Fast subband filtering in MPEG audio coding," *IEEE Sig. Proc. Letters*, vol. 1, pp. 26–28, Feb. 1994.

[Kost95] B. Kostek, "Feature extraction methods for the intelligent processing of musical signals," in *Proc. 99th Conv. Aud. Eng. Soc.,* preprint #4076, Oct. 1995.

[Kost96] B. Kostek and M. Szczerba, "MIDI database for the automatic recognition of musical phrases," in *Proc. 100th Conv. Aud. Eng. Soc.,* preprint #4169, May 1996.

[Kova95] J. Kovacevic, "Subband coding systems incorporating quantizer models," *IEEE Trans. Image Proc.*, vol. 4, no. 5, pp. 543–553, May 1995.

[Krah85] D. Krahe, *"New Source Coding Method for High Quality Digital Audio Signals,"* NTG Fachtagung Hoerrundfunk, Mannheim, 1985.

[Krah88] D. Krahe, "Grundlagen eines Verfahrens zur Dataenreduktion bei Qualitativ Hochwertigen, digitalen Audiosignalen auf Basis einer Adaptiven Transformationscodierung unter Berucksightigung Psychoakustischer Phanomene," Ph.D. Thesis, Duisburg, 1988.

[kroo86] P. Kroon, E. Deprettere, and R.J. Sluyeter, "Regular-pulse excitation – a novel approach to effective and efficient multi-pulse coding of speech," *IEEE Trans. on Acoustics, Speech, and Signal Proc.*, vol. 34, no. 5, pp. 1054–1063, Oct. 1986.

[Kroo90] P. Kroon and B. Atal, "Pitch predictors with high temporal resolution," in *Proc. IEEE ICASSP-90*, pp. 661–664, Albuquerque, NM, April 1990.

[Kroo95] P. Kroon and W.B. Kleijn, "Linear-prediction based analysis-synthesis coding," in *Speech Coding and Synthesis*, Eds., W.B. Kleijn and K.K. Paliwal, Kluwer Academic Publishers, Norwell, MA, Nov 1995.

[Krug88] E. Kruger and H. Strube, "Linear prediction on a warped frequency scale," *IEEE Trans. on Acoustics Speech and Signal Proc.*, vol. 36, no. 9, pp. 1529–1531, Sept. 1988.

[Kudu95a] P. Kudumakis and M. Sandler, "On the performance of wavelets for low bit rate coding of audio signals," in *Proc. IEEE ICASSP-95*, pp. 3087–3090, Detroit, MI, May 1995.

[Kudu95b] P. Kudumakis and M. Sandler, "Wavelets, regularity, complexity, and MPEG-Audio," in *Proc. 99th Conv. Aud. Eng. Soc.*, preprint #4048, Oct. 1995.

[Kudu96] P. Kudumakis and M. Sandler, "On the compression obtainable with four-tap wavelets," *IEEE Sig. Proc. Let.*, pp. 231–233, Aug. 1996.

[Kuma00] A. Kumar, "Low complexity ACELP coding of 7 kHz speech and audio at 16 kbps," *IEEE Int. Conf. on Personal Wireless Communications*, pp. 368–372, Dec. 2000.

[Kunt80] M. Kunt and O. Johnsen, "Block coding of graphics: A tutorial review," in *Proc. of the IEEE,* vol. 7, pp. 770–786, 1980.

[Kuo02] S.S. Kuo, J.D. Johnston, W. Turin, and S.R. Quackenbush, "Covert audio watermarking using perceptually tuned signal independent multiband phase modulation," in *Proc. IEEE ICASSP-02,* vol. 2, pp. 1753–1756, Orlando, FL, May 2002.

[Kurt02] F. Kurth, A. Ribbrock, and M. Clausen, "Identification of highly distorted audio material for querying large scale data bases," in *Proc. 112th Conv. Aud. Eng. Soc.*, preprint #5512, Munich, May 2002.

[Kuzu03] B.H. Kuzuki, N. Fuchigami, and J.R. Stuart, "DVD-audio specifications," in *IEEE Sig. Proc. Mag.*, pp. 72–82, July 2003.

[Lacy98] J. Lacy, S.R. Quackenbush, A.R. Reibman, D. Shur, and J.H. Snyder, "On combining watermarking with perceptual coding," in *Proc. IEEE ICASSP-98,* vol. 6, pp. 3725–3728, Seattle, WA, May 1998.

[Lafl90] C. Laflamme et al., "On reducing computational complexity of codebook search in CELP coder through the use of algebraic codes," in *Proc. IEEE ICASSP-90,* vol. 1, pp. 177–180, Albuquerque, NM, Apr. 1990.

[Lafl93] C. Laflamme, R. Salami, and J.-P Adoul, "9.6 kbit/s ACELP coding of wideband speech," in *Speech and Audio Coding for Wireless and Network Applications,* Eds., B. Atal, V. Cuperman, and A. Gersho, Kluwer Academic Publishers, Norwell, MA, 1993.

[Laft03] C. Laftsidis, A. Tefas, N. Nikolaidis, and I. Pitas, "Robust multibit audio watermarking in the temporal domain," in *Proc. IEEE ISCAS-03*, vol. 2, pp. 944–947, May 2003.

[Lanc02] R. Lancini, F. Mapelli, and S. Tubaro, "Embedding indexing information in audio signal using watermarking technique," *Int. Symp. on Video/Image Proc. and Mult. Comm.*, pp. 257–261, June 2002.

[Laub98] P. Lauber and N. Gilchrist, "ATLANTIC audio: switching layer 3 signals," in *Proc. 104th Aud. Eng. Soc. Conv.*, preprint 4738, Amsterdam, May 1998.

[Lean00] C.T.G. de Leandro, M. Bonnet, and N. Moreau, "Cyclostationarity-based audio watermarking with private and public hidden data," in *Proc. 109th Conv. Aud. Eng. Soc.,* preprint #5258, Los Angeles, Sept. 2000.

[Leand01] C.T.G. de Leandro, E. Gomez, and N. Moreau, "Resynchronization methods for audio watermarking an audio system," in *Proc. 111th Conv. Aud. Eng. Soc.*, preprint 5441, New York, Dec. 2001.

[Lee90] J.I. Lee et al., "On reducing computational complexity of codebook search in CELP coding," *IEEE Trans. on Comm.,* vol. 38, no. 11, pp. 1935–1937, Nov. 1990.

[Lee00a] S.K. Lee and Y.S. Ho, "Digital audio watermarking in the cepstrum domain," *Int. Conf. on Consumer Electronics (ICCE-00),* pp. 334–335, June 2000.

[Lee00b] S.K. Lee and Y.S. Ho, "Digital audio watermarking in the cepstrum domain," *Trans. on Consumer Electronics,* vol. 46, no. 3, pp. 744–750, Aug. 2000.

[Leek84] M. Leek and C. Watson, "Learning to detect auditory pattern components," *J. Acoust. Soc. Am.*, vol. 76, pp. 1037–1044, 1984.

[Lefe94] R. Lefebvre, R. Salami, C. Laflamme, and J.-P. Adoul, "High quality coding of wideband audio signals using transform coded excitation (TCX)," in *Proc. IEEE ICASSP-94*, vol. 1, pp. 93–96, Adelaide, Australia, 1994.

[LeGal92] D.J. Le Gall, "The MPEG video compression algorithm," *J. Sig. Proc.: Image Comm.,* vol. 4, no. 2, pp. 129–140, 1992.

[Lehr93] P. Lehrman and T. Tully, *MIDI for the Professional*, Music Sales Corp, 1993.

[Lemm03] A.N. Lemma, J. Aprea, W. Oomen, and L.V. de Kerkhof, "A temporal domain audio watermarking technique," *Trans. on Sig. Proc.*, vol. 51, no. 4, pp. 1088–1097, Apr. 2003.

[Levi98a] S. Levine and J. Smith, "A sines+transients+noise audio representation for data compression and time/pitch scale modifications," in *Proc. 105th Conv. Audio Eng. Soc.*, preprint #4781, Sep. 1998.

[Levi99] S.N. Levine and J.O. Smith, "A switched parametric and transform audio coder," in *Proc. IEEE ICASSP-99*, vol. 2, pp. 985–988, Phoenix, AZ, Mar. 1999.

[Lie01] W-N. Lie and L-C. Chang, "Robust and high-quality time-domain audio watermarking subject to psychoacoustic masking," in *Proc. IEEE ISCAS-01,* vol. 2, pp. 45–48, May 2001.

[Lin82] S. Lin and D.J. Costello, *Error Control Coding: Fundamentals and Applications*, Englewood Cliffs, NJ, Prentice Hall, 1982.

[Lin91] X. Lin, R.A. Salami, and R. Steele, "High quality audio coding using analysis-by-synthesis technique," in *Proc. IEEE ICASSP-91*, vol. 5, pp. 3617–3620, Toronto, Canada, Apr. 1991.

[Lina94] J. Linacre, *Many-Facet Rasch Measurement.* MESA Press, Chicago, 1994.

[Lind80] Y. Linde, A. Buzo, and A. Gray, "An algorithm for vector quantizer design," *IEEE Trans. Commun.,* vol. 28, no. 1, pp. 84–95, Jan. 1980.

[Lind99] A. T. Lindsay and W. Kriechbaum, "There's more than one way to hear it: multiple representations of music in MPEG-7," *J. New Music Res.,* vol. 28, no. 4, pp. 364–372, 1999.

[Lind00] A. T. Lindsay, S. Srinivasan et al., "Representation and linking mechanism for audio in MPEG-7," *Signal Proc.: Image Commun.,* vol. 16, pp. 193–209, 2000.

[Lind01] A.T. Lindsay and J. Herre, "MPEG-7 and MPEG-7 audio – an overview," *J. Audio Eng. Soc.,* vol. 49, no. 7/8, pp. 589–594, July/Aug. 2001.

[Link93] M. Link, "An attack processing of audio signals for optimizing the temporal characteristics of a low bit-rate audio coding system," in *Proc. 95th Conv. Audio Eng. Soc.*, preprint #3696, 1993.

[Liu98] C-M. Liu and W-C. Lee, "A unified fast algorithm for cosine modulated filter banks in current audio coding standards," in *Proc. 104th Conv. Audio Eng. Soc.*, preprint #4729, 1998.

[Liu99] F. Liu, J. Shin, C-C. J. Kuo, and A.G. Tescher, "Streaming of MPEG-4 speech/audio over Internet," *Int. Conf. on Consumer Electronics,* pp. 300–301, June 1999.

[LKur96] E. Lukac-Kuruc, "Beyond MIDI: XMidi, an affordable solution," in *Proc. 101st Conv. Aud. Eng. Soc.,* preprint #4348, Nov. 1996.

[Lloy82] S. P. Lloyd, "Least squares quantization in PCM," originally published in 1957 and republished in *IEEE Trans. Inform. Theory*, vol. 28, pt. II, pp.127–135, Mar. 1982.

[Lokh91] G.C.P. Lokhoff, "DCC – digital compact cassette," in *IEEE Trans. on Consumer Electronics,* vol. 37, no. 3, pp. 702–706, Aug. 1991.

[Lokh92] G.C.P. Lokhoff, "Precision adaptive sub-band coding (PASC) for the digital compact cassette (DCC)," *IEEE Trans. Consumer Electron.*, vol. 38, no. 4, pp. 784–789, Nov. 1992.

[Lu98] Z. Lu and W. Pearlman, "An efficient, low-complexity audio coder delivering multiple levels of quality for interactive applications," in *Proc. IEEE Sig. Proc. Soc. Workshop on Multimedia Sig. Proc.*, Dec. 1998.

[Luft83] R. Lufti, "Additivity of simultaneous masking," *J. Acoust. Soc. Am.*, vol. 73, pp. 262–267, 1983.

[Luft85] R. Lufti, "A power-law transformation predicting masking by sounds with complex spectra," *J. Acoust. Soc. of Am.*, vol. 77, pp. 2128–2136, 1985.

[Maco97] M.W. Macon, L. Jensen-Link, J. Oliverio, M.A. Clements, and E.B. George, "Concatenation-based MIDI-to-singing voice synthesis," in *Proc. 101st Conv. Aud. Eng. Soc.,* preprint #4591, Nov. 1997.

[Madi97] V. Madisetti and D.B. Williams, *The Digital Signal Processing Handbook*, CRC Press, Boca Raton, FL, Dec. 1997.

[Mahi89] Y. Mahieux, et al., "Transform coding of audio signals using correlation between successive transform blocks," in *Proc. IEEE ICASSP-89*, pp. 2021–2024, Glasgow, Scotland, May 1989.

[Mahi90] Y. Mahieux and J. Petit, "Transform coding of audio signals at 64 kbits/sec," in *Proc. IEEE Globecom '90*, vol. 1, pp. 518–522, Nov. 1990.

[Mahi94] Y. Mahieux and J.P. Petit, "High-quality audio transform coding at 64 kbps," *IEEE Trans. Comm.*, vol. 42, no. 11, pp. 3010–3019, Nov. 1994.

[Main96] L. Mainard and M. Lever, "A bi-dimensional coding scheme applied to audio bitrate reduction," in *Proc. IEEE ICASSP-96*, pp. 1017–1020, Atlanta, GA, May 1996.

[Main96] L. Mainard, C. Creignou, M. Lever, and Y.F. Dehery, "A real-time PC-based high-quality MPEG layer II codec," in *Proc. 101st Conv. Aud. Eng. Soc.,* preprint #4345, Nov. 1996.

[Makh75] J. Makhoul, "Linear prediction: A tutorial review," in *Proc. of the IEEE*, vol. 63, no. 4, pp. 561–580, Apr. 1975.

[Makh78] J. Makhoul et al., "A mixed-source model for speech compression and Synthesis," *J. Acoust. Soc. Am.*, vol. 64, pp. 1577–1581, Dec. 1978.

[Makh85] J. Makhoul et al., "Vector quantization in speech coding," *Proc. of the IEEE*, vol. 73, no. 11, pp. 1551–1588, Nov. 1985.

[Maki00] K. Makino and J. Matsumoto, "Hybrid audio coding for speech and audio below medium bit rate," *Int. Conf. on Consumer Electronics (ICCE-00)*, pp. 264–265, June 2000.

[Malv90a] H. Malvar, "Lapped transforms for efficient transform/subband coding," *IEEE Trans. Acous., Speech, and Sig. Process.*, vol. 38, no. 6, pp. 969–978, June 1990.

[Malv90b] H. Malvar, "Modulated QMF filter banks with perfect reconstruction," *Electronics Letters*, vol. 26, pp. 906–907, June 1990.

[Malv91] H.S. Malvar, *Signal Processing with Lapped Transforms*, Artech House, NorWood, MA, 1991.

[Malv98] H. Malvar, "Biorthogonal and nonuniform lapped transforms for transform coding with reduced blocking and ringing artifacts," *IEEE Trans. Sig. Proc.*, vol. 46, no. 4, pp. 1043–1053, Apr. 1998.

[Malv98] H. Malvar, "Enhancing the performance of subband audio coders for speech signals," in *Proc. IEEE ISCAS-98,* vol. 5, pp. 98–101, May 1998.

[Mana02] B. Manaris, T. Purewal, and C. McCormick, "Progress towards recognizing and classifying beautiful music with computers – MIDI-encoded music and the Zipf-Mandelbrot law," in *Proc. of Southeast Conf.* pp. 52–57, Apr. 2002.

[Manj02] B.S. Manjunath, P. Salembier, and T. Sikora, editors, *Introduction to MPEG-7: Multimedia Content Description Language*, John Wiley & Sons, New York, 2002.

[Mans01a] M.F. Mansour and A.H. Tewfik, "Audio watermarking by time-scale modification," in *Proc. IEEE ICASSP-01,* vol. 3, pp. 1353–1356, Salt Lake City, UT, May 2001.

[Mans01b] M.F. Mansour and A.H. Tewfik, "Efficient decoding of watermarking schemes in the presence of false alarms," *IEEE 4th Workshop on Mult. Sig. Proc.*, pp. 523–528, Oct. 2001.

[Mans02] M.F. Mansour and A.H. Tewfik, "Improving the security of watermark public detectors," *Int. Conf. on DSP-02,* vol. 1, pp. 59–66, July 2002.

[Mao03] W. Mao, *Modern Cryptography: Theory and Practice*, Prentice Hall PTR, July 2003.

[Mark76] J. Markel and A. Gray, Jr., *Linear Prediction of Speech*, Springer-Verlag, New York, 1976.

[Marp87] S. Marple, *Digital Spectral Analysis with Applications*, Prentice-Hall, Englewood Cliffs, NJ, 1987.

[Marp89] L. Marple, *Digital Spectral Analysis with Applications,* 2nd edition, Prentice Hall, Englewood Cliffs, NJ, 1989.

[Mart02] L.G. Martins, A.J.S. Ferreira, "PCM to MIDI transposition," in *Proc. 112th Conv. Aud. Eng. Soc.,* preprint #5524, May 2002.

[Mass85] J. Masson, and Z. Picel, "Flexible design of computationally efficient nearly perfect QMF filter banks," in *Proc. IEEE ICASSP-85*, vol. 10, pp. 541–544, Tampa, FL, Mar. 1985.

[Matv96] G. Matviyenko, "Optimized Local Trigonometric Bases," *Appl. Comput. Harmonic Anal.*, v. 3, n. 4, pp. 301-323–, 1996.

[McAu86] R. McAulay and T. Quatieri, "Speech analysis synthesis based on a sinusoidal representation," *IEEE Trans. Acous. Speech, and Sig. Proc.*, vol. 34, no. 4, pp. 744–754, Aug. 1986.

[McAu88] R. McAulay and T. Quatieri, "Computationally efficient sine-wave synthesis and its application to sinusoidal transform coding," in *Proc. IEEE ICASSP-88*, New York, NY, 1988.

[McCr91] A. McCree and T. Barnwell III, "A new mixed excitation LPC vocoder," in *Proc. IEEE ICASSP-91*, vol. 1, pp. 593–596, Toronto, Canada, Apr. 1991.

[McCr93] A. McCree and T. Barnwell III, "Implementation and evaluation of a 2400 BPS mixed excitation LPC vocoder," in *Proc. IEEE ICASSP*-93, vol. 2, p. 159, Minneapolis, MN, Apr. 1993.

[Mear97] D.J. Meares and G. Theile, "Matrixed surround sound in an MPEG digital world," in *Proc. 102th Conv. Aud. Eng. Soc.*, preprint #4431, Mar. 1997.

[Mein01] N. Meine, B. Edler, and H. Purnhagen, "Error protection and concealment for HILN MPEG-4 parametric audio coding," in *Proc. 110th Conv. Aud. Eng. Soc.,* preprint #5300, May 2001.

[Merh00] N. Merhav, "On random coding error exponents of watermarking systems," *Trans. on Info. Theory,* vol. 46, no. 2, pp. 420–430, Mar. 2000.

[Merm88] P. Mermelstein, "G.722, A new CCITT coding standard for digital transmission of wideband audio signals," *IEEE Comm. Mag.*, vol. 26, pp. 8–15, Feb. 1988.

[Mesa99] V.Z. Mesarovic, M.V. Dokic, and R.K. Rao, "DTS multichannel audio decoder on a 24-bit fixed-point dual DSP," in *Proc. 106th Conv. Aud. Eng. Soc.*, preprint #4964, May 1999.

[MID03] "MIDI and musical instrument control," in the *Journal of AES*, vol. 51, no. 4, pp. 272–276, Apr. 2003.

[MIDI] The MIDI Manufacturers Association Official home page, available online at http://www.midi.org.

[Mill47] G. Miller, "Sensitivity to changes in the intensity of white noise and its relation to masking and loudness," *J. Acoust. Soc. Am.*, vol. 19, pp. 609–619, 1947.

[Miln92] A. Milne, "New test methods for digital audio data compression algorithms," in *Proc. 11th Int. Conf. Aud. Eng. Soc.*, pp. 210–215, May 1992.

[Mitc97] J.L. Mitchell, W.B. Pennebaker, C.E. Fogg and D.J. LeGall, *MPEG Video Compression Standard*, in *Digital Multimedia Standards Series*, Chapman & Hall, New York, NY, 1997.

[Mizr02] S. Mizrachi and D. Malah, "Parameter identification of a class of nonlinear systems for robust audio watermarking verification," in *Proc. 22nd Conv. Electrical and Electronics* Engineers, pp. 13–15, Dec. 2002.

[Mode98] T. Modegi and S-I, Iisaku, "Proposals of MIDI coding and its application for audio authoring," in *Proc. Int. Conf. on Mult. Comp. and Syst.* pp. 305–314, June–July 1998.

[Mode00] T. Modegi, "MIDI encoding method based on variable frame-length analysis and its evaluation of coding precision," in *IEEE Int. Conf. on Mult. and Expo.* ICME-00, vol. 2, pp. 1043–1046, July–Aug. 2000.

[Mont94] P. Monta and S. Cheung, "Low rate audio coder with hierarchical filterbanks," in *Proc. IEEE ICASSP-94*, vol. 2, pp. 209–212, Adelaide, Australia, May 1994.

[Moor77] B.C.J. Moore, *Introduction to the Psychology of Hearing*, University Park Press, London 1977.

[Moor78] B.C.J. Moore, "Psychophysical tuning curves measured in simultaneous and forward masking," *J. Acoust. Soc. Am.*, vol. 63, pp. 524–532, 1978.

[Moor83] B.C.J. Moore and B. Glasberg, "Suggested formulae for calculating auditory-filter bandwidths and excitation patterns," *J. Acous. Soc. Am.*, vol. 74, pp. 750–753, 1983.

[Moor96] B.C.J. Moore, "Masking in the human auditory system," in *Collected Papers on Digital Audio Bit-Rate Reduction*, N. Gilchrist and C. Grewin, Eds., Aud. Eng. Soc., pp. 9–19, 1996.

[Moor96] J.A. Moorer, "Breaking the sound barrier: mastering at 96 kHz and beyond," in *Proc. 101st Conv. Aud. Eng. Soc.*, preprint #4357, Los Angeles, Nov. 1996.

[Moor97] B.C.J. Moore, B. Glasberg, and T. Baer, "A model for the prediction of thresholds, loudness, and partial loudness," *J. Audio Eng. Soc.*, vol. 45, no. 4, pp. 224-240–, Apr. 1997.

[Moor03] B.C.J. Moore, *An Introduction to the Psychology of Hearing*, Fifth Edition, Academic Press, San Diego Jan. 2003.

[More95] F. Moreau de Saint-Martin et al., "A measure of near orthogonality of PR biorthogonal filter banks," in *Proc. IEEE ICASSP-95*, pp. 1480–1483, Detroit, MI, May 1995.

[Mori96] T. Moriya, et al., "Extension and complexity reduction of TWINVQ audio coder," in *Proc. IEEE ICASSP-96*, vol. 2, pp. 1029–1032, Atlanta, GA, May 1996.

[Mori97] T. Moriya, Y. Takashima, T. Nakamura, and N. Iwakami, "Digital watermarking schemes based on vector quantization," in *IEEE Workshop on Speech Coding for Telecommunications,* pp. 95–96, 1997.

[Mori00] T. Moriya, N. Iwakami, A. Jin, and T. Mori, "A design of lossy and lossless scalable audio coding," in *Proc. IEEE ICASSP 2000*, vol. 2, pp. 889–892, Istanbul, Turkey, June 2000.

[Mori00a] T. Moriya, A. Jin, N. Iwakami, and T. Mori, "Design of an MPEG-4 general audio coder for improving speech quality," in *Proc. of IEEE Workshop on Speech* Coding, pp. 139–141, Sept. 2000.

[Mori00b] T. Moriya, T. Mori, N. Iwakami, and A. Jin, "A design of error robust scalable coder based on MPEG-4/Audio," in *Proc. IEEE ISCAS-2000,* vol. 3, pp. 213–216, May 2000.

[Mori02a] T. Moriya, "Report of AHG on issues in lossless audio coding," ISO/IEC JTC1/SC29/WG11 M7955, Mar. 2002.

[Mori02b] T. Moriya, A. Jin, T. Mori, K. Ikeda, and T. Kaneko, "Lossless scalable audio coder and quality enhancement," in *Proc. IEEE ICASSP-02,* vol. 2, pp. 1829–1832, Orlando, FL, 2002.

[Mori03] T. Moriya, A. Jin, T. Mori, K. Ikeda, and T. Kaneko, "Hierarchical lossless audio coding in terms of sampling rate and amplitude resolution," in *Proc. IEEE ICASSP-03,* vol. 5, pp. 409–412, Hong Kong, 2003.

[Moul98a] D. Moulton and M. Moulton, "Measurement of small impairments of perceptual audio coders using a 3-facet Rasch model," in *Proc. 104th Conv. Aud. Eng. Soc.*, preprint #4709, May 1998.

[Moul98b] D. Moulton and M. Moulton, "Codec 'transparency,' listener 'severity,' program 'intolerance:' suggestive relationships between Rasch measures and some background variables," in *Proc. 105th Conv. Aud. Eng. Soc.*, preprint #4843, Sept. 1998.

[MPEG] The MPEG workgroup web-page, available online at http://www.mpeg.org.

[Munt02] T. Muntean, E. Grivel, and M. Najim "Audio digital watermarking based on hybrid spread spectrum," in *Proc. 2nd Int. Conf. WEBDELMUSIC-02*, pp. 150–155, Dec. 2002.

[Nack99a] F. Nack and A.T. Lindsay, "Everything you wanted to know about MPEG-7, Part-I," *IEEE Multimedia*, vol. 6, no. 3, pp. 65–71, July–Sept. 1999.

[Nack99b] F. Nack and A.T. Lindsay, "Everything you wanted to know about MPEG-7, Part-II," *IEEE Multimedia*, vol. 6, no. 4, pp. 64–73, Oct.–Dec. 1999.

[Naja00] H. Najafzadeh and P. Kabal, "Perceptual bit allocation for low rate coding of narrowband audio," in *Proc. IEEE ICASSP-00*, vol. 2, pp. 5–9, Istanbul, Turkey, June 2000.

[Neub00a] C. Neubauer and J. Herre, "Audio watermarking of MPEG-2 AAC bit streams," in *Proc. 108th Conv. Aud. Eng. Soc.*, preprint #5101, Paris, Feb. 2000.

[Neub00b] C. Neubauer and J. Herre, "Advanced watermarking and its applications," in *Proc. 109th Conv. Aud. Eng. Soc.*, preprint #5176, Los Angeles, Sept. 2000.

[Neub02] C. Neubauer, F. Siebenhaar, J. Herre, and R. Bauml, "New high data rate audio watermarking based on SCS (scalar costa scheme)," in *Proc. 113th Conv. Aud. Eng. Soc.*, preprint #5645, Los Angeles, Oct. 2002.

[Neub98] C. Neubauer and J. Herre, "Digital watermarking and its influence on audio quality," in *Proc. 105th Conv. Aud. Eng. Soc.*, preprint #4823, San Francisco, Sept. 1998.

[NICAM] NICAM 728, *"Specification for Two Additional Digital Sound Channels with System I Television,"* Independent Broadcasting Authority (IBA), British Radio Equipment Manufacturers' Association (BREMA), and British Broadcasting Corporation (BBC), Aug. 1988.

[Ning02] D. Ning and M. Deriche, "A new audio coder using a warped linear prediction model and the wavelet transform," in *Proc. IEEE ICASSP-02*, vol. 2, pp. 1825–1828, Orlando, FL, 2002.

[Nish96a] A. Nishio et al, "Direct stream digital audio system," in *Proc. 100th Conv. Aud. Eng. Soc.*, preprint #4163, May 1996.

[Nish96b] A. Nishio et al, "A new CD mastering processing using direct stream digital," in *Proc. 101st Conv. Aud. Eng. Soc.*, preprint #4393, Los Angeles, Nov. 1996.

[Nish99] M. Nishiguchi, "MPEG-4 speech coding," in *Proc. AES 17th Int. Conf.*, Sept. 1999.

[Nish01] A. Nishio and H. Takahashi, "Investigation of practical 1-bit delta-sigma conversion for professional audio applications," in *Proc. 110th Conv. Aud. Eng. Soc.*, preprint #5392, May 2001.

[Nogu97] M. Noguchi, G. Ichimura, A. Nishio, and S. Tagami, "Digital signal processing in a direct stream digital editing system," in *Proc. 102nd Conv. Aud. Eng. Soc.*, Preprint #4476, May 1997.

[Noll93] P. Noll, "Wideband speech and audio coding," *IEEE Comm. Mag.*, pp. 34–44, Nov. 1993.

[Noll95] P. Noll, "Digital audio coding for visual communications," *Proc. of the IEEE*, pp. 925–943, June 1995.

[Noll97] P. Noll, "MPEG digital audio coding," *IEEE Sig. Proc. Mag.*, pp. 59–81, Sept. 1997

[Nomu98] T. Nomura et al., "A bitrate and bandwidth scalable CELP coder," in *Proc. IEEE ICASSP-1998*, vol. 1, pp. 341–344, Seattle, WA, May 1998.

[Nors92] S.R. Norsworthy, "Effective dithering of sigma-delta modulators," in *Proc. IEEE ISCAS-92*, vol. 3, pp. 1304–1307, 1992.

[Nors93] S.R. Norsworthy and D.A. Rich, "Idle channel tones and dithering in delta-sigma modulators," in *Proc. 95th Conv. Aud. Eng. Soc.*, preprint #3711, New York, Oct. 1993.

[Nors97] S.R. Norsworthy, R. Schreier, and G.C. Temes, *Delta-Sigma Converters, Theory, Design and Simulation*, New York, IEEE Press, 1997.

[Nuss81] Nussbaumer, H. J., "Pseudo QMF filter bank," *IBM Tech. Disclosure Bulletin*, vol. 24, pp. 3081–3087, Nov. 1981.

[Ogur97] Y. Ogura, T. Sugihara, S. Okada, H. Yamauchi, and A. Nishio, "One-bit two-channel recorder system," in *Proc. 103rd Conv. Aud. Eng. Soc.,* preprint #4564, New York, Sept. 1997.

[Oh01] H.O. Oh, J.W. Seok, J.W. Hong, and D.H. Youn, "New echo embedding technique for robust and imperceptible audio watermarking," in *Proc. IEEE ICASSP-01,* vol. 3, pp. 1341–1344, Salt Lake City, UT, May 2001.

[Oh02] H.O. Oh, D.H. Youn, J.W. Hong, J.W. Seok, "Imperceptible echo for robust audio watermarking," in *Proc. 113th Conv. Aud. Eng. Soc.*, preprint #5644, Los Angeles, Oct. 2002.

[Ohma97] I. Ohman, "The placement of one or several subwoofers," in *Music och Ljudteknik*, no. 1, Sweden, 1997.

[Ojan99] J. Ojanpera and M. Vaananen, "Long-term predictor for transform domain perceptual audio coding," in *Proc. 107th Conv. Aud. Eng. Soc.*, preprint #5036, New York, 1999.

[Oliv48] B.M. Oliver, J. Pierce, and C.E. Shannon, "The philosophy of PCM," *Proc. IRE,* vol. 36, pp. 1324–1331, Nov. 1948.

[Onno93] P. Onno and C. Guillemot, "Tradeoffs in the design of wavelet filters for image compression," in *Proc. VCIP*, pp. 1536–1547, Nov. 1993.

[Oom03] W. Oomen, E. Schuijers, B. den Brinker, and J. Breebaart, "Advances in parametric coding for high-quality audio," in *Proc. 114th Conv. Aud. Eng. Soc.,* preprint #5852, Mar. 2003.

[Oome96] A.W.J.L. Oomen, A.A.M. Bruekers, R.J. van der Vleuten, and L.M. Van de Kerkhof, "Lossless coding for DVD audio," in *101st Conv. Aud. Eng. Soc.*, preprint #4358, Nov. 1996.

[Oppe71] A. Oppenheim et al., "Computation of spectra with unequal resolution using the fast Fourier transform," *Proc. of the IEEE*, vol. 59, pp. 299–301, Feb. 1971.

[Oppe72] A. Oppenheim and D. Johnson, "Discrete representation of signals," *Proc. of the IEEE*, vol. 60, pp. 681–691, June 1972.

[Oppe99] A.V. Oppenheim, R.W. Schafer, J.R. Buck, *Discrete-Time Signal Processing,* 2nd edition, Prentice Hall, Englewood Cliffs, NJ, 1999.

[Orde91] E. Ordentlich and Y. Shoham, "Low-delay code excited linear predictive coding of wideband speech at 32 kb/s," in *Proc. IEEE ICASSP-93*, vol. 2, pp. 9–12, Minneapolis, MN, 1993.

[Pail92] B. Paillard, et al., "PERCEVAL: perceptual evaluation of the quality of audio signals," *J. Aud. Eng. Soc.*, vol. 40, no. 1/2, pp. 21–31, Jan./Feb. 1992.

[Pain00] T. Painter and A. Spanias, "Perceptual coding of digital audio," *Proc. of the IEEE*, vol. 88, no. 4, pp. 451–513, Apr. 2000.

[Pain00a] T. Painter and A. Spanias, "Perceptual Coding of Digital Audio," *Proc. of the IEEE*, vol. 88, no. 4, pp. 451–513, April 2000.

[Pain01] T. Painter and A. Spanias, "Perceptual component selection in sinusoidal coding of audio," *IEEE 4th Workshop on Multimedia Signal Processing*, pp. 187–192, Oct. 2001.

[Pain05] T. Painter and A. Spanias, "Perceptual segmentation and component selection for sinusoidal representations of audio," *IEEE Trans. Speech Audio Proc.*, vol. 13, no. 2, pp. 149–162, Mar. 2005.

[Pali91] K.K. Paliwal and B. Atal, "Efficient vector quantization of LPC parameters at 2.4 kb/s," in *Proc. IEEE ICASSP-91*, pp. 661–663, Toronto, ON, Canada, 1991.

[Pali93] K.K. Paliwal and B.S. Atal, "Efficient vector quantization of LPC parameters at 24 bits/frame," *IEEE Trans. Speech Audio Process.*, vol. 1, no. 1, pp. 3–14, 1993.

[Pan93] D. Pan, "Digital audio compression," *Digital Tech. J.*, vol. 5, no. 2, pp. 28–40, 1993.

[Pan95] D. Pan, "A tutorial on MPEG/audio compression," *IEEE Mult.*, vol. 2, no. 2, pp. 60–74, Summer 1995.

[Papa95] P. Papamichalis, "MPEG audio compression: algorithms and implementation," in *Proc. DSP 95 Int. Conf. on DSP*, pp. 72–77, June 1995.

[Papo91] A. Papoulis, *Probability, Random Variables, and Stochastic Processes*, 3rd edition, McGraw-Hill, New York, 1991.

[Para95] M. Paraskevas and J. Mourjopoulos, "A differential perceptual audio coding method with reduced bitrate requirements," *IEEE Trans. Speech and Audio Proc.*, vol. 3, no. 6, pp. 490–503, Nov. 1995.

[Park97] S.H. Park et al., "Multi-layered bit-sliced bit-rate scalable audio coding," in *Proc. 103rd Conv. Aud. Eng. Soc.*, preprint #4520, 1997.

[Paul82] D. Paul, "A 500–800 b/s adaptive vector quantization vocoder using a perceptually motivated distance measure," in *Proc. Globecom* 1982 (Miami, FL, Dec. 1982), p. E6.3.1.

[PEAQ] The PEAQ webpage available at http://www.peaq.org/.

[Peel03] C.B. Peel, "On 'dirty-paper coding'," in *IEEE Sig. Proc. Mag.* vol. 20, no. 3, pp. 112–113, May 2003.

[Pena95] A. Pena, "A suggested auditory information environment to ease the detection and minimization of subjective annoyance in perceptive-based systems," in *Proc. 98th Conv. Aud. Eng. Soc.*, preprint #4019, Paris, 1995.

[Pena96] A. Pena et al., "ARCO (Adaptive Resolution COdec): a hybrid approach to perceptual audio coding," in *Proc. 100th Conv. Aud. Eng. Soc.*, preprint #4178, May 1996.

[Pena97a] A. Pena et al., "New improvements in ARCO (Adaptive Resolution Codec)," in *Proc. 102nd Conv. Aud. Eng. Soc.*, preprint #4419, Mar. 1997.

[Pena97b] A. Pena et al., "A flexible tiling of the time axis for adaptive wavelet packet decompositions," in *Proc. IEEE ICASSP-97*, pp. 2137–2140, Munich, Germany, Apr. 1997.

[Pena01] A. Pena, E. Alexandre, B. Rivas, R. Perez, and A. Duenas, "Realtime implementations of MPEG-2 and MPEG-4 natural audio coders," in *Proc. 110th Conv. Aud. Eng. Soc.*, preprint #5302, May 2001.

[Penn80] M. Penner, "The coding of intensity and the interaction of forward and backward masking," *J. Acoust. Soc. of Am.*, vol. 67, pp. 608–616, 1980.

[Penn95] B. Pennycook, "Audio and MIDI Markup Tools for the World Wide Web," in *Proc. 99th Conv. Aud. Eng. Soc.*, preprint #4058, Oct. 1995.

[Peti99] F.A.P. Petitcolas, R.J. Anderson, and M.G. Kuhn, "Information hiding – a survey," in *Proc. of the IEEE*, vol. 87, no. 7, pp. 1062–1078, July 1999.

[Peti02] F.A.P. Petitcolas, and D. Kirovski, "The blind pattern matching attack on watermark systems," in *Proc. IEEE ICASSP-02*, vol. 4, pp. 3740–3743, Orlando, FL, May 2002.

[Petr01] R. Petrovic, "Audio signal watermarking based on replica modulation," it Int. Conf. on Telecomm. in Modern Satellite, Cable and Broadcasting, vol. 1, pp. 227–234, Sept. 2001.

[Phil95a] P. Philippe et al., "A relevant criterion for the design of wavelet filters in high-quality audio coding," in *Proc. 98th Conv. Aud. Eng. Soc.*, preprint #3948, Feb. 1995.

[Phil95b] P. Philippe, et al., "On the Choice of Wavelet Filters for Audio Compression," in *Proc. IEEE ICASSP-95*, pp. 1045–1048, Detroit, MI, May 1995.

[Phil96] P. Philippe et al., "Optimal wavelet packets for low-delay audio coding," in *Proc. IEEE ICASSP-96*, pp. 550–553, Atlanta, GA, May 1996.

[PhSo82] Philips and Sony, *Audio CD System Description*, Philips Licensing, Eindhoven, The Netherlands, 1982.

[Pick82] R. Pickholtz, D. Schilling, and L. Milstein, "Theory of spread-spectrum communications – a tutorial," *IEEE Trans. on Comm.*, vol. 30, no. 5, pp. 855–884, May 1982.

[Pick88] J. Pickles, *An Introduction to the Physiology of Hearing*. Academic Press, New York, 1988.

[Podi01] C.I. Podilchuk and E.J. Delp, "Digital watermarking: algorithms and applications," *IEEE Sig. Proc. Mag.*, vol. 18, no. 4, pp. 33–46, July 2001.

[Port80] M.R. Portnoff, "Time-frequency representation of digital signals and systems based on short-time fourier analysis," *IEEE Trans. ASSP*, vol. 28, no. 1, pp. 55–69, Feb. 1980.

[Port81a] M.R. Portnoff, "Short-time fourier analysis of sampled speech," *IEEE Trans. ASSP*, vol. 29, no. 3, pp. 364–373, June 1981.

[Prec97] K. Precoda and T. Meng, "Listener differences in audio compression Evaluations," *J. Aud. Eng. Soc.*, vol. 45, no. 9, pp. 708–715, Sept. 1997.

[Prel96a] N. Prelcic and A. Pena, "An adaptive tree search algorithm with application to multiresolution based perceptive audio coding," in *Proc. IEEE Int. Symp. on Time-Freq. and Time-Scale Anal.*, 1996.

[Prel96b] N. Prelcic et al., "Considerations on the performance of filter design methods for wavelet packet audio decomposition," in *Proc. 100th Conv. Aud. Eng. Soc.*, preprint #4235, May 1996.

[Pres89] W. Press, et al., *Numerical Recipes*, Cambridge Univ. Press, 1989.

[Prin86] J. Princen and A. Bradley, "Analysis/synthesis filter bank design based on time domain aliasing cancellation," *IEEE Trans. ASSP*, vol. 34, no. 5, pp. 1153–1161, Oct. 1986.

[Prin87] J. Princen et al., "Subband/transform coding using filter bank designs based on time domain aliasing cancellation," in *Proc. IEEE ICASSP-87*, vol. 12, pp. 2161–2164, Dallas, TX, May 1987.

[Prin94] J. Princen, "The design of non-uniform modulated filterbanks," in *Proc. IEEE Int. Symp. on Time-Frequency and Time-Scale Analysis*, Oct. 1994.

[Prin95] J. Princen and J. Johnston, "Audio coding with signal adaptive filter banks," in *Proc. IEEE ICASSP-95*, pp. 3071–3074, Detroit, MI, May 1995.

[Prit90] W.L. Pritchard and M. Ogata, "Satellite direct broadcast," *Proc. of the IEEE*, vol. 78, no. 7, pp. 1116–1140, July 1990.

[Pura95] M. Purat and P. Noll, "A new orthonormal wavelet packet decomposition for audio coding using frequency-varying modulated lapped transforms," in *IEEE ASSP Workshop on Applic. of Sig. Proc. to Aud. and Acous.*, Oct. 1995.

[Pura96] M. Purat and P. Noll, "Audio coding with a dynamic wavelet packet decomposition based on frequency-varying modulated lapped transforms," in *Proc. IEEE ICASSP-96*, pp. 1021–1024, Atlanta, GA, May 1996.

[Pura97] M. Purat, T. Liebchen, and P. Noll, "Lossless transform coding of audio signals," in *Proc. 102nd Conv. Aud. Eng. Soc.*, Munich, preprint #4414, Mar. 1997.

[Purn00a] H. Purnhagen, N. Meine, "HILN-the MPEG-4 parametric audio coding tools," in *Proc. IEEE ISCAS-2000*, vol. 3, pp. 201–204, May 2000.

[Purn00b] H. Purnhagen, N. Meine, and B. Edler, "Speeding up HILN-MPEG-4 parametric audio encoding with reduced complexity," in *Proc. 109th Conv. Aud. Eng. Soc.*, preprint #5177, Sept. 2000.

[Purn02] H. Purnhagen, N. Meine, and B. Edler, "Sinusoidal coding using loudness-based component selection," in *Proc. IEEE ICASSP-02*, vol. 2, pp. 1817–1820, Orlando, FL, 2002.

[Purn97] H. Purnhagen et al., "Proposal of a Core Experiment for extended 'Harmonic and Individual Lines plus Noise' Tools for the Parametric Audio Coder Core," ISO/IEC JTC1/SC29/WG11 MPEG97/2480, July 1997.

[Purn98] H. Purnhagen et al., "Object-based analysis/synthesis audio coder for very low bit rates," in *Proc. 104th Conv. Aud. Eng. Soc.*, preprint #4747, May 1998.

[Purn99a] H. Purnhagen, "Advances in parametric audio coding," in *Proc. IEEE WASPAA*, New Paltz, NY, Oct. 17–20, 1999.

[Purn99b] H. Purnhagen, "An overview of MPEG-4 audio version-2," in *Proc. AES 17th Int. Conf.*, Florence, Italy, Sept. 1999.

[Purn03] H. Purnhagen, et al., "Combining low complexity parametric stereo with high efficiency AAC," ISO/IEC JTC1/SC29/WG11 MPEG2003/M10385, Dec. 2003.

[Qiu01] T. Qiu, "Lossless audio coding based on high order context modeling," in *IEEE Fourth Workshop on Multimedia Signal Processing*, pp. 575–580, Oct. 2001.

[Quac01] S. Quackenbush and A.T. Lindsay, "Overview of MPEG-7 Audio," *IEEE Trans. on Circuits and Systems for Video Technology*, vol. 2, no. 6, pp. 725–729, June 2001.

[Quac97] S. Quackenbush, "Noiseless coding of quantized spectral components in MPEG-2 advanced audio Coding," *IEEE ASSP Workshop on Apps. of Sig. Proc. to Aud. And Acous.*, Mohonk, NY, 1997.

[Quac98a] S. Quackenbush and Y. Toguri, ISO/IEC JTC1/SC29/WG11 N2005, "Revised Report on Complexity of MPEG-2 AAC Tools," Feb. 1998.

[Quac98b] S. Quackenbush, "Coding of Natural Audio in MPEG-4," in *Proc. IEEE* ICASSP-98, Seattle, WA May 1998.

[Quat02] T. Quatieri, *Discrete-Time Speech Signal Processing,* Prentice Hall PTR, Upper Saddle River, NJ, 2002.

[Quei93] R. de Queiroz, "Time-varying lapped transforms and wavelet packets," *IEEE Trans. Sig. Proc.*, vol. 41, pp. 3293–3305, 1993.

[Raad01] M. Raad and I.S. Burnett, "Audio coding using sorted sinusoidal parameters," in *Proc. IEEE ISCAS-02,* vol. 2, pp. 401–404, May 2001.

[Raad02] M. Raad and A. Mertins, "From lossy to lossless audio coding using SPIHT," in *Proc. of the 5th Int. Conf. on Digital Audio Effects,* Hamburg, Germany, Sept. 26–28, 2002.

[Rabi78] L.R. Rabiner and R.W. Schafer, *Digital Processing of Speech Signals*, Prentice-Hall, Englewood Cliffs, NJ, 1978.

[Rabi89] L.R. Rabiner, "A tutorial on hidden markov models and selected applications in speech recognition," *Proc. of the IEEE,* vol. 77, no. 2, pp. 257–286, Feb. 1989.

[Rabi93] L.R. Rabiner and B.H. Juang, *Fundamentals of Speech Recognition*. Prentice Hall, Englewood Cliffs, NJ, 1993.

[Rahm87] M.D. Rahman and K. Yu, "Total least squares approach for frequency estimation using linear Prediction," *IEEE Trans. ASSP*, vol. 35, no. 10, pp. 1440–1454, Oct. 1987.

[Ramp98] S.A. Ramprashad, "A two stage hybrid embedded speech/audio coding structure," in *Proc. IEEE ICASSP-98*, vol. 1, pp. 337–340, Seattle, WA, May 1998.

[Ramp99] S.A. Ramprashad, "A multimode transform predictive coder (MTPC) for speech and audio," *IEEE Workshop on Speech Coding,* pp. 10–12, June 1999.

[Rams82] T. Ramstad, "Sub-band coder with a simple adaptive bit-allocation algorithm a possible candidate for digital mobile telephony," in *Proc. IEEE ICASSP-82*, vol. 7, pp. 203–207, Paris, France, May 1982.

[Rams86] T. Ramstad, "Considerations on quantization and dynamic bit-allocation in subband coders," in *Proc. IEEE ICASSP-86*, vol. 11, pp. 841–844, Tokyo, Japan, Apr. 1986.

[Rams91] T. Ramstad, "Cosine modulated analysis-synthesis filter bank with critical sampling and perfect reconstruction, in *Proc. IEEE ICASSP-91*, pp. 1789–1792, Toronto, Canada, May 1991.

[Rao90] K. Rao and P. Yip, *The Discrete Cosine Transform: Algorithm, Advantages, and Applications*, Academic Press, Boston 1990.

[Rasc80] G. Rasch, *Probabilistic Models for Some Intelligence Attainment Tests*, University of Chicago Press, Chicago, 1980.

[Reef01a] D. Reefman and E. Janssen, "White paper on signal processing for SACD," Philips IP&S, [Online] Available at http://www.superaudiocd.philips.com.

[Reef01b] D. Reefman and P. Nuijten, "Why direct stream digital (DSD) is the best choice as a digital audio format," in *Proc.* 110th *Conv. Aud. Eng. Soc.*, preprint #5396, Amsterdam, May 2001.

[Reef01c] D. Reefman and P. Nuijten, "Editing and switching in 1-bit audio streams," in *Proc. 110th Conv. Aud. Eng. Soc.*, preprint #5399, Amsterdam, May 12–15, 2001.

[Reef02] D. Reefman and E. Janssen, "Enhanced sigma delta structures for super audio CD application," in *Proc. 112th Conv. Aud. Eng. Soc.,* preprint #5616, Munich, Germany, May 2002.

[Rice79] R. Rice, "Some practical universal noiseless coding techniques," *Jet Propulsion Laboratory,* JPL Publication, Pasadena, CA, Mar. 1979.

[Riou91] O. Rioul and M. Vetterli, "Wavelets and signal processing," *IEEE SP Mag.*, pp. 14–38, Oct. 1991.

[Riou94] O. Rioul and P. Duhamel, "A Remez exchange algorithm for orthonormal wavelets," *IEEE Trans. Circ. Sys. II*, Aug. 1994.

[Riss79] J. Rissanen and G. Langdon, "Arithmetic coding," *IBM J. Res. Develop.* vol. 23, no. 2, pp. 146–162, Mar. 1979.

[Rits96] S. Ritscher and U. Felderhoff, "Cascading of Different Audio Codecs," in *Proc. 100th Audio Eng. Soc. Conv.*, preprint 4174, Copenhagen, May 1996.

[Rix01] Rix et al., "Perceptual evaluation of speech quality (PESQ) – a new method for speech quality assessment of telephone networks and codecs," in *Proc. IEEE ICASSP-01*, vol. 2, pp. 749–752, Salt Lake City, UT, 2001.

[Robi94] T. Robinson, "SHORTEN: Simple lossless and near-lossless waveform compression," *Technical Report 156*, Engineering Department, Cambridge University, Dec. 1994.

[Roch98] S. Roche and J.-L Dugelay, "Image watermarking based on the fractal transform," in *Proc. Workshop Multimedia Signal Processing*, pp. 358–363, Los Angeles, CA, 1998.

[Rode92] X. Rodet and P. Depalle, "Spectral envelopes and inverse FFT synthesis," in *Proc. 93rd Conv. Audio Eng. Soc.*, preprint #3393, Oct. 1992.

[Rong99] Y. Rongshan and K.C. Chung, "High quality audio coding using a novel hybrid WLP-subband coding algorithm," *5th Asia-Pacific Conf. on Comm.,* vol. 2, pp. 952–955, Oct. 1999.

[Rose01] B. Rosenblatt, B. Trippe, and S. Mooney, *Digital Rights Management: Business and Technology*, New York, John Wiley & Sons, Nov. 2001.

[Roth83] J.H. Rothweiler, "Polyphase quadrature filters – a new subband coding technique," in *Proc. ICASSP-83*, pp. 1280–1283, May 1983.

[Ruiz00] F.J. Ruiz and J.R. Deller, "Digital watermarking of speech signals for the National Gallery of the Spoken Word," in *Proc. IEEE ICASSP-00,* vol. 3, pp. 1499–1502, Istanbul, Turkey, June 2000.

[Ryde96] T. Ryden, "Using listening tests to assess audio codecs," in *Collected Papers on Digital Audio Bit-Rate Reduction*, N. Gilchrist and C. Grewin, Eds., Aud. Eng. Soc., pp. 115–125, 1996.

[Sabi82] M. Sabin and R. Gray, "Product code vector quantizers for speech waveform coding," in *Proc. Globecom* 1982 (Miami, FL, Dec. 1982), p. E6.5.1.

[SACD02] Philips and Sony, "Super Audio CD System Description," Philips Licensing, Eindhoven, The Netherlands, 2002.

[SACD03] Philips Super Audio CD, 2003. Available at http://www.superaudiocd.philips.com/.

[Sait02a] S. Saito, T. Furukawa, and K. Konishi, "A data hiding for audio using band division based on QMF bank," in *Proc. IEEE ISCAS-02,* vol. 3, pp. 635–638, May 2002.

[Sait02b] S. Saito, T. Furukawa, and K. Konishi, "A digital watermarking for audio data using band division based on QMF bank," in *Proc. IEEE ICASSP-02,* vol. 4, pp. 3473–3476, Orlando, FL, May 2002.

[Saka00] T. Sakamoto, M. Taruki, and T. Hase, "MPEG-2 AAC decoder for a 32-bit MCU," *Int. Conf. on Consumer Electronics,* ICCE-00, pp. 256–257, June 2000.

[Sala98] R. Salami, et al., "Design and description of CS-ACELP: A toll quality 8 kb/s speech coder," *IEEE Trans. on Speech and Audio Processing,* vol. 6, no. 2, pp. 116–130, Mar. 1998.

[Sand01] F.E. Sandnes, "Efficient large-scale multichannel audio coding," in *Proc. of Euromicro Conference*, pp. 392–399, Sept. 2001.

[Saun96] J. Saunders, "Real time discrimination of broadcast speech/music," in *Proc. IEEE ICASSP-96*, pp. 993–996, Atlanta GA, May 1996.

[SBC90] Swedish Broadcasting Corporation, "ISO MPEG/Audio Test Report," Stokholm, July 1990.

[Scha70] B. Scharf, "Critical bands," in *Foundations of Modern Auditory Theory*, Academic Press, New York, 1970.

[Scha75] R. Schafer and L. Rabiner, "Digital representations of speech signals," *Proc. IEEE*, vol. 63, no. 4, pp. 662–677, Apr. 1975.

[Scha79] R. Schafer and J. Markel, *Speech Analysis*, IEEE Press, New York, 1979.

[Scha95] R. Schäfer and T. Sikora, "Digital video coding standards and their role in video communications," in *Proc. of the IEEE*, vol. 83, no. 6, June 1995.

[Sche98a] E. Scheirer, "The MPEG-4 Structured Audio Standard," in *Proc. IEEE* ICASSP-98, vol. 6, pp. 3801–3804, Seattle, WA May 1998.

[Sche98b] E. Scheirer and M. Slaney, "Construction and Evaluation of a Robust Multifeature Speech/Music Discriminator," in *Proc. IEEE ICASSP-98*, Seattle, WA, May 1998.

[Sche98c] E. Scheirer and L. Ray, "Algorithmic and wavetable synthesis in the MPEG-4 multimedia standard," in *Proc. 105th Conv. Aud. Eng. Soc.*, preprint #4811, Sept. 1998.

[Sche98d] E. Scheirer, "The MPEG-4 structured audio orchestra language," in *Proc. ICMC*, Oct. 1998.

[Sche98e] E. Scheirer et al., "AudioBIFS: the MPEG-4 standard for effects processing," in *Proc. DAFX98 Workshop on Digital Audio Effects Processing*, Nov. 1998.

[Sche99a] E. Scheirer, "Structured audio and effects processing in the MPEG-4 multimedia standard," *ACM Multimedia Sys.*, vol.7, no. 1, pp. 11–22, 1999.

[Sche99b] E. Scheirer and B.L. Vercoe, "SAOL: The MPEG-4 structured audio orchestra language," *Comput. Music J.*, vol. 23, no. 2, pp. 31–51, 1999.

[Sche01] E. Scheirer, "Structured audio, Kolmogorov complexity, and generalized audio coding," in *IEEE Trans. on Speech and Audio Proc.*, vol. 9, no. 8, pp. 914–931, Nov. 2001.

[Schi81] S. Schiffman et al., *Introduction to Multidimensional Scaling: Theory, Method, and Applications*, Academic Press, New York, 1981.

[Schr79] M.R. Schroeder, B.S. Atal, and J.L. Hall, "Optimising digital speech coders by exploiting masking properties of the human ear," *J. Acoust. Soc. Am.*, vol. 66, 1979.

[Schr79] M. Schroeder et al., "Optimizing digital speech coders by exploiting masking properties of the human ear," *J. Acoust. Soc. Am.*, pp. 1647–1652, Dec. 1979.

[Schr85] M.R. Schroeder and B. Atal, "Code-excited linear prediction (CELP): high quality speech at very low bit rates," in *Proc. IEEE ICASSP-85*, vol. 10, pp. 937–940, Tampa, FL, Apr. 1985.

[Schr85] M. Schroeder and B. Atal, "Code-excited linear prediction (CELP): high quality speech at very low bit rates," in *Proc. IEEE ICASSP-85*, pp. 937–940, Tampa, FL, May 1985.

[Schr86] E. Schroder, and W. Voessing, "High quality digital audio encoding with 3.0 bits/sample using adaptive transform coding," in *Proc. 80th Conv. Aud. Eng. Soc.*, preprint #2321, Mar. 1986.

[Schu96] D. Schulz, "Improving audio codecs by noise substitution," *J. Audio Eng. Soc.*, pp. 593–598, July/Aug. 1996.

[Schu98] G. Schuller, "Time-varying filter banks with low delay for audio coding," in *Proc. 105th Conv. Audio Eng. Soc.*, preprint #4809, Sept. 1998.

[Schu00] G. Schuller and T. Karp, "Modulated filter banks with arbitrary system delay: efficient implementations and the time-varying case," *IEEE Trans. Speech Audio Proc.*, vol. 48, no. 3, pp. 737–748, Mar. 2000.

[Schu01] G. Schuller, Y. Bin, and D. Huang, " Lossless coding of audio signals using cascaded prediction," in *Proc. IEEE ICASSP-01*, vol. 5, pp. 3273–3276, Salt Lake City, UT, 2001.

[Schu02] G. Schuller and A. Harma, "Low delay audio compression using predictive coding," in *Proc. IEEE ICASSP-02*, vol. 2, pp. 1853–1856, Orlando, FL, 2002.

[Schu04] E. Schuijers, et al., "Low complexity parametric stereo coding," in *116th Conv. Aud. Eng. Soc.*, preprint # 6073, Berlin, Germany, May 2004.

[Schu05] G. Schuller, J. Kovacevic, F. Masson, and V. Goyal, "Robust low-delay audio coding using multiple descriptions," *IEEE Trans. Speech Audio Proc.*, vol. 13, no. 5, pp. 1014–1024, Sep. 2005.

[Sega76] A. Segall, "Bit allocation and encoding for vector sources," *IEEE Trans. Info. Theory*, vol. 22, no. 2, pp. 162–169, Mar. 1976.

[Seok00] J-W. Seok, and J-W. Hong, "Prediction-based audio watermark detection algorithm," in *Proc. 109th Conv. Aud. Eng. Soc.*, preprint 5254, Los Angeles, Sept. 2000.

[Sera97] C. Serantes et al., "A fast noise-scaling algorithm for uniform quantization in audio coding schemes," in *Proc. IEEE ICASSP-97*, pp. 339–342, Munich, Germany, Apr. 1997.

[Serr89] X. Serra, "A System for Sound Analysis/Transformation/Synthesis Based on a Deterministic Plus Stochastic Decomposition," Ph.D. Thesis, Stanford University, 1989.

[Serr90] X. Serra and J.O. Smith III, "Spectral modeling and synthesis: a sound analysis/synthesis system based on a deterministic plus stochastic decomposition," *Comput. Mus. J.*, pp 12–24, Winter 1990.

[Sevi94] D. Sevic and M. Popovic, "A new efficient implementation of the oddly-stacked Princen-Bradley filter bank," *IEEE Sig. Proc. Lett.*, vol. 1, no. 11, Nov. 1994.

[Shan48] C.E. Shannon, "A mathematical theory of communication," *Bell Sys. Tech. J.*, vol. 27, pp. 379–423, pp. 623–656, July/Oct., 1948.

[Shao02] S. Shaopeng, Y. Junxun, Y. Yongcong, and A-M. Raed, "A low bit-rate audio coder based on modified sinusoidal model," in *IEEE Int. Conf. on Comm. Circuits and Systems and West Sino Expositions*, vol. 1, pp. 648–652, June 2002.

[Shap93] J. Shapiro, "Embedded image coding using zerotrees of wavelet coefficients," *IEEE Trans. Sig. Proc.*, vol. 41, no. 12, pp. 3445–3462, Dec. 1993.

[Shin02] S. Shin, O. Kim, J. Kim, and J. Choil, "A robust audio watermarking algorithm using pitch scaling," *Int. Conf. on DSP-02,* vol. 2, pp. 701–704, July 2002.

[Shli94] S. Shlien, "Guide to MPEG-1 audio standard," *IEEE Trans. Broadcast.*, pp. 206–218, Dec. 1994.

[Shli96] S. Shlien and G. Soulodre, "Measuring the characteristics of 'expert' listeners," in *Proc. 101st Conv. Aud. Eng. Soc.*, preprint #4339, Nov. 1996.

[Shli97] S. Shlien, "The modulated lapped transform, its time-varying forms, and its applications to audio coding standards," *IEEE Trans. on Speech and Audio Proc.*, vol. 5, no. 4, pp. 359–366, July 1997.

[Shoh88] Y. Shoham and A. Gersho, "Efficient bit allocation for an arbitrary set of quantizers," *IEEE Trans. ASSP,* vol. 36, pp. 1445–1453, 1988.

[Siko97a] T. Sikora, "The MPEG-4 Video Standard Verification Model," in *IEEE Trans. CSVT*, vol.7, no. 1, Feb.1997.

[Siko97b] T. Sikora, "MPEG digital video-coding standards," in *IEEE Sig. Proc. Mag.*, vol. 14, no. 5, pp. 82–100, Sept. 1997.

[Silv74] H.F. Silverman and N.R. Dixon, "A parametrically controlled spectral analysis system for speech," *IEEE Trans. ASSP*, vol. 22, no. 5, pp. 362–381, oct. 1974.

[Silv01] G.C.M. Silvestre, N.J. Hurley, G.S. Hanau, and W.J. Dowling, "Informed audio watermarking scheme using digital chaotic signals," in *Proc. IEEE ICASSP-01,* vol. 3, pp. 1361–1364, Salt Lake City, UT, May 2001.

[Sing84] S. Singhal and B. Atal, "Improving performance of multi-pulse LPC coders at low bit rates," in *Proc. IEEE ICASSP-84*, vol. 9, pp. 9–12, San Diego, CA, Mar. 1984.

[Sing90] S. Singhal, "High quality audio coding using multipulse LPC," in *Proc. IEEE ICASSP-90*, vol. 2, pp. 1101–1104, Albuquerque, NM, Apr. 1990.

[Sinh93a] D. Sinha and A. Tewfik, "Low bit rate transparent audio compression using a dynamic dictionary and optimized wavelets," in *Proc. IEEE ICASSP-93*, pp. I-197–I-200–, Minneapolis, MN, May 1993.

[Sinh93b] D. Sinha and A. Tewfik, "Low bit rate transparent audio compression using adapted wavelets," *IEEE Trans. Sig. Proc.*, vol. 41, no. 12, pp. 3463–3479, Dec. 1993.

[Sinh96] D. Sinha and J. Johnston, "Audio compression at low bit rates using a signal adaptive switched filter bank," in *Proc. IEEE ICASSP-96*, pp. 1053–1056, Atlanta, GA, May 1996.

[Sinh98a] D. Sinha et al., "The perceptual audio coder (PAC)," in *The Digital Signal Processing Handbook*, V. Madisetti and D. Williams, Eds., CRC Press, Boca Raton, FL, pp. 42.1–42.18, 1998.

[Sinh98b] D. Sinha and C.E.W. Sundberg, "Unequal error protection (UEP) for perceptual audio coders," in *Proc. 104th Conv. Aud. Eng. Soc.*, preprint 4754, May 1998.

[Smar95] G. Smart and A. Bradley, "Filter bank design based on time-domain aliasing cancellation with non-identical windows," in *Proc. IEEE ICASSP-94*, vol. 3, no. 185–188, Detroit, MI, May 1995.

[Smit86] M.J.T. Smith and T. Barnwell, "Exact reconstruction techniques for tree-structured subband coders," *IEEE Trans. Acous., Speech, and Sig. Process.*, vol. 34, no. 3, pp. 434–441, June 1986.

[Smit95] J.O. Smith and J. Abel, "The Bark bilinear transform," in *Proc. IEEE Workshop on App. Sig. Proc. to Audio and Electroacoustics*, Oct. 1995.

[Smit99] J.O. Smith and J.S. Abel, "Bark and ERB bilinear transforms," *IEEE Trans. on Speech and Audio Proc.,* vol. 7, no. 6, pp. 697–708 Nov. 1999.

[SMPTE99] SMPTE 320M-1999, *Channel Assignments and Levels on Multichannel Audio Media*, Society of Motion Picture and Television Engineers, White Plains, NY, 1999.

[SMPTE02] SMPTE RP 173–2002, Loudspeaker Placements for Audio Monitoring in High-Definition Electronic Production, Society of Motion Picture and Television Engineers, White Plains, NY, 2002.

[Smyt96] S. Smyth, W.P. Smith, M. Smyth, M. Yan, and T. Jung, "DTS coherent acoustics delivering high quality multichannel sound to the consumer," in *Proc. 100th Conv. Aud. Eng. Soc.*, May 11–14, preprint #4293, 1996.

[Smyt99] M. Smyth, "White paper: An Overview of the Coherent Acoustics Coding System," June 1999. Available on the Internet at www.dtsonline.com/media/uploads/pdfs/whitepaper.pdf.

[Soda94] I. Sodagar et al., "Time-varying filter banks and wavelets," *IEEE Trans. Sig. Proc.*, vol. 42, pp. 2983–2996, Nov. 1994.

[Sore96] H.V. Sorensen, P.M. Azziz, and J. van der Spiegel, "An overview of sigma-delta converters," *IEEE Sig. Proc. Mag.*, vol. 13, pp 61–84, Jan. 1996.

[Soul98] G. Soulodre et al., "Subjective evaluation of state-of-the-art two-channel audio codecs," *J. Aud. Eng. Soc.*, vol. 46, no. 3, pp. 164–177, Mar. 1998.

[Span91] A. Spanias, "A hybrid transform method for speech analysis and synthesis," *Signal Processing*, Vol. 24, pp. 217–229, 1991.

[Span92] A. Spanias, M. Deisher, P. Loizou, and G. Lim, "Fixed-point implementation of the VSELP algorithm," ASU-TRC Technical Report, TRC-SP-ASP-9201, July 1992.

[Span94] A. Spanias, "Speech coding: A tutorial review," in *Proc. of the IEEE*, vol. 82, no. 10, pp. 1541–1582, Oct. 1994.

[Span97] A. Spanias and T. Painter, "Universal Speech and Audio Coding Using a Sinusoidal Signal Model," ASU-TRC Technical Report 97–001, Jan. 1997.

[Spec98] "Special issue on copyright and privacy protection," *IEEE J. Select. Areas Commun.*, vol. 16, May 1998.

[Spec99] "Special issue on identification and protection of multimedia information," in *Proc. of the IEEE* vol. 87, July 1999.

[Spen01] G. Spenger, J. Herre, C. Neubauer, and N. Rump, "MPEG-21 – What does it bring to Audio?," in *Proc. 111th Conv. Aud. Eng. Soc.*, preprint #5462, Nov–Dec. 2001.

[Sper00] R. Sperschneider, "Error resilient source coding with variable length codes and its application to MPEG advanced audio coding," in *Proc. 109th Conv. Aud. Eng. Soc.,* preprint #5271, Sept. 2000.

[Sper01] R. Sperschneider and P. Lauber, "Error concealment for compressed digital audio," in *Proc. 111th Conv. Aud. Eng. Soc.,* preprint #5460, Nov–Dec. 2001.

[Sper02] R. Sperschneider, D. Homm, and L-H. Chambat, "Error resilient source coding with differential variable length codes and its application to MPEG advanced audio coding," in *Proc. 112th Conv. Aud. Eng. Soc.,* preprint #5555, May 2002.

[Spor95a] T. Sporer and K. Brandenburg, "Constraints of filter banks used for perceptual measurement," *J. Aud. Eng. Soc.*, vol. 43, no. 3, pp. 107–115, Mar. 1995.

[Spor95b] T. Sporer et al., "Evaluating a measurement system," *J. Aud. Eng. Soc.*, vol. 43, no. 5, pp. 353–363, May 1995.

[Spor96] T. Sporer, "Evaluating small impairments with the mean opinion scale – reliable or just a guess?," in *Proc. 101st Conv. Aud. Eng. Soc.*, preprint #4396, Nov. 1996.

[Spor97] T. Sporer, "Objective audio signal evaluation – applied psychoacoustics for modeling the perceived quality of digital audio," in *Proc. 103rd Conv. Aud. Eng. Soc.*, preprint #4512, Sept. 1997.

[Spor99a] T. Sporer, web site on TG 10/4 and other perceptual measurement standardization activities, available at http://www.lte.e-technik.uni-erlangen.de/~spo/tg104/index.html.

[SQAM88] Sound Quality Assessment Material: Recordings for Subjective Tests," EBU Doc. Tech. 3253 (includes SQAM Compact Disc), 1988.

[Sree98a] T. Sreenivas and M. Dietz, "Vector quantization of scale factors in advanced audio coder (AAC)," in *Proc. IEEE ICASSP-98*, vol. 6, pp. 3641–3644, Seattle, WA, May 1998.

[Sree98b] T. Sreenivas and M. Dietz, "Improved AAC Performance @<64 kb/s using VQ," in *Proc. 104th Conv. Audio Eng. Soc.*, preprint 4750, Amsterdam, 1998.

[Srin97] P. Srinivasan, *Speech and Wideband Audio Compression Using Filter Banks and Wavelets*, Ph. D. Thesis, Purdue University, May 1997.

[Srin98] P. Srinivasan, and L. Jamieson, "High quality audio compression using An adaptive wavelet packet decomposition and psychoacoustic modeling," *IEEE Trans. Sig. Proc.*, pp. 1085–1093, Apr. 1998. 'C' Libraries and examples available on http://www.wavelet.ecn.purdue.edu/~ speechg.

[Stau92] J. Stautner, "Physical testing of psychophysically-based digital audio coding systems," in *Proc. 11th Int. Conf. Aud. Eng. Soc.*, pp. 203–209, May 1992.

[Stol93a] G. Stoll et al., "Extension of ISO/MPEG-audio layer II to multi-channel coding: the future standard for broadcasting, telecommunication, and multimedia applications," in *Proc. 94th Conv. Aud. Eng. Soc.*, preprint #3550, Mar. 1993.

[Stol96] G. Stoll, et al., "ISO-MPEG-2 Audio: A generic standard for the coding of two-channel and multi-channel Sound," in *Collected Papers on Digital Audio Bit-Rate Reduction*, N. Gielchrist and C. Grewin, Eds., 1996.

[Stoll88] G. Stoll, et al., "Masking-pattern adapted subband coding: use of the dynamic bit-rate margin," in *Proc. 84th Conv. Aud. Eng. Soc.*, preprint #2585, Mar. 1988.

[Stor97] Storey, R. "ATLANTIC: Advanced Television at Low Bitrates and Networked Transmission over Integrated Communication Systems," *ACTS Common European Newsletter*, Feb. 1997.

[Stra96] G. Strang and T. Nguyen, *Wavelets and Filter Banks*, Wellesley-Cambridge Press, Wellesley, MA, 1996.

[Stru80] H. Strube, "Linear prediction on a warped frequency scale," *J. Acoust. Soc. Am.*, vol. 68, no. 4, pp. 1071–1076, Oct. 1980.

[Stua93] J. Stuart, "Noise: methods for estimating detectability and threshold," in *Proc. 94th Audio Eng. Soc. Conv.*, preprint 3477, Berlin, Mar. 1993.

[Sugi90] A. Sugiyama, et al., "Adaptive transform coding with an adaptive block size (ATC-ABS)," in *Proc. IEEE ICASSP-90*, pp. 1093–1096, Albuquerque, NM, May 1990.

[Swan98a] M. Swanson, B. Zhu, A. Tewfik, and L. Boney, "Robust audio watermarking using perceptual masking," *Elsevier Signal Processing, Special Issue on Copyright Protection and Access Control,* vol. 66, no. 3, pp. 337–355, May 1998.

[Swan98b] M.D. Swanson, M. Kobayashi, and A.H. Tewfik, "Multimedia data-embedding and watermarking technologies," *Proc. of the IEEE*, vol. 86, no. 6, pp. 1064–1087, June 1998.

[Swan99] M. Swanson, B. Zhu, and A. Tewfik, "Current state of the art, challenges and future directions for audio watermarking," in *Proc. ICMCS,* vol. 1, pp. 19–24, Florence, Italy, June 1999.

[Swee02] P. Sweeney, *Error Control Coding: From Theory to Practice,* John Wiley & Sons, New York, May 2002.

[Taka88] M. Taka et al., "DSP implementations of sophisticated speech codecs," *IEEE Trans. on Selected Areas in Communications*, vol. 6, no. 2, pp. 274–282, Feb. 1988.

[Taka97] Y. Takamizawa, "An efficient tonal component coding algorithm for MPEG-2 audio NBC," in *Proc. IEEE ICASSP-97*, pp. 331–334, Munich, Germany, Apr. 1997.

[Taka01] Y. Takamizawa and T. Nomura, "Processor-efficient implementation of a high quality MPEG-2 AAC encoder," in *Proc. 110th Conv. Aud. Eng. Soc.*, preprint #5294, May 2001.

[Taki95] Y. Takishima, M. Wada, and H. Murakami, "Reversible variable length codes," *IEEE Trans. Commun.*, vol. 43, pp. 158–162, Feb.–Apr. 1995.

[Tan89] E. Tan and B. Vermuelen, "Digital audio tape for data storage," *IEEE Spectrum*, vol. 26, no. 10, pp. 34–38, Oct. 1989.

[Tang95] B. Tang et al., "Spectral analysis of subband filtered signals," in *Proc. IEEE ICASSP-95*, pp. 1324–1327, Detroit, MI, May 1995.

[Taor99] R. Taori and R.J. Sluijter, "Closed-loop tracking of sinusoids for speech and audio coding," *IEEE Workshop on Speech Coding Proceedings*, pp. 1–3, 1999.

[Tefa03] A. Tefas, A. Nikolaidis, N. Nikolaidis, V. Solachidis, S. Tsekeridou, and I. Pitas, "Performance analysis of correlation-based watermarking schemes employing Markov chaotic sequences," *Trans. on Sig. Proc.*, vol. 51, no. 7, pp. 1979–1994, July 2003.

[Teh90] D. Teh et al., "Subband coding of high-fidelity quality audio signals at 128 kbps," in *Proc. IEEE ICASSP-92*, vol. 2, pp. 197–200, Albuquerque, NM, May 1990.

[tenK92] W. ten Kate et al., "Matrixing of bit-rate reduced signals," in *Proc. IEEE ICASSP*, vol. 2, pp. 205–208, San Francisco, CA, May 1992.

[tenK94] W. ten Kate et al., "Compatibility matrixing of multi-channel bit rate reduced audio signals," in *Proc. 96th Conv. Aud. Eng. Soc.*, preprint #3792, 1994.

[tenK96b] W. ten Kate, "Maintaining audio quality in cascaded psychoacoustic coding," in *Proc. 101st Conv. Aud. Eng. Soc.*, preprint #4387, Nov. 1996.

[tenK97] R. tenKate, "Disc-technology for super-quality audio applications," in *Proc. 103rd Conv. Aud. Eng. Soc.*, preprint #4565, Sept. 1997, New York.

[Terh79] Terhardt, E., "Calculating virtual pitch," *Hearing Research*, vol. 1, pp. 155–182, 1979.

[Terh82] E. Terhardt et al., "Algorithm for extraction of pitch and pitch salience from complex tonal signals," *J. Acous. Soc. Am.*, vol. 71, pp. 679–688, Mar. 1982.

[Terr96] K. Terry and J. Seaver, "A real-time, multichannel Dolby AC-3 audio encoder implementation," in *Proc. 101st Conv. Aud. Eng. Soc.*, preprint #4363, Nov. 1996.

[Tewf93] A. Tewfik and M. Ali, "Enhanced wavelet based audio coder," in *Conf. Rec. of the 27th Asilomar Conf. on Sig. Sys., and Comp.*, pp. 896–900, Nov. 1993.

[Thei87] G. Theile, et al., "Low-bit rate coding of high quality audio signals," in *Proc. 82nd Conv. Aud. Eng. Soc.*, preprint #2432, Mar. 1987.

[Thie96] T. Thiede and E. Kabot, "A new perceptual quality measure for bit rate reduced audio," in *Proc. 100th Conv. Aud. Eng. Soc.*, preprint #4280, May 1996.

[Thom82] D. Thomson, "Spectrum estimation and harmonic analysis," *Proc. of the IEEE*, pp. 1055–1096, Sep. 1982.

[Thor00] N.J. Thorwirth, P. Horvatic, R. Weis, and J. Zhao, "Security methods for MP3 music delivery," in *34th ASILOMAR Signals, Systems, and Computers,* vol. 2, pp. 1831–1835, Nov. 2000.

[Todd94] C. Todd et al., "AC-3: flexible perceptual coding for audio transmission and storage," in *Proc. 96th Conv. Aud. Eng. Soc.*, preprint #3796, Feb. 1994.

[Toh03] C.-K. Toh et al., "Transporting audio over wireless ad hoc networks: experiments and new insights," in *Proc. of Personal, Indoor and Mobile Radio Communications (IMRC-2003)*, vol. 1, pp. 772–777, Sept. 2003.

[Tran90] I. Trancoso and B. Atal, "Efficient search procedures for selecting the optimum innovation in stochastic coders," *IEEE Trans. ASSP*, vol. 38, no. 3, pp. 385–396, Mar. 1990.

[Trap02] W. Trappe and L.C. Washington, *Introduction to Cryptography with Coding Theory*, Prentice Hall, Englewood Cliffs, NJ, Jan. 2002.

[Trem82] T.E. Tremain, "The government standard linear predictive coding algorithm: LPC-10," *Speech Technology Magazine*, pp. 40–49, Apr. 1982.

[Treu96] W. Treurniet, "Simulation of individual listeners with an auditory model," in *Proc. 100th Conv. Aud. Eng. Soc.*, preprint #4154, May 1996.

[Tsai01] C.W. Tsai and J.L. Wu, "On constructing the Huffman-code based reversible variable length codes," *IEEE Trans. Commun.,* vol. 49, pp. 1506–1509, Sept. 2001.

[Tsai02] T-H. Tsai and J-N. Liu, "Architecture design for MPEG-2 AAC filterbank decoder using modified regressive method," in *Proc. IEEE ICASSP*, vol. 3, pp. 3216–3219, Orlando, FL, May 2002.

[Tsut96] K. Tsutsui et al., "ATRAC: Adaptive Transform Acoustic Coding for MiniDisc," in *Collected Papers on Digital Audio Bit-Rate Reduction*, N. Gilchrist and C. Grewin, Eds., Aud. Eng. Soc., pp. 95–101, 1996.

[Tsut98] K. Tsutsui, "ATRAC (Adaptive Transform Acoustic Coding) and ATRAC 2," in *The Digital Signal Processing Handbook*, V. Madisetti and D. Williams, Eds., CRC Press, Boca Raton, FL, pp. 43.16–43.20, 1998.

[Twve99] On-line animations of cochlear traveling waves: a) Boys Town National Research Hospital, Communication Engineering Laboratory, http://www.btnrh.boystown.org/cel/waves.htm; b) Department of Physiology at the University of Wisconsin, Madison, http://www.neurophys.wisc.edu/animations/; c) Scuola Internazionale Superiore di Studi Avanzati/International School for Advanced Studies (SISSA/ISAS) http://www.sissa.it/bp/Cochlea/twlo.htm; and d) Ear Lab at Boston University http://earlab.bu.edu/physiology/mechanics.html.

[USAT95a] United States Advanced Television Systems Committee (ATSC), Doc. A/53, "Digital Television Standard," Sep. 1995. Available online at http://www.atsc.org/Standards/A53/.

[USAT95b] United States Advanced Television Systems Committee (ATSC), Doc. A/52, "Digital Audio Compression Standard (AC-3)," Dec. 1995. Available online at http://www.atsc.org/Standards/A52/.

[Vaan00] R. Väänänen, "Synthetic audio tools in MPEG-4 standard," in *Proc. 108th Conv. Aud. Eng. Soc.,* preprint #5080, Feb. 2000.

[Vaid87] P.P. Vaidyanathan, "Quadrature mirror filter banks, *M*-band extensions, and perfect-reconstruction techniques," *IEEE ASSP Mag.*, vol. 4, no. 3, pp. 4–20, July 1987.

[Vaid88] P.P. Vaidyanathan and P.Q. Hoang, "Lattice structures for optimal design and robust implementation of two-channel perfect reconstruction QMF banks," *IEEE Trans. Acous., Speech, and Sig. Process.*, vol. 36, no. 1, pp. 81–94, Jan. 1988.

[Vaid90] P.P. Vaidyanathan, "Multirate digital filters, filter banks, polyphase networks, and applications: a tutorial," *Proc. of the IEEE*, vol. 78, no. 1, pp. 56–93, Jan. 1990.

[Vaid93] P.P. Vaidyanathan, *Multirate Systems and Filter Banks*, Prentice-Hall, Englewood Cliffs, NJ, 1993.

[Vaid94] P.P. Vaidyanathan and T. Chen, "Statistically optimal synthesis banks for subband coders," in *Proc. 28th Asilomar Conf. on Sig., Sys., and Comp.*, vol. 2, pp. 986–990, Nov. 1994.

[vand98] O. van der Vrecken et al., "A new subband perceptual audio coder using CELP," in *Proc. IEEE ICASSP-98*, pp. 3661–3664, Seattle, WA, May 1998.

[Vande02] S. van de Par, A. Kohlrausch, G. Charestan, and R. Heusdens, "A new psychoacoustical masking model for audio coding applications," in *Proc. IEEE ICASSP-02*, vol. 2, pp. 1805–1808, Orlando, FL, 2002.

[Vasi02] N. Vasiloglou, R.W. Schafer, and M.C. Hans, "Lossless audio coding with MPEG-4 structured audio," *2nd Int. Conf. on WEBDELMUSIC-02*, pp. 184–191, Dec. 2002.

[Vaup91] T. Vaupel, "Ein Beitrag zur Transformationscodierung von Audiosignalen unter Verwendung der Methode der 'Time Domain Aliasing Cancellation (TDAC)' und einer Signalkompandierung in Zeitbereich," Ph.D. Dissertation Duisburg, Germany: Univ. of Duisburg, Apr. 1991.

[Veen03] M.V. Veen, A.V. Leest, and F. Bruekers, "Reversible audio watermarking," in *Proc. 114th Conv. Aud. Eng. Soc.*, preprint 5818, Amsterdam, Mar. 2003.

[Veld89] R.N.J. Veldhuis, "Subband coding of digital audio signals without loss of quality," in *Proc. IEEE ICASSP-89*, pp. 2009–2012, Glasgow, Scotland, May 1989.

[Verc95] B.L. Vercoe, *Csound: A Manual for the Audio-Processing System*, MIT Media Lab, Cambridge 1995.

[Verc98] B.L. Vercoe et al., "Structured audio: creation, transmission, and rendering of parametric sound representations," *Proc. IEEE*, vol. 85, no. 5, pp. 922–940, May 1998.

[Verm98a] T. Verma and T. Meng, "An analysis/synthesis tool for transient signals that allows a flexible sines+transients+noise model for audio," in *Proc. IEEE ICASSP-98*, Seattle, WA, May 1998.

[Verm98b] T. Verma et al., "transient modeling synthesis: a flexible analysis/synthesis tool for transient signals," in *Proc. Int. Comp. Mus. Conf.*, pp. 164–167, Greece, 1997.

[Vern93] S. Vernon et al., "A single-chip DSP implementation of a high-quality, low bit-rate multi-channel audio coder," in *Proc. 95th Conv. Aud. Eng. Soc.*, 1993.

[Vern95] S. Vernon, "Design and implementation of AC-3 coders," *IEEE Trans. Consumer Elec.*, vol. 41, no. 3, Aug. 1995.

[Vett92] M. Vetterli and C. Herley, "Wavelets and filter banks," *IEEE Trans. Sig. Proc.*, vol. 40, no. 9, pp. 2207–2232, Sept. 1992.

[Vier86] M.A. Viergever, "Cochlear macromechanics – a review," in *Peripheral Auditory Mechanisms.* J. Allen et al., Eds., Springer, Berlin, 1986.

[Volo01] S. Voloshynovskiy, S. Pereira, T. Pun, J.J. Eggers, and J.K. Su, "Attacks on digital watermarks: classification, estimation based attacks, and benchmarks," *IEEE Comm. Mag.*, vol. 39, no. 8, pp. 118–126, Aug. 2001.

[Vora97] S. Voran, "Perception-based bit-allocation algorithms for audio coding," in *IEEE App. of Sig. Proc. to Audio and Acoustics (ASSP) Workshop*, 19–22 Oct. 1997.

[Voro88] P. Voros, "High-quality sound coding within 2×64 kbit/s using instantaneous dynamic bit-allocation," in *Proc. IEEE ICASSP-88*, pp. 2536–2539, New York, NY, May 1988.

[Wang98] Y. Wang, "A new watermarking method of digital audio content for copyright protection," in *Proc. ICSP-98,* vol. 2, pp. 1420–1423, Oct. 1998.

[Wang01] Y. Wang, J. Ojanpera, M. Vilermo, and M. Vaananen, "Schemes for re-compressing MP3 audio bitstreams," in *Proc. 111th Conv. Aud. Eng. Soc.,* preprint #5435, Nov.–Dec. 2001.

[Watk88] J. Watkinson, *Art of Digital Audio,* Focal Press, London and Boston, 1988.

[Wege97] A. Wegener, "MUSICompress: lossless, low-MIPS audio compression in software and hardware," in *Proc. of the Int. Conf. on Sig. Proc. App. and Tech.*, Sept. 1997.

[Wein96] M. Weinberger, G. Seroussi, and G. Sapiro, "LOCO-I: a low complexity, context-based, lossless image compression algorithm," *Proc. Data Compression Conf.*, 1996.

[Welc84] T. Welch, "A technique for high performance data compression," *IEEE Comp.*, vol. 17, no. 6, pp. 8–19, June 1984.

[Wen98] J. Wen and J. Villasenor, "Reversible variable length codes for efficient and robust image and video coding," in *Proc. IEEE Data Compression Conf.,* pp. 471–480, 1998.

[West88] P. H. Westerink, et al., "An optimal bit allocation algorithm for sub-band coding," in *Proc. IEEE ICASSP-88*, pp. 757–760, New York, NY, May 1988.

[Whit00] P. White, *Basic MIDI*, Sanctuary Press, Feb. 2000.

[Wick94] M. Wickerhauser, *Adapted Wavelet Analysis from Theory to Software*, A.K. Peters, Wellesley, MA, 1994.

[Wick95] S.B. Wicker, *Error Control Systems for Digital Communication and Storage*, Prentice Hall, Englewood Cliffs, NJ, 1995.

[Widr85] B. Widrow and S. Stearns, *Adaptive Signal Processing*, Prentice Hall, Englewood Cliffs, NJ, 1985.

[Wies90] D. Wiese and G. Stoll, "Bitrate reduction of high quality audio signals by modelling the ear's masking thresholds," in *Proc. 89th Conv. Aud. Eng. Soc.*, preprint #2970, Sep. 1990.

[Wind98] B. Winduratna, "FM analysis/synthesis based audio coding," *Proc. 104th Conv. Audio Eng. Soc.*, preprint #4746, May, 1998.

[Witt87] I.H. Witten, R.M. Neal, and J.G. Cleary, "Arithmetic Coding for data compression," *Comm. ACM*, vol. 30, no. 6, pp. 520–540, June 1987.

[Wolt03] M. Wolters, et al., "A closer look into MPEG-4 high efficiency AAC," in *115th Conv. Aud. Eng. Soc.,* preprint # 5871, New York, NY, Oct. 2003.

[Wu97] X. Wu, "High-order context modeling and embedded conditional entropy coding of wavelet coefficients for image compression," in *Proc. of 31st ASILOMAR Conf. on Signals, Systems, and Computers,* pp. 1378–1382, Nov., 1997.

[Wu98] X. Wu and Z. Xiong, "An empirical study of high-order context modeling and entropy coding of wavelet coefficients," ISO/IEC JTC-1/SC -29/WG-1. No. 771, Feb. 1998.

[Wu01] M. Wu, S. Craver, E.W. Felten, and B. Liu, "Analysis of attacks on SDMI audio watermarks," in *Proc. IEEE ICASSP-01,* vol. 3, pp. 1369–1372, Salt Lake City, UT, May 2001.

[Wu03] M. Wu and B. Liu, *Multimedia Data Hiding,* Springer Verlag, New York, Jan. 2003.

[Wust98] U. Wustenhagen et al., "Subjective listening test of multichannel audio codecs," in *Proc. 105th Conv. Aud. Eng. Soc.*, preprint #4813, Sept. 1998.

[Wyli96b] F. Wylie, "APT-x100: low-delay, low-bit-rate-subband ADPCM digital audio coding," in *Collected Papers on Digital Audio Bit-Rate Reduction*, N. Gilchrist and C. Grewin, Eds., Aud. Eng. Soc., pp. 83–94, 1996.

[Xu99a] C. Xu, J. Wu, and Q. Sun, "A robust digital audio watermarking technique," in *Proc. of the Int. Symposium on Signal Processing and Its Applications, (ISSPA-99),* vol. 1, pp. 95–98, Aug. 1999.

[Xu99b] C. Xu, J. Wu, and Q. Sun, "Digital audio watermarking and its application in multimedia databasc," in *Proc. of the Int. Symposium on Signal Processing and Its Applications, (ISSPA-99),* vol. 1, pp. 91–94, Aug. 1999.

[Xu99c] C. Xu, J. Wu, Q. Sun, and K. Xin "Applications of watermarking technology in audio signals," *J. Aud. Eng. Soc.*, vol. 47, no. 10, pp. 805–812, Oct. 1999.

[Yama98] H. Yamauchi et al., "The SDDS system for digitizing film sound," in *The Digital Signal Processing Handbook*, V. Madisetti and D. Williams, Eds., CRC Press, Boca Ration, FL pp. 43.6–43.12, 1998.

[Yang02] D. Yang, A. Hongmei, and C-C.J. Kuo, "Progressive multichannel audio codec (PMAC) with rich features," in *Proc. IEEE ICASSP-02,* vol. 3, pp. 2717–2720, Orlando, FL, May 2002.

[Yang03] C-H. Yang and H-M. Hang, "Efficient bit assignment strategy for perceptual audio coding," in *Proc. IEEE ICASSP*-03, vol. 5, pp. 405–408, Hong Kong, 2003.

[Yatr88] P. Yatrou and P. Mermelstein, "Ensuring predictor tracking in ADPCM speech coders under noisy transmission conditions," *IEEE J. Selected Areas Commun*. (Special Issue on Voice Coding for Communications), vol. 6, no. 2, pp. 249–261, Feb. 1988.

[Yeo01] I-K. Yeo and H.J. Kim, "Modified patchwork algorithm: a novel audio watermarking scheme," *Int. Conf. on Inf. Tech.: Coding and Computing*, pp. 237–242, Apr. 2001.

[Yeo03] I-K. Yeo and H.J. Kim, "Modified patchwork algorithm: a novel audio watermarking scheme," in *Trans. on Speech and Aud. Proc.*, vol. 11, no. 4, pp. 381–386, July 2003.

[Yin97] L. Yin, M. Suonio, and M. Vaananen, "A new backward predictor for MPEG audio coding," in *Proc. 103th Conv. Aud. Eng. Soc.*, preprint #4521, Sept. 1997.

[Yosh94] T. Yoshida, "The rewritable MiniDisc System," *Proc. of the IEEE*, vol. 82, no. 10, pp. 1492–1500, Oct. 1994.

[Yosh97] S. Yoshikawa et al., "Does high sampling frequency improve perceptual time-axis resolution of a digital audio signal?" *103rd Conv. Aud. Eng. Soc.*, preprint # 4562, New York, Sept. 1997.

[Zara02] R.H. Morelos-Zaragoza, *The Art of Error Correcting Coding*, John Wiley & Sons, New York, Apr. 2002.

[Zhan00] Q. Zhang, Y-Q. Zhang, and W. Zhu, "Resource allocation for audio and video streaming over the Internet," in *Proc. IEEE ISCAS-00*, vol. 4, pp. 21–24, May 2000.

[Zhen03] L. Zheng and A. Inoue, "Audio watermarking techniques using sinusoidal patterns based on pseudorandom sequences," in *Trans. Circuits and Systems for Video Technology*, vol. 13, no. 8, pp. 801–812, Aug. 2003.

[Zhou01] J. Zhou, Q. Zhang, Z. Xiong, and W. Zhu, "Error resilient scalable audio coding (ERSAC) for mobile applications," in *IEEE 4th Workshop on Multimedia Signal Processing*, pp. 307–312, Oct. 2001.

[Zieg02] T. Ziegler, et al., "Enhancing MP3 with SBR: Features and capabilities of the new mp3PRO algorithm," in *112th Conv. Aud. Eng. Soc.*, preprint # 5560, Munich, Germany, May 2002.

[Ziv77] J. Ziv and A. Lempel, "A universal algorithm for sequential data compression," *IEEE Trans. Inform. Theory*, vol. 23, no. 3, pp. 337–343, May 1977.

[Zwic67] E. Zwicker and R. Feldtkeller, *Das Ohr als Nachrichtenempfinger*. Hirzel, Stuttgart, Germany, 1967.

[Zwic77] E. Zwicker, "Procedure for calculating loudness of temporally variable sounds," *J. Acoust. Soc. of Am.*, vol. 62, pp. 675–682, 1977.

[Zwic90] E. Zwicker and H. Fastl, *Psychoacoustics Facts and Models*, Springer-Verlag, New York, 1990.

[Zwic91] E. Zwicker and U. Zwicker, "Audio engineering and psychoacoustics: matching signals to the final receiver, the human auditory system," *J. Audio Eng. Soc.*, pp. 115–126, Mar. 1991.

[Zwis65] J. Zwislocki, "Analysis of some auditory characteristics," in *Handbook of Mathematical Psychology*, R. Luce, et al., Eds., John Wiley and Sons, New York, 1965.

INDEX